Polar Microbes and Climate Change

Alysson Wagner Fernandes Duarte
Michel Rodrigo Zambrano Passarini
Josef Elster • Prashant Kumar Singh
Ajay Kumar
Editors

Polar Microbes and Climate Change

A Molecular Understanding for Sustainable Future

Springer

Editors
Alysson Wagner Fernandes Duarte
Laboratório de Microbiologia
Imunologia e Parasitologia. Complexo de
Ciências Médicas e Enfermagem
Universidade Federal de Alagoas
Campus Arapiraca
Arapiraca, AL, Brazil

Josef Elster
Centre for Polar Ecology
University of South Bohemia
Ceske Budejovice, Czech Republic

Ajay Kumar
Amity Institute of Biotechnology
Amity University
Noida, Uttar Pradesh, India

Michel Rodrigo Zambrano Passarini
Laboratório de Biotecnologia Ambiental
Federal University of Lat Am Integ.
Foz do Iguacu, Paraná, Brazil

Prashant Kumar Singh
Department of Biotechnology
Pacchunga University College
Aizawl, Mizoram, India

ISBN 978-3-032-07095-1 ISBN 978-3-032-07096-8 (eBook)
https://doi.org/10.1007/978-3-032-07096-8

This Springer imprint is published by the registered company Springer Nature Switzerland AG
The registered company address is: Gewerbestrasse 11, 6330 Cham, Switzerland

Contents

About the Editors

Alysson Wagner Fernandes Duarte is an Adjunct Professor at the Federal University of Alagoas, Arapiraca Campus. He graduated in Biological Sciences from the Federal University of Alagoas (2007), Master in Human Nutrition from Universidade Federal de Alagoas (UFAL) (2009), and PhD in Biotechnology from Universidade de São Paulo (USP) (2015). He completed a Postdoctoral internship at the Microbial Resources Division Laboratory at the Multidisciplinary Center for Chemical, Biological and Agricultural Research, University of Campinas (UNICAMP), in 2015–2016. He participated as a researcher in November/December 2013 of the XXXII OPERANTAR (Brazilian Antarctic Research Expedition). He has experience in microbiology, with emphasis on food microbiology, microbiology of the Antarctic environment, microbial ecology, taxonomic evaluation of yeasts, enzyme prospecting using cultivable (culture-dependent) and non-cultivable microorganisms (metagenomic libraries), construction of metagenomic libraries, heterologous expression of microbial enzymes, and experimental planning aiming at the optimization of processes and products.

Michel Rodrigo Zambrano Passarini graduated and completed his PhD in Genetics and Molecular Biology of Microorganisms at the State University of Campinas. He has vast research experience in biotechnology, emphasizing molecular and microorganism genetics. His current research areas are microbial ecology and microbial bioprospecting of extreme environments in the search for enzymes and metabolites of industrial interest applicable to the production of biofuel and degradation processes of environmental pollutants, as well as metagenomic and metabarcoding. He has published more than 60 research articles and book chapters of international repute.

Josef Elster is a full professor at the Centre for Polar Ecology at the University of South Bohemia. His research focuses on the polyphasic diversity of phototrophic microorganisms from cold environments and their bioprospection potential. He has published over 200 research articles and completed many cryobiology-related projects. He is also a fellow and member of various scientific societies.

Prashant Kumar Singh works at the Department of Biotechnology, Mizoram University (A Central University), Pachhunga University College Campus, Aizawl, Mizoram. He completed his PhD degree at Banaras Hindu University, Varanasi, India. In his 14 years of research career, he has published more than 80 articles and book chapters and four books. He has also been exposed to various national and international institutes. Young Scientist Award from Anusandhan National Research Foundation (ANRF; formerly known as Science and Engineering Research Board). He is a lifetime member of different scientific societies and reviewer of various peer-reviewed journals.

Ajay Kumar is a visiting scientist in the Agriculture Research Organization, Volcani Centre, Rishon Leziyon, Ministry of Agriculture and Rural Development, Israel. He completed his doctoral research in Botany, Institute of Science, Banaras Hindu University, Varanasi, India. He has published over 75 research papers, including research articles, review papers, and book chapters in leading international and national journals or books. He has a wide area of research experience, especially in plant microbiome, plant-microbe interactions, and microbe-mediated environmental contamination management. In addition, he is an invited member of the editorial committee of *The International Journal of Current Microbiology and Applied Sciences* and *Journal of Plant Science and Agriculture Research*. He is also a reviewer of journals such as *PLOS ONE*, *Postharvest Biology and Technology*, *Agriculture, Ecosystem and Environment*, *Biological Control*, *Plant Biochemistry and Biotechnology*, *3Biotech*, etc. He has edited several books with Elsevier such as *PGPR Amelioration in Sustainable Agriculture*, *Climate Change and Agriculture Ecosystem*, *Role of Plant Growth-Promoting Bacteria in Sustainable Agriculture and Nanotechnology*, *Microbial Endophytes: Prospects for Sustainable Agriculture*, *Abatements of Environmental Pollutants*, and *Fungi Bio-prospects in Sustainable Agriculture, Environment and Nanotechnology* (volumes 1–3), and several are under production.

Chapter 1
The Enigma of Photopsychrophily and Adaptation to Polar Habitats

Norman P. A. Hüner, Rachael M. Morgan-Kiss, David R. Smith, Alexander G. Ivanov, and Beth Szyszka-Mroz

Abstract Oceans cover about 70% of the Earth's surface and exhibit an average temperature of about 5 °C. Thus, Earth may be considered a rather cold place. The coldest ecosystems on Earth are represented by the Antarctic and Arctic polar regions, which constitute approximately 20% of the Earth's surface area and account for the annual sequestration of about 32% of the anthropogenic CO_2 through carbon capture by the process of photosynthesis. The photosynthetic microorganisms that dominate these cold, polar aquatic habitats include cyanobacteria and green algae that provide an essential ecosystem service for the planet. Many of these polar microorganisms are psychrophiles and thus are adapted to low temperatures. However, since cyanobacteria and algae are photosynthetic, they are considered photopsychrophiles, which represent some of the least studied organisms on Earth. We review the physiological, biochemical, molecular, and genomic adaptations that permit aquatic, polar photopsychrophiles to survive their severe habitats with respect to persistent cold temperatures, low light availability, and extreme photoperiod. These adaptations are discussed first, with respect to the observed biodiversity of polar photopsychrophiles; second, the extent of and the role of gene duplication in the evolution of polar photopsychrophily, and finally, the mechanisms that govern photosynthetic energy capture and utilization through carbon metabolism, growth, and development. Although the maintenance of energy balance (photostasis) appears to be a critical feature of adaptation of photopsychrophiles to polar environments, the enigma is that there does not appear to be an obvious, single molecular or

N. P. A. Hüner (✉) · D. R. Smith · B. Szyszka-Mroz
Department of Biology, Western University, London, Canada
e-mail: nhuner@uwo.ca

R. M. Morgan-Kiss
Department of Microbiology, Miami University of Ohio, Oxford, OH, USA

A. G. Ivanov (✉)
Department of Biology, Western University, London, Canada

Institute of Biophysics and Biomedical Engineering, Bulgarian Academy of Sciences, Sofia, Bulgaria
e-mail: aivanov@uwo.ca

A. W. Fernandes Duarte et al. (eds.), *Polar Microbes and Climate Change*,
https://doi.org/10.1007/978-3-032-07096-8_1

cellular mechanism that is sufficient to explain photopsychrophily. Rather, we conclude that evolution in polar environments appears to provide the "possibility space" for the generation of adaptive mechanisms as diverse and varied as the extreme environments in which they have evolved. This fuels a "dynamic persistence" that is the product of natural selection during the evolution of photopsychrophiles in a harsh, thermodynamically challenging polar environment.

Keywords Photopsychrophiles · Polar ecosystems · Carbon sequestration · Photosynthetic microorganisms · Cold adaptation · Energy balance (photostasis)

1.1 Introduction

Oceans cover about 70% of the Earth's surface with temperatures ≤5 °C. Consequently, Earth generally can be considered a rather cold place. The northern Arctic and southern Antarctic Oceans together constitute approximately 20% of the Earth's surface (Casanueva et al. 2010). The extreme environments of these aquatic polar regions provide an exceptional opportunity to unravel the basis of evolution and successful photosynthetic adaptation to polar life (Chown et al. 2015; Priscu et al. 1998; Bielewicz et al. 2011; Morgan-Kiss et al. 2006; Bakermans 2012; Zakhia et al. 2008; Vincent 2000, 2007; Dolhi et al. 2013; Kennicutt II et al. 2014; Kennicutt et al. 2015; Xavier et al. 2016; Yin et al. 2023). Persistent low temperature is a predominant characteristic of Antarctic and Arctic environments, which is coupled with variation in light intensities and predictable annual photoperiod as well as varied salt and nutrient levels depending on the specific polar habitat. Since most model prokaryotic and eukaryotic micro-organisms and plants with which we are familiar cannot survive under these extreme polar conditions, non-model biological species endemic to natural, seemingly inhospitable Antarctic and Arctic ecosystems allow one to preview the evolution of populations of myriad species that have survived and adapted to life at the edge (Kennicutt II et al. 2014; Kennicutt et al. 2015; Xavier et al. 2016; Yin et al. 2023). What are the characteristics of a polar photoautrophic lifestyle that differentiate it from photoautrophic life in warmer, more hospitable climates?

Because of the extremely limited biodiversity of terrestrial angiosperms in the Antarctic (Bravo et al. 2001, 2007), primary terrestrial production in this perpetually cold environment is largely dependent on mosses and lichens (Chown et al. 2015; Yin et al. 2023). In contrast, aquatic biodiversity is much more extensive with cyanobacteria and eukaryotic algae as the dominant photosynthetic primary producers not only of Antarctic and Arctic Oceans but also the water column below perennially ice-covered polar lakes such as Lake Bonney (Bielewicz et al. 2011; Morgan-Kiss et al. 2006; Priscu et al. 1998; Chrismas et al. 2015), meltwater pools of the McMurdo ice shelf (Sutherland et al. 2020), sea ice (Arrigo et al. 1997; Thomas and Dieckmann 2002; Liu et al. 2006; Zhang et al. 2019) and snowfields (Rogers 2007; ; Morgan-Kiss et al. 2006; Zhang et al. 2019). The study of these non-model organisms from such extreme and varied environments provides surprising evidence of what is not only biologically feasible but also biologically possible through the slow but inexorable process of evolution in polar environments over the past 100 M years (Rogers 2007; Cvetkovska et al. 2017; Yin et al. 2023). First, we summarize our present knowledge of the observed biodiversity and evolution of polar photopsychrophiles. Second, the role of gene duplication in the evolution of polar photopsychrophily through comparative genomics is addressed. Finally, we address the adaptations that govern photosynthetic energy capture and utilization through carbon metabolism, growth, and development with a focus on the Antarctic photopsychrophile, *Chlamydomonas priscuii*. It is tempting to consider this green alga as a model biological system for the study of polar photopsychrophily. However, we conclude that an enigma of polar photopsychrophily is that there may not be a single species that can be considered a model organism for study, given the diverse, extreme environments in which they have evolved and to which they have adapted.

1.1.1 Photostasis

The second law of thermodynamics dictates that a constant influx of energy from the environment is an absolute requirement to counteract the constant, emergent disorder called entropy to maintain the order necessary for life on Earth (Schroedinger 1947; Prigogine and Stengers 1984; Atkins 2010; England 2020). Thus, Schroedinger (1947) concluded that life may be defined as "negative entropy." Photosynthesis is the process that provides access to a relatively constant and reliable external energy source in the form of sunlight, either directly, in the case of all photoautrophs, or indirectly for all other forms of heterotrophic life on our planet through their diet of photosynthetic organisms (Nelson 1807; Judson 2017). The first photoautrophic eukaryotic cell evolved as a consequence of successive rounds of endosymbiosis between a cyanobacterium and a eukaryotic heterotrophic host (Sagan 1967; Gray et al. 2020).

Through oxygenic photosynthesis, cyanobacteria, green algae, and terrestrial plants transform absorbed light energy into electrons consumed in the reduction of CO_2 via the photosynthetic apparatus, ensuring a constant supply of energy through the biosynthesis of carbohydrates and the generation of oxygen consumed through aerobic respiration in most living organisms on our planet. Thus, a major challenge for all endemic polar photoautotrophs is to balance the energy flux between extremely rapid, temperature-insensitive light absorption and photochemistry (Nelson 1807) versus temperature-sensitive, energy consumption through the slower, enzyme-catalyzed biochemical reactions essential for CO_2 fixation, respiration, growth, development, and reproduction in an extremely inhospitable, and thermodynamically challenging environment. The maintenance of such cellular energy balance in photoautotrophs is called photostasis (Melis 1998; Hüner et al. 2003; Hollis and Hüner 2014). Under thermodynamically challenging environments, photostasis requires a regulated reorganization of the structure, composition, and function of the photosynthetic apparatus (Fig. 1.1). This is reflected in the phenotypic plasticity observed in photoautotrophs either during short-term acclimation or long-term adaptation to low temperature (Hüner et al. 1998, 2003, 2020).

Microbes dominate the Antarctic and Arctic Oceans, and it has been estimated that these polar oceans sequester about 32% of the global anthropogenic CO_2, which makes these polar aquatic basins major sinks for the sequestration of atmospheric CO_2 (Sabine et al. 2004). Thus, these polar marine ecosystems contribute significantly not only to the governance of global carbon balance and the mitigation of anthropogenically-induced climate change (Sabine et al. 2004) but also to global biodiversity (Saikonen et al. 2023). Due to present space limitations, we suggest several excellent books (Whitton and Potts 2000; Margesin et al. 2008; Vincent 2007; Seckbach 2007) and reviews (Morgan-Kiss et al. 2006; Dolhi et al. 2013; Vincent 2000, 2007; Zakhia et al. 2008; Priscu et al. 1998; Mock and Thomas 2008; Lyon and Mock 2014; Kennicutt II et al. 2014; Kennicutt et al. 2015; Xavier et al. 2016; Yin et al. 2023) for readers requiring further information on the adaptation of eukaryotic and prokaryotic organisms to polar environments.

1.2 Evolution and Biodiversity of Photopsychrophiles

The fragmentation of Earth's original supercontinent, Gondwana, was accompanied by successive widespread glaciation events (Hoffman et al. 1998). It has been hypothesized that this extensive glaciation led to a "snowball Earth" in which the oceans were ice covered (Kirschvink et al. 2000). Successive glaciation events interspersed with extreme warm environments afforded selective pressure on the evolution of life on Earth (Hoffman et al. 1998; Kirschvink et al. 2000). Prokaryotes dominated the Neoproterozoic era (1000 to 500 million years ago) and survived these successive glaciation events during which red and green algae evolved (Hofmann et al. 1990). Recently, Krupovic et al. (2020) proposed that LUCA (Last Universal Cellular Ancestor) represents the most recent population of cellular organisms from which all life on Earth descended. It is proposed that the succession of glacial events on Earth resulted in a series of genetic "bottleneck and flush cycles" (Hofmann et al. 1990), which may have established demes or small populations of organisms (Carson 1987) that led to the evolution of psychrophiles and photosynthetic psychrophiles, generally termed photopsychrophiles (Morgan-Kiss et al. 2008).

Organisms obligately adapted to cold temperatures are defined as psychrophiles, which typically include the Bacteria and Archaea as well as eukaryotes such as fungi, including yeasts, and snow molds (Newsted and Hüner 1987, 1988), as well as micro- and macro-algae (Morita 1975). Psychrophiles typically have optimal growth temperatures of ≤ 15 °C but do not survive at temperatures exceeding 18–20 °C. In contrast, micro-organisms that exhibit maximum growth rates ≥ 20 °C but are also able to acclimate, grow, and survive at low temperatures (0–5 °C) are considered psychrotolerant (Morita 1975). Stenothermic organisms can grow and

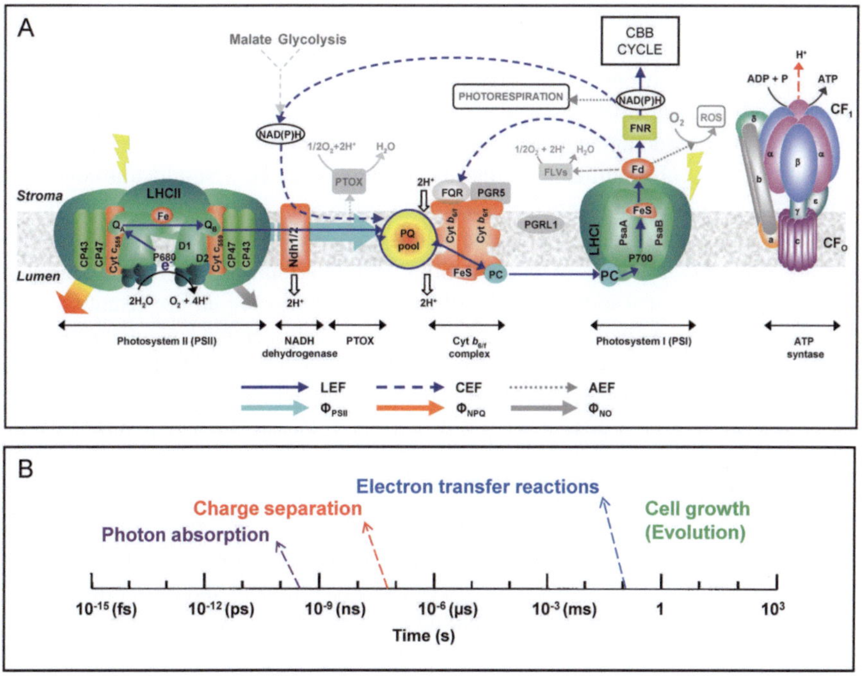

Fig. 1.1 Schematic representations of linear photosynthetic electron transport (LEF), PSI-driven cyclic electron transport (CEF), and alternative electron transport (AEF) pathways in higher plants/green algae thylakoid membranes. During growth and development under optimal conditions, the plastoquinone (PQ) pool remains preferentially oxidized because the rate of consumption of photosynthetic electrons through major metabolic sinks such as carbon fixing reactions in the Calvin-Benson-Bassham (CBB) cycle, as well as N and S reduction, keeps pace with the rate at which PSII undergoes charge separation to reduce the PQ pool. PSI and PSII are linked in series by an intersystem photosynthetic electron transport chain (PETC) consisting of coupled redox carriers, plastoquinone (PQ), the cytochrome b_6/f complex and plastocyanin (PC), all localized within the plane of the thylakoid membrane (**a**). Coordination of these events results in the production of the strongest oxidizing agent in nature associated with PSII ($P680^+$) with the concomitant production of O_2 and the strongest reducing agent in nature (reduced Fd) as a consequence of the photooxidation of P700 to $P700^+$ and the storage of reductant in the form of NADPH coupled to the generation of a proton motive force (pmf) which the chloroplast ATPase converts into ATP. The photosynthetic reductant and ATP are consumed in CO_2 fixation by the Calvin-Benson-Bassham cycle (**a**). Under normal conditions, linear photosynthetic electron flow (LEF, dark blue solid arrows) from PSII (water splitting) to PSI ($NADP^+$ generation) dominates, although additional PSI cyclic electron flow (CEF, blue dashed lines) and alternative electron transport pathways (gray dashed arrows) may also be induced. When under various environmental stress conditions, such as low temperature, the rate of electron consumption governed by the metabolic sink capacity of the CBB-cycle may be insufficient to consume all of the photochemically-generated electrons. Consequently, the excess photosynthetic electrons are diverted at ferredoxin (Fd) serving as the major electron distribution hub, and re-circulated by NDH and FQR-dependent PSI-driven CEFs and/or safely utilized through oxygen-dependent electron sinks by PTOX- and/or FLVs-alternative electron pathways. Regulation of the circulation of electron flow contributes to the protection of the photosynthetic apparatus from photodamage due to exposure to excess excitation energy (EEE). *PSII* photosystem II, *PSI* photosystem I, *PQ* plastoquinone pool, *Cytb6/f* cytochrome b6/f complex, *PTOX* plastid terminal oxidase, *PC* plastocyanin, *Cyt oxidase* cytochrome oxidase, *Fd* ferredoxin, *NDH* NADPH dehydrogenase, *FNR* ferredoxin-$NADP^+$ reductase, *FLVs* flavodiuron proteins. (**b**). Time scales of photosynthetic events beginning with light absorption and extending to long-term photosynthetic adaptation to the environment. (Chow 2003)

survive only over a narrow temperature range, whereas eurythermal organisms can grow and survive over a broad temperature range (Bligh and Johnson 1973). Consequently, psychrophiles are stenotherms in contrast to psychrotolerant organisms, which are considered eurytherms. A eurythermal, psychrotolerant survival strategy may provide an adaptative advantage in environments that fluctuate daily. In contrast, a psychrophilic, stenothermal strategy may be a significant benefit in long-term, relatively constant seasonal environments characteristic of the polar regions (Tang and Vincent 1999; Nadeau and Castenholz 2000). Thermal limits for life are not only relevant in understanding biogeography over varying thermal regimes: physiological thermal limits of microorganisms are an important component of survival in a changing climate. Studies have shown that organisms adapted to either stable (stenothermal) or highly variable (eurythermal) lifestyles are highly affected by climate change, while organisms from moderately variable thermal environments exhibit the highest ability to acclimate to environmental change (Tomanek 2008).

There are ~100 isolated algal photopsychrophilic species in algal strain collections and laboratories around the world: these include both unicellular and multicellular representatives of green algae, rhodophytes, stramenopiles, and dinoflagellates (Cvetkovska et al. 2017). However, most of the physiological and molecular studies have focused on diatoms (stramenopiles) and green algae (chlorophytes). Chlorophyta represent more than a third of all confirmed photopsychrophiles of which 23 species can be placed in the order Chlamydomonadales (Cvetkovska et al. 2017). Isolates of photopsychrophilic *Chlamydomonas* spp. have been cultivated from polar and alpine habitats, with representatives from both marine and inland water bodies (Fig. 1.2). Photopsychrophily has probably evolved independently several times during algal evolution. However, it is difficult to determine whether the shift to cold tolerance over evolutionary time was gained independently in multiple lineages or whether it represents an ancestral state. Given the small number of photopsychrophilic species within a given family, it is unlikely that algal psychrotolerance and photopsychrophily are ancestral.

The global biogeography of psychrophilic microorganisms is extensive: in permanently cold ecosystems that receive sufficient light levels, photopsychrophilic microalgae and phototrophic bacteria (cyanobacteria, anoxygenic phototrophic bacteria) contribute significantly to primary production. Thus, polar and alpine ecosystems harbor a massive diversity of microalgal species, many of which are likely photopsychrophilic. Global aquatic habitats for photopsychrophiles include marine ecosystems in the Arctic and Antarctic, as well as freshwater inland bodies of water in polar and alpine environments (Fig. 1.3). The Tibetan Plateau (often referred to as the Earth's "third pole") represents a vast high-altitude permanently cold ecosystem containing thousands of seasonally and perennially ice-covered lakes (Zhang et al. 2015). Specific lineages of photopsychrophiles, including chlorophytes and diatoms, specialize in colonizing the surface of snow and glaciers (Hoham and Remias 2020) and brine channels within sea ice (Arrigo 2014). While all habitats for photopsychrophiles share some characteristics, such as permanent low

Fig. 1.2 Global distribution of isolates of psychrophilic *Chlamydomonas* spp. Antarctic isolates: *C. priscuii**. (CCMP1619, Neale and Priscu 1995), *C.* sp. ICE-MDV. (Li et al. 2016), *C.* sp. WAP05, *C.* sp. WAP30 (unpublished), C. sp. ICE-L*, C. sp. ICE-W; Arctic isolates: *C. malina.* (RCC2488, Morales-Sáchez et al. 2020), *C. klinobasis** (CCCryo 050-99; unpublished), *C.* sp. ARC. (Eddie et al. 2008), *C.* sp. KNM0029C. (Kim et al. 2020). Alpine isolates: *C. nivalis.* (Remias et al. 2005), *C. yellowstonesis.* (Hoham 1975)*, sequenced genomes available

temperatures, there are multiple site-specific environmental variables that also influence survival, growth, and photosynthesis.

Antarctica harbors many inland lakes that share physicochemical characteristics such as permanent chemical stratification (meromixis) and extreme light limitation. Antarctic lakes often exhibit extremes in salinity (freshwater to hypersaline) and nutrient status (ultra-oligotrophic to hypereutrophic). In continental Antarctica, most inland aquatic ecosystems represent the last permanently ice-covered lakes on our planet, while maritime Antarctic and sub-Antarctic regions contain many lakes that experience seasonal meltout and ice-free summers. Several lakes within the McMurdo Dry Valleys (Victoria Land) and the Vestfold Hills (Princess Elizabeth Land) are sites of over three decades of long-term research on the ecology of photopsychrophiles and their functional roles in Antarctic lake food webs (Bell 1998; Bielewicz et al. 2011; Izaguirre et al. 2021; Perriss and Laybourn-Parry 1996; Priscu et al. 1998) (Fig. 1.4).

Antarctic lake food webs are dominated by the microbial loop, with many phytoplankton communities dominated by nanoflagellates, that is, flagellated microorganisms within the size range of 2–20 μm. Chlorophyte species (*Chlamydomonas* spp., *Chlorogonium* spp.) have been reported to be highly abundant in productive lakes and ponds from Maritime Antarctica (Laybourn-Parry et al. 1996), which are typically mesotrophic to hypereutrophic due to proximity of natural eutrophication from sea animals and birds (Hawes 1990). *Chlamydomonas* spp. from the McMurdo Dry Valley (MDV) lakes have been the subject of numerous studies, including field-based (Bielewicz et al. 2011; Kong et al. 2012; Li and Morgan-Kiss 2019;

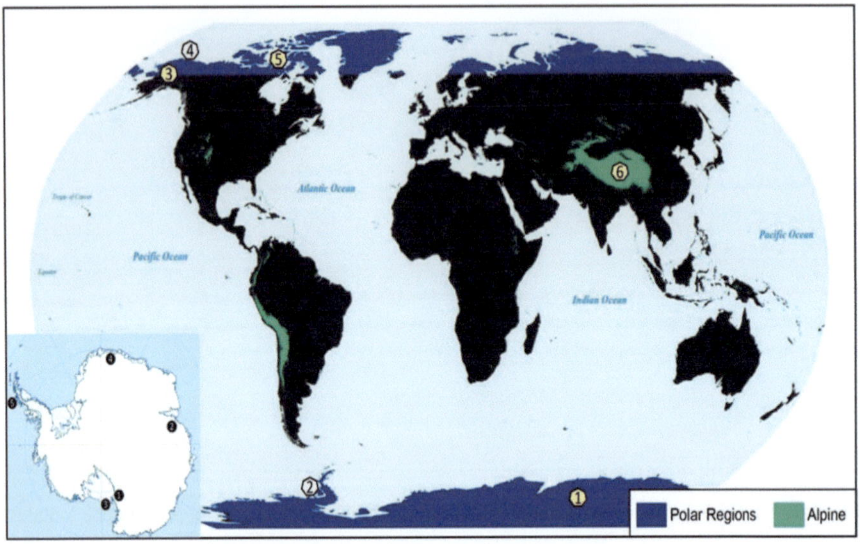

Fig. 1.3 Global locations of habitats for photopsychrophiles. Numbers indicate examples of long-term ecological research projects: 1—McMurdo Long Term Ecological Research. (MCM LTER, https://lternet.edu/site/mcmurdo-dry-valleys-lter/); 2—Palmer LTER; 3—Arctic LTER. (ARC, https://lternet.edu/site/arctic-lter/); 4—Beaufort Lagoon Ecosystems LTER. (https://lternet.edu/site/beaufort-lagoon-ecosystem/); 5—Canadian High Arctic Research Station. (CHARS, https://www.canada.ca/en/polar-knowledge/CHARScampus.html); 6—Institute of Tibetan Plateau Research. (ITP; http://english.itpcas.cas.cn/). Yellow sites, freshwater; grey sites, marine. Inset: Sites of intensive study of Antarctic photopsychrophiles. 1—McMurdo Dry Valleys, Victoria Land; 2 —Vestfold Hills, Princess Elizabeth Land; 3—Bratina Island, Brown Peninsula; 4—Lake Untersee, Queen Maud Land; 5—western Antarctic Peninsula (WAP)

Morgan-Kiss et al. 2016; Sherwell et al. 2022) and laboratory-controlled experiments (reviewed in: Cvetkovska et al. 2017; Hüner et al. 2020). *Chlamydomonas* spp. have been detected in only one MDV lake, Lake Bonney (east and west lobes). Sequences of phylogenetic (18S rRNA) and photosynthetic genes (*rbcL, psbA*) for *Chlamydomonas* spp. have been detected throughout the photic zone of Lake Bonney, but are most abundant within the shallow layers of the water column, directly beneath the ice-cover (Kong et al. 2012). A distinct consortium of *Chlamydomonas* spp. dominate the deep photic zone, where light levels diminish to near the compensation point for photosynthesis and salinity rises to above seawater levels. The conditions within the upper water column of Lake Bonney would appear to be at odds with the typical conditions in which *Chlamydomonas* thrives. Lake Bonney is oligotrophic, and the shallow layers are particularly phosphorus-limited. It is not clear why the Lake Bonney Chlamydomonads thrive in the oligotrophic environment; however, from studies on isolates such as *C. priscuii* and *Chlamydomonas* sp. ICE-MDV (both Lake Bonney isolates) have shown that they possess a robust toolbox of physiological and metabolic strategies to thrive under permanent stress (see sections below). Moreover, recent field studies indicate that

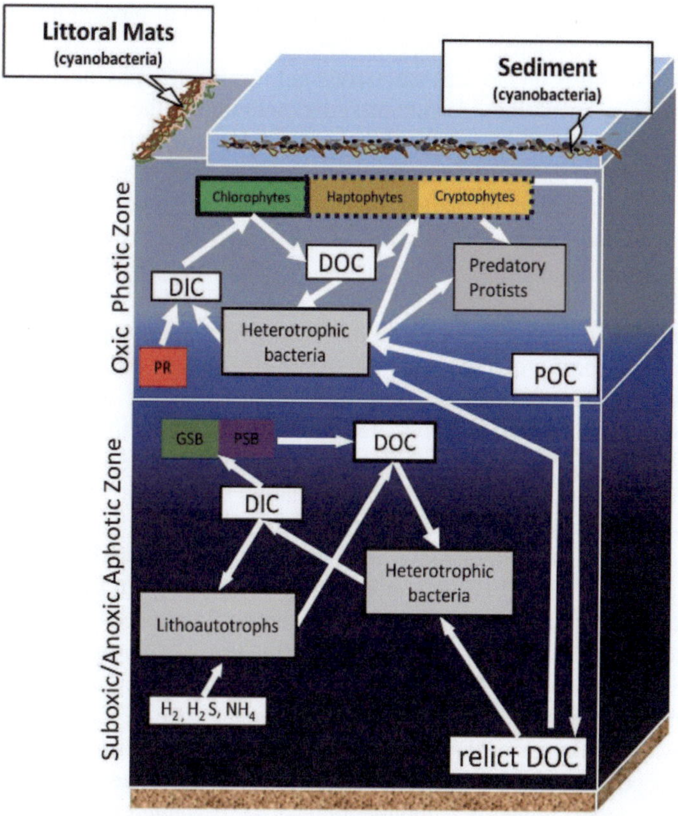

Fig. 1.4 Role of photopsychrophilic micro-organisms in the Antarctic Lake food web. Solid square, obligate photoautotroph; dashed square, mixotrophs

Chlamydomonads are competitive under climate-driven change in the MDVs. In a long-term study, Patriarche et al. (2021) observed that phytoplankton standing biomass has gradually increased in Lake Bonney over the past two decades: chlorophytes appear to be the major contributors to this long-term positive trend in phytoplankton biomass. In addition, Teufel et al. (2017) monitored phytoplankton community composition and growth in a nutrient bioassay experiment: growth of chlorophyte populations was stimulated under either nitrogen or phosphorus amendment, displacing other phytoplankton species.

Mixotrophic phytoplankton are also abundant in Antarctic inland lakes, including the MDV lakes (e.g., Lake Bonney, Lake Fryxell) (Bielewicz et al. 2011; Li and Morgan-Kiss 2019) and other meromictic lakes within the Vestfold Hills (e.g., Ace Lake) (Bell 1998; Bell and Laybourn-Parry 2003). The permanent chemocline of Lake Bonney is dominated by a psychrophilic haptophyte species, *Isochrysis* sp. MDV (Li et al. 2016; Li and Morgan-Kiss 2019). Although there is an isolate of

Isochrysis in culture, we anxiously await the publication of physiological information regarding this isolate.

A year-long study on the seasonal trends in Lake Bonney phytoplankton discovered a surprising result that, despite total darkness during the polar winter, chlorophyll a accumulated during early winter within the chemocline (Patriarche et al. 2021). Further analysis of data from a spectral fluorometer revealed that the winter-associated accumulation in phytoplankton biomass was due to the growth of a "brown alga" spectral group. In temperate systems, the "brown" group is often dominated by diatoms; however, previous research has shown that this spectral group is largely composed of the *Isochrysis* sp. in Lake Bonney (Dolhi et al. 2015). Thus, while the obligate photoautotrophic Chlamydomonads appear to be competitive during the short growing season, photopsychrophiles with mixed metabolism can continue to grow through the long, dark winter.

A second abundant group of psychrophilic mixotrophs in the Antarctic lakes is the cryptophytes. Cryptophyta is a globally important group of mixotrophic flagellates, often detected at high abundances in marine systems as well as forming the deep chlorophyll maxima in freshwater ecosystems. Cryptophytes have been reported as abundant phytoplankton in both Lakes Bonney and Fryxell, as well as Ace Lake in the Vestfold Hills (van den Hoff et al. 2020). In Lakes Bonney and Fryxell, cryptophyte populations appear to be largely dominated by a phytoplankton related to a marine photopsychrophile, *Gemingera cryophila*. The 18S rRNA sequences of *G. cryophila* from Lakes Bonney and Fryxell are identical; however, cryptophyte populations occupy the shallow photic zone of Lake Bonney (5 m depth), while Lake Fryxell cryptophytes dominate the permanent chemocline (9–11 m depth) (Li and Morgan-Kiss 2019). In Ace Lake, the permanent chemocline (15 m depth) is nearly a monoculture of *G. cryophila* (van den Hoff et al. 2020). The MDV *G. cryophila* has proven elusive to capture in isolation; however, a marine isolate has been studied for several years (McKie-Krisberg et al. 2015; Trimborn et al. 2019; Camoying and Trimborn 2023). *G. cryophila* is also a major prey species for non-constitutive mixotrophic ciliate species, such as *Myrionecta rubra* in the Antarctic marine ecosystem (Johnson et al. 2007; van den Hoff et al. 2020) and *Mesodinium* in the MDV Lake Fryxell (Laybourn-Parry et al. 2000).

Photopsychrophiles are critical members of polar marine food webs. In the Southern Ocean, sea ice diatoms are a dominant algal group. Diatoms are a major primary producer in this ecosystem, where iron levels are extremely low and photosynthetic organisms must survive seasonal fluctuations in temperature, light, and sea ice extent (Strzepek et al. 2012). Small grazers such as krill, which link carbon fixed by primary producers with upper trophic levels, exhibit a strong preference for a diatom-dominated diet (Haberman et al. 2003). On the other hand, smaller marine phytoplankton, including cryptophytes, are not consumed by krill (Pauli et al. 2021). The pennate diatom *Fragilariopsis cylindrus* represents a common diatom within polar marine ecosystems, especially within the marginal ice-edge zones of the Southern Ocean (Kang and Fryxell 1992): it is considered to be an indicator species for polar marine systems (von Quillfeldt 2004) and a model for cold adaptation of photosynthesis (Mock et al. 2017). Recent studies in the western Antarctic

Peninsula (WAP) have reported significant shifts in the phytoplankton communities in favor of cryptophytes dominated by a marine *G. cryophila* species at the expense of the dominant sea ice diatoms (Mendes et al. 2023). Experiments on isolates of *G. cryophila* have found that it exhibits high physiological plasticity under variable temperature and CO_2. The WAP has experienced significant increases in both temperature and dissolved CO_2 levels over the past three decades: it is hypothesized that these climate-related changes in the WAP ecosystem are allowing cryptophytes to gradually outcompete diatoms (Mendes et al. 2023). Given the prey-specificity of grazers, these shifts in the phytoplankton community to small, less palatable phytoplankton prey could have significant impacts on the Southern Ocean marine food web.

The Arctic region harbors an enormous diversity and abundance of lakes and ponds, which support diverse phytoplankton communities (Schartau et al. 2022). Common phytoplankton detected in Arctic lakes are cryptophytes, chlorophytes, diatoms, and chrysophytes; however, there is great variability in community composition and biomass across these lakes (Schartau et al. 2022). Meromictic lakes are a common feature in the high Arctic and are considered close analogues to the ice-covered, saline Antarctic lakes, such as Lakes Bonney and Vanda (Markager et al. 1999). However, unlike their Antarctic counterparts, Arctic lakes have experienced increasing ice-free events during the summer (Walsh 2009), which has resulted in wind-driven mixing of water columns, extreme increases in underwater radiation, and significant changes in allochthonous inputs from streams (Veillette et al. 2010). There appears to be no photopsychrophilic isolates from Arctic lakes that have been studied in detail. Phytoplankton blooms in the Arctic marine system are dominated by pico-phytoeukaryotes, including the psychrophilic picoprasinophyte *Micromonas* (Bolaños et al. 2020).

1.2.1 Gene Duplication and Photopsychrophily

The arrival of a newly completed genome sequence from a land plant or alga used to be big news. But today, with the help of third-generation sequencing technologies (Kovaka et al. 2023), even the smallest research groups can tackle the largest of genomes. Thus, the challenge for plant genomicists is not generating data, but rather, it is choosing which species to concentrate on. Generally, they have focused on genomes of model organisms. Indeed, a quick scan of the Joint Genome Institute's PhycoCosm database (Grigoriev et al. 2021) reveals only a handful of genomes from extremophiles (e.g., *Dunaliella salina*), with fewer than five coming from *bona fide* photopsychrophiles. We are at a precipice. Soon, hundreds—maybe even thousands—of whole genomes from photopsychrophiles will be making their way into the scientific database, and this will undoubtedly usher in a new chapter of comparative plant genomics. For the moment, however, we are left with the data at hand, which, although not huge, still include an eclectic mix of species, such as *Fragilariopsis cylindrus* (Mock et al. 2017), *Chlamydomonas* sp. ICE-L (Zhang et al. 2020), *Microglena* sp. YARC (Zhang et al. 2023) and *C. priscuii* (Zhang et al. 2021), as well as various whole transcriptome analyses from psychrophiles (Leebens-Mack et al. 2019).

What have we learned about the evolution of photopsychrophily from deciphering the DNA and RNA sequences of these organisms? As so often is the case in genomics, there is no obvious genetic feature (or features) that someone could point to and say, "That's what allows this species to be a psychrophile…" In fact, it would be nearly impossible to establish if an organism is a photopsychrophile from the genome or transcriptome sequence alone, phylogenetic placement notwithstanding. Even features like multicellularity, for example, are not necessarily obvious from the nucleotide data, as was so clearly illustrated from the genomes of *Volvox carteri* (Prochnik et al. 2010) and *Ulva mutabilis* (De Clerck et al. 2018). Still, some of the secrets of photopsychrophily can be gleaned from genomes and transcriptomes of land plants and algae, and they tend to revolve around three key themes: gene duplication, gene family expansion, and horizontal gene transfer (HGT). In addition to genomics, epigenetics also contributes significantly to the regulation of plant responses to cold stress and cold acclimation (Eichten et al. 2014; Cavalli and Heard 2019; Hemenway and Gehring 2023). However, there is minimal published information regarding the potential contribution of epigenetics to the regulation of photopsychrophily, cold acclimation, and adaptation to extreme polar environments.

Genetic redundancy has long been acknowledged as a potential means for coping with, adapting to, and thriving in extreme and tumultuous environments (Bomblies 2020). This redundancy can manifest itself in various forms, including localized gene duplications, whole genome duplications, and/or gene family expansion, and is not necessarily constrained to coding regions. Because these phenomena are near-ubiquitous across the eukaryotic domain, it can be challenging to tweeze out if genetic redundancy is directly and positively associated with an extremophilic lifestyle or is merely the consequence of random events, like genetic drift. And, admittedly, some still hold that large-scale gene duplication events are deleterious, no matter the environmental conditions (Comai 2005). However, it is hard to ignore that psychrophiles of all types appear to be especially prone to genetic redundancy (Zhang et al. 2020, 2021; Kim et al. 2023). The genomes of *Chlamydomonas* sp. ICE-L and *C. priscuii* exemplify this point, showing thousands of gene duplication/expansion events, spanning a plethora of functional categories, and even involving noncoding regions (Zhang et al. 2020, 2021). Similar observations have come from the transcriptome analyses of photopsychrophiles as well, including *Chlamydomonas* sp. ICE-MDV (Raymond and Morgan-Kiss 2017). However, presently, there is little published data for the potential contribution of alternative splicing (Kjer-Hansen and Weatheritt 2023) in the generation of the redundancy apparent in the proteomes of photopsychrophiles.

How could this redundancy help psychrophiles survive a cold, harsh environment? Some maintain that it can improve mutational and environmental robustness because of an increase in genetic variation (Posadas-García and Espinosa-Soto 2022). Recently, it was argued that populations with duplicated gene regulatory networks show higher signal output variation than those without such duplications, leading to better niche expansion and, thus, improved ability to withstand environmental turmoil (Ebadi et al. 2023). Another explanation is that if your proteins are working at a reduced capacity, because of cold temperatures, for instance, having more of them can simply help compensate for this limitation. This can at least partly explain some of the redundancy in *C. priscuii* (Cvetkovska et al. 2018; Zhang et al. 2021). Perhaps the most striking observation of genetic redundancy in photopsychrophiles is the expansions of protein-coding genes containing an ICE-binding domain. The genomes of *Chlamydomonas* sp. ICE-L, ICE-MDV, and *C. priscuii* harbor at least a dozen or more ice-binding proteins (IBPs), with the latter two species potentially encoding >35 IBPs (Zhang et al. 2020, 2021). Data from other photopsychrophilic algae also suggest multiple copies of genes for IBPs (Sorhannus 2011; Dorrell et al. 2023). It is believed that the IBPs in these various species originate through one or multiple instances of horizontal gene transfer, followed by gene duplication and gene family expansion, and aid with survival in cold habitats.

As we all eagerly await the genome sequences of more photopsychrophiles, it is important to remember that there are still a lot of "old" data in NCBI waiting to be explored in greater detail. The "One thousand plant transcriptomes" project (Leebens-Mack et al. 2019) is a perfect example of this. These freely available datasets include several psychrophilic/psychrotolerant algal species from across the eukaryotic domain, most of which have only been analyzed in a cursory manner. Similarly, the *Tara Oceans* metagenomes and meta-transcriptomes project (Sunagawa et al. 2020) explored a diversity of protists from the Arctic and Antarctic regions. Recent analyses of these data have uncovered, for example, the presence of IBPs in a range of Arctic algae and suggested widespread horizontal transfer of these IBPs among species (Dorrell et al. 2023). Clearly, there is a lot more work needed to be done in the field of comparative genomics of photopsychrophiles.

1.3 Energy Balance and Photopsychrophily

Light absorption and energy transfer associated with photosynthesis entail very rapid (femtosecond to microsecond time scale), temperature-insensitive photobiophysical processes associated with the energy transfer in light-harvesting complexes coupled to the photochemistry associated with PSI and PSII reaction centers of the chloroplast thylakoid membranes. Evolution has integrated this temperature-insensitive, extremely rapid photochemistry with the much slower and highly temperature-sensitive, biochemical redox reactions associated with the intersystem photosynthetic electron transfer chain (PETC) of the thylakoid membranes and the soluble, stromal enzymes of the Calvin-Benson-Bassham Cycle (CBB) which utilize the NADPH and ATP to reduce C, N and S required either for energy storage or for growth, development, and reproduction (Fig. 1.1). The conversion of energy in the form of sunlight ultimately into biomass energy is the "magic of photosynthesis" on which all life located on Earth's surface is dependent. Given the extensive research published on photosynthetic adaptation in *Chlamydomonas priscuii* over the past thirty years (Morgan-Kiss et al. 2006; Cvetkovska et al. 2022), it may represent a useful template to elucidate the mechanisms of photosynthetic adaptation to maintain photostasis and energy balance in polar photosychrophiles and psychrotolerant species in general.

An astonishing feature of *C. priscuii* is that it exhibits maximum light- and CO_2-saturated rates of O_2 evolution at its optimum growth temperature (8–12 °C), which are similar to the optimum rates of photosynthesis in *C. reinhardtii* but measured at its optimal growth temperature of 28 °C (Pocock et al. 2007). These data are consistent with the comparable light response curves for excitation pressure, defined as the relative redox state of the intersystem photosynthetic electron transport chain (Hüner et al. 1998), for *C. priscuii* and *C. reinhardtii* measured at their optimal growth temperatures (Szyszka et al. 2007). Thus, the photopsychrophile, *C. priscuii*, maintains a comparable energy balance as that of the model mesophile, *C. reinhardtii*, albeit under drastically different thermodynamic constraints. This is partly accounted for by the significant genetic redundancy within its genome (Zhang et al. 2021), which translates into significant changes in the structure, function, and organization of the photosynthetic apparatus of this photopsychrophile relative to the model photomesophile, *C. reinhardtii* (Szyszka-Mroz et al. 2015, 2019; Cvetkovska et al. 2018, 2019; Kalra et al. 2020, 2023).

A reorganization of the photosynthetic apparatus in *C. priscuii* compared with C. *reinhardtii* can be detected *in vivo* by 77 K fluorescence emission spectroscopy combined with *in vitro* separation of the major thylakoid pigment-protein complexes (Szyszka-Mroz et al. 2019). As expected, the mesophile exhibits two major emission bands associated with PSII: one at about 680 nm associated with LHCII and another at about 685 nm associated with the PSII core complexes. A third major emission band is typically detected at around 715 nm, associated with PSI. In contrast, *C. priscuii* consistently exhibits a very distinctive and unusual 77 K emission spectrum characterized by prominent emission bands between 680 and 685 nm associated with LHCII and PSII core complexes but minimal emission of a prominent PSI emission band at 715 nm, even though PSI is present and functional in this psychrophile (Kalra et al. 2020; Morgan et al. 1998; Morgan-Kiss et al. 2005; Szyszka et al. 2007, Szyszka-Mroz et al. 2015, 2019; Cook et al. 2019). The absence of a major PSI 77 K emission band in this psychrophile is, at least in part, due to the following: (1) a reduction in the level of 7 of the 11 Lhca polypeptides normally associated with PSI (Morgan et al. 1998); (2) the presence of a novel PSI-Cytb6/f supercomplex which appears to be important in the regulation of PSI cyclic electron flow (CEF) (Kalra et al. 2020; Szyszka-Mroz et al. 2015) and (3) a major reorganization in the distribution of PSI-LHCI and PSII-LHCII complexes within thylakoid membranes based on digitonin fractionation experiments combined with immunoblot analyses (Szyszka-Mroz et al. 2019). We note with interest that the Arctic photopsychrophile, *Chlamydomonas malina,* exhibits comparable 77 K emission spectra to that of the Antarctic psychrophile, *C. priscuii* (Hüner et al. 2020).

1.3.1 State Transitions

State transitions are considered essential for acclimation to a fluctuating light environment for all known green algae and terrestrial plants. This process is governed by the reversible phosphorylation of LHCII that is regulated by the thylakoid protein kinases, Stt7 and Stl1, in *C. reinhardtii* and by STN7 and STN8, in *A. thaliana* (Rochaix 2014; Wunder et al. 2013). These protein kinases sense and respond to the reduction state of the intersystem PETC (Allen et al. 1981; Oxborough et al. 1987; Allen 1992; Zer and Ohad 2003) in which the Cyt b6/f complex is an obligatory electron transfer and proton-translocating complex in all oxygenic photosynthetic organisms. This complex couples the oxidation of the plastoquinol (PQH2) pool with the reduction of plastocyanin (PC), the electron donor to $P700^{+}$. The pmf generated between the stroma and the lumen of thylakoid membranes by the PETC is governed by the Q cycle associated with the Cyt b6/f complex. This contributes to the formation of the proton gradient required for ATP biosynthesis through either linear electron flow (LEF) or PSI CEF (Fig. 1.1). An enhanced reduction state of the intersystem PETC induces the phosphorylation of a mobile population of LHCII which disengages from PSII with its subsequent migration within the plane of the thylakoid membrane to become physically associated with PSI (Fig. 1.5). Thus, state transitions govern the energy distribution between PSII and PSI in a coordinated manner under fluctuating light conditions to maximize rates of photosynthetic electron transport.

C. priscuii's natural environment in Lake Bonney, Antarctica, consists of a relatively constant but low temperature and low light conditions enriched in the blue region of the spectrum during 6 months of austral summer and complete darkness for 6 months of the austral winter (Morgan-Kiss et al. 2006). Even though this polar photopsychrophile can maintain high rates of photosynthesis at low growth temperature (Pocock et al. 2007), it is deficient in the ability to undergo state transitions to regulate energy distribution between PSII and PSI (Morgan-Kiss et al. 2002a, b, 2005; Szyszka et al. 2007; Szyszka-Mroz et al. 2019; Gudynaite-Savitch 2006, 2007; Kalra et al. 2023). This is a novel feature of this photopsychrophile. Concomitantly, *C. priscuii* exhibits an unusual Cytf/b6 complex in which the Cyt f is 7kD smaller than that of *C. reinhardtii* based on SDS-PAGE, and it is less heat stable than that of *C. reinhardtii*. We hypothesized that the inability to undergo state transitions in *C. priscuii* was due to the presence of its unique Cyt f. However, transformation of the *ΔpetA* mutant of *C. reinhardtii* with the *petA* gene from the psychrophile retained the capacity to undergo state transitions, indicating that the novel Cyt f of *C. priscuii* did not explain the inability of this psychrophile to undergo classic state transitions (Gudynaite-Savitch 2006, 2007). Nevertheless, *C. priscuii* retained the capacity for long-term adjustment in energy distribution between PSI and PSII by differentially adjusting the stoichiometry of PSI relative to PSII at the protein level in response to red light (Morgan-Kiss et al. 2005). Thus, it appears that the maintenance of the capacity to perform short-term state transitions is not essential for the survival of *C. priscuii* in Lake Bonney. In contrast to *C. priscuii*,

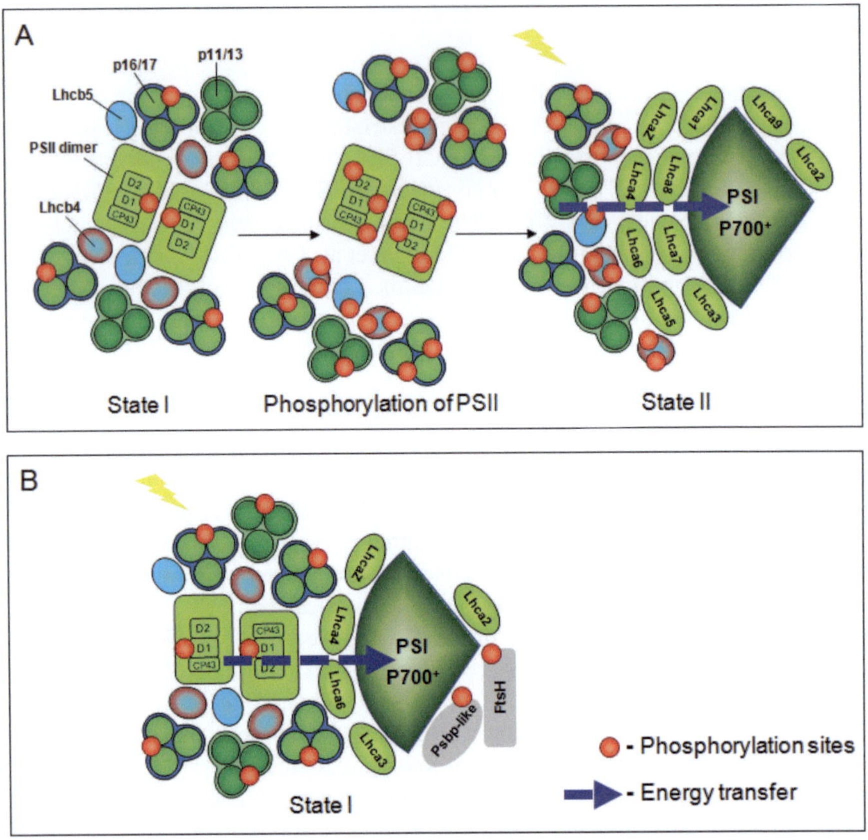

Fig. 1.5 (**a**) Hypothetical model for state transitions from state I to state II, in the model green alga, *Chlamydomonas reinhardtii. C.reinhardtii* LHCII consists of two minor (monomeric) LHCIIs, Lhcb4 and Lhcb5, in addition to ten major (trimeric) Lhcbm proteins (Lhcbm1-10). Photosystem I of *C. reinhardtii* contains 9–10 light-harvesting proteins (Lhca) located on one side of the core. (Drop et al. 2011). Major Lhcbm proteins are organized in trimers, which are connected to the PSII core complex through monomeric Lhcb4 and Lhcb5. It has been proposed that both minor LHCIIs, Lhcb4, Lhcb5, and one major LHCII, Lhcbm5, physically shuttle between PSII and PSI during state transitions, and their affinity for either photosystem may depend on phosphorylation state. During state I conditions, these three LHCII proteins may act as linker proteins as they are located at the interface between the PSII core and major LHCII trimers. Therefore, it is likely that dissociation of Lhcb4, Lhcb5, and Lhcbm5 results in the release of major (trimeric) LHCII. (**b**). In contrast to *C. reinhardtii* thylakoid membranes *of C. priscuii* exhibit lower phosphorylation of PSII light-harvesting proteins and much lower abundance of Lhca polypeptides, which effectively restrict their capacity to perform state transitions. Instead, formations of PSII-LHCII-PSI-LHCI megacomplexes are involved in spillover-type light energy transfer from PSII to PSI and subsequent quenching of excess energy by the oxidized reaction center of PSI ($P700^+$)

psychrophilic *Chlorella* sp. strain BI as well as *Chlamydomonas* sp. ICE-MVD do undergo short-term state transitions (Morgan-Kiss et al. 2008; Kalra et al. 2023). Clearly, the presence or absence of short-term state transitions is not a prerequisite for photopsychrophily in eukaryotic algae. This may reflect, in part, the very different habitats and ecosystems occupied by *C. priscuii*, *Chlorella* sp. strain BI, and *Chlamydomonas* sp. ICE-MVD.

Despite the deficiency to undergo classic state transitions, *C. priscuii* exhibits the presence of *Stt7* and *Stl1* at the genome (Zhang et al. 2021) as well as at the protein level (Szyszka-Mroz et al. 2019). Consistent with adaptation to its low-temperature environment, the extent of thylakoid protein phosphorylation in *C. priscuii* was highest at 5 °C but inhibited at 25 °C, whereas the opposite temperature dependence for protein phosphorylation was observed for the mesophile, *C. reinhardtii* (Szyszka-Mroz et al. 2019). The kinase domain of Stt7 in *C. priscuii* exhibited significant structural alterations, which were interpreted to indicate its predisposition to function maximally at low temperature and exhibit minimal sensitivity to the known protein kinase inhibitor, staurosporine, compared to that of *C. reinhardtii* (Szyszka-Mroz et al. 2019). Furthermore, the phosphorylation of the complement of light-harvesting polypeptides, as well as PSII and PSI core polypeptides such as subunits of the PSI-Cytf/b6 complex, the PsbP subunit as well as the FtsH metalloprotease subunit was distinct in *C. priscuii*. Functionally, the phosphorylation status of these subunits, combined with the requirement for high salt, governed the stability of the PSI-Cytf/b6 supercomplex and the rates of CEF in *C. priscuii* (Kalra et al. 2020, 2023; Szyszka-Mroz et al. 2019).

The Stt7/Stl1 kinases from *C. priscuii* (Szyszka-Mroz et al. 2019) are examples of membrane-associated, cold-adapted enzymes that function optimally at cold temperatures but are inhibited at high temperatures (Aquist et al. 2017). Rather than governing state transitions, we suggested thylakoid protein phosphorylation may regulate thylakoid membrane reorganization as reflected in the distribution of PSII, PSI, and Cytf/b6 complexes coupled with a marked energy quenching capacity (Szyszka et al. 2007; Szyszka-Mroz et al. 2019) controlling a proposed PSII-PSI spillover quenching mechanism similar to the spillover-mediated PSI-dependent energy quenching reported in cold-adapted pine (Bag et al. 2020), Arabidopsis (Velitchkova et al. 2020), and other organisms exposed to harsh environmental conditions (Slavov et al. 2013, 2016).

This conclusion remains equivocal since further experimentation is required to differentiate the contribution of the LHCII phosphorylation-dependent mechanism for state transitions from its potential in the PSII-PSI spillover mechanism in *C. priscuii*. However, the regulation of state transitions through LHCII phosphorylation and granal thylakoid stacking appears to act synergistically to govern CEF versus LEF in *Arabidopsis thaliana* in response to changes in irradiance (Hepworth et al. 2021). When low temperatures limit CO_2 assimilation, both NADPH levels and reduced ferredoxin (Fd) accumulate, which enhance the reduction state of the intersystem PETC, which is manifested by increased excitation pressure (Hüner et al. 1998). This activates the STN7 kinase, stimulates LHCII phosphorylation and the concomitant decrease in granal stacking in *Arabidopsis thaliana,* and increases excitation of PSI compared to PSII. This results in a functional transition from predominantly LEF to CEF (Hepworth et al. 2021). However, since green algae, including photopsychrophilic *C. priscuii* and mesophilic *C. reinhardtii,* exhibit granal stacking that is much less extensive than that observed in *Arabidopsis thaliana,* this mechanism to govern the ratio of LEF/CEF may not apply to green algae. Consequently, Hepworth et al. (2021) suggest that the novel PSI-supercomplexes observed in *C. reinhardtii* (Takahashi et al. 2013; Iwai et al. 2010) and *C. priscuii* (Szyszka-Mroz et al. 2019; Kalra et al. 2020, 2023) may govern the ratio of CEF/LEF in green algae.

General models of plant thylakoid membranes assume lateral heterogeneity with respect to the asymmetric distributions of PSII and PSI supercomplexes, with PSII largely localized to the granal stacked regions and PSI associated with granal margins and stromal lamellae (Anderson and Anderson 1982; Anderson et al. 1995; Anderson 1986, 2012). In contrast, recent data indicate that PSI and PSII can be found physically associated with one another in mega-PSI-PSII- pigment protein complexes (Jarvi et al. 2011; Yokono et al. 1865, 2019; Gao et al. 2019; Kim et al. 2023) that increase the propensity for energy spillover from PSII to PSI. Thus, energy spillover represents a mechanism whereby PSI quenches EEE associated with PSII, thereby protecting PSII from photodamage (Yokono et al. 2019). Using digitonin fractionation of the thylakoid membranes, we reported that *C. priscuii*, exhibits a significant alteration in the distribution of PSI and PSII pigment-protein complexes between grana and stromal lamellae that is associated with altered protein phosphorylation patterns (Szyzska-Mroz et al. 2019) compared to the model photomesophile, *C. reinhardtii*. Since the photopsychrophile exhibits minimal capacity for state transitions, we suggest that these changes in thylakoid organization and distribution of PSI and PSII supercomplexes in *C. priscuii* support the thesis that PSII-PSI energy spillover regulates photosystem energy partitioning and quenching of EEE. This may account for the absence of state transitions in *C. priscuii*. However, this hypothesis requires validation through the isolation of PSI-PSII megacomplexes from *C. priscuii*, which is presently under investigation.

1.3.2 Ferredoxins and Energy Balance

An important role of photosynthetic ferredoxins (Fd) is to regulate the distribution of electrons from PSI to various essential metabolic reactions in cyanobacteria, green algae, and plants (Knaff 1996; McKay et al. 1999; Boehm et al. 2015; Schorsch et al. 2018). A global interaction Fd network has been reported for *C. reinhardtii*, providing putative roles for this protein in redox metabolism, carbohydrate modification, fatty acid biosynthesis, hydrogen production, nitrogen and sulfur metabolism, state transitions, and dark anoxia (Peden et al. 2013).

The *PETF* isoform (Fd1) of the family of photosynthetic ferredoxins represents about 98% of all transcribed Fd genes in *C. reinhardtii* (Terauchi et al. 2009). Psychrophilic PETF from *C. priscuii* exhibited high structural stability and activity at 10 °C but is more labile at temperatures greater than 40 °C than that of *C. reinhardtii* (Cvetkovska et al. 2018). In contrast to other photoautotrophs, transcriptome coupled with polypeptide analyses indicated that *C. priscuii* expressed and accumulated two forms of ferredoxin, Fd-1A and Fd-1B (Cvetkovska et al. 2018). Detailed genomic comparisons of *Fd-1A* and *Fd-1B* from *C. priscuii* with 21 other *Fd* genes from green algae, diatoms, plants, and cyanobacteria indicated that all Fd isoforms were highly conserved at the amino acid level. Photomesophilic *C. reinhardtii* and the photopsychrophilic *C. priscuii* Fd differed by 11 amino acids. Furthermore, these differences occur in regions distant from the active site and appear to be involved in Fe binding and protein-protein interactions (Cvetkovska et al. 2018). It is interesting to note that such differences in predicted amino acid sequences can lead to such significant differences in temperature stability and activity. The observation that the small changes in amino acid sequence of *C. priscuii* Fd can lead to significant changes in temperature stability is consistent with those of other psychrophilic as well as thermophillic enzymes examined to date (Feller and Gerday 2003; D'Amico et al. 2006; Aquist et al. 2017; Rainer 2000; Pinney et al. 2021). This apparent enigma may, in part, reflect epistatic effects of amino acid residues on protein structure (Starr and Thornton 2016). Regardless of the molecular mechanism, ferredoxin from *C. priscuii* is adapted to function at cold temperatures consistent with the psychrophilic nature of *C. priscuii* (Cvetkovska et al. 2018). However, Fd-1 from *C. priscuii* appears to be unique among psychrophilic proteins since it appears not only to exhibit enhanced activity and structural stability at low temperature, but it also accumulates to levels that are double those observed in mesophilic *C. reinhardtii* (Cvetkovska et al. 2018). This combination may reflect the inherent redundancy observed in the genome of *C. priscuii* (Zhang et al. 2021; Hüner et al. 2020) and/or the contributions of alternative splicing (Kjer-Hansen and Weatheritt 2023) to the higher capacity for photosynthetic electron transport and the maintenance of photostasis in its natural, cold environment (Hüner et al. 2020).

1.3.3 Adaptation to an Extreme Photoperiod

In addition to adaptation to low temperature and constant low light levels in Lake Bonney during austral summer, *C. priscuii* is also adapted to 6 months of darkness during austral winter. Consequently, *C. priscuii* exhibits a shade-adapted state but maintains normal Chl levels during the dark austral winter months (Morgan-Kiss et al. 2006, 2016). Most terrestrial plants and green algae become etiolated when exposed to prolonged darkness (Nemhauser and Chory 2009). In contrast, *C. priscuii* maintains high levels of Chl a and Chl b during the austral winter period despite the prolonged darkness (Morgan-Kiss et al. 2016). In eukaryotic photoautrophs, the rate-limiting step in chlorophyll biosynthesis is the conversion of protochlorophyllide to chlorophyllide catalyzed by protochlorophyllide oxidoreductase (Reinbothe and Reinbothe 1996). However, in most eukaryotes, two distinct forms of this enzyme catalyze this reaction: LPOR is the light-dependent form of protochlorophyllide oxidoreductase that catalyzes this reaction in the light, whereas DPOR is the light-independent protochlorophyllide oxidoreductase that catalyzes this reaction in the dark (Reinbothe and Reinbothe 1996). The three genes (*chlL*, *chlN*, and *chlB*) that encode DPOR are absent not only from the genome of *C. priscuii* but are also absent from its plastome and mitochondrial genomes (Cvetkovska et al. 2019). In contrast, two other duplicated genes, *GUN4* and *CAO*, that govern Chl biosynthesis and retrograde signaling (Chory and Wu 2001; Nott et al. 2006; Tanaka and Tanaka 2007) are present in the genome of *C. priscuii*. Therefore, this photopsychrophile appears to have a functional Chl biosynthetic pathway that is totally dependent on LPOR and has lost DPOR, even though this alga is exposed to prolonged annual seasonal darkness.

This astonishing enigma may be rationalized by the fact that at 17 m below the bottom of the ice cover, where *C. priscuii* exists, Lake Bonney exhibits extremely high O_2 concentrations (Morgan-Kiss et al. 2006). DPOR is very sensitive to O_2 concentrations due to the presence of an essential Fe-S cluster, which is not present in LPOR. Consequently, it has been hypothesized that the evolutionary elimination of DPOR from *C. priscuii* would not result in a deleterious effect on the survival of this photopsychrophile, which may explain the absence of *chlL*, *chlN*, and *chlB* genes in *C. priscuii* (Cvetkovska et al. 2019). However, to validate this hypothesis with respect to O_2 concentrations and the absence of DPOR, more phototrophs from Lake Bonney must be examined. Furthermore, the mechanism by which *C. priscuii* represses the normal process of etiolation in the dark and maintains normal Chl levels has yet to be elucidated.

What physiological and biochemical adaptations allow *C. priscuii* to persist through such a prolonged dark period? An elegant field experiment was performed by Morgan-Kiss and colleagues (2016) to address this question. Laboratory cultures of *C. priscuii* were placed in dialysis bags containing water from Lake Bonney and subsequently transferred back into Lake Bonney and suspended at a depth of 17 m to allow the photopsychrophile to respond to its natural environment. Physiological and molecular analyses of these transplanted cultures were subsequently compared with the natural phytoplankton community of Lake Bonney during the natural transition from the light of austral summer to the darkness of austral winter (Morgan-Kiss et al. 2016). During this transition, *C. priscuii* ceased CO_2 fixation associated with a down-regulation of *rbcL* and *psbA* expression. It was concluded that *C. priscuii* transplanted cultures shifted from a light-adapted state to a shade-adapted state during the transition to polar night. This is consistent with the previously proposed model for dark adaptation in this polar alga (Morgan-Kiss et al. 2006, 2016; Mock and Thomas 2008). The processes involved in the reactivation and the induction of photosynthesis in the transition from the darkness of austral winter to the light conditions of austral summer to maintain photostasis remains to be elucidated.

1.3.4 Colony Formation and Photoprotection

C. priscuii cultured under continuous light at low, permissive temperatures (4–10 °C) typically exhibit 20% nonmotile palmelloid colonies and 80% motile single cells (Fig. 1.6) (Szyszka-Mroz et al. 2022). Palmelloids of *C. priscuii* show decreased levels of the PSII light-harvesting complex II (LHCII) proteins, consistent with a twofold higher Chl a/b ratio during growth at 16 °C than growth at 8 °C, compared to single cells which reduces light absorption efficiency (Szyszka-Mroz et al. 2022). This is reflected in comparative differences in 77 K fluorescence emission spectra of *C. priscuii* single cells relative to palmelloids (Szyszka-Mroz et al. 2022), which, in turn, reflects changes in the pigment and polypeptide compositions as well as the organization of PSII and PSI and their functional interactions (Krause and Weis 1991; Baker 2008; Swoczyna et al. 2022). Thus, surprisingly, the transition from single cells to multicellular palmelloids appears to result in detectable alterations in the organization and composition of the PSI and PSII pigment-protein complexes (Szyszka-Mroz et al. 2022).

Although the maximum PSII photochemical efficiency, measured as Fv/Fm, was comparable for isolated single cells and isolated palmelloids, the reduction state of the intersystem electron transport chain, measured as 1-qL, was lower in palmelloids than in single cells (Szyszka-Mroz et al. 2022), consistent with an increased capacity to dissipate EEE in palmelloids compared to single cells. This differential reduction state was associated with 50% higher light saturated rates of ETR in palmelloids relative to single cells exposed to high light at low temperature. The decreased sensitivity to high light was, in part, due to enhanced non-regulated dissipation of EEE (ΦNO) presumably through reaction center quenching (Szyszka et al. 2007; Ivanov et al. 2008) and/or PSII-PSI energy spillover rather than normal antenna quenching via the xanthophyll cycle (Adams and Demmig-Adams 1992, 1993; Adams et al. 2007; Hüner et al. 2003; Verhoeven 2014). Functionally, the temperature-induced morphological transition of *C. priscuii* from a single to a multicellular state alters the composition and organization of the photosynthetic

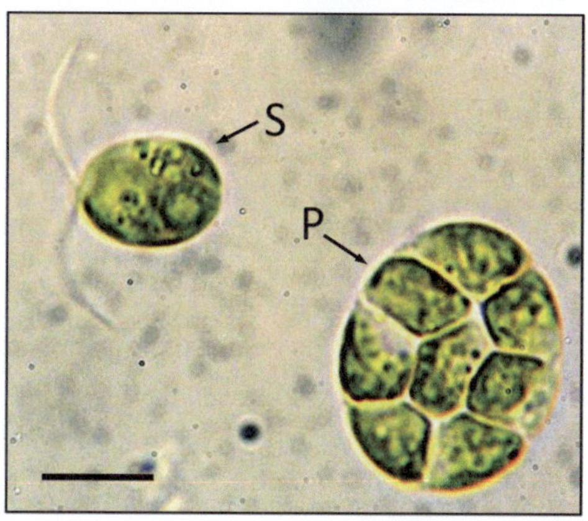

Fig. 1.6 Light microscope image of *C. priscuii*, indicating a single cell (S) and palmelloid (P). Magnification 100 x. Scale bar, 5 µm

apparatus, which predisposes it to dissipate EEE via non-regulated dissipation of excess absorbed energy (Szyszka-Mroz et al. 2022). Thus, surprisingly, single cells are morphologically and photosynthetically distinct from the single cells enclosed within a palmelloid. Although palmelloid formation may be an additional selective criterion for multicellularity in some green algae, the enigma is that there is no evidence that palmelloid accumulation occurs in natural Antarctic and Arctic populations of photopsychrophiles. Regardless, the molecular mechanism underlying the observed alterations in the organization of the photosynthetic apparatus in response to the transition of cultured single cells to multicellular palmelloids remains to be elucidated.

1.4 Conclusions

A major challenge for all endemic polar photoautotrophs is to maintain photostasis, the balance of energy flux between rapid, temperature-insensitive light absorption and photochemistry versus temperature-sensitive, energy consumption through the slower, enzyme-catalyzed biochemical reactions essential for light-dependent CO_2 assimilation, respiration, growth, development, and reproduction in an extremely inhospitable, and thermodynamically challenging environment. This is associated with reorganization of the structure, composition, and function of the photosynthetic apparatus, which is reflected in the phenotypic plasticity observed in photoautotrophs either during short-term acclimation or long-term adaptation to low temperature. Functionally, *C. priscuii* exhibits high rates of CEF but appears to regulate energy distribution between PSI and PSII through PSI-PSII energy spillover rather than state transitions, which is associated with distinct protein phosphorylation patterns for PSII and PSI. This polar photopsychrophile exhibits a light-adapted state during austral summer, which is converted to a shade-adapted state during the dark austral winter months. However, unlike most terrestrial plants and aquatic green algae, *C. priscuii* does not etiolate during prolonged darkness during austral winter, but rather, maintains high levels of Chl a and Chl b. Single cells of *C. priscuii* are morphologically and photosynthetically distinct from the single cells enclosed within a palmelloid membrane. Although palmelloid formation may be a developmental transition to multicellularity, its associated resistance to photoinhibition may be considered an evolutionary advantage not previously considered. However, the molecular mechanism underlying the observed alterations in the organization of the photosynthetic apparatus in response to the transition of single cells to multicellular palmelloids remains an enigma. Despite the novel insights that the genomics of *C. priscuii* has provided, no single genetic feature is evident to indicate the determining factor that confers photopsychrophily. This is yet another enigma of photopsychrophily and its inherent complexity at all levels of cellular organization. However, an essential physiological feature of *C. priscuii* and perhaps photopsychrophily in general is the capacity to maintain energy balance and photostasis under such a thermodynamically challenging environment. Clearly, photopsychrophily is a very complex phenomenon that requires continued fundamental controlled experimentation validated by field research to untangle the complexity of photostasis, adaptation, and survival in the extreme polar environment. The extensive biodiversity of Arctic and Antarctic photopsychrophiles, coupled with their wide-ranging environments to which they are adapted, reflects an enormous "possibility space" (Wong et al. 2023) for the evolution of photopsychrophily. Thus, it is unlikely that photopsychrophily can be explained by any single genetic, molecular, or morphological trait. Rather, we suggest that it is the range of cellular processes essential to maintain photosynthetic energy balance in the thermodynamically challenging polar environments that best explains photopsychrophily. This range of processes reflects the evolutionary "possibility space" available that leads to "dynamic persistence" (Wong et al. 2023), the evolutionary selection criterion necessary for polar adaptation of all photopsychrophilic and photopsychrotolerant species. We suggest that this "dynamic persistence" is functionally reflected in the maintenance of photostasis, which functionally balances the ratio of photosynthetic efficiency versus photosynthetic capacity. We suggest that the search for a model polar photopsychrophile may be illusory. Rather, what is needed is greater emphasis on field research combined with controlled, laboratory experimentation focused on the ecology, physiology, and molecular biology of the myriad species that have evolved in thermodynamically challenging and diverse polar environments.

References

Adams WW III, Demmig-Adams B (1992) Operation of the xanthophyll cycle in higher plants in response to diurnal changes in incident sunlight. Planta 186:390–398

Adams WW III, Demmig-Adams B (1993) Energy dissipation and photoprotection in leaves of higher plants. In: Yamamoto S (ed) Photosynthetic responses to the environment, vol 8. American Society of Plant Physiology, pp 27–38

Adams WW, Watson AM, Mueh KE et al (2007) Photosynthetic acclimation in the context of structural constraints to carbon export from leaves. Photosynth Res 94:455–466

Allen JF (1992) Protein phosphorylation in regulation of photosynthesis. Biochim Biophys Acta 1098:275–335

Allen JF, Bennett J, Steinback et al (1981) Chloroplast protein phosphorylation couples plastoquinone redox state to distribution of excitation energy between photosystems. Nature 291:21–25

Anderson JM (1986) Photoregulation of the composition, function and structure of thylakoid membranes. Ann Rev Plant Physiol 37:93–136

Anderson JM (2012) Lateral heterogeneity of plant thylakoid protein complexes: early reminiscences. Philos Trans Royal Soc B: Biol Sci 367:3384–3388

Anderson JM, Andersson B (1982) The architecture of photosynthetic membranes: lateral and transverse organization. Trends Biochem Sci 7:288–292

Anderson JM, Chow WS, Park Y-I (1995) The grand design of photosynthesis: acclimation of the photosynthetic apparatus to environmental cues. Photosynth Res 46:129–139

Aquist J, Isaksen GV, Brandsdal BO (2017) Computation of enzyme cold adaptation. Nat Rev Chem 1:1–14

Arrigo KR (2014) Sea ice ecosystems. Annu Rev Mar Sci 6:439–467

Arrigo KR, Worthen DL, Lizotte MP et al (1997) Primary production in Antarctic sea ice. Science 276:394. https://doi.org/10.1126/science.276.5311.394

Atkins P (2010) The laws of thermodynamics. A very short introduction. Oxford University Press
Bag P, Chukhutsina V, Zhang Z et al (2020) Direct energy transfer from photosystem II to photosystem I confers winter sustainability in Scots pine. Nat Commun 11:6388
Baker NR (2008) Chlorophyll fluorescence: a probe of photosynthesis in vivo. Annu Rev Plant Biol 59:89–113
Bakermans C (2012) Psychrophiles: life in the cold. In: Extremophiles: microbiology and biotechnology. Caister Academic Press, pp 53–76
Bell EM (1998) Plankton dynamics in the saline lakes of the Vestfold Hills, Eastern Antarctica. University of Nottingham
Bell EM, Laybourn-Parry J (2003) Mixotrophy in the Antarctic phytoflagellate, Pyramimonas gelidicola (Chlorophyta: Prasinophyceae). J Phycol 39(4):644–649
Bielewicz S, Bell E, Kong W et al (2011) Protist diversity in a permanently ice-covered Antarctic Lake during the polar night transition. ISME J 5:1559–1564
Bligh J, Johnson KG (1973) Glossary of terms for thermal physiology. J Appl Physiol 35:941–961
Boehm M, Alahuhta M, Mulder DW et al (2015) Crystal structure and biochemical characterization of Chlamydomonas FDX2 reveal two residues that, when mutated, partially confer FDX2 the redox potential and catalytic properties of FDX1. Photosynth Res 128:45–57
Bolaños LM, Lee K-B, Chang JC et al (2020) Small phytoplankton dominate western North Atlantic biomass. ISME J 14(7):1663–1674
Bomblies K (2020) When everything changes at once: finding a new normal after genome duplication. Proc R Soc B 287:20202154
Bravo LA, Ulloa N, Zuniga GE et al (2001) Cold resistance in Antarctic angiosperms. Physiol Plant 111:55–65
Bravo LA, Saavedra-Mella FA, Vera F et al (2007) Effect of cold acclimation on the photosynthetic performance of two ecotypes of Colobanthus quitensis (Kunth) Bartl. J Exp Bot 58:3581–3590
Camoying MG, Trimborn S (2023) Physiological response of an Antarctic cryptophyte to increasing temperature, CO_2, and irradiance. Limnol Oceanogr 68:1880–1894
Carson HL (1987) The genetic system, the deme, and the origin of species. Annu Rev Genet 21:405–423
Casanueva A, Tuffin M, Cary C et al (2010) Molecular adaptations to psychrophily: the impact of 'omic' technologies. Trends Microbiol 18:374–381
Cavalli G, Heard E (2019) Advances in epigenetics link genetics to the environment and disease. Nature 571:489–499
Chory J, Wu D (2001) Weaving the complex web of signal transduction. Plant Physiol 125:77–80
Chow WS (2003) Photosynthesis: From natural towards artificial. J Biol Phys 29:447–459
Chown SL, Clarke A, Fraser CI et al (2015) The changing form of Antarctic biodiversity. Nature 522:431–438
Chrismas N, Anesio A, Sanchez-Baracaldo P (2015) Multiple adaptations to polar and alpine environments within cyanobacteria: a phylogenomic and Bayesian approach. Front Microbiol 6:01070
Comai L (2005) The advantages and disadvantages of being polyploid. Nat Rev Genet 6:836–846
Cook G, Teufel A, Kalra I, Wang X et al (2019) The Antarctic psychrophiles Chlamydomonas spp. UWO241 and ICE-MDV exhibit differential restructuring of photosystem I in response to iron. Photosynth Res 141:209–228
Cvetkovska M, Hüner NPA, Smith DR (2017) Chilling out: the evolution and diversification of psychrophilic algae with a focus on Chlamydomonadales. Polar Biol 40:1169–1184
Cvetkovska M, Szyszka-Mroz B, Possmayer M et al (2018) Characterization of photosynthetic ferredoxin from the Antarctic alga Chlamydomonas sp. UWO241 reveals novel features of cold adaptation. New Phytol 219:588–604
Cvetkovska M, Orgnero S, Hüner NPA et al (2019) The enigmatic loss of light-independent chlorophyll biosynthesis from an Antarctic green alga in a light-limited environment. New Phytol 222:651–656

Cvetkovska M, Vakulenko G, Smith DR et al (2022) Temperature stress in psychrophilic green microalgae: mini-review. Physiol Plant 174:e13811
D'Amico S, Marx J-C, Feller et al (2006) Psychrophilic microorganisms: challenges for life. EMBO Rep 7:385–389
De Clerck O, Kao SM, Bogaert KA et al (2018) Insights into the evolution of multicellularity from the sea lettuce genome. Curr Biol 28:2921–2933
Dolhi JM, Maxwell DP, Morgan-Kiss RM (2013) Antarctic Chlamydomonas raudensis: an emerging model for cold adaptation of photosynthesis. Extremophiles 17:711–722
Dolhi JM, Teufel AG, Kong W et al (2015) Diversity and spatial distribution of autotrophic communities within and between ice-covered Antarctic lakes (McMurdo Dry Valleys). Limnol Oceanogr 60:977–991
Dorrell RG, Kuo A, Füssy Z et al (2023) Convergent evolution and horizontal gene transfer in Arctic Ocean microalgae. Life Sci Alliance 6:e202201833
Drop B, Webber-Birungi M, Fusetti F et al (2011) Photosystem I of Chlamydomonas reinhardtii contains nine light-harvesting complexes (Lhca) located on one side of the core. J Biol Chem 286:44878–44887
Ebadi M, Bafort Q, Mizrachi E et al (2023) The duplication of genomes and genetic networks and its potential for evolutionary adaptation and survival during environmental turmoil. Proc Natl Acad Sci USA 120:2307289120
Eddie B, Krembs C, Neuer S (2008) Characterization and growth response to temperature and salinity of psychrophilic, halotolerant Chlamydomonas sp. ARC isolated from Chukchi Sea ice. Mar Ecol Progr Ser 354:107–117
Eichten SR, Schmitz RJ, Springer NM (2014) Epigenetics: beyond chromatin modifications and complex genetic regulation. Plant Physiol 165:933–947
England J (2020) Every life is on fire- how thermodynamics explains the origins of living things. Basic Books, New York
Feller G, Gerday C (2003) Psychrophilic enzymes: hot topics in cold adaptation. Nat Rev Microbiol 1:200–208
Gao S, Chi Z, Chen H et al (2019) A supercomplex, of approximately 720 kDa and composed of both photosystem reaction centers, dissipates excess energy by PSI in green macroalgae under salt stress. Plant Cell Physiol 60:166–175
Gray A, Krolikowski M, Fretwell et al (2020) Remote sensing reveals Antarctic green snow algae as important terrestrial carbon sink. Nat Commun 11:2527
Grigoriev IV, Hayes RD, Calhoun S et al (2021) PhycoCosm, a comparative algal genomics resource. Nucleic Acids Res 49:D1004–D1011
Gudynaite-Savitch L, Gretes M, Morgan-Kiss RM et al (2006) Cytochrome f from the Antarctic psychrophile, Chlamydomonas raudensis UWO 241: structure, sequence, and complementation in the mesophile, Chlamydomonas reinhardtii. Mol Genet Genomics 275:387–398
Gudynaite-Savitch L, Loiselay C, Savitch LV et al (2007) The small domain of cytochrome f from the psychrophile, Chlamydomonas raudensis UWO 241, modulates the apparent molecular mass and decreases the accumulation of cytochrome f in the mesophile, Chlamydomaonas reinhardtii. Biochem Cell Biol 85:616–627
Haberman KL, Ross RM, Quetin LB (2003) Diet of the Antarctic krill (Euphausia superba Dana): II. Selective grazing in mixed phytoplankton assemblages. J Exp Mar Biol Ecol 283:97–113
Hawes I (1990) Eutrophication and vegetation development in a maritime Antarctic lake. In: Antarctic ecosystems: ecological change and conservation, pp 83–90
Hemenway EA, Gehring M (2023) Epigenetic regulation during plant development and the capacity for epigenetic memory. Annu Rev Plant Biol 74:87–109
Hepworth C, Wood WHJ, Emrich-Mills et al (2021) Dynamic thylakoid stacking and state transitions work synergistically to avoid acceptor-side limitation of photosystem I. Nat Plants 7:87–93
Hoffman PF, Kaufman AJ, Halverson GP et al (1998) A neoproterozoic snowball earth. Science 281:1342. https://doi.org/10.1126/science.281.5381.1342

Hofmann HJ, Narbonne GM, Aitken JD (1990) Ediacaran remains from intertillite beds in northwestern Canada. Geology 18:1199–1202
Hoham RW (1975) Optimum temperatures and temperature ranges for growth of snow algae. Arctic Alpine Res 7:13–24
Hoham RW, Remias D (2020) Snow and glacial algae: a review. J Phycol 56:264–282
Hollis L, Hüner NPA (2014) Retrograde operational sensing and signalling pathways maintain photostasis in green algae, cyanobacteria and terrestrial plants. Trends Photochem Photobiol 16:47–61
Hüner NPA, Öquist G, Sarhan F (1998) Energy balance and acclimation to light and cold. Trends Plant Sci 3:224–230
Hüner NPA, Öquist G, Melis A (2003) Photostasis in plants, green algae and cyanobacteria: the role of light harvesting antenna complexes. In: Advances in photosynthesis and respiration, Light harvesting antennas in photosynthesis, vol 13. Kluwer Academic, Dordrecht, pp 401–421
Hüner NPA, Ivanov AG, Cvetkovska M et al (2020) Photosynthetic acclimation and adaptation to cold ecosystems. In: Climate change, photosynthesis and advanced biofuels: the role of biotechnology in the production of value-added plant bio-products. Springer, Singapore, pp 159–201
Ivanov AG, Sane PV, Hurry V et al (2008) Photosystem II reaction centre quenching: mechanisms and physiological role. Photosynth Res 98:565–574
Iwai M, Takizawa K, Tokutsu R et al (2010) Isolation of the elusive supercomplex that drives cyclic electron flow in photosynthesis. Nature 464:1210–1213
Izaguirre I, Allende L, Schiaffino MR (2021) Phytoplankton in Antarctic lakes: biodiversity and main ecological features. Hydrobiologia 848:177–207
Järvi S, Suorsa M, Paakkarinen V (2011) Optimized native gel systems for separation of thylakoid protein complexes: novel super- and mega-complexes. Biochem J 439:207–214
Johnson MD, Oldach D, Delwiche CF et al (2007) Retention of transcriptionally active cryptophyte nuclei by the ciliate Myrionecta rubra. Nature 445:426–428
Judson OP (2017) The energy expansions of evolution. Nat Ecol Evol 1:0138
Kalra I, Wang X, Cvetkovska M et al (2020) Chlamydomonas sp. UWO 241 exhibits high cyclic electron flow and rewired metabolism under high salinity. Plant Physiol 183:588–601
Kalra I, Wang X, Zhang R et al (2023) High salt-induced PSI-supercomplex is associated with high CEF and attenuation of state transitions. Photosynth Res 157:65–84
Kang S-H, Fryxell GA (1992) Fragilariopsis cylindrus (Grunow) Krieger: the most abundant diatom in water column assemblages of Antarctic marginal ice-edge zones. Polar Biol 12:609–627
Kennicutt MC, Chown SL et al (2014) Polar research: six priorities for Antarctic science. Nature 512:523–525
Kennicutt MC, Chown SL, Cassano JJ et al (2015) A roadmap for Antarctic and Southern Ocean science for the next two decades and beyond. Antarct Sci 27:3–18
Kim EJ, Kim S, Choi H-G et al (2020) Co-production of biodiesel and bioethanol using psychrophilic microalga Chlamydomonas sp. KNM0029C isolated from Arctic sea ice. Biotechnol Biofuels 13:1–13
Kim E, Yokono M, Tsugane K et al (2023) Formation of a stable PSI–PSII megacomplex in rice that conducts energy spillover. Plant Cell Physiol 64:858–865
Kirschvink JL, Eric JG, Bertani LE et al (2000) Paleoproterozoic snowball earth: extreme climatic and geochemical global change and its biological consequences. Proc Natl Acad Sci USA 97:1400–1405
Kjer-Hansen P, Weatheritt RJ (2023) The function of alternative splicing in the proteome: rewiring protein interactomes to put old functions into new contexts. Nat Struct Mol Biol 30:1844–1856
Knaff DB (1996) Ferredoxin and ferredoxin-dependent enzymes. In: Advances in photosynthesis, Oxygenic photosynthesis, the light reactions, vol 4. Kluwer Academic, Dordrecht, pp 333–361
Kong W, Ream D, Priscu JC et al (2012) Diversity and expression of RubisCO genes in a perennially ice-covered Antarctic lake during the polar night transition. Appl Environ Microbiol 78:4358–4366

Kovaka S, Ou S, Jenike KM et al (2023) Approaching complete genomes, transcriptomes and epi-omes with accurate long-read sequencing. Nat Methods 20:12–16
Krause GH, Weis E (1991) Chlorophyll fluorescence and photosynthesis: the basics. Annu Rev Plant Physiol Plant Mol Biol 42:313–349
Krupovic M, Dolja VV, Koonin EV (2020) The LUCA and its complex virome. Nat Rev Microbiol 18:661–670
Laybourn-Parry JJ, Roberts EC, Bell EM et al (1996) Microbial dynamics during the summer ice-loss phase in maritime Antarctic lakes. J Plankton Res 18:495–511
Laybourn-Parry J, Roberts EC, Bell EM (2000) Mixotrophy as a survival strategy in Antarctic lakes. In: Antarctic ecosystems. Caxton Press, Caldwell, pp 33–40
Leebens-Mack JH, Barker MS, Carpenter EJ et al (2019) One thousand plant transcriptomes and the phylogenomics of green plants. Nature 574:679–685
Li W, Morgan-Kiss RM (2019) Influence of environmental drivers and potential interactions on the distribution of microbial communities from three permanently stratified Antarctic lakes. Front Microbiol 10:1067
Li W, Podar M, Morgan-Kiss RM (2016) Ultrastructural and single-cell-level characterization reveals metabolic versatility in a microbial eukaryote community from an ice-covered Antarctic lake. Appl Environ Microbiol 82:3659–3670
Liu C, Huang X, Wang X (2006) Phylogenetic studies on two strains of Antarctic ice algae based on morphological and molecular characteristics. Phycologia 45:190–198
Lyon BR, Mock T (2014) Polar microalgae: new approaches to an extreme and changing environment. Biology 3:56–80
Margesin R, Schinner F, Marx J-C et al (2008) Psychrophiles: from biodiversity to biotechnolgy. Springer, Berlin
Markager S, Vincent WF, Tang EPY (1999) Carbon fixation by phytoplankton in high Arctic lakes: implications of low temperature for photosynthesis. Limnol Oceanogr 44:597–607
McKay RML, La Roche J, Yakunin AF et al (1999) Accumulation of ferredoxin and flavodoxin in a marine diatom in response to Fe. J Phycol 35:510–519
McKie-Krisberg ZM, Gast RJ, Sanders RW (2015) Physiological responses of three species of Antarctic mixotrophic phytoflagellates to changes in light and dissolved nutrients. Microb Ecol 70:21–29
Melis A (1998) Photostasis in plants. In: Photostasis and related phenomena. Plenum Press, New York, pp 207–220
Mendes CRB, Mendes CR, Borges C et al (2023) Cryptophytes: an emerging algal group in the rapidly changing Antarctic Peninsula marine environments. Glob Chang Biol 29:1791–1808
Mock T, Thomas DN (2008) Microalgae in polar regions: linking functional genomics and physiology with environmental conditions. In: Psychrophiles: from biodiversity to biotechnolgy. Springer, Berlin, pp 285–312
Mock T, Otillar RP, Strauss J et al (2017) Evolutionary genomics of the cold-adapted diatom Fragilariopsis cylindrus. Nature 541:536–540
Morales-Sánchez D, Schulze PSC, Kiron V et al (2020) Production of carbohydrates, lipids and polyunsaturated fatty acids (PUFA) by the polar marine microalga Chlamydomonas malina RCC2488. Algal Res 50:102016
Morgan RM, Ivanov AG, Priscu JC et al (1998) Structure and composition of the photochemical apparatus of the Antarctic green alga, Chlamydomonas subcaudata. Photosynth Res 56:303–314
Morgan-Kiss RM, Ivanov AG, Huner NPA (2002a) The Antarctic psychrophile, Chlamydomonas subcaudata, is deficient in state I – state II transitions. Planta 214:435–445
Morgan-Kiss R, Ivanov AG, Williams J et al (2002b) Differential thermal effects on the thylakoid pigment-protein complexes in a psychrophilic and mesophilic algae. Biochim Biophys Acta 1561:251–265

Morgan-Kiss RM, Ivanov AG, Pocock T et al (2005) The Antarctic psychrophile, Chlamydomonas raudensis Ettl (UWO241) (Chlorophyceae, Chlorophyta), exhibits a limited capacity to photoacclimate to red light. J Phycol 41:791–800

Morgan-Kiss RM, Priscu JC, Pocock T et al (2006) Adaptation and acclimation of photosynthetic microorganisms to permanently cold environments. Microbiol Mol Biol Rev 70:222–252

Morgan-Kiss RM, Ivanov AG, Modla S et al (2008) Identity and physiology of a new psychrophilic eukaryotic green alga, Chlorella sp., strain BI, isolated from a transitory pond near Bratina Island, Antarctica. Extremophiles 12:701–711

Morgan-Kiss R, Lizotte MP, Kong W et al (2016) Photoadaptation to the polar night by phytoplankton in a permanently ice-covered Antarctic lake. Limnol Oceanogr 61:3–13

Morita RY (1975) Psychrophilic bacteria. Bacteriol Rev 39:144–167

Nadeau TL, Castenholz RW (2000) Characterization of psychrophilic oscillatorians (cyanobacteria) from Antarctic meltwater ponds. J Phycol 36:914–923

Neale PJ, Priscu JC (1995) The photosynthetic apparatus of phytoplankton from a perennially ice-covered Antarctic lake: acclimation to an extreme shade environment. Plant Cell Physiol 36:253–263

Nelson N (1807) Photosystems and global effects of oxygenic photosynthesis. Biochim Biophys Acta 2011:856–863

Nemhauser JL, Chory J (2009) Photomorphogenesis. In: The arabidopsis book. American Society of Plant Biologists, Rockville, pp 1–12

Newsted WJ, Hüner NPA (1987) Major polypeptides associated with differentiation in psychrophilic fungi. Can J Bot 65(2):1755–1761

Newsted WJ, Hüner NPA (1988) Major sclerotial polypeptides of psychrophilic pathogenic fungi: intracellular localization and antigenic relatedness. Protoplasma 147(2):162–169

Nott A, Jung H-S, Koussevitzky et al (2006) Plastid-to-nucleus retrograde signalling. Annu Rev Plant Biol 57:739–759

Oxborough K, Lee P, Horton P (1987) Regulation of thylakoid protein phosphorylation by high-energy-state quenching. FEBS Lett 221:211–214

Patriarche J, Priscu JC, Takacs-Vesbach C et al (2021) Year-round and long-term phytoplankton dynamics in Lake Bonney, a permanently ice-covered Antarctic lake. J Geophys Res Biogeosci 126:e2020JG005925

Pauli N-C, Flintrop CM, Konrad C et al (2021) Selective feeding in Southern Ocean key grazers – diet composition of krill and salps. Commun Biol 4:1061

Peden EA, Boehm M, Mulder DW et al (2013) Identification of global ferredoxin interaction networks in Chlamydomonas reinhardtii. J Biol Chem 288:35192–35209

Pinney MM, Mokhtari DA, Akiva E et al (2021) Parallel molecular mechanisms for enzyme temperature adaptation. Science 371:eaay2784

Pocock TH, Koziak A, Rosso D et al (2007) Chlamydomonas raudensis (UWO 241), Chlorophyceae, exhibits the capacity for rapid D1 repair in response to chronic photoinhibition at low temperature. J Phycol 43:924–936

Posadas-García YS, Espinosa-Soto C (2022) Early effects of gene duplication on the robustness and phenotypic variability of gene regulatory networks. BMC Bioinformatics 23:1–23

Prigogine I, Stengers I (1984) Order out of chaos. Bantam Books, Toronto

Priscu JC, Fritsen CH, Adams EE et al (1998) Perennial Antarctic lake ice: an oasis for life in a polar desert. Science 280:2095–2098

Prochnik SE, Umen J, Nedelcu AM et al (2010) Genomic analysis of organismal complexity in the multicellular green alga Volvox carteri. Science 329:223–226

Rainer J (2000) Do ultrastable proteins from hyperthermophiles have high or low conformational rigidity? Proc Natl Acad Sci USA 97:2962–2964

Raymond JA, Morgan-Kiss R (2017) Multiple ice-binding proteins of probable prokaryotic origin in an Antarctic lake alga, Chlamydomonas sp. ICE-MDV (Chlorophyceae). J Phycol 53:848–854

Reinbothe S, Reinbothe C (1996) Regulation of chlorophyll biosynthesis in angiosperms. Plant Physiol 111:1–7
Remias D, Lütz-Meindl U, Lütz C (2005) Photosynthesis, pigments and ultrastructure of the alpine snow alga Chlamydomonas nivalis. Eur J Phycol 40(3):259–268
Rochaix J-D (2014) Regulation and dynamics of the light-harvesting system. Annu Rev Plant Biol 65:287–309
Rogers AD (2007) Evolution and biodiversity of Antarctic organisms: a molecular perspective. Philos Trans R Soc B 362:2191–2214
Sabine CL, Feely RA, Gruber N et al (2004) The oceanic sink for anthropogenic CO_2. Science 305:367
Sagan L (1967) On the origin of mitosing cells. J Theor Biol 14:225–274
Saikonen K, Birge T, Fuchs B et al (2023) Toward an integrated understanding of how extreme polar light regimes, light sensitive microbes and hybridization shape global biodiversity. One Earth. in press
Schartau AK, Mariash HL, Christoffersen KS et al (2022) First circumpolar assessment of Arctic freshwater phytoplankton and zooplankton diversity: spatial patterns and environmental factors. Freshw Biol 67:141–158
Schorsch M, Kramer M, Goss T et al (2018) A unique ferredoxin acts as a player in the low-iron response of photosynthetic organisms. Proc Natl Acad Sci USA 115:E12111–E12120
Schroedinger E (1947) What is life? The physical aspect of the living cell. Cambridge University Press, London
Seckbach J (2007) Algae and cyanobacteria in extreme environments. Springer, Dordrecht
Sherwell S, Kalra I, Li W et al (2022) Antarctic lake phytoplankton and bacteria from near-surface waters exhibit high sensitivity to climate-driven disturbance. Environ Microbiol 24:6017–6032
Slavov C, Schrameyer V, Reus M et al (2016) "Super-quenching" state protects symbiodinium from thermal stress – implications for coral bleaching. Biochim Biophys Acta 1857:840–847
Slavov C, Reus M, Holzwarth AR (2013) Two different mechanisms cooperate in the desiccation-induced excited state quenching in Parmelia lichen. J Phys Chem B 117:11326–11336
Sorhannus U (2011) Evolution of antifreeze protein genes in the diatom genus Fragilariopsis: evidence for horizontal gene transfer, gene duplication and episodic diversifying selection. Evol Bioinformatics Online 7:S8321
Starr TN, Thornton JW (2016) Epistasis in protein evolution. Protein Sci 25:1204–1218
Strzepek RF, Hunter KA, Frew RD et al (2012) Iron-light interactions differ in Southern Ocean phytoplankton. Limnol Oceanogr 57:1182–1200
Sunagawa S, Acinas SG, Bork P et al (2020) Tara Oceans: towards global ocean ecosystems biology. Nat Rev Microbiol 18:428–445
Sutherland DL, Howard-Williams C, Ralph P et al (2020) Environmental drivers that influence microalgal species in meltwater pools on the McMurdo Ice Shelf, Antarctica. Polar Biol 43:467–482
Swoczyna T, Kalaji HM, Bussotti F et al (2022) Environmental stress- what can we learn from chlorophyll a fluorescence analysis in woody plants? A review. Front Plant Sci 13:1048582
Szyszka B, Ivanov AG, Hüner NPA (2007) Psychrophily is associated with differential energy partitioning, photosystem stoichiometry and polypeptide phosphorylation in Chlamydomonas raudensis. Biochim Biophys Acta Bioenerg 1767:789–800
Szyszka-Mroz B, Pittock P, Ivanov AG et al (2015) The Antarctic psychrophile Chlamydomonas sp. UWO 241 preferentially phosphorylates a photosystem I-cytochrome b6/f supercomplex. Plant Physiol 169(2022):717–736
Szyszka-Mroz B, Cvetkovska M, Ivanov AG et al (2019) Cold-adapted protein kinases and thylakoid remodeling impact energy distribution in an Antarctic psychrophile. Plant Physiol 180:1291–1309
Szyszka-Mroz B, Ivanov AG, Trick CG et al (2022) Palmelloid formation in the Antarctic psychrophile, Chlamydomonas priscuii, is photoprotective. Front Plant Sci 13:911035

Takahashi H, Clowez S, Wollman F-A et al (2013) Cyclic electron flow is redox-controlled but independent of state transition. Nat Commun 4:2954

Tanaka R, Tanaka A (2007) Tetrapyrrole biosynthesis in higher plants. Annu Rev Plant Biol 58:321–346

Tang EPY, Vincent WF (1999) Strategies of thermal adaptation by high-latitude cyanobacteria. New Phytol 142:315–323

Terauchi AM, Lu S-F, Zaffagnini M et al (2009) Pattern of expression and substrate specificity of chloroplast ferredoxins from Chlamydomonas reinhardtii. J Biol Chem 284:25867–25878

Teufel AG, Li W, Kiss AJ et al (2017) Impact of nitrogen and phosphorus on phytoplankton production and bacterial community structure in two stratified Antarctic lakes: a bioassay approach. Polar Biol 40:1007–1022

Thomas DN, Dieckmann GS (2002) Antarctic Sea Ice-a habitat for extremophiles. Science 295:641

Tomanek L (2008) The importance of physiological limits in determining biogeographical range shifts due to global climate change: the heat-shock response. Physiol Biochem Zool 81:709–717

Trimborn S, Thoms S, Karitter P et al (2019) Ocean acidification and high irradiance stimulate the photo-physiological fitness, growth and carbon production of the Antarctic cryptophyte Geminigera cryophila. Biogeosciences 16:2997–3008

van den Hoff J, Bell E, Whittock L (2020) Dimorphism in the Antarctic cryptophyte Geminigera cryophila (Cryptophyceae). J Phycol 56:1028–1038

Veillette J, Martineau M-J, Antoniades D et al (2010) Effects of loss of perennial lake ice on mixing and phytoplankton dynamics: Insights from High Arctic Canada. Ann Glaciol 51:56–70

Velitchkova M, Popova AV, Faik A et al (2020) Low temperature and high light dependent dynamic photoprotective strategies in Arabidopsis thaliana. Physiol Plant 170:93–108

Verhoeven AS (2014) Sustained energy dissipation in winter evergreens. New Phytol 201:57–65

Vincent W (2000) Cyanobacterial dominance in the polar regions. In: The ecology of cyanobacteria. Kluwer Academic, Dordrecht, pp 321–340

Vincent WF (2007) Cold tolerance in cyanobacteria and life in the cryosphere. In: Algae and cyanobacteria in extreme environments. Springer, Dordrecht, pp 289–301

von Quillfeldt CH (2004) The diatom Fragilariopsis cylindrus and its potential as an indicator species for cold water rather than for sea ice. Vie Milieu 54(2–3):137–143

Walsh JE (2009) A comparison of Arctic and Antarctic climate change, present and future. Antarct Sci 21:179–188

Whitton BA, Potts M (2000) The ecology of cyanobacteria: their diversity in time and space. Kluwer Academic, Dordrecht

Wong ML, Cleland CE, Arend D et al (2023) On the roles of function and selection in evolving systems. Proc Natl Acad Sci USA 120:e2310223120

Wunder T, Xu W, Liu Q (2013) The major thylakoid protein kinases STN7 and STN8 revisited: effects of altered STN8 levels and regulatory specificities of the STN kinases. Front Plant Sci 4:417

Xavier JC, Brandt A, Ropert-Coudert et al (2016) Future challenges in Southern Ocean ecology research. Front Mar Sci 3:article 94

Yin H, Perera-Castro AV, Randall KL et al (2023) Basking in the sun: how mosses photosynthesise and survive in Antarctica. Photosynth Res 158:151–169

Yokono M, Noda C, Minagawa J (1865) Spillover in the direct-type PSI-PSII megacomplex isolated from Arabidopsis thaliana is regulated by pH. Biochim Biophys Acta Bioenerg 2024:149012

Yokono M, Takabayashi A, Akimoto S et al (2019) A megacomplex composed of both photosystem reaction centres in higher plants. Nat Commun 6:6675

Zakhia F, Jungblut A-D, Taton A et al (2008) Cyanobacteria in cold ecosystems. In: Psychrophiles from biodiversity to biotechnology. Springer, Berlin, pp 121–135

Zer H, Ohad I (2003) Light, redox state, thylakoid-protein phosphorylation and signaling gene expression. Trends Biochem Sci 28:467–470

Zhang G, Yao T, Xie H et al (2015) An inventory of glacial lakes in the Third Pole region and their changes in response to global warming. Glob Planet Chang 131:148–157

Zhang Z, Qu C, Yao R et al (2019) The parallel molecular adaptations to the Antarctic cold environment in two psychrophilic green algae. Genome Biol Evol 11:1897–1908

Zhang Z, Qu C, Zhang K et al (2020) Adaptation to extreme Antarctic environments revealed by the genome of a sea ice green alga. Curr Biol 30:3330–3341

Zhang X, Cvetkovska M, Morgan-Kiss R et al (2021) Draft genome sequence of the Antarctic green alga Chlamydomonas sp. UWO241. iScience 24:102084

Zhang X, Han W, Fan X et al (2023) Gene duplication and functional divergence of new genes contributed to the polar acclimation of Antarctic green algae. Mar Life Sci Technol. https://doi.org/10.1007/s42995-023-00203-z

Chapter 2
Impact of Microplastics on Microbial Communities in Polar Environments

Dell'Agnese, S. Cappello, S. Patania, A. Lunetta, G. Caruso, and A. Sherry

Abstract Plastic pollution has been detected in several ecosystems and has reached remote areas in the Arctic and Antarctic environments. These regions are unique ecosystems and have been reported as the most susceptible habitats due to climatic changes and pollution events, especially with recalcitrant compounds. Microorganisms from the plastisphere, involving microbial communities growing on plastic debris surfaces, could be a bioremediation resource to mitigate and restore affected areas. Bioremediation is especially challenging in cold environments, where temperatures can be a limiting parameter for microbial performance, thereby hindering an efficient clean-up process. Synthetic microplastics (MP) (≤5 mm fragments) are emerging contaminants that have been found to accumulate in aquatic environments and soils. Microfibres (MF) are derived from the degradation of textiles, with major environmental concerns associated with synthetic materials, such as polyester and nylon. Research into cold-adapted microorganisms capable of degrading synthetic recalcitrant MP and MF is essential to discover new catalysts for low-temperature process applications. The present chapter aims to describe the impact of synthetic microfibres and microplastic pollution on cold-adapted microbial communities, as well as microorganisms capable of plastic degradation occurring in polar and subpolar ecosystems.

Keywords Microplastics · Arctic · Antarctic · Biodegradation

Dell'Agnese (✉) · A. Sherry
Hub for Biotechnology in the Built Environment, Department of Applied Sciences, Northumbria University, Newcastle upon Tyne, UK

S. Cappello · S. Patania · A. Lunetta
Section of Messina, Institute for Biological Resources and Marine Biotechnologies (IRBIM), National Research Council, Institute of Polar Sciences (CNR-ISP), Messina, Italy

G. Caruso
Institute of Polar Sciences (CNR-ISP), National Research Council (CNR), Messina, Italy

A. W. Fernandes Duarte et al. (eds.), *Polar Microbes and Climate Change*,
https://doi.org/10.1007/978-3-032-07096-8_2

2.1 Introduction

The global plastic products market grew at a compound annual growth rate (CAGR) of 7.7%, reaching $1066.84 billion in 2023, with projections to grow 6.9% until 2027. The main types of plastic products are plastic packaging materials, plastic bottles and polystyrene foam, among other plastic products (Research and Market, 2023). Along with increasing consumption of plastics, the textile industry is considered one of the most polluting, globally responsible for 5–10% of CO_2 emissions and 20% of wastewater effluent, mostly compounded by complex and carcinogenic substances (Anacleto et al. 2022; Ribul et al. 2021). Associated with textile production, modern consumption within the fashion industry has changed. In the last 15 years, the production of clothes has doubled, but the time of reusing before being discarded has decreased by 36% (MacArthur Foundation 2017). This way of consumption, called "fast fashion," has as a consequence a high amount of textile waste generation, becoming an environmental issue for several countries, where about 60% end up incinerated or in landfill within less than one year of use (The Price of Fast Fashion 2018). Synthetic textiles or fibre blended textiles (i.e. polycotton) present beneficial features to consumers such as softness, lightweight and waterproof, and more durability. Due to preference in the market, synthetic textiles have increased in recent years, reaching 65% of global fibre production in 2022, mainly polyester (Textile Exchange Report 2023). However, one of the main drawbacks of synthetic and/or mixed-blend textiles is their recalcitrancy in the environment and ability to form microfibres (Zambrano et al. 2019).

Microplastics and microfibres have been detected globally in different environments such as soils (Machado et al. 2018), lake sediments (Silva Cavalcanti et al. 2023; Gonzales-Pleiter et al. 2021), seawater (Gago et al. 2018) and have reached cold regions, even in considered pristine areas (Lo Giudice et al. 2019; Bergmann et al. 2022). Cold environments are more susceptible to climate changes and pollution events since temperature is intrinsically involved in the capacity of self-healing. The presence of microplastics and synthetic microfibres has been detected in different habitats in Arctic and Antarctic regions, such as alpine lakes, seawater, sea ice and snow (Obbard et al. 2014; Bergmann et al. 2019; González-Pleiter et al. 2021; Velasco et al. 2020; Caruso et al. 2022a, b). Despite the concerns about environmental disruptions in Arctic or Antarctic regions, these environments harbour potential microorganisms, with intrinsic mechanisms to mitigate pollution sources under cold temperatures.

The detection of microplastic/microfibre pollution in polar regions has been increasingly reported, whereas studies elucidating the role and structure of the psychrophiles associated with plastic, called the *plastisphere*, are still scarce. Plastisphere composition varies according to the environment, which complex microbial assemblages, including bacteria, fungi, diatoms, algae and some invertebrates may be associated with (Lacerda et al. 2019). This chapter will focus on bacterial characterisation colonising plastic debris from several matrices in polar environments, including sub-Arctic and sub-Antarctic regions. Moreover, the term

"MP" will be used to define both plastic and synthetic textiles' microfibres. It will briefly present the sources and transport pathways in both regions, followed by a section that describes a few studies reporting the bacterial plastisphere identification, and lastly, the description of the biotechnological potential of isolated cold-adapted bacteria to plastic degradation/bioremediation.

2.2 How Do Microplastics Reach and Affect Cold Regions?

Microplastics (MPs) are defined as solid particles or polymeric matrix, from 1 nm to 5 mm (Frias and Nash 2019; Mishra et al. 2020). Microplastics (MP) can be produced by two main sources: (i) manufactured products containing plastics used for personal care and cosmetics, known as primary microplastics, and (ii) detrition of large plastic debris through environmental factors such as water, wind and sunlight, which are defined as secondary microplastics, generating fragments of <5 mm (Horton et al. 2017; Sun et al. 2019). The major types include the petroleum-based plastics as polypropylene (PP), polyethylene (PE), polyvinyl-chloride (PVC), polystyrene (PS) and polyethylene terephthalate (PET), which are known to be highly persistent, resulting in the global plastics waste problem. On the other hand, a sustainable alternative is the biodegradable bioplastics (BBP), less recalcitrant in the environment, synthesised from biomass and renewable resources such as poly-lactic acid (PLA) and polyhydroxyalkanoate (PHA) or polybutylene succinate (PBS), which can also be utilised as a substrate by microorganisms (Emadian et al. 2017). Textile microfibres are tiny threads that can be classified as natural (cotton, wool) or synthetic, as polyester, nylon, acrylic and other synthetic textiles (Henry et al. 2019).

Generally, MPs are released from domestic use of products containing microplastics, domestic laundering and weathering from plastic litter from industrialised and urban areas, reaching aquatic systems mainly via insufficient wastewater treatment plants (WWTPs) (Kay et al. 2018; Uddin et al. 2020). From contaminated rivers, plastic debris can enter the oceans after being widely transported by ocean currents. Atmospheric distribution, through wind transportation, has been considered another pathway for these microparticles to reach remote areas (Evangeliou et al. 2020).

By their physicochemical composition, MP can adsorb various hydrophobic and toxic chemical contaminants during their long residence time in the water column (Uddin et al. 2020). Cell attachment and biofilm formation are the crucial steps for microbial colonisation on MP surfaces and can be influenced by several factors: i) microplastic characteristics (polymer type, morphology), ii) period/succession ("time-specific"), iii) "location-specific" factors (microbial community and environmental conditions) (Yang et al. 2020).

The harmful effects of MPs entering the environment include their ingestion, leading to the accumulation throughout the food web from invertebrates to mammals (Setälä et al. 2014; Eckert et al. 2018). By their ingestion, animals like birds indirectly contribute to the transport of MP to other areas, which the persistence of

MPs can lead to exposure risk for wild organisms and human populations. Nanoplastics accumulation in mammalian and human tissues can lead to diseases linked with hormonal disruption, reproductive problems, nervous tissue, liver and kidney damage (Mishra et al. 2019). In addition, microplastics can function as vessels for opportunistic pathogenic bacteria and antibiotic resistance genes (ARGs) in marine and terrestrial environments (Keswani et al. 2016; Pedrotti et al. 2022; Zhu et al. 2022a, b).

2.2.1 Transport of Microparticles in Antarctic and Arctic

The geographic structure of the Arctic Circle is land, such as the Bering Strait and Alaska Archipelago, surrounded by adjacent seas, where the propagation of plastic litter through ocean currents can extend into the Arctic. Macro-litter, as floating plastic, reaches the Arctic Ocean mainly brought from the Atlantic Ocean current. MP can undergo a weathering process through sunlight, temperature or wave actions and be transported to and/or within the Arctic via atmosphere, rivers, ocean currents, sea ice and thawed permafrost (Bergmann et al. 2022). Plastics reach the Arctic mainly from: (i) local industrial activities (fishing, ship traffic, hydrocarbon exploration, aquaculture and tourism), (ii) domestic sources (e.g. washing of synthetic textile clothing, personal care products containing microplastics, iii) agriculture, wastewater treatment and landfills. After entering the marine environment, these MPs can precipitate to the bottom of deep sediments, and sea ice are areas where plastic accumulation is observed to occur, considered as an MP sink (Cózar et al. 2017; Bergmann et al. 2017; Peeken et al. 2018). MP input varies along the Arctic coasts, expecting high quantities in increasing urban population zones, such as some areas in Europe, Canada and Alaska. Another influencing factor is the infrastructure for waste disposal and wastewater treatment plants, often absent in remote areas (Halsband and Herzke 2019; Bergmann et al. 2022).

In contrast with the geographical aspects of the Arctic, Antarctica is an isolated continent, covered by an ice cap and surrounded by sea ice and the Southern Ocean (NOAA 2023). Plastic debris has been found in several locations and matrices, as sediment, soil, sea ice and mostly in seawater, in which similar MP sources and transportation paths by atmospheric and oceanic currents can occur in Antarctica (Caruso et al. 2022a, b). The detection of exogenous waste in the Southern Ocean indicates its origin can be from lower latitudes, mainly from the Atlantic Ocean. A study performed at Weddel Sea found 47 different categories of plastics from several matrices, as sediment, air and seawater samples, concluding that fibres can reach Antarctica by air masses from southern South America (Cunningham et al. 2022).

Compared to surface and subsurface from the Southern Ocean, microplastic pollution is higher in coastal areas, near research stations, where macro and microplastics are present and could impact endemic biota. Still, the distribution of MP in Antarctica remains uncertain since available data have been collected mostly from the West Antarctic Peninsula and the Ross Sea regions (Rota et al. 2022; Cincinelli et al. 2017). MP pollution was evaluated in surface and subsurface seawater in 28 stations, in four locations in the Ross Sea. Generally, the authors found microplastic abundance in subsurface water was higher than that in surface water and dominant microplastics in surface and subsurface waters differed. Among the samples, surface waters from the Ross Sea presented the highest abundance of MP. This work can corroborate that MP debris in Antarctic seawater comes from local input as well as from the ocean at mid- and low latitudes (Zhang et al. 2022a, b).

Polymer type and amounts vary according to the sampling location and matrices surveyed. MP and synthetic textile fibres have been reported in different locations in the Arctic, such as marine sediments in Canada (Adams et al. 2021), seawater in Artic Ocean, Greenland and Barent Sea (Cozar et al. 2017), deep sea sediments in Norway, HAUSGARTEN Observatory area (Bergmann et al. 2017), Finland (Scopetani et al. 2019), Siberia (Materic et al. 2022) and in different coastal areas in Antarctica (Caruso et al. 2022a, b), including freshwater systems (González-Pleiter 2020; Citterich et al. 2023), sea ice (Kelly et al. 2020). The most common plastic type reported in polar regions is polyethylene (PE), followed by polyester (PET), which could be associated with global high demand and production of this material (Mishra et al. 2021).

2.3 Microbial Communities Associated with Microplastics

The identification of microbial groups growing on plastic debris from the Arctic or Antarctic is reported mainly based on microbial isolation/cultivation, while few studies identifying the plastisphere through 16sRNA gene massive sequencing are available. Thus, this section will describe the plastisphere diversity based on both dependent and independent culture methods.

Pollution sources are often reported as a disrupting factor for microbial communities in pristine environments, causing a decrease in diversity and selection and thriving of a few well-adapted groups (Yakimov et al. 2007; Hazen et al. 2010). The effects of plastic pollution and microbial response have shown the same profile, where few bacterial groups inhabiting plastic surfaces differ from surrounding matrices (González-Pleiter et al. 2021; Ruthi et al. 2023). The term "plastisphere" was defined by Zettler and collaborators in 2013, in a study that identified microbial communities in plastic marine debris from seawater samples from the North Atlantic Ocean Gyre. The plastisphere community consisted of a morphologically diverse range of microbes comprising prokaryotes such as bacteria, eukaryotes, fungi, diatoms and dinoflagellates. They observed plastic-associated communities differed from those present in surrounding surface water, indicating the plastic material could select microorganisms, especially those capable of producing biofilm. Carson et al. (2013) evaluated plastic-associated microorganisms from North Pacific Gyre samples based on morphological characterisation, suggesting the microbial colonisation could be influenced by plastic size, polymer type and surface roughness, mainly for eukaryotes. Nonetheless, sampling geographic origin could be the major determinant of the plastisphere composition (Amaral-Zettler et al. 2015).

In the marine environment, opportunistic and pathogenic bacteria, such as *Vibrio* spp. were found in the plastisphere. Due to intrinsic transportation by currents, these pathogenic groups could be introduced in other areas, posing risks to human health and negative effects on the quality of bathing water in beaches (Keswani et al. 2016). MPs have been found to be a hotspot for antimicrobial resistance genes in aquatic and soil matrices (Zhu et al. 2022a, b; Junaid et al. 2022). Gene exchange between the plastisphere and planktonic microbes might contribute to the growing antimicrobial resistance (AMR). Other factors can affect the antimicrobial resistance genes (ARG) abundance in the plastisphere, including types of microplastics and environmental conditions, such as salinity and coexisting contaminants (Luo et al. 2023).

Laganà et al. (2019) investigated 27 isolated strains associated with polystyrene macroplastic from King George Island (South Shetlands, Antarctica), in which seven were able to produce biofilm and resistance to several antibiotics, like cephalosporins, quinolones and beta-lactams. Bacteria were affiliated with *Pseudomonas, Alteromonas, Thalassopira* and *Halomonas* genera.

Polar plastisphere (bacteria, eukaryotes and fungi) from a freshwater lake (Svalbard Archipelago, Norway) was identified growing on three types of MPs: poly-3-hydroxybutyrate (PHB), a biodegradable plastic and two non-biodegradable plastics, low-density polyethylene (LDPE) and high-density polyethylene (HDPE), spheres with a 5 mm diameter. MPs were deployed into a shallow freshwater lake for 11 days. Microbial diversity from MPs, surrounding water and lake sediment was assessed, where core microbiomes were assigned to each substrate. Phyla *Proteobacteria, Cyanobacteria, Bacteroidetes, Actinobacteria* and *Verrucomicrobia* dominated MPs biofilms. Moreover, genera *Mycoplana, Erythromicrobium*, and *Rhodoferax* were found in high abundance in MP compared to the surrounding water and sediment, and are described as involved in the degradation of recalcitrant chemicals (González-Pleiter et al. 2021).

The effects of plastics were evaluated in Arctic and alpine soils. Polylactic acid (PLA) as plastic bag, polybutylene adipate terephthalate (PBAT) mulch foil and polyethylene (PE) waste bag were buried in two different soils, in mesocosms experiments from Swiss Alps and Northern Greenland soils, at 15 °C for 8 weeks. The authors worked with dependent and independent cultivation methods, and microbial diversity was lower in the plastisphere than in bulk soil. Prominent differences were found in both soil and plastic microbiomes, mainly in PLA, the most degradable material, which could explain the community shift in a short time. Taxa with a known potential to degrade xenobiotics, including *Burkholderiales, Caulobacterales, Pseudomonas, Rhodococcus* and *Streptomyces*, thrived in the plastisphere of the Alpine and Arctic soils (Rüthi et al. 2020).

Following a similar approach, the same group evaluated the effect of plastic in soil microbiome, through mesocosm experiments, using the same plastic typology: Polylactic acid (PLA) and polybutylene adipate terephthalate (PBAT) as biodegradable. Metagenomic analysis revealed differences between biodegradable plastics and bulk soil, with an increase of Proteobacteria phyla (orders *Rhizobiales,* α-Proteobacteria and *Burkholderiales*, β-Proteobacteria) in the plastisphere. Genes related to α/β-hydrolases synthesis and other enzymes involved in cellular processes and signalling, as well as metabolism (carbohydrate, lipid and secondary metabolism), were enriched in the plastisphere (Ruthi et al. 2023).

Polystyrene and polyethylene, as macro and mesoplastic fragments, were collected in distinct land and sea environments, from coastal areas in King George Island, Antarctica. Microbial diversity was evaluated through enzymatic profile, isolation and 16S amplicon sequencing. Sequences analysis revealed the class Betaproteobacteria as the most abundant (21%, followed by Alphaproteobacteria (13%) and the presence of hydrocarbon-degrading genera *Marinobacter* and *Alcanivorax* in polystyrene. Polyethylene showed the highest diversity index, harbouring more groups, whose sequences showed Gamma- and Betaproteobacteria (31% and 28%, respectively) as dominant groups and *Variovorax* (22%) and *Marinobacter* (14%) as the most abundant genera. These differences suggest plastic type may influence biofilm composition and metabolic functions. Moreover, the presence of hydrocarbonoclastic groups may indicate pollution disturbance in this area in Antarctica (Cappello et al. 2021).

Bacterial strains were isolated from water and biofilm formed on polyvinyl chloride (PVC) and polyethylene (PE) panels placed in different depths in the ice pack, in two areas in Ross Bay, Antarctica. Phenotypic and enzymatic assays showed higher diversity in PVC than PE, mostly assigned to Gram-negative bacteria, such as *Vibrio/Aeromonas/Photobacterium* genera. Pigmented strains were classified as belonging to the *Flavobacterium* genus, whereas others were identified as *Pseudomonas* sp. and *Micrococcus* sp. (Caruso et al. 2022a, b). These few studies aiming strictly the plastisphere identification from Arctic and Antarctic areas are summarised in Table 2.1.

These studies were performed in situ and could demonstrate the impact of MP in the autochthonous microbiome, in which: (i) plastisphere communities will vary according to the geographical location; (ii) the diversity observed from the plastisphere microorganisms differs from free-living in the surrounding environment; (iii) biofilm composition can vary according to time, initial vs. mature. Primarily, a plastic surface offers a physical substrate for microbial attachment, on which bacteria can attach and form biofilms. Moreover, microorganisms able to degrade xenobiotic compounds, as hydrocarbons, were found in high abundance in the plastisphere, which could be explained by the presence of these groups in the sampled matrix, likely being an indicator of pollution and the attraction for these substrates as a carbon source. A study involving a meta-analysis of the 16S rRNA gene of global plastisphere composition from 2229 samples within 35 studies in multiple matrices, marine, freshwater, other aquatic and terrestrial environments also verified the presence of hydrocarbonoclastic *Oceanospirillales* and *Alteromonadales* groups consistently more abundant in plastic than control biofilm samples (Wright et al. 2021).

Table 2.1 Plastic-associated microorganisms occurring in polar regions

Microbial community	Sample matrix	Plastic type	Location	References
Proteobacteria, Cyanobacteria, Mycoplana, Erythromicrobium and *Rhodoferax*	Shallow freshwater lake	Poly-3-hydroxybutyrate (PHB), polyethylene (LDPE) and (HDPE)	Ny-Ålesund, Svalbard Archipelago, Arctic	González-Pleiter et al. (2021)
Burkholderiales, Caulobacterales, Pseudomonas, Rhodococcus	Soil mesocosms	Polylactic acid (PLA), polybutylene adipate terephthalate (PBAT) and polyethylene (PE)	Swiss Alps and from Northern Greenland, Arctic	Ruthi et al. (2020)
Proteobacteria Rhizobiales, Burkholderiales	Soil mesocosms	Polylactic acid (PLA), polybutylene adipate terephthalate (PBAT) and polyethylene (PE)	Eastern Swiss Alps	Ruthi et al. (2023)
Marinobacter, Alcanivorax Variovorax	Coastal environment (land and sea)	Polystyrene (PS), polyethylene (HDPE)	King George Island, Antarctica	Cappello et al. (2021)
Flavobacterium Pseudomonas Micrococcus	Ice pack	Polyvinyl chloride (PVC), Polyethylene (PE)	Ross Bay, Antarctica	Caruso et al. (2022a, b)

2.4 Cold-Adapted Microorganisms Capable of Plastic Degradation

Cold environments represent more than 70% of the Earth's surface, comprising aquatic environments, including deep-sea water, polar and high alpine soils, and glaciers. Psychrophilic microorganisms can grow in the temperature range from −20 to 20 °C, with an optimal growth temperature below 15 °C. The diversity of psychrophilic bacteria comprises genera as *Pseudomonas, Psychrobacter, Pseudoalteromonas, Colwellia* and *Arthrobacter* (Atanasova et al. 2021). Physiological adaptations are known in these microorganisms, like membrane structure (high unsaturated fatty acids and short-chain fatty acids in membranes), to keep membrane fluidity. Synthesis of cold-shock and anti-freeze proteins, regulating vital cellular processes, enables bacterial thriving in extreme conditions; accumulation of cryo-protectant molecules to prevent cell damage. Enzymes are structurally adapted to activity at low temperature (Lo Giudice et al. 2019; Atanasova et al. 2021).

Although works employ cultivation for functional degradation assays, culture-dependent methods are a bottleneck for microbial diversity studies, since many taxa remain unclassified and uncultivated. The biotechnological potential of novel enzymes aiming at plastic degradation has been investigated by a metagenomics approach (Zhu et al. 2022a, b). Oxidative enzymes, such as monooxygenases, peroxidases and alkane hydroxylases and extracellular hydrolytic esterases, cutinases, lipases and proteases, are involved in plastic degradation (Ruthi et al. 2023). Cold enzymes have been optimised mostly for bacteria isolated from abiotic matrices such as Antarctic soil and sediments, Arctic and Subarctic glaciers, deep sea, permafrost soils and actives layer (Rizzo and Lo Giudice 2022).

Some works aiming at plastic degradation from Arctic or Antarctic strains, involving culture-dependent and independent approaches, are summarised in Table 2.2.

To identify potential PET hydrolases from glacial environments, 208 metagenomes and 952 cultivated bacterial genomes from global glaciers were assessed in databases. Analysis showed 414 putative PET hydrolase hits were obtained using a collection of curated PET hydrolase sequences. In total, 12 of them were expressed, one of which was demonstrated to show exceptional activity against PET nanoplastics, mainly at 30 °C and 37 °C. These putative PET hydrolases were affiliated mostly with Proteobacteria (67%), followed by *Actinobacteria* (17%), *Gemmatimonadetes* (5%) and *Deinococcus-Thermus* (4%) (Qi et al. 2023).

The degradation of polyethylene (PE) was determined using the cold-adapted *Psychrobacter* sp. NJ228, isolated from Antarctic Sea ice. Bacterial laccase was heterologously expressed, purified and characterised, resulting in a new recombinant laccase enzyme (psLac), cold-resistant and salt-tolerant. PE-degradation experiments in mesophilic conditions at 30 °C indicated changes in weight and hydrophobicity (Zhang et al. 2022a, b).

Another cold-adapted enzyme (MoPE) from the Antarctic bacterium *Moraxella* sp. was cloned and expressed in a heterologous vector for the PET model substrate. The amino acid composition and structure found on MoPE were similar to PET-degrading enzymes. Tests showed the ability to metabolise biodegradable compounds and non-biodegradable plastics, such as polyethylene terephthalate and polyurethane. MoPE was characterised as a versatile psychrophilic polyesterase demonstrating a broad-range plastics degradation potential (Nikolaivitis et al. 2022).

A total of 34 soil bacteria isolated from Greenwich Island, Antarctica, were isolated and tested for polystyrene (PS) degradation. A preliminary screening and tests showed that the isolate AYDL1, identified as *Brevundimonas* sp., was the most efficient in utilising PS microplastics. PS structure morphology was changed, indicating the bacteria altered the chemical structure of PS after incubation for 40 days (Tang et al. 2023).

Microbial isolates from buried plastics buried in alpine and Arctic soils were identified. The experiments were conducted at 15 °C with conventional polyethylene (PE) and biodegradable plastics, polyester-polyurethane (PUR) and plastic films made of polybutylene adipate-co-terephthalate (PBAT) and polylactic acid (PLA). The biodegradable plastics were modified by about one-third of the tested strains, identified as belonging almost exclusively to the phyla Actinobacteria and Proteobacteria (Ruthi et al. 2023).

In the subarctic region of Siberia, microbial communities of cryogenic soils were able to degrade microbial bioplastic (polyhydroxyalkanoates, PHAs) poly-3-hydroxybutyrate [P(3HB)]. The study consisted of in-situ experiments, with plastic burial in cold soils for 15 months. Biofilms were removed from the plastic surface and cultivated in agar plates. The microbial community from biofilm formed on the

Table 2.2 Isolated cold-adapted microorganisms able to biodegrade plastic

Microorganism	Sample matrix	Plastic type and temperature	Origin	References
Proteobacteria Actinobacteria, Gemmatimonadetes	Glacier	Polyethylene terephthalate (PET) 30 °C	208 metagenomes (global glacier) genomes of 952 isolates (snow, ice, cryoconite)	Qi et al. (2023)
Psychrobacter sp. NJ228	Sea ice	Polyethylene (PE) 30 °C	Antarctica	Zhang et al. (2022a, b)
Brevundimonas sp. AYDL1	Soil	Polystyrene (PS) 17 °C	Soil from Greenwich Island, Antarctica	Tang et al. (2023)
Neodevriesia and *Lachnellula*	Soil	Polyester-polyurethane (PUR) and compostable PBAT and PLA 15 °C	Swiss Alps and Greenland	Ruthi et al. (2023)
Bacillus pumilus, Paraburkholderia sp., *Pseudomonas* sp., *Rhodococcus* sp., *Stenotrophomonas rhizophila,*	Soil	Poly-3-hydroxybutyrate [P(3HB)] 4.5–8.3 °C	Siberia	Prudnikova et al. (2021)
Rhodococcus sp. *Streptomyces* sp.	Soil, sediment and water	Mesophilic degradation 22 °C	Norway	Charnock (2021)
Pseudomonas and *Rhodococcus*	Soil	Microbial plastics PBS, PBSA and PCL (20 °C and 28 °C)	Svalbard Archipelago, Arctic	Urbanek et al. (2017)
Pseudomonas and *Bacillus* species	Soil	Microbial plastics PBS, PBSA and PCL (20 °C and 28 °C)	King George Island, Antarctica	Urbanek et al. (2021)
Pseudomonas sp. *ADL15* and *Rhodococcus* sp. *ADL36*	Soil	Polypropylene (PP) 10 °C	Victoria Land (Ross Sea sector Antarctica)	Habib et al. (2020)
Roseovarius pelagicus sp.	Surface seawater	Polypropylene (PP) 15 °C	Arctic Ocean	Hong et al. (2023)
Moraxella sp. TA144	Seawater	Polyethylene terephthalate (PET) derivatives	Antarctic	Nikolaivitis et al. (2022)

surface of P(3HB) films differed from the background soil in concentrations and composition of microorganisms. Moreover, screening of cryogenic soil was carried out for plastic degradation. Sequences analysis of rRNA genes identified the P(3HB)-degrading species: *Bacillus pumilus, Paraburkholderia* sp., *Pseudomonas* sp., *Rhodococcus* sp., *Stenotrophomonas rhizophila, Streptomyces prunicolor,* and

Variovorax paradoxus bacteria and the *Penicillium thomii, P. arenicola, P. lanosum, Aspergillus fumigatus,* and *A. niger fungi* (Prudnikova et al. 2021).

Plastic-degrading bacteria in soils, water and sediments were investigated through isolation, enumeration and polyphasic characterisation of present in Norway. Isolates were tested against 8 types of plastics: poly[(R)-3-hydroksybutyrate], PHB; poly(3-hydroksybutyrate-co-3- hydroksyvalerate), PHBV; poly(ethylene succinate), PES; poly(1,4-butylene succinate), polycaprolactone, PCL, poly(L-lactide), PLLA (Poly(D,L-Lactide-co-Glycolide), RES polyethylene terephthalate, PET. These bacteria were present in relatively high numbers in plastic samples, and plastic-degrading capabilities are found in several taxa, most especially *Streptomyces*. Some isolates were able to degrade the biocompostable as PHB and PHBV. Notably, a *Rhodococcus* sp. and a *Streptomyces* sp. degraded, respectively, four and six of the eight plastics investigated (Charnock 2021).

Urbanek et al. (2017) evaluated around 300 microorganisms, bacteria and fungi from 52 soil samples from the Arctic region (Spitsbergen) through screening for degrading bioplastics, poly(butylene succinate-co-adipate) (PBSA), poly(butylene succinate) (PBS) and poly(ε-caprolactone) (PCL). Almost 40% of microorganisms were able to degrade plastic, mostly PBSA. Based on the 16S rRNA sequence, 10 bacterial strains with the highest ability for biodegradation were identified as species belonging to *Pseudomonas* sp. and *Rhodococcus* sp. A similar approach was applied to investigate 161 bacteria isolated from 22 Antarctic soil samples regarding the potential for degradation of these bioplastics (PBS, PBSA and PCL) at low temperatures. For Antarctic microorganisms, screening temperature was tested at 28 °C, and fungal plastic degradation was carried out at 20 °C. Among them, the largest percentage of bacteria with a halo zone was observed on PCL medium. Based on the 16S sequences, bacterial strains were identified as species belonging to *Pseudomonas* and *Bacillus* (Urbanek et al. 2021).

Pseudomonas sp. ADL15 and *Rhodococcus* sp. ADL36 from Antarctic soil were tested against the polypropylene (PP) microplastics in liquid medium at 10 °C. Structural changes on the polypropylene surface and weight loss percentage showed higher activity from *Pseudomonas* ADL15 after 40 days (Habib et al. 2020). A bacterium isolated and characterised from Arctic seawater, affiliated to the genus *Roseovarius* (family *Rhodobacteraceae* within the class *Alphaproteobacteria*), was observed to use polypropylene as the sole carbon source after two years of incubation (Hong et al. 2023).

2.5 Conclusion

The detection and impacts of MP in both the Arctic and Antarctic indicate that plastic pollution is currently ubiquitous. The heterogeneity of methods in monitoring and access to remote areas could influence the development of proper measures of risk assessment plans, since MP input can directly affect endemic species, accumulating in trophic webs.

Microorganisms naturally inhabiting the poles possess intrinsic and potential metabolism for biotechnological applications, such as cold-adapted enzymes, exopolymers substances (EPS) and antimicrobial compounds. Microbial response and the role of the plastisphere are not fully elucidated and are scarce in these environments. It is known that geographical location, physicochemical/abiotic factors and polymer type could influence the microbial composition in plastic surfaces. Studies gathered showed psychrophiles' diversity could be similar to that occurring in other biogeographic regions, composed mostly by Proteobacteria and Actinobacteria phyla, including groups capable of utilising these polymers as a carbon source, which could be harnessed for bioremediation means in polluted areas. Nonetheless, the ecological aspects need further investigation to clarify the existence of core groups related to the autochthonous microbiome and the differences from other latitudes. Since plastic debris can often be dispersed to other areas, microbial diversity found in the plastisphere may not reflect the indigenous microbiome inhabiting matrices where they are detected. Undoubtedly, research in cold-adapted plastisphere is important to search for new catalysts, especially at low temperatures, where both dependent and independent culture methods may be employed to optimise the enzymatic process from psychrophiles.

References

Adams JK et al (2021) Anthropogenic particles (including microfibers and microplastics) in marine sediments of the Canadian Arctic. Sci Total Environ 784:147155. https://doi.org/10.1016/j.scitotenv.2021.147155

Amaral-Zettler LA et al (2015) The biogeography of the plastisphere: implications for policy. Front Ecol Environ 13(10):541–546. https://doi.org/10.1890/150017

Anacleto TM, Kozlowsky-Suzuki B, Wilson AE, Enrich-Prast A (2022) Comprehensive meta-analysis of pathways to increase biogas production in the textile industry. Energies 15:5574. https://doi.org/10.3390/en15155574

Atanasova N, Stoitsova S, Paunova-Krasteva T, Kambourova M (2021) Plastic degradation by extremophilic bacteria. Int J Mol Sci 22:5610. https://doi.org/10.3390/ijms22115610

Bergmann M, Wirzberger V, Krumpen T et al (2017) High quantities of microplastic in arctic deep-sea sediments from the HAUSGARTEN observatory. Environ Sci Technol 51:11000–11010. https://doi.org/10.1021/acs.est.7b03331

Bergmann M, Mützel S, Primpke S, Tekman MB, Trachsel J, Gerdts G (2019) White and wonderful? Microplastics prevail in snow from the Alps to the Arctic. Sci Adv 5:2019

Bergmann M et al (2022) Plastic pollution in the Arctic. Nat Rev Earth Environ 3. https://doi.org/10.1038/s43017-022-00279-8

Cappello S, Caruso G, Bergami E et al (2021) New insights into the structure and function of the prokaryotic communities colonizing plastic debris collected in King George Island (Antarctica): preliminary observations from two plastic fragments. J Hazard Mater 414:125586. https://doi.org/10.1016/j.jhazmat.2021.125586

Carson HS, Nerheim MS, Carroll KA, Eriksen M (2013) The plastic-associated microorganisms of the North Pacific Gyre. Mar Pollut Bull 75:126–132. https://doi.org/10.1016/j.marpolbul.2013.07.054

Caruso G, Dell'Acqua O, Caruso R, Azzaro M (2022a) Phenotypic characterization of bacterial isolates from marine waters and plastisphere communities of the Ross Sea (Antarctica). J Clin Microbiol Biochem Technol 8(1):001–009. https://doi.org/10.17352/jcmbt.000048

Caruso G, Bergami E, Singh N, Corsi I (2022b) Plastic occurrence, sources, and impacts in Antarctic environment and biota. Water Biol Security 1(2):100034. https://doi.org/10.1016/j.watbs.2022.100034

Charnock C (2021) Norwegian soils and waters contain mesophilic, plastic-degrading bacteria. Microorganisms 9(94):2021. https://doi.org/10.3390/microorganisms9010094

Cincinelli A, Scopetani C, Chelazzi D et al (2017) Microplastic in the surface waters of the Ross Sea (Antarctica): Occurrence, distribution and characterization by FTIR. Chemosphere 175:391e400. https://doi.org/10.1016/j.chemosphere.2017.02.024

Citterich F, Lo Giudice A, Azzaro M (2023) A plastic world: a review of microplastic pollution in the freshwaters of the Earth's poles. Sci Total Environ 869:161847. https://doi.org/10.1016/j.scitotenv.2023.161847

Cózar A, Elisa Martí E, Duarte CM et al (2017) The Arctic Ocean as a dead end for floating plastics in the North Atlantic branch of the Thermohaline Circulation. Sci Adv 3:1600582

Cunningham EM, Rico Seijo N, Altieri KE, Audh RR, Burger JM, Bornman TG, Fawcett S, Gwinnett CMB, Osborne AO, Woodall LC (2022) The transport and fate of microplastic fibres in the Antarctic: the role of multiple global processes. Front Mar Sci 9:1056081. https://doi.org/10.3389/fmars.2022.1056081

Eckert EM, Di Cesare A, Kettner MT, Arias-Andresc M, Fontaneto D, Grossart HP, Corno G (2018) Microplastics increase impact of treated wastewater on freshwater microbial community. Emerging Contam 5:308e318. https://doi.org/10.1016/j.emcon.2019.11.001

Emadian SM, Onay TT, Demirel B (2017) Biodegradation of bioplastics in natural environments. Waste Manag 59:526–536

Evangeliou N, Grythe H, Klimont Z (2020) Atmospheric transport is a major pathway of microplastics to remote regions. Nat Commun 11:3381. https://doi.org/10.1038/s41467-020-17201-9

Frias JPGL, Nash R (2019) Microplastics: finding a consensus on the definition. Mar Pollut Bull 138:145–147. https://doi.org/10.1016/j.marpolbul.2018.11.022

Gago J, Carretero O, Filgueiras AV, Viñas L (2018) Synthetic microfibers in the marine environment: a review on their occurrence in seawater and sediments. Mar Pollut Bull 127:365–376. https://doi.org/10.1016/j.marpolbul.2017.11.070

González-Pleiter M et al (2020) Fibers spreading worldwide: microplastics and other anthropogenic litter in an Arctic freshwater lake. Sci Total Environ 722:137904. https://doi.org/10.1016/j.scitotenv.2020.137904

González-Pleiter M, Velázquez D, Casero MC, Tytgat B, Verleyen E, Leganés F, Fernandez-Pinas F (2021) Microbial colonizers of microplastics in an Arctic freshwater lake. Sci Total Environ 795:148640. https://doi.org/10.1016/j.scitotenv.2021.148640

Habib S, Iruthayam A, Shukor MYA, Siti Aisyah Alias AS, Smykla J, Yasid NA (2020) Biodeterioration of untreated polypropylene microplastic particles by Antarctic bacteria. Polymers 12:2616. https://doi.org/10.3390/polym12112616

Halsband C, Herzke D (2019) Plastic litter in the European Arctic: what do we know? Emerging Contam 5:308–318. https://doi.org/10.1016/j.emcon.2019.11.001

Hazen et al (2010) Deep-sea oil plume enriches indigenous oil-degrading bacteria. Science 2010. https://doi.org/10.1126/science.1195979

Henry B, Laitala K, Klepp I G (2019) Microfibres from apparel and home textiles: Prospects for including microplastics in environmental sustainability assessment. Science of the Total Environment 652, 483–494. https://doi.org/10.1016/j.scitotenv.2018.10.166

Hong YW, Jang GI, Kim SY et al (2023) *Roseovarius pelagicus* sp. nov., a facultatively anaerobic bacterium with potential for degrading polypropylene, isolated from Arctic seawater. Int J Syst Evol Microbiol 73:005999. https://doi.org/10.1099/ijsem.0.005999

Horton AA, Walton A, Spurgeon DJ, Lahive E, Svendsen C (2017) Microplastics in freshwater and terrestrial environments: Evaluating the current understanding to identify the knowledge gaps and future research priorities. Sci Total Environ 586:127–141

Junaid M, Liu X, Wu Y, Wang J (2022) Selective enrichment of antibiotic resistome and bacterial pathogens by aquatic microplastics. J Hazard Mater Adv 7:100106. https://doi.org/10.1016/j.hazadv.2022.100106

Kay P, Hiscoe R, Moberley I, Bajic L, McKenna N (2018) Wastewater treatment plants as a source of microplastics in river catchments. Environ Sci Pollut Res 25:20264–20267

Kelly A, Lannuzel D, Rodemann T, Meiners KM, Auman HJ (2020) Microplastic contamination in east Antarctic sea ice. Sci Total Environ 869:161847

Keswani A, Oliver DM, Gutierrez T, Quilliam RS (2016) Microbial hitchhikers on marine plastic debris: Human exposure risks at bathing waters and beach environments. Mar Environ Res 118:10–19. https://doi.org/10.1016/j.marenvres.2016.04.006

Lacerda ALF, Rodrigues LS, Sebille E, Rodrigues FL, Ribeiro L, Secchi ER, Kessler F, Proietti MC (2019) Plastics in sea surface waters around the Antarctic Peninsula. Sci Rep 9:3977

Laganà P, Caruso G, Corsi I (2019) Do plastics serve as a possible vector for the spread of antibiotic resistance? First insights from bacteria associated to a polystyrene piece from King George Island (Antarctica). Int J Hyg Environ Health 222:89–100. https://doi.org/10.1016/j.ijheh.2018.08.009

Lo Giudice A, Caruso G, Rizzo C, Papale M, Azzaro M (2019) Bacterial communities versus anthropogenic disturbances in the Antarctic coastal marine environment. Environ Sustain 2:297–310. https://doi.org/10.1007/s42398-019-00064-2

Luo G et al (2023) Determining the contribution of micro/nanoplastics to antimicrobial resistance: challenges and perspectives. Environ Sci Technol 2023(57):12137–12152. https://doi.org/10.1021/acs.est.3c01128

MacArthur Foundation (2017) A new textiles economy: redesigning fashion's future. http://www.ellenmacarthurfoundation.org/publications

Machado AAS, Kloas W, Zarfl C, Hempel S, Rillig MC (2018) Microplastics as an emerging threat to terrestrial ecosystems. Glob Chang Biol 24:1405–1416. https://doi.org/10.1111/gcb.14020

Materic D, Peacock M, Dean J (2022) Presence of nanoplastics in rural and remote surface waters. Environ Res Lett 17:054036. https://doi.org/10.1088/1748-9326/ac68f7

Mishra S, Rath CC, Das AP (2019) Marine microfiber pollution: a review on present status and future challenges. Mar Pollut Bull 140:188–197. https://doi.org/10.1016/j.marpolbul.2019.01.039

Mishra S, Singh RP, Rath CC, Das AD (2020) Synthetic microfibers: source, transport and their remediation. J Water Process Eng 38:101612

Mishra AK, Singh J, Mishra PP (2021) Microplastics in polar regions: an early warning to the world's pristine ecosystem. Sci Total Environ 784:147149. https://doi.org/10.1016/j.scitotenv.2021.147149

National Oceanic and Atmospheric Administration (NOAA) (2023) https://www.climate.gov/newsfeatures/understanding-climate/understanding-climate-antarctic-sea-ice-extent(acessed in Jan 2024)

Nikolaivits E, Taxeidis G, Gkountela C et al (2022) A polyesterase from the Antarctic bacterium *Moraxella* sp. degrades highly crystalline synthetic polymers. J Hazard Mater 434:128900

Obbard RW, Sadri S, Wong YQ, Khitun AA, Baker I, Thompson RC (2014) Global warming releases microplastic legacy frozen in Arctic Sea ice. Earth's Future 2:315–320. https://doi.org/10.1002/2014EF000240

Pedrotti ML, de Figueiredo Lacerda AL, Petit S, Ghiglione JF, Gorsky G (2022) Vibrio spp and other potential pathogenic bacteria associated to microfibers in the North-Western Mediterranean Sea. PLoS One 17(11):e0275284. https://doi.org/10.1371/journal.pone.0275284

Peeken I et al (2018) Arctic sea ice is an important temporal sink and means of transport for microplastic. Nat Commun 9:1505. https://doi.org/10.1038/s41467-018-03825-5

Prudnikova SV, Evgrafova SY, Volova TG (2021) Metabolic activity of cryogenic soils in the subarctic zone of Siberia towards "green" bioplastics. Chemosphere 263:128180. https://doi.org/10.1016/j.chemosphere.2020.128180

Qi X, Ji M, Yin C-F, Zhou N-Y, Liu Y (2023) Glacier as a source of novel polyethylene terephthalate hydrolases. Environ Microbiol 25(12):2822–2833. https://doi.org/10.1111/1462-2920.16516

Research and Market (2023) (https://www.researchandmarkets.com/report/plastic-product), accessed in Feb 2024

Ribul M et al (2021) Mechanical, chemical, biological: Moving towards closed-loop bio-based recycling in a circular economy of sustainable textiles. J Clean Prod 326:129325. https://doi.org/10.1016/j.jclepro.2021.129325

Rizzo C, Lo Giudice A (2022) The variety and inscrutability of polar environments as a resource of biotechnologically relevant molecules. Microorganisms 8:1422. https://doi.org/10.3390/microorganisms8091422

Rota E, Bergami E, Corsi I, Bargagli R (2022) Macro-microplastics in the antarctic environment: ongoing assessment and perspectives. Environments 9:93. https://doi.org/10.3390/environments9070093

Rüthi J, Bölsterli D, Pardi-Comensoli L, Brunner I, Frey B (2020) The "plastisphere" of biodegradable plastics is characterized by specific microbial taxa of Alpine and Arctic soils. Front Environ Sci 8:562263. https://doi.org/10.3389/fenvs.2020.562263

Rüthi J, Rast BM, Qi W, Perez-Mon C, Pardi-Comensoli L, Brunner I, Frey B (2023) The plastisphere microbiome in alpine soils alters the microbial genetic potential for plastic degradation and biogeochemical cycling. J Hazard Mater 441:129941. https://doi.org/10.1016/j.jhazmat.2022.129941

Scopetani C, Chelazzi D, Cincinelli A, Esterhuizen-Londt M (2019) Assessment of microplastic pollution: occurrence and characterisation in Vesijärvi lake and Pikku Vesijärvi pond. Finland Environ Monit Assess. https://doi.org/10.1007/s10661-019-7843-z

Setälä O, Fleming-Lehtinen V, Lehtiniemi M (2014) Ingestion and transfer of microplastics in the planktonic food web. Environ Pollut 185:77e83. https://doi.org/10.1016/j.envpol.2013.10.013

Silva-Cavalcanti JS, Silva JCP, de Andrade FM, Brito AMSS, da Costa MF (2023) Microplastic pollution in sediments of tropical shallow lakes. Sci Total Environ 855:158671. https://doi.org/10.1016/j.scitotenv.2022.158671

Sun J, Dai X, Wang Q, van Loosdrecht MCM, Ni BJ (2019) Microplastics in wastewater treatment plants: detection, occurrence and removal. Water Res 152:21e37

Tang PM, Habib S, Shukor MYA, Alias SA, Smykla J, Yasid NA (2023) Evaluation of the deterioration of untreated commercial polystyrene by psychrotrophic Antarctic bacterium. Polymers 15:1841. https://doi.org/10.3390/polym15081841

Textile Exchange, Materials market report. (2023). https://textileexchange.org/knowledge-center/reports/materials-market-report-2023

The price of fast fashion (2018) Nat Clim Change 8:1. https://doi.org/10.1038/s41558-017-0058-9

Uddin S, Fowler SW, Behbehani M (2020) An assessment of microplastic inputs into the aquatic environment from wastewater streams. Mar Pollut Bull 160:111538. https://doi.org/10.1016/j.marpolbul.2020.111538

Urbanek AK, Rymowicz W, Strzelecki MC, Kociuba W, Franczak L, Mirończuk AM (2017) Isolation and characterization of Arctic microorganisms decomposing bioplastics. AMB Express 7:148. https://doi.org/10.1186/s13568-017-0448-4

Urbanek AK, Strzelecki MC, Mironczuk AM (2021) The potential of cold-adapted microorganisms for biodegradation of bioplastics. Waste Manag 119:72–81. https://doi.org/10.1016/j.wasman.2020.09.031

Velasco AJN, Rard L, Blois W (2020) Microplastic and fibre contamination in a remote mountain lake in Switzerland. Water 12(9):2410. https://doi.org/10.3390/w12092410

Wright RJ, Langille MGI, Walker TR (2021) Food or just a free ride? A meta-analysis reveals the global diversity of the Plastisphere. ISME J 15:789–806. https://doi.org/10.1038/s41396-020-00814-9

Yakimov MM, Timmis KN, Golyshin PN (2007) Obligate oil-degrading marine bacteria. Current Opinion in Biotechnology, 18:257–266. https://doi.org/10.1016/j.copbio.2007.04.006

Yang Y, Liu W, Zhang Z, Grossart HP, Gadd GM (2020) Microplastics provide new microbial niches in aquatic environments. Appl Microbiol Biotechnol 104:6501–6511. https://doi.org/10.1007/s00253-020-10704-x

Zambrano MC, Pawlak JJ, Daystar J, Ankeny M, Cheng JJ, Venditti RA (2019) Microfibers generated from the laundering of cotton, rayon and polyester based fabrics and their aquatic biodegradation. Mar Pollut Bull 142:394–407. https://doi.org/10.1016/j.marpolbul.2019.02.062

Zettler ER, Mincer TJ, Amaral-Zettler LA (2013) Life in the "plastisphere": microbial communities on plastic marine debris. Environ Sci Technol 47:7137–7146. https://doi.org/10.1021/es401288x

Zhang S, Zhang W, Ju M, Ling Qu L, Chu X, Huo C, Wang J (2022a) Distribution characteristics of microplastics in surface and subsurface Antarctic seawater. Sci Total Environ 838:156051

Zhang A, Hou Y, Wang Q, Wang Y (2022b) Characteristics and polyethylene biodegradation function of a novel cold-adapted bacterial laccase from Antarctic sea ice psychrophile *Psychrobacter* sp. NJ228. J Hazard Mater 439:129656. https://doi.org/10.1016/j.jhazmat129656

Zhu D, Ma J, Li G, Rillig MC, Zhu YG (2022a) Soil plastispheres as hotspots of antibiotic resistance genes and potential pathogens. ISME J 16:521–532. https://doi.org/10.1038/s41396-021-01103-9

Zhu B, Wang D, Wei N (2022b) Enzyme discovery and engineering for sustainable plastic recycling. Trends Biotechnol 40(1):22–37. https://doi.org/10.1016/jtibtech.2021.02.008

Chapter 3
Antarctic and Arctic Seaweeds: Reassessing Diversity, Ecophysiological Adaptations, Connections and Changes in Biogeographical Distribution as a Survival Strategy Facing Meteorological and Oceanographic Changes

Franciane Pellizzari, Mariana Cabral de Oliveira, Nair Sumie Yokoya, Andres Mansilla, Johanna Marambio, Frithjof C. Küpper, and Peter Convey

Abstract Diversity, distribution and biogeographical aspects of polar macroalgae must be reassessed as they face contemporary environmental changes. Macroalgae are the basis of the marine food web, efficient carbon sinks, biogenic habitats and promoters of homeostasis in the oceans. Additionally, algae synthesise bioactive compounds that are used or potentially usable industrially. Algal growth can be limited and sensitive to thermohaline, nutrient, light and pH patterns. Furthermore,

F. Pellizzari (✉)
Department of Biological Sciences, Universidade Estadual do Paraná, Paranaguá, Brazil
e-mail: franciane.pellizzari@unespar.edu.br

M. C. de Oliveira
Universidade de São Paulo, São Paulo, Brazil

N. S. Yokoya
Instituto de Pesquisas Ambientais, São Paulo, Brazil

A. Mansilla · J. Marambio
Universidad de Magalhanes, Punta Arenas, Chile

Cape Horn International Center (CHIC), University of Magallanes, Punta Arenas, Chile

F. C. Küpper
University of Aberdeen, School of Biological Sciences, Aberdeen, Scotland, UK

P. Convey
British Antarctic Survey, Cambridge, UK

Department of Zoology, University of Johannesburg, Johannesburg, South Africa

Millennium Institute – Biodiversity of Antarctic and Sub-Antarctic Ecosystems, Santiago, Chile

A. W. Fernandes Duarte et al. (eds.), *Polar Microbes and Climate Change*,
https://doi.org/10.1007/978-3-032-07096-8_3

as the polar regions drive global climate and oceanographic circulations and as their ecosystems are highly sensitive to changes, their monitoring is essential. In 2012, Wiencke and Amsler published the first compilation of polar macroalgal studies. In this chapter, after a further decade and the intensification of global changes, we revisit this subject, hypothesising that polar biogeographic isolation is being broken, driven by meteorological and oceanographic changes and/or extreme events which alter algal dispersal patterns, resulting in more connected ecosystems. We focus on species richness and diversity, cryptic, cryptogenic and endemic taxa, setting out to reassess the distributions and connections of Antarctic and Arctic seaweed assemblages and species. Current abiotic data are associated with species lists in order to address macroecological aspects, focusing on endemism, ecotones, connections and changes in the biogeographical distribution of these seaweed communities. This data compilation is relevant for the selection of key areas for monitoring in Antarctica and the Arctic that will be fundamental for the identification and interpretation of the impacts of environmental changes on this biota, as well as in supporting conservation actions and advancing polar ocean science.

Keywords Climate change · Polar biodiversity · Algae · Seaweed assemblages · Ecosystem services

3.1 How Seaweeds Survive in Polar Regions: Physiological Adaptations of Antarctic and Arctic Assemblages

3.1.1 Antarctic and Arctic Ecosystem Features, Oceanographic Concepts and Climate

Contemporary Antarctic and Arctic meteorological and oceanographic processes and changes present distinct patterns and processes. Although both are polar regions, they show clear differences regarding their genesis and the physicochemical features of their water masses. Antarctica is a unique continent, geographically isolated by a combination of the vast area of the Southern Ocean (SO), particularly the Antarctic Circumpolar Current (ACC), as well as the circumpolar atmospheric circulation and strong westerly winds. The continent has never had a native human population and has experienced only a few and transient visitors over the last century. In contrast, the Arctic is a seasonally frozen ocean surrounded by the large northern continental landmasses and islands, the latter forming part of eight nations (Fig. 3.1). The Arctic is less isolated in geographic and human terms and faces greater anthropogenic impacts, both through the direct impacts of populations and activities within the region and indirectly originating in the densely populated and industrially active Northern Hemisphere continents.

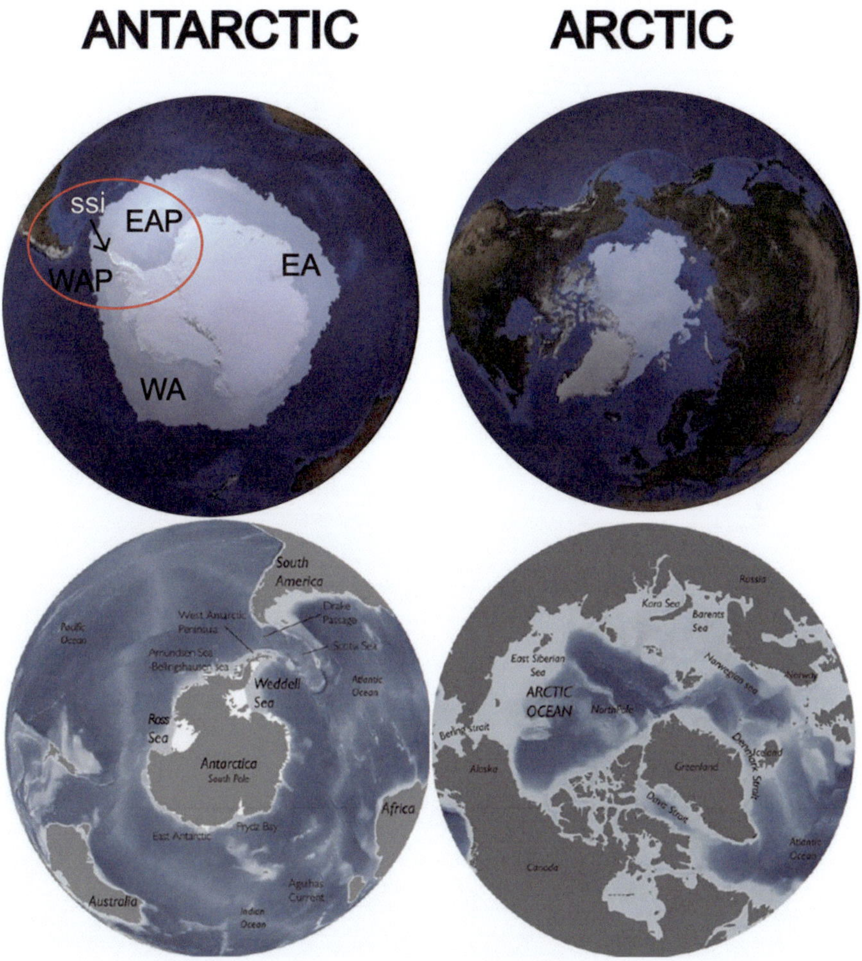

Fig. 3.1 Comparative overview of Antarctica (A, C) and the Arctic (B, D). Antarctica is a frozen continent surrounded by extensive and cold oceanic waters, while the Arctic is a seasonally frozen ocean surrounded by largely ice-free continents. A. *SSI* South Shetland Islands, *EAP* Eastern Antarctic Peninsula, *WAP* Western Antarctic Peninsula, *EA* East Antarctica, *WA* West Antarctica. (Image credits: A/B. NASA Goddard Space Flight Center, climate.nasa.gov; C/D. NOAA, www.climate.gov)

Southern Ocean (SO)

The ACC connects the Atlantic, Indian and Pacific Oceans, and defines key physico-chemical properties of water masses, such as temperature, salinity, pH and nutrients. It includes three main circumpolar fronts, from north to south the Sub-Antarctic Front (SAF), the Antarctic Polar Front (APF), and the Southern Antarctic Circumpolar Front (SACF) (Fig. 3.2a). Between the latter two polar fronts lies the Antarctic Circumpolar Current (ACC), which is the world's largest and strongest ocean current, flowing continuously eastward around Antarctica. Cold Antarctic Bottom water

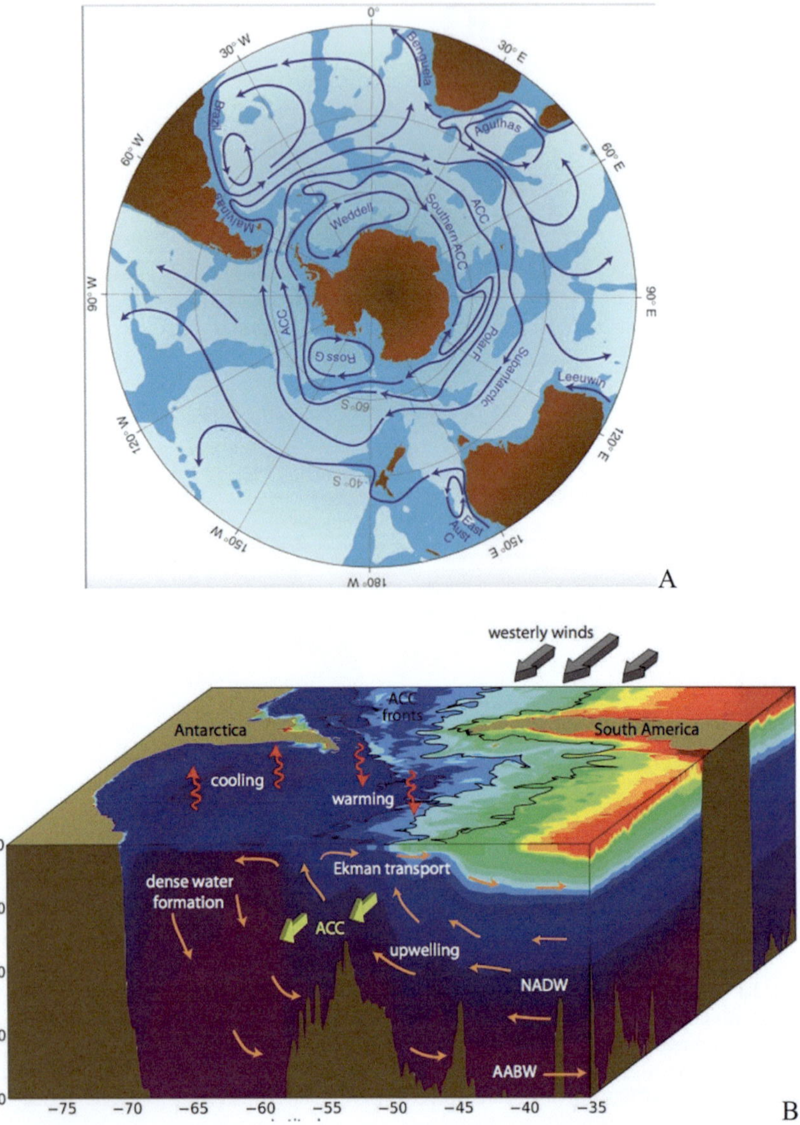

Fig. 3.2 (**a**) Schematic view of the major ocean currents of the Southern Ocean (SO) and their connections. Depths shallower than 3500 m are shaded. *ACC* Antarctic Circumpolar Current. (Source: H. Grobe, Alfred Wegener Institute, adapted from Rintoul, 2009). (**b**) Surface and deep circulation in the SO, between the Antarctic Peninsula (EAP/WAP) and southern South America. (Source: Australian National University. https://earthsciences.anu.edu.au/study/student-projects/southern-ocean-circulation)

flows northwards from Antarctica into the abyssal depths of the other major oceans, providing an important driver of the 'global conveyor belt' or global overturning circulation, which can only have been initiated once the tectonic movements separating Antarctica from South America had completed, allowing the ACC to develop. As oceanic water travels the globe in these currents, it carries dissolved gases such as oxygen and carbon dioxide, influencing the natural uptake and release of CO_2 from the oceans, contributing to the absorption of anthropogenic CO_2 and modulating the transport of heat energy globally. These functions underpin Antarctica's central role in regulating climate and ocean circulation globally (Fig. 3.2b).

Given its large geographical scale, temperatures in the SO range between −1.9 °C and +10 °C, and salinity ranges from 33.5 to 37.5 psu (Rintoul and Garabato 2013). However, even small differences in thermohaline patterns are sufficient to set up powerful currents. Sea ice forming in winter in the SO and remaining present year-round in parts of it causes surface water to sink vertically, due to the colder and saltier waters being denser than the water below. When this dense water reaches the ocean floor, it spreads out to fill the depths of the world's oceans. The existence of a circumpolar oceanic channel in the latitude band of the Drake Passage between southern South America and the Antarctic Peninsula is fundamental for the existence of the ACC and has profound implications for the global ocean circulation and climate (Rintoul and Garabato 2013). Given the influence of SO processes on the global ocean circulation, climate and biogeochemical cycles, any changes in the circulation of the SO might have widespread consequences.

These connections are highlighted by features such as the upwelling of deep waters in the Antarctic Divergence (AD), which originated in the North Atlantic. Part of the Antarctic Intermediate Water spreads northward until it reaches northern temperate latitudes, influencing thermohaline patterns and large-scale climate (Sloyan and Rintoul 2001). The other part is upwelled deep around the Antarctic continental margins by advection, rotating clockwise in the Ross and Weddell gyres. Heat losses to the atmosphere, interactions with the ice shelves and salinity enrichment result in very dense water masses which flow down the continental slope towards the deep-sea floor. The newly formed and highly O_2-rich water is driven by the bottom topography, carrying Antarctic water into the deep basins of the global oceans (Orsi et al. 2002). At the same time, AD deep water upwelling in Antarctica is rich in nutrients and gases, which may be released to the atmosphere (Lumpkin and Speer 2007), providing further positive feedback to global climate change.

Algal primary production generates organic carbon compounds through photosynthesis by sequestering CO_2. Part of this organic matter sinks and carries carbon and nutrients downwards to the deep ocean layers (Ganachaud and Wunsch 2000) in a process called the biological pump. Whether the SO acts as a source or a sink of CO_2 for the atmosphere depends on the balance between the physical carbon pump coupled to the overturning circulation and the biological pump. Controversially discussed among the researchers is whether climatic and oceanographic changes observed over recent decades will result in a stronger overturning circulation or enhanced eddy activity (Ganachaud 2003). This remains a subject of intense debate, as does uncertainty over which direction this delicate balance will shift in future.

Seaweed habitats in Antarctica are distributed in the following major areas (Fig. 3.1): 1. South Shetland Islands (SSI), 2. Eastern Antarctic Peninsula (EAP, Weddell Gyre), 3. Western Antarctic Peninsula (WAP, bordering the Bellingshausen Sea), 4. East Antarctica (EA) and 5. West Antarctica (WA), with the western side of the Ross Sea bordering EA. The SSI comprises 11 main islands, each with geomorphological and thermohaline characteristics, that lie approximately 110 km northwest of the AP, separated by the Bransfield Strait. In turn, the AP and SSI are separated from South America by the Drake Passage. The northern-most of the SSI is Elephant Island, located 885 km from Cape Horn, forming a transitional area between the EAP and South America (Sanches et al. 2016; Pellizzari et al. 2023). Another area of particular interest in the SSI is Deception Island, which is a volcanically active island whose flooded caldera forms a circular semi-enclosed bay, which is one of the safest and most protected anchorages in Antarctica. The island last erupted in the late 1960s. It hosted a shore-based whaling station in the first half of the twentieth century and currently hosts two active summer-only national research stations (with two further being heavily damaged and abandoned in the 1960s eruptions. It is also an important tourist destination. Pellizzari et al. (2017, 2023) described the island's algal assemblages and the thermohaline patterns of seawater around this island, recording thermal anomalies of up to +8 °C in the shallow water column, in addition to higher salinities and lower pH values compared to other SSI islands.

The EAP borders the Weddell Sea, reaching latitude 75°S. The southern sector of the EAP is heavily glaciated and adjoined by permanent ice shelves (including the Larsen Ice Shelf), which continue southwards to EA. It has an area of approximately 2.8 million km^2 and is considered one of the areas in Antarctica that is most vulnerable to ice losses due to climate change (Turner et al. 2020; Pehlke et al. 2022). The Weddell Sea is one of the very few areas among the global oceans that produce the very deep and cold-water masses that make a fundamental contribution to global thermohaline circulation (Vernet et al. 2019). Weddell Sea circulation is dominated by a cyclonic current, whose western part flows northwards along the EAP (Weddell Gyre; Fig. 3.2a). This northwards part of the Gyre is also the primary source of water export from the Weddell Sea and is an essential mixing and exchange zone for oceanic water. In this area, strong surface winds blow northwards parallel to the Antarctic Peninsula (AP), carrying cold air towards lower latitudes and forming cold fronts that affect the Southern Hemisphere (SH). These meteoceanographic processes are essential not only because they influence the temperature regime in the EAP but also as a driver of ice release into the South Atlantic sector of the SO (Pellizzari et al. 2023).

The Western Antarctic Peninsula (WAP) has a northern opening in Gerlache Strait, a channel forming the boundary between the Palmer Archipelago and the AP. To its south, the Gerlache Strait leads eventually to Marguerite Bay, connecting with the Bellingshausen Sea north and west of Alexander Island, which, in turn, connects with the Amundsen Sea and western Ross Sea, in West Antarctica (WA). Coastal landscapes in the northern part of the WAP are montane/alpine and heavily glaciated, with some ice-free rocky substrates (~3% of the entire Antarctic Peninsula region is ice-free; the majority of this is in the northern part). The Bellingshausen Sea reaches depths of up to 4500 m, with its geomorphology associated with the collision of an asteroid at the end of the Pliocene, the only known case globally of this type of ocean basin geoformation (Porter et al. 2016). In the northern part of the WAP (including the SSI), and to a lesser extent in the WAP in general, there are locally significant concentrations of national operator research and logistic activities and tourism, although the influence of ice limiting access by vessels can be significant through much of the year, especially in the south of the region.

Within the main body of the continent, the 2000 km mountain range of the Transantarctic Mountains forms a dividing line across the continent, separating EA and WA. EA is considerably larger than WA. The ice sheet covering EA, reaching over 5 km thick in some regions and on average more than 2 km thick, includes the most climatically extreme areas in Antarctica and the coldest and some of the driest areas on the planet. The heavy mass of ice presses down on the continental bedrock, depressing the entire continent several hundred meters into the Earth's crust, as a result of which the surrounding continental shelf is the world's deepest. Some large areas under Antarctica's ice sheets are below the mean sea level. As these low-lying areas are near the edge of the ice sheet, as is the case in large parts of WA and the southern part of WAP, ocean water extends under the ice, in some circumstances speeding up ice flow rates and mass loss. Ice flow rates in WA are generally much greater than in EA, and Porter et al. (2016) suggested that recent declines in ice in this region are practically irreversible.

Despite the preponderance of ice in EA, in parts of the continental coastline such as Wilkes Land (66°S 110°E), islands and shallow bays are present, providing a range of physical conditions that may allow and influence the distribution of benthic organisms (Johnston et al. 2007; Irving et al. 2005). Bays are generally protected areas subject to prolonged ice-cover, while islands and coastal headlands can experience ice breakout earlier in the season, releasing some hard substrate to light. During the austral summer, benthic communities in such areas are exposed to continuous solar radiation, due to the high latitudes, concurrently with little sea-ice cover. In these areas, subtidal habitats are the only substrates available to macroalgae, in contrast with SSI, and northern parts of the EAP and WAP, where intertidal colonisation can also be possible due to sheltered ice-free areas being present.

West Antarctica is covered by the WA Ice Sheet. Parts of this region have experienced some of the most rapid rates of atmospheric warming on the planet (Turner et al. 2009; Chown et al. 2022a, b). The geological history of WA was summarised by Jordan et al. (2020), who noted that, while EA has been continuously ice-covered for ~15 million years, albeit with fluctuations with depth, WA has shown repeated large-scale changes in ice over climatic cycles. Current accelerating ice loss in this region may signify an important tipping change initiating, with the primary cause being warming of ocean water, which then enters beneath the WA ice sheet and glaciers, where they are below sea level, accelerating bottom melting and increasing rates of ice flow.

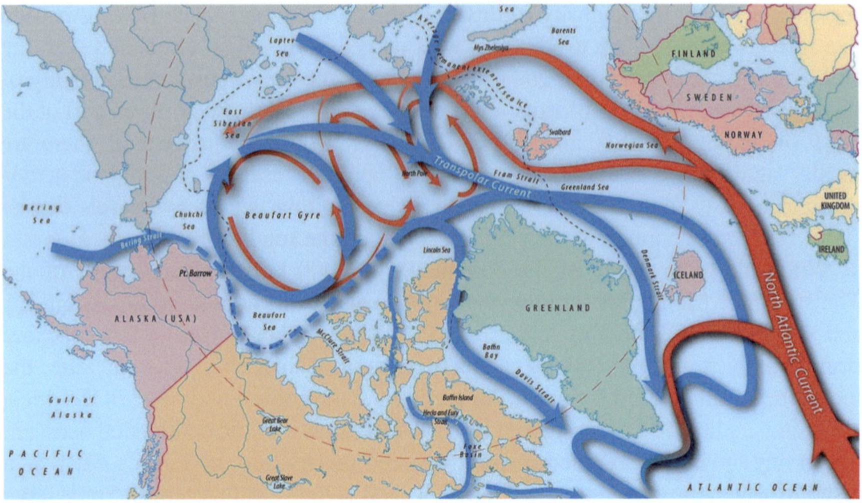

Fig. 3.3 The complex circulation patterns of the Arctic Ocean (in red, warm currents, and in blue, cold). (Source: J. Cook, Woods Hole Oceanographic Institution. (Retrieved from: https://www.whoi.edu/know-your-ocean/ocean-topics/how-the-ocean-works/ocean-circulation/arctic-ocean-circulation/))

Arctic Ocean (AO)

The complex Arctic Ocean circulation system (Fig. 3.3) rules the AO's entire food web in a delicate balance. In recent decades, changes in several areas of the Arctic have been recorded, including a dramatic reduction in seasonal sea ice cover and a weakening of the Beaufort Gyre, which are attributed to climate change. The changes affect not only biota, but the Arctic's indigenous human communities, as well as the wider populations of Europe and North America. The AO, surrounded by land, is covered entirely by a thin layer of annual and multi-year sea ice, which forms during winter (average of 2 m thickness), and can shrink by up to two-thirds every summer (Timmermans and Marshall 2020). Arctic summer sea ice extent has declined rapidly in recent decades (Perovich et al. 2019), coincident with unparalleled air temperature increases, warming ocean temperatures and melting ice and permafrost. Northern polar regions are warming faster than the global mean in a phenomenon known as Arctic amplification (Overland et al. 2019). In the imminent future, an entirely new regime may come into place, in which the interior Arctic Ocean is entirely ice-free in summer and sea ice will become thinner and more mobile in winter (Haine and Martin 2017), increasing the availability of consolidated substrates for colonisation by benthic organisms such as macroalgae. A seasonally ice-free Arctic will have major implications for oceanography, marine food webs and the larger-scale climate (Collins et al. 2013), as well as for regional human populations, economies, and trade.

The AO receives inflows from the Atlantic and Pacific oceans, as well as considerable freshwater inputs from the major North American and Siberian rivers.

According to Timmermans and Marshall (2020), there is a prominent halocline (rather than thermoclines, as also observed in some areas of Antarctica), and its stratification is predominantly driven by melting and freezing of sea ice and through the mediation of surface wind stress. Unlike in Antarctica, the Sverdrup balance is not applicable in the Arctic, where the north-south gradient of the Coriolis effect is too small to influence the dynamics (Timmermans and Marshall 2020). Rapid contemporary changes have raised many questions about the AO's future dynamics, as well as those of its biota. The AO, along with the Greenland, Iceland and Norwegian seas, has been referred to as the 'Arctic Mediterranean' because it is basically a large deep basin of seawater surrounded by land and shallower channels (Fig. 3.3), marked mainly by the Greenland-Scotland Ridge (Timmermans and Marshall 2020). Relatively warm and saline North Atlantic water flows across this ridge into the Nordic Seas. Atlantic Water enters the Arctic via the Fram Strait and the Barents Sea opening. The only oceanic gateway between the Pacific and Arctic oceans is the Bering Strait, where Pacific Water inflow provides a source of lower salinity water and heat to the AO. Water leaves the AO via straits in the Canadian Arctic Archipelago and in the East Greenland current that flows south on the west side of the Fram Strait (Timmermans and Marshall 2020, and citations therein).

Arctic Ocean ice has a strong seasonal cycle, with summer sea-ice extent, in recent years generally around one third of the winter extent. The winter maximum extent occurs in March, while the sea-ice minimum is usually in September. Since 1979 (the first satellite record), there has been a linear trend of summer sea ice decline, with a rate of about one million km^2 per decade. Minimum sea-ice cover has been about 4.5 million km^2 in September in recent years (Perovich et al. 2019). Declining sea-ice extent is accompanied by changes in marine benthic communities, including migration of seaweeds (Lebrun et al. 2022). The availability of inorganic nutrients in the AO can limit the biomass or composition of macroalgal communities, where concentrations of nutrients can become depleted seasonally, or may even be highly available in certain areas due to organic input. This contrasts with Antarctica, where nutrients are rarely limiting for the well-adapted primary producers, despite being considered oligotrophic in many areas (Zacher et al. 2009; Deppeler and Davidson 2017).

Despite being at opposite poles globally, Antarctica and the Arctic are connected oceanographically by a system known as the Thermohaline Conveyor (Fig. 3.4), a complex process that occurs around the globe. Warm ocean currents (in red) are those which flow near the surface of the ocean. These warmer and less dense waters transport heat from the equator to the polar regions. Cold, more saline and denser currents (in blue) flow at greater depths and are mixed with freshwater from melting. There is a constant flow of warm currents flowing to higher latitudes, where they cool and sink in downwelling areas and return towards the equator (Fig. 3.3).

The climate between 60 and 70°N varies from temperate maritime oceanic during the summer, to sub-polar and polar during the winter, and average surface seawater temperature (SST) varies from 1 °C during winter to 13 °C in summer. In the past decade, extreme air temperatures have been recorded, with a higher temperature of +28 °C in July, and a lower record of −23 °C in January (Timmermans and

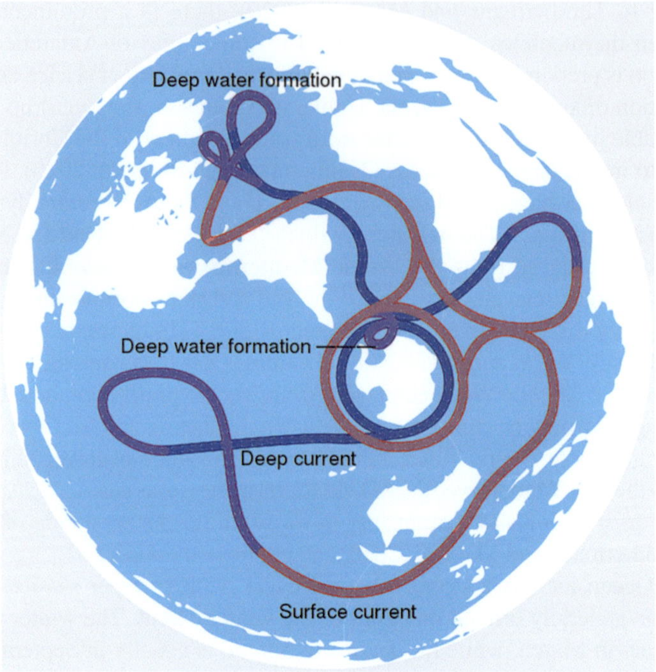

Fig. 3.4 The Antarctic Circumpolar Current (ACC), surrounding the continent in the middle of the map, and its connection (cold waters in blue, and warm in red) to the wider Thermohaline Conveyor. (https://www.antarcticglaciers.org/glaciers-and-climate/ocean-circulation/)

Marshall 2020), although the extreme temperatures vary broadly across different parts of the Arctic. The general climate is windy and cloudy (low irradiance), and average annual precipitation is ~1000 mm, with snowfall concentrated from November to February (Timmermans and Marshall 2020). Fog is frequent during summer due to the cooling effect of the sea interacting with warmer atmospheric flows from the south. Regionally, within the Arctic, the geology is complex, with rugged and montane coastlines. The main economic activities are current fishing and oil and gas extraction, with an expected drastic expansion in shipping using the 'northern sea route' as Arctic sea-ice extent declines or is lost altogether. It is also an environmentally sensitive area with various protected areas declared within the territories of the Arctic nations. The Svalbard Archipelago (between 74°N and 81°N) is the northern-most permanently inhabited location, experiencing a High Arctic climate but with temperate influence by North Atlantic currents originating from the Gulf Stream. Annual precipitation here, generally in the form of snow in winter but increasingly of rain in both summer and winter, ranges from 200 to 600 mm. Svalbard geology is varied, and currently, around 60% of the archipelago's area is permanently ice-covered (https://www.thearcticinstitute.org/). As with all areas north (south) of the Arctic (Antarctic) Circle, Svalbard experiences periods of continuous daylight in summer and darkness in winter.

With these brief and inevitably selective introductions to parts of Antarctica and the Arctic, it is already clear that there is a great distinction among their ecosystems, as well as the levels of protection and conservation already in place and their vulnerability. There is an urgent need to compare aspects of the diversity and ecology of polar algae, particularly relating to their great potential as indicators of meta-oceanographic changes, which this chapter focuses on. Compared to Antarctica, the Arctic currently faces higher anthropogenic impacts and has been the subject of a greater number of studies, including climatological and biological modelling on large spatial and temporal scales, giving assistance in understanding anomalous patterns, generating models predicting biodiversity responses and incorporating sensitive bioindicators such as the macroalgae.

3.1.2 Physiological Survival Strategies

Seaweeds, as benthic primary producers, underpin the marine food web, as well as providing biogenic habitat for marine organisms, contributing to seawater homeostasis, pH balance, carbon stocks and nutrient cycles. They are highly sensitive to abrupt changes in the physical and chemical parameters of their environment, being considered excellent bioindicators of climate change. As monitoring proxies, compiling the 'state of the art' of knowledge of algal communities in the polar regions will aid in both prediction and the development of mitigation strategies towards changes in biotic and abiotic patterns, in the context of climatological and oceanographic changes. As noted in Sect. 3.1.1, the polar regions have key influences on the global climate and oceanographic circulation patterns. Changes occurring in the polar regions affect the entire planet and must be monitored, described and understood.

Antarctic ecosystems are characterised by abiotic conditions with wide seasonal ranges, dominated by continental-scale ice and its dynamics, as well as the consequences of extreme geographical isolation and particular geomorphologies (Convey et al. 2020). Over evolutionary time, polar macroalgae have developed survival strategies to combat these various challenges. Survival of the multiple extreme environmental stresses presented by the Antarctic and Arctic involves (a) synthesis of specific chemical compounds and their use at often higher concentrations than those reported for temperate and tropical relatives; (b) distinct and more effective routes of assimilation of inorganic macro- and micronutrients from seawater; (c) higher and more efficient synthesis of specialised organic compounds, associated with algal secondary metabolism.

The growth, development, reproduction and spatial distribution of seaweeds in polar zones are strongly driven by environmental parameters (Wiencke and Amsler 2012). At regional scales, the dynamics of ice formation, the type of substrate and seasonality are further important factors. Biotic and abiotic factors act in combination to drive the development and growth of algal assemblages, including temperature and its variation on different spatial and temporal scales, salinity, nutrients, pH, abrasion, photoperiod, irradiance, UV levels, shading, competition and herbivory (Zacher et al. 2009; Mystikou et al. 2014). Sea ice has a strong influence on the life history, ecology, morphology, and physiology of seaweeds. In two known cases, one from the Arctic (*Platysiphon glacialis*; Kawai et al. 2015) and one from the Antarctic (*Desmarestia menziesii*; Küpper et al. 2019), sea ice cover results in a strong change from juvenile and adult morphology.

Polar seaweeds are adapted to the strong seasonality and a pronounced seasonal light cycle, 24 h of darkness during winter and 24 h of light during summer, reaching their peak during mid-summer (Pavlov et al. 2019), developing strategies to tolerate high light after ice break-up (Zacher et al. 2009). Therefore, they have developed mechanisms for adjusting the photosynthetic apparatus by altering the content and composition of photochemically active and accessory pigments (Brey 2009). The effects of different wavelengths of UV radiation on polar seaweeds have received scientific attention due to the increased UV levels (specifically UV-B) experienced in these regions as a result of stratospheric ozone depletion (Bischof et al. 2006; Karsten et al. 2009; Gómez and Huovinen 2015).

Temperature is an important abiotic factor influencing the growth and reproduction of polar seaweeds (Wiencke et al. 2009). They are well adapted to low temperatures, showing high ratios of photosynthesis: respiration (P:R), which underlie their high productivity and growth at low temperatures (Goméz et al. 2009). Antarctic endemic species (e.g. representatives of the Desmarestiales) have the lowest temperature demands, while Arctic seaweeds are comparatively less well adapted to low temperatures (Goméz et al. 2009). Low temperatures affect photosynthetic activity (as enzymatic reactions are temperature-dependent), membrane fluidity and protein function (Becker et al. 2009). Intertidal seaweeds are frequently exposed to air temperatures below 0 °C in polar regions, as well as to solar warming to temperatures well above that of seawater, and few species can tolerate the combination of low temperatures and high irradiance in the intertidal.

The extreme polar conditions challenge survival, in which capturing light to fuel photosynthesis emerges as a vital process, as seaweeds must simultaneously address the risks associated with the induction of reactive oxygen species (ROS), as an indicator of emerging stresses, such as high irradiance, some algal species react by generating ROS ultimately generating intracellular stress in the organism. ROS generation promotes the non-photochemical quenching of excited chlorophyll molecules in photosystems or light-harvesting complexes (Ruban 2016). In response to ROS generation, seaweeds (mainly brown) can synthesise phlorotannins, which act as antioxidants, mainly protecting photosynthetic tissue, making them important photoprotective compounds (Schoenwaelder and Wiencke 2000; Amsler 2008).

The interaction between the need to absorb solar radiation and defend against environmental adversities highlights the remarkable resilience and adaptation of algal species in some of the most hostile and challenging environments on the planet.

Polar macroalgae have developed distinct biochemical and physiological strategies to enable survival under extreme conditions. Although well adapted to low-light conditions (Marambio and Bischof 2021), they also face radiation stress (Noctor et al. 2019). For example, since seawater is very clear after sea-ice break-up in spring, solar radiation penetrates deeper into the water column than it does elsewhere (Gómez and Huovinen 2015). Consequently, exposure to high levels of both photosynthetically active radiation (PAR: 400–700 nm) and ultraviolet radiation (UV-A: 315–400 nm; UV-B: 280–315 nm) affects the physiological processes of these algae (Aguilera et al. 2002; Huovinen and Gómez 2013).

When macroalgae are exposed to both UV stress and high temperatures, their primary metabolism can generate an imbalance in the REDOX system, and generate an accumulation of EROX, such as the superoxide anion radical (O_2•−), hydrogen peroxide (H_2O_2) and singlet oxygen ($^1O_2$ = oxygen with all electrons paired, unstable and at 760 nm becomes a triplet), which are continuously produced at basal levels as part of the normal metabolic process and are eliminated by various antioxidant mechanisms (Kaur et al. 2022). In polar macroalgae, the induction of ROS can pose a significant challenge as these reactive molecules have the potential to damage vital cellular components such as nucleic acids, proteins and lipids (Riaz et al. 2022). To minimise the cost of damage and repair to photosystems and other cellular components due to oxidative stress, a key mechanism is the activation of the antioxidant system. Furthermore, mycosporin-like amino acids (MAAs) act as natural solar filters, absorbing ultraviolet (UV) radiation, and are found in higher concentrations and diversity in temperate and polar red algae (Sáez et al. 2023). In polar environments where UV exposure is high, and further exacerbated by the anthropogenic ozone hole, the potential to mitigate damage to photosynthetic apparatus by synthesising and accumulating MAAs responsively indicates a high degree of physiological plasticity and resistance.

Antarctica

The SO, whose natural northern boundary is the Antarctic Convergence, includes some of the most extreme marine environments on Earth, resulting in wide temporal variability in productivity and biomass, as well as a large diversity of primary producers (Sanches et al. 2016). Deregibus et al. (2023) suggested that, under sustained climate change, the processes of colonisation and expansion of benthic assemblages will lead to significant transformations in Antarctic coastal ecosystems through increasing primary production and energy flow through food webs, creating habitats giving physical refuge to fauna and capturing and storing more carbon. Antarctica experiences a strong seasonal light regime, especially at latitudes beyond the polar circle, with periods of complete darkness in winter and daylight in summer, facing up to 1700μmol m^{-2} s^{-1} PAR, 44 W m^{-2} UV-A, and 2.3 W m^{-2} UV-B (Huovinen and Gómez 2020). Antarctic seaweeds are adapted to survive under low light conditions and through extended darkness, and their high efficiency of photosynthesis permits colonisation of deeper habitats (Gómez et al. 1997, 2009, 2011; Gómez and Huovinen 2015). Growth and photosynthesis of some endemic brown seaweeds are not light-limited even at depths of 30–40 m (Weykam et al. 1996a, b; Schwarz et al. 2003). Antarctic seaweeds have also protective mechanisms to tolerate high levels of solar radiation during the spring-summer period, through photoinhibition and down-regulation or feedback regulation of photosynthesis (i.e. when non-structural carbohydrates accumulate and photosynthesis may be suppressed), dissipating the excess energy absorbed by photosystem II (Adams et al. 2008). In this context, the photosynthetic pigments of macroalgae can regulate the process of acclimatisation and adaptation, as well as generate physiological responses in cold-water algae (Plastino and Mansilla 2004). An example is the physiological regulation through photosynthetic pigments in the genus *Desmarestia*, a genus studied in a latitudinal gradient from the sub-Antarctic to the Antarctic region, which has shown how the high concentration of Chlorophyll-a is key in the acclimation of polar algae (Mansilla et al. 2016).

When PAR levels are high, Antarctic seaweeds can also be exposed to deleterious UV radiation (Huovinen et al. 2016), which can affect cellular components, biomolecules such as DNA and proteins (e.g. D1 protein in PSII) and photosynthetic pigments, and cause oxidative stress by the production of reactive oxygen species (ROS) (Huovinen and Gómez 2020; Sáez et al. 2023). Protective mechanisms include the biosynthesis of mycosporine-like amino acids (MAAs) in red algae, phlorotannins in brown algae (Gómez and Huovinen 2020a, b) and antioxidants (e.g. carotenoids, enzymes such as superoxide dismutase and glutathione peroxidase). Other metabolites also have protective roles against low temperatures and freezing, e.g. floridean starch, low molecular weight carbohydrates and polysaccharides in the Antarctic endemic red alga *Sarcopeltis antarctica* (Pasqualetti et al. 2023).

Phenolic compounds, specifically phlorotannins, play a photoprotective role by acting as UV-shielding substances, efficient antioxidants and herbivory deterrents. The concentration of soluble phlorotannins is positively correlated with antioxidant capacity in several species of Antarctic brown algae (Gómez and Huovinen 2020a, b). In *D. anceps*, phlorotannins, although not induced by UV radiation, increase significantly at higher temperatures, positively correlating with ROS scavenging (Rautenberger et al. 2013). Furthermore, they are fundamental to several morpho-functional processes (Gómez et al. 2019), allowing Antarctic seaweeds to thrive under extreme conditions.

A study of the Antarctic red macroalgae *Curdiea racovitzae* and *Iridaea cordata* showed that extracts from these algae have high concentrations of MAAs, mainly palithine, demonstrating significant photoprotective properties, absorbing UVA and UVB radiation, being photostable and showing antioxidant activity against superoxide radicals (Rangel et al. 2020). Crustose calcareous algae from Antarctica are highly sensitive to increased light levels (Irving et al. 2005) and sensitive bioindicators of environmental changes at small scales.

Arctic

Arctic coastal environments are similarly characterised by strong seasonal changes in light levels. In spring, day length and temperature increase, and ice begins to melt. Thawing land and sea ice decreases salinity and affects the light regime by delivering dissolved and particulate matter, which increases turbidity. In autumn and winter, as day length shortens and temperature decreases, ice re-forms on the sea surface and land (Lebrun et al. 2022).

The AO warms in summer via surface-water heating from solar radiation absorbed in ice-free regions (Timmermans and Marshall 2020). The net surface heat flux is the sum of incoming shortwave radiation, longwave emission, and sensible plus latent heat fluxes. Throughout the year over most of the AO, vertical sensible and latent heat fluxes are small ($\lesssim$10 W m^{-2}). Net longwave flux is greater (~50 W m^{-2}) and remains approximately constant throughout the year. The net shortwave component has a strong seasonal cycle, dominating in summer when average values over the AO are ~150 W m^{-2}. Incoming solar radiation is effectively zero between October and March. The AO also receives heat via warm inflows, as well as UV radiation, from the Atlantic and Pacific oceans (Timmermans and Marshall 2020). While ocean temperature may have only a weak influence on ocean dynamics, it has a strong influence on the fate of Arctic sea-ice cover when warmer water is mixed to the surface.

Interactions of certain environmental contaminants (such as PAHs) with UV radiation can lead directly or via ROS to their photo-modification (photo-oxidation, photolysis), resulting in photodegradation or the formation of photoproducts with different characteristics and effects (which can include lower or higher toxicity) to the original compound (Huovinen and Gómez 2020; Sulzberger et al. 2019).

Species present in intertidal and shallow subtidal zones are generally more tolerant to high irradiances of PAR and UV and exhibit faster recovery and repair mechanisms than those from deeper zones (Wiencke et al. 2000; Karsten et al. 2001). However, species such as *Palmaria palmata* and *Desmarestia aculeata* inhabiting shallow waters in Kongsfjorden, Svalbard, responded with a decrease in quantum yield in the presence of high irradiance (Marambio and Bischof 2021), showing a short-term effect. Some studies suggest that the sensitivity to UV radiation of early life stages, propagules and adult stages is distinct, reflecting on the recruitment and succession of species (Zacher et al. 2007; Campana et al. 2009). Summers et al. (2023) studied the photophysiological compared activity among green (*Ulva* sp.), red (*Palmaria palmata*), and brown macroalgae (*Laminaria digitata, Alaria esculenta, and Saccharina latissima*) living in the Arctic Polar night along Svalbard fiords. The authors report that even during winter, seaweeds appeared healthier (morphologically) and had longer lamina, revealing that some perennial species grow new tissue in darkness. Also, light saturation parameter (Ek) and pigment composition were higher with lower maximum quantum yield of PS II fluorescence (Fv/Fm) for red and brown algae, suggesting that the photosynthetic apparatus is active, although slow, during Arctic winter.

The Arctic brown alga *Fucus distichus* is one of the very few intertidal algae present along the Svalbard coastline (Wiencke et al. 2004). The photosynthetic activity of *F. distichus* specimens submitted to low temperatures showed decreased quantum yield, which recovered as soon as the temperature increased, indicating that, under the experimental conditions, photodamage did not occur (Becker et al. 2009). On the other hand, the combination of factors such as temperature and salinity shows a negative effect on the response of species such as *Laminaria solindungula* and *P. palmata*, both of which show an increase in antioxidants when affected by both environmental factors in the Arctic zone (Diehl et al. 2020; Marambio et al. 2022a, b).

In other parts of the Arctic, the ecological success of macroalgae in the supra- and mid-upper littoral is supported by an enhanced reactive oxygen scavenging system, allowing rapid acclimation to changing environmental radiation conditions (Aguilera et al. 2002). The recovery of photosynthesis after photo-inhibition in species from shallower areas is more rapid, indicating their adaptation to varying light levels. *Fucus distichus* and *F. vesiculosus*, dominant species in Arctic intertidal zones, do not exhibit photo-inhibition at high light levels, recovering completely after thawing. Similarly, only small changes in ROS-mediated MDA formation (= malondialdehyde, metabolised lipid peroxides) were induced by the experimental treatments, suggesting high levels of protection from free radical formation, or that the radiation conditions experienced were too low to induce significant oxidative stress.

Clearly, the antioxidant defence strategies of these algae are efficient. Furthermore, Arctic species such as *Undaria pinnatifida, Arthrothamnus bifidus, Thalassiophyllum clathrus* and *Agarum turneri* are potential sources of natural antioxidants, and the synthesis of photoprotective compounds such as MAAs and phenolics is also a vital strategy (Aminina et al. 2020). An inverse correlation between MAA content and reduction in photosynthetic electron transport following UV exposure has been found in the Arctic species *Devaleraea ramentacea* and *Palmaria palmata*, indicating the potential of MAAs to partially protect photosynthesis against harmful radiation (Stévant et al. 2023). Phenolic compounds such as mannitol are part of the carbon storage during the photosynthetic process in brown algae, also acting as a protector against osmotic stress such as hyposalinity a condition that is becoming increasingly prevalent due to rising temperatures and thus increased meltwater in polar regions. In the species *Desmarestia aculeata*, high mannitol concentration has been observed at low temperature; this response shows how this compound also acts as an antifreeze in polar algae (Monteiro et al. 2021; Marambio et al. 2022a, b).

F. vesiculosus also accumulate a wide variety of phenolic compounds, such as phloroglucinol, gallic acid and catechin; and their preventive roles against lipid peroxidation are due to radical scavenging properties, protecting the cell from the formation of ROS (Sánchez-Bonet et al. 2021). Most studies on the response of Arctic macroalgae to climate change have been performed ex situ, due to the multiple challenges in carrying out in situ experiments (Müller et al. 2009; Wilson et al. 2018). Warming, changes in underwater light regime, local hyposalinity and ocean acidification can all have large effects on biological processes, including reproduction, primary production and respiration (Müller et al. 2009; Bonsell and Dunton 2018).

3.1.3 Morphological and Ecological Survival Strategies

The distinct morphological and ecological features of polar macroalgae can be related to the geological and biogeographic histories of the polar regions, with seasonality, biomass, dispersal, succession processes and zonation and distribution being influenced by the interaction of multiple abiotic and biotic factors, including the ability to resist and survive herbivory (Wulff et al. 2009).

In the classification of algal functional groups (AFG) proposed by Steneck and Dethier (1994), Antarctica and the Arctic host the following AFGs: Filamentous, foliose, corticated foliose, corticated terete (fleshy), leathery, calcareous crustose, non-calcareous crustose and vesicular (or balloon). While algal richness is lower in the polar compared to temperate and tropical zones, this approach provides an alternative for studying seaweed communities, with broad applicability. AFGs provide a simple approach to predict algal community composition, not requiring detailed taxonomic skills, with different AFGs showing similar characteristics in areas such as productivity and growth rates, life cycles, thallus longevity and canopy height. Herbivore-induced disturbances have functionally similar impacts on most morphologically and anatomically similar algae regardless of their taxonomic or geographic affinities. The use of AFGs for the description of algal communities in polar coastal zones provides a consistent and rapid outcome, giving it a practical advantage over more detailed taxonomic approaches at the species level. Seaweed biomasses in both polar regions primarily consist of robust species, mainly fleshy and leathery functional groups.

In marine ecosystems dominated by macroalgae, the presence of classic AFGs or morpho-functional models may suggest features of physiological performance and physical gradients, assuming that macroalgal species with comparable gross morphology share similar ecophysiological features. For example, algae with sheet-like morphology are predicted to have higher growth rates, photosynthesis and nutrient uptake, lower resistance to herbivory and lower competitive capacity than algae with more robust morphologies (reviewed by Gómez et al. 2019).

Polar seaweeds have developed two different life history strategies to synchronise their growth and reproduction to seasonal variations in abiotic factors: season anticipators and season responders (Kain 1989). Season anticipators grow and reproduce during the winter, under short day or dark conditions during the winter-spring, in a life history strategy similar to K-strategists and triggered by day length. In Antarctica, Robinson et al. (2022) reported *P. decipiens* as a seasonal anticipator, developing new blades in August following circannual rhythms and preparing to grow and reproduce in late winter/spring, as well as many other Antarctic species, while in the Arctic, only *Laminaria solidungula* is a seasonal anticipator (Table 3.1). In contrast, season responders grow and reproduce in summer when conditions of temperature, light and nutrients are suitable. This more opportunistic life history strategy is similar to R-strategists and several Antarctic species, and the Arctic brown alga *Saccharina latissima* are examples (Table 3.1). The majority of endemic Antarctic and Arctic seaweeds are season anticipators, showing physiological

Table 3.1 Life history strategies, *sensu* Kain (1989), of Antarctic and Arctic seaweeds based on Wiencke et al. (2009) and Navarro et al. (2020)

	Antarctica	Arctic
Life strategy	Species	Species
Season anticipators (K-strategists)	Phaeophyceae	
	Ascoseira mirabilis	*Laminaria solidungula*
	Desmarestia anceps	
	Desmarestia antarctica	
	Desmarestia menziesii	
	Himantothallus grandifolius	
	Phaeurus antarcticus	
	Rhodophyta	
	Gymnogongrus antarcticus	
	Gymnogongrus turquetii	
	Hymenocladiopsis crustigena	
	Kallymenia antarctica	
	Palmaria decipiens	
	Paraglossum salicifolia	
	Phyllophora ahnfeltioides	
	Trematocarpus antarcticus	
Season responders (R-strategists)	Phaeophyceae	
	Adenocystis utricularis	*Saccharina latissima*
	Geminocarpus geminatus	
	Pylaiella littoralis	
	Rhodophyta	
	Iridaea cordata	
	Sarcopeltis antarctica (formerly *Gigartina skottsbergii*)	
	Ulvophyceae	
	Acrosiphonia arcta	
	Ulva hookeriana	

adaptation to the seasonal changes of polar environments and occurring in the subtidal zone, while season responders have wider distributions extending into cold-temperate regions as well as wider vertical distribution, growing in both intertidal and subtidal zones (Wiencke et al. 2009).

A small number of studies have addressed succession in polar seaweed assemblages (see Campana et al. (2009) for an overview). Succession commences with rapid colonisers, e.g. diatoms and ephemeral small thallus seaweeds, followed by recruitment of annual and perennial species in late winter–spring. Colonisation is severely limited by physical disturbance (e.g. ice scouring) and seasonal changes in abiotic conditions, including thermohaline patterns and UV stress (Karsten et al. 2001), as well as by biotic factors such as grazing (Zacher et al. 2007).

Antarctica

In the SO, the presence and magnitude of scouring by ice shelves and icebergs have driven the evolution of a distinct biota adapted to both conditions south of the ACC and by extreme geographical isolation (Barnes et al. 2006). However, Sanches et al. (2016) and Fraser et al. (2018, 2020) reported a higher connectivity among seaweed assemblages across the ACC, and on both sides of Drake Passage, suggesting that it may not be considered a watertight barrier anymore, as previously reported. According to (Oliveira et al. 2020), morphological, reproductive and physiological characteristics, which determine species tolerances to environmental conditions, are susceptible to evolutionary pressure and, ultimately, reflected in phylogenetic relationships.

The seaweed drifted biomass in the SSI is dominated during summer mainly by *Desmarestia* spp. and *Palmaria decipiens* (Oliveira et al. 2009). Gomez et al. (2019) studied the photosynthetic and bio-optical properties of 29 seaweeds from King George Island (SSI) to document their different functional traits, grouped according to gross morphology, bathymetric distribution, taxonomic category, and biogeographic affinity, and assess whether these could explain patterns of vertical distribution and ecosystem services. Based on analysis of light requirements for photosynthesis, thallus (absorbance and thickness), photosynthetic efficiency and tolerance to stress (UV and temperature), the results obtained suggested that these algae showed high tolerance to stress caused by short-term exposure to UV radiation and enhanced temperature (7 °C). The stress tolerance was independent of gross morphology, with the greatest differences found along the depth gradient and between algae of different biogeographic affinity. Thus, the studied traits were strongly limited by photosynthetic and bio-optical properties, concluding that three major clusters of species could be defined based on the morphological framework: (1) delicate (filamentous and foliaceous) green algae, present mostly in the intertidal zone and with wide geographic distribution; (2) thick leathery Rhodophyta with Antarctic-sub-Antarctic origin, found mostly in shallow-sub-tidal locations; and (3) endemic and/or conspicuous robust fleshy and/or coriaceous Antarctic brown algae inhabiting the lower sub-tidal zone. Overall, these results highlight the contribution of AFGs in explaining algal distribution across the depth gradient, providing insights into their potential resilience in a changing polar environment. This is particularly relevant in Antarctica as macroalgae underpin biogeochemical processes in coastal ecosystems.

Goméz and Huovinen (2020a, b) considered a total of 131 species of Antarctic seaweeds, concluding they were distributed in the following functional groups: filamentous and finely-branched (45), foliose (9), coarsely-branched, including corticated species (48), thick leathery, including terete forms (11) and prostrate species (18). Coarsely-branched and leathery species, the most robust and large-sized forms, represented 49% of the total number of species, mostly being endemic brown and red algal species growing in the subtidal zone. Filamentous, finely-branched and foliose species (41%) represented mainly green and red algae, growing in shallow sub- and intertidal zones, with wide geographic distributions.

Irving et al. (2005) reported that *Himantothallus grandifolius* covered approximately 80% of the seafloor at 12 m depth in a study in EA, with the red seaweed, *Palmaria decipiens,* covering nearly all the remainder. In another area of EA, Johnston et al. (2007) reported 80–90% cover of *Desmarestia* sp. and *P. decipiens* at 6 m depth, and approximately 80% cover of *H. grandifolius* at 12 m. In the western Ross Sea, the southern-most WA seaweed habitats are dominated by the frondose red seaweed *Iridaea* sp. (formerly known as *I. cordata*; see Guiry and Guiry (2025)) and *Phyllophora antarctica* and the crustose coralline alga *Phymatolithon foecundum* in Terra Nova Bay (Cormaci et al. 1992). Miller and Pearse (1991), in an ecological study at McMurdo Sound (WA), listed three species of seaweeds as the main components in a quantitative evaluation demonstrating a depth-related distribution pattern. *Iridaea* sp. again dominated in shallow water, *P. antarctica* were abundant and fertile at intermediate depths, and *Leptophytum coulmanicum* dominated below 20 m. The vertical distribution of species was related to irradiance levels. Miller and Pearse (1991) suggested that disturbance caused by ice scour and anchor ice determined the upper limit of algal distribution, where herbivory was apparently absent. The SO is the only ocean globally with no record of kelps, which reach their latitudinal southern limit at 55°S. Ecologically, Laminariales are replaced in Antarctica by Desmarestiales, including *Himantothallus grandifolius* and *Desmarestia* spp. This latter order possibly originated in Antarctica or the sub-Antarctic, subsequently spreading widely to other locations, including the Northern Hemisphere, based on phylogenetic, morphological and ecophysiological studies (Peters et al. 1997).

Sea ice is a major environmental driver of marine ecological processes in Antarctica. Sea ice exerts direct (through export of sympagic primary production) and indirect (through coupling with oceanographic processes) influence on the nature and abundance of food available for benthic organisms. In contrast with the WAP, sea ice cover along the coasts of EA has tended to increase (Michel et al. 2019), and the ecological impacts of the increase in sea ice cover on benthic ecosystems are still poorly understood, as effects of sea ice loss have received more attention than those of expansion. Feeding habits of benthic consumers vary between locations featuring different sea ice extent and persistence. Moreover, seasonal sea-ice breakup causes strong and predictable food pulses which benthic organisms can take advantage of by altering their dietary niche. Antarctic coastal ecosystems commonly feature co-occurrence of three groups of producers, each associated with a specific ecosystem compartment: sympagic algae (blooming in or on ice), pelagic algae (mainly phytoplankton) and benthic producers (mainly seaweeds). In a comparison, Michel et al. (2019) mention that sympagic algae in the Arctic are readily consumed by benthic invertebrates, and may be preferentially selected by some species, possibly due to their high essential fatty acid content and palatability; while in Antarctica, they found the ratio between elemental contents of carbon and nitrogen (C/N ratio), a commonly used proxy for nutritional quality of a food item, very high for *H. grandifolius* blades. However, *H. grandifolius* has robust blades, and large Phaeophyceae commonly contain high amounts of refractive chemical compounds, as well as herbivore deterrents. Regardless of the underlying factors, the authors highlighted that the predominance of sympagic algae in benthic invertebrate diets is not a common feature of coastal Antarctic food webs, and that, especially in the AP, as well in EA (Ross Sea), a mixed diet consisting of the three groups of producers is typical, which provide the main sources fuelling food webs in these regions.

Changes in food web structure are mostly associated with shifts in lower trophic levels. The structure in more intensely ice-covered regions involves more generalist feeders with a loss of specialists, consistent with an increased reliance on more reworked detrital food sources (Michel et al. 2019). Recently, Iken et al. (2023) studied the influence of macroalgal abundance and diversity changes on coastal food web structure along an annual sea ice cover gradient in the WAP. The authors measured the stable carbon and nitrogen isotope values of primary producers and observed that these changed in relation to the sea ice cover gradient, while the isotope values of most invertebrate feeding groups increased with sea ice cover. Food web length became shorter and consumer trophic niche width smaller in regions with higher sea ice cover. These results suggest that some benthic invertebrates can respond to differences in basal energy sources. Conversely, these food webs dominated by generalist feeders are likely less efficient in energy transfer, which can create fewer stable systems with lower adaptive capacity to disturbance (Iken et al. 2023). The predicted sea ice loss along the WAP may ultimately lead to a longer food web with higher macroalgal abundance, more specialist species, and wider consumer trophic niches than are currently present in more ice-covered regions.

Clark et al. (2015) surveyed benthic communities in Commonwealth Bay (EA) 3 years after continuous sea-ice cover began and found algal bed communities in severe decline, suggesting the vulnerability of polar ecosystems to such an environmental change. The majority (~75%) of large macroalgae were decomposing, and the remainder were discoloured or bleached, while approximately 40% of encrusting coralline algae were bleached. In recent years, and in contrast with other Antarctic regions, sea ice cover along the coasts of EA has tended to increase. Michel et al. (2019) studied the ecological implications of these environmental changes on benthic food web structure during an event of unusually high sea ice cover using integrative trophic markers (stable isotope ratios of carbon, nitrogen and sulphur) to build ecological models and explore feeding habits of macroinvertebrates, many of them grazers of seaweeds. The results obtained indicated that local trends of sea ice increase in some areas of Antarctica have the potential to cause drastic changes in food web structure, and therefore to impact benthic communities, including seaweeds.

Decreases in sea ice cover predicted by the end of the twenty-first century are likely to greatly increase overall macroalgal cover in AP (Amsler et al. 2023a) and in the central WAP (Amsler et al. 2023b), while glacier retreat driven by climate change is also likely to allow macroalgae to expand their distribution (Pellizzari et al. 2020, 2023; Deregibus et al. 2023). Both types of change have important implications not only with respect to change zonation and expansion of benthic food webs, but also to understanding the potential for the WAP to contribute to blue carbon sequestration (Deregibus et al. 2023; Iken et al. 2023). Amsler et al. (2023b) suggested that the distribution of conspicuous fleshy macroalgae across depths in these expanded future communities will converge with those of the more widely studied northern WAP communities. Consequently, current knowledge of the northern WAP should be relevant to predictions about food web structure and blue carbon sequestration under future Antarctic Peninsula conditions.

Thus, macroalgal functional forms are well distributed across the major groups of Antarctic seaweeds: coarsely-branched and leathery species, the most robust and large-sized forms, represent 49% of the total number of species. In this group, endemic brown and red algae dominate, mainly growing in the subtidal zone. Filamentous, finely-branched, and foliose species (41%) representing mostly green and red algae are common at shallow and intertidal sites and are geographically widely distributed. Each of these groups is characterised by different life history strategies and hence distinct ecosystem functions. For example, perennial canopy-forming species show good competitive abilities for light and substrate, but are generally less prevalent in sites subject to strong physical perturbation. Here, small colonisers and opportunistic species dominate by virtue of their rapid metabolic responsiveness and turnover rates.

Arctic

Key features of the Arctic seaweed flora have been reviewed by Wilce (2016), one of the pioneers of marine botany in the American and Canadian Arctic (Küpper 2022). Similar to the Antarctic, in the intertidal and the first upper subtidal zones. In some parts of the Arctic, seaweeds are subject to wave action, drying, freezing, and scouring from both sea ice and smaller icebergs. Furthermore, there may be sediment input (mud to fine sand) affecting seaweeds, often near freshwater outflows. Hop et al. (2002) described the intertidal zone, between maximum spring and minimum neap tides on Svalbard and noted that benthic organisms, regardless of AFG, must be able to tolerate osmotic stress, sediment accumulation and displacement. Intertidal hard substrata include bedrock and stable rocks or large boulders. Level or sloping bedrock surfaces are commonly overgrown by macroalgae, which are dominated in the central and inner areas by *Fucus evanescens* and *Pylaiella littoralis* and in the exposed outer areas by *Chordaria flagelliformis*. Recent studies in western Spitsbergen (Svalbard) have shown that the diversity and biomass of algal species have increased in recent decades, particularly the biomass and occurrence of *Fucus* and some species of kelp, with new species being recorded in this area (Malavenda 2021; Weslawski et al. 1997).

Arctic macroalgal assemblages present on hard-bottomed coastal ecosystems comprise fucoid species, kelps, and coralline algae (Lebrun et al. 2022). Fucoids dominate intertidal assemblages, while kelps dominate the subtidal domain until they reach their minimum light threshold. The greatest depth kelps have been recorded at is 61 m in the Disko Bay region (Greenland; Krause-Jensen et al. 2012). Coralline algae form the understory of kelp forests and are found to 77 m depth. Climate change alters the zonation and spatial distribution of species by expanding or contracting suitable habitats, while species invasions, declines or extinctions are projected to increase in the future (Wiencke et al. 2014). Half of all benthic taxa are projected to experience significant changes in the extent of their suitable habitat (> 10% increase or decrease), including the dominant and/or the few endemic species of Arctic shelf areas (Lebrun et al. 2022).

In contrast with the patterns found in Antarctica, where green seaweeds are opportunistic and show higher richness compared to Phaeophyceae in some stressed areas (Pellizzari et al. 2017, 2023), in the Arctic, some brown algae appear to tolerate broader abiotic variation and take advantage of their robust thalli and physiological plasticity to invest in rapid growth, mainly during the boreal summer. The biomass (wet weight) of *Laminaria saccharina* in Svalbard may vary between 15 and 60 kg m^{-2} depending on the respective sectors of the fjords surveyed, with an overall diversity of 107 species (Malavenda 2021); and some areas near the mouths of fjords are also experiencing a significant increase in the inflow of Atlantic Water, resulting in persistent positive temperature anomalies and pronounced increases in SST, as well as broader variation in salinity patterns.

Malavenda (2021) reported that, among brown algae, the richness zonation ratio of sublittoral/midlittoral/common species was 19/13/3 and for red algae, 12/4/1, respectively. In the sublittoral communities of Svalbard, the most abundant species were *Fucus vesiculosus,* associated with *Ceramium deslongchampsii,* and *Chaetopteris plumosa*, and their maximum biomasses were 275.0 ± 125.2, 10.0 ± 5.1, and 166.4 ± 84.4 g m^{-2}, respectively. Absolute values of biomass were generally higher in the sublittoral zone compared with those in the littoral zone. Maximum biomass values reported for sublittoral communities were those of *Laminaria digitata, Alaria esculenta* and *Saccharina latissima*, with an overall average biomass of 15,100 g m^{-2}, on boulders at depths of 0.5–5.0 m. This study of *F. vesiculosus* zonation in Svalbard reported that the species prefers the sublittoral in some regions and the littoral in others, with the reasons for this not being clear. In Grønfjorden, kelp grows at a depth range of 5–10 m, and biomass and cover depend primarily on the availability of light and the presence of solid substrate. Favourable conditions and highly productive sites were present at the mouth of the fjord, more distant from sources of freshwater run-off. Thus, Malavenda (2021) characterised the littoral and sublittoral zones of Svalbard mostly by benthic algal communities with limited diversity and low biomass, compared to other Arctic areas.

3.2 Diversity of Antarctic and Arctic macroalgae

3.2.1 Comparative Richness and Diversity of Polar Seaweeds

Wulff et al. (2009) were pioneers in bipolar studies of seaweeds, reviewing composition, biogeography and zonation, followed by Wiencke and Clayton (2002). A general characteristic of the macrophycoflora of both polar regions is their low species richness, compared to temperate and tropical zones. Wulff et al. (2009) summarised that macroalgal richness was higher in Antarctica than in the Arctic, with a greater number of endemic species, the differences being explained by the differing biogeographical histories of isolation of both regions. They also concluded that the zonation or local distribution of polar macroalgae is influenced by physiological, morphological, chemical and ecological characteristics that determine responses to a range of environmental factors, and the ability to resist grazing. More recent syntheses indicate that the Arctic hosts higher macroalgal richness, estimated between 160 and 210 species in the Canadian Arctic alone (Archambault et al. 2010; Bluhm et al. 2011), compared to 151 species in Antarctica (Oliveira et al. 2020).

Polar seaweeds are uniquely adapted to the extreme environmental features of their respective regions; however, these assemblages show high sensitivity to abrupt changes in abiotic factors, making them vulnerable to environmental changes (Schoenrock et al. 2015; Vernet et al. 2019; Pellizzari et al. 2020). Changes in polar ecosystems lead to disruption in the structure and functioning of the entire food web. The importance of knowledge of the diversity and richness of seaweeds in their use as bioindicators of climatic and oceanographic shifts lies in the detection of changes in phycofloristic community composition and distribution. Although our understanding of the physical processes driving ecological change in polar ecosystems has improved in recent decades, robust monitoring programs are required to develop conservation guidelines and then assess their effectiveness.

Antarctica

Antarctic seaweed assemblages (Figs. 3.5 and 3.6) have long been considered to be highly endemic (37–40% at the species level; Wiencke and Clayton 2002), attributed to Antarctica's geographical isolation. Historically, Antarctic algal communities were characterised by high biomass and low richness (Lamb 1977; Clayton et al. 1997). Wiencke and Clayton (2002) reported a complete diversity for Antarctica of 120 species. However, Oliveira et al. (2020), in a recent review of available macrophycological data, reported an increase in species richness to 151 species and a reduction in the degree of endemism (to 27%), the latter in part due to increased knowledge of the wider biogeographic distribution of some taxa previously thought to occur only in Antarctica (Pellizzari et al. 2017, 2020, 2023).

Macroalga from King George Island (SSI), the largest island in the archipelago, were studied by Zielinski (1990), Quartino et al. (2005), Oliveira et al. (2009) and Valdivia et al. (2014). The diversity present on the Deception and Livingston Islands was recorded by Gallardo et al. (1999). Phycological studies carried out before 2000 in the SSI were generally specific for each island, and with the lack of application of molecular approaches at that time, focused on morphological aspects of vegetative and reproductive structures. Pellizzari et al. (2017) published the first comparative study across eight islands in the SSI, reporting a total of 104 species of macroalgae, a richness representing approximately 82% of all algal taxa listed for Antarctica. They identified King George and Livingston Islands as hotspots and reported six new species records, previously listed only at lower latitudes, and the presence of four species was confirmed using molecular analyses, many years after their first report. The algal assemblages of Deception Island were first studied by Gallardo et al. (1999). Pellizzari et al. (2017, 2023) studied the spatial distribution

Fig. 3.5 Seaweeds from South Shetland Island (SSI), Antarctica. **a**. *Iridaea* sp. and *Adenocystis utricularis* in a tidal pool; **b**. *A. utricularis;* **c**. *Monostroma hariotii*; **d**. *Desmarestia* holdfast; **e**. *Palmaria decipien*; **f**. *Cystosphaera jacquinotii.* (Images: F. Pellizzari 2022)

Fig. 3.6 Seaweeds from SSI, Antarctica. **a**. King George Island; **b, c**. *P. decipiens* bed at Deception Is.; **d**. Gigartinales; **e**. *Monostroma* sp.; **f**. *Clathromorphum* sp., crustose calcareous algae (Rhodophyta) and *A. utricularis.* (Images: F. Pellizzari 2022)

of macroalgal assemblages across different locations on Deception Island, reporting higher richness of Rhodophyta and Ulvophyceae, as well as pH and thermal anomalies in the shallow water column of the island's flooded caldera (up to 8°C compared to other areas of Antarctica), proposing the area as a sensitive natural laboratory facing both climatic changes and the threat of anthropogenic impacts.

Considerable climatic and oceanographic changes have taken place over recent decades in the AP region, including the SSI, EAP and WAP. These areas are considered transitional ecosystems, and abiotic changes have caused changes in the diversity, richness and degree of endemism of algal populations (Ducklow et al. 2013; Pellizzari et al. 2017, 2020, 2023). In contrast to SSI and WAP, very few macrophycological studies have been carried out in the EAP, largely due to logistical difficulties of access. Pellizzari et al. (2023) reported new occurrences of seaweed species on Elephant, Deception and Half Moon Islands (SSI) and for the first time from Vega Island (EAP). A total of 48 macroalgal species were identified, including eight new records (NR) for the EAP and five NR from the SSI. Comparison of these data with the existing literature from the WAP provided insights into macroecology, ecoregional connectivity and potential changes in thermohaline circulation patterns. Richness and diversity differences among the assemblages from the SSI, WAP and EAP were identified, being greatest in the SSI, followed by the WAP and then the EAP. This approach is relevant for ongoing monitoring, to support conservation aims, as new dispersal pathways potentially become available, supporting the transfer of currently non-native species.

Macroalgal inventories conducted in the WAP (Gulhemin et al. 2020) have highlighted that those changes in thermohaline patterns, UV exposure, ice cover and dynamics of sediments/substrates, combined with competition and herbivory, will affect the composition and richness of macroalgal communities in the WAP (Saiz et al. 2008; Clarke et al. 2012; Mystikou et al. 2014; Deregibus et al. 2016). Wiencke et al. (2014) considered that the limits of seaweed abundance and distribution in the southern part of the WAP indicate a decline in species richness and biomass southwards from Anvers Island. They associated these biomasses, and diversity decreases with an increase in ice cover. Decrease in algal coverage with progression southwards is also described at other Antarctic locations (Johnston et al. 2007).

In terms of biomass, a key genus in the WAP is *Desmarestia*, represented by the species *D. antarctica*, *D. menziesii* and *D. anceps*, which represents a high percentage of the algal biomass (Quartino et al. 2001, 2005, 2013). This species provides habitats for other species of invertebrates and algae to inhabit (Mansilla et al. 2016). On the other hand, its importance at an ecological level could be comparable to that of the species *Macrocystis pyrifera* present in cold temperate waters (Mansilla et al. 2005; Ríos et al. 2007). Due to these characteristics, both species can be considered 'sentinel of climate change' (Marambio et al. 2023).

It is recognised that the general WAP richness is low compared with the northern AP and SSI (Pellizzari et al. 2017, 2023), as well as the sub-Antarctic (Sanches et al. 2016). Ducklow et al. (2013) further confirmed, based on statistical and ecological models, that the southern AP has lower seaweed biomass and diversity compared with the northern WAP. They also noted that the higher richness of macroalgal communities typical of the northern AP appeared to be expanding southwards as annual sea ice cover averages declined. Mystikou et al. (2014) analysed algal diversity in part of the southern WAP over 35 years. Based on the diving surveys of F. Küpper in 2010–11 around Rothera Point, Adelaide Island, these authors studied the diversity of seaweed in southern Adelaide Island using as a baseline the previous checklist from the northern Marguerite Bay. Furthermore, the authors combined the records of both campaigns, with data obtained during Moe's and De Laca's work in 1973–1975 and Küpper's work in 2010–2011, overall resulting in a checklist of 41 species (7 brown, 27 red, 6 green, 1 chrysophyte).

The first macroalgal collections from Antarctica were obtained in the Ross Sea (RS) (e.g. Levring 1945; Skottsberg 1953; Neushul 1965; Zaneveld 1966a, b; Zaneveld and Sanford 1980, among others). Cormaci et al. (1992) reported 17 taxa (including 9 Rhodophyta, 4 Fucophyceae and 4 Ulvophyceae) of benthic macroalgae from Terra Nova Bay (RS), as well as giving brief descriptions and phytogeographic data. Wiencke and Clayton (2002) recorded 30 species from the northern RS. Nelson et al. (2022) published a spatial checklist for the RS recording 30 species, including 27 from the Balleny Islands (northern RS) and 18 from the western RS. Fifteen taxa were recorded from both locations, and 3 were exclusive to the RS. Wiencke et al. (2014) emphasised that much of the EA coast (between 45°E and 160°E) remains undersampled. Furthermore, many records are based on dredged or drift specimens, which are of limited usefulness, with only very few records and possible confusion with morphologically similar species.

Antarctic species richness is highest in the AP region and lowest in the southernmost part of the Ross Sea (approx. 77°S). Oliveira et al. (2020) compiled records of macroalgae for Antarctica, incorporating previous lists (including Wiencke and Clayton 2002; Hommersand et al. 2009; Quartino et al. 2013; Wiencke et al. 2014; Mystikou et al. 2014, 2016; Pellizzari et al. 2017), recording a total of 151 species (85 Rhodophyta, 34 Phaeophyceae, 32 Ulvophyceae). Additional new records of green and red macroalgae have been reported by Guillemin et al. (2018), Dubrasquet et al. (2018) from the SSI and WAP, and Pellizzari et al. (2023) from the SSI and the EAP. Pellizzari et al. (2023) compared three major areas of the AP, concluding that the EAP hosted the lowest richness, followed by the WAP, and then the SSI, with 21, 82 and 112 taxa, respectively. Approximately 30 spp. are known from WA, with a similar or lower richness for EA, each representing approximately 20% of the total Antarctic phycoflora (151 spp.).

Arctic

Benthic macroalgal communities (Figs. 3.7 and 3.8) are an essential component of hard-bottomed shallow water ecosystems in the Arctic. Fucoid species and kelps are large brown seaweeds that can dominate Arctic rocky intertidal and sub-tidal areas and exhibit a high biomass (Lebrun et al. 2022). Arctic kelp forests provide a food source, substrate, shelter and nursery for many species (Filbee-Dexter et al. 2019). Arctic kelp forests are hotspots of polar marine biodiversity. Fucoids and kelps play an important role in the transfer of carbon (Filbee-Dexter et al. 2022) through high production and export of dissolved organic matter and detritus, as well as carbon sequestration (Krause-Jensen and Duarte 2014).

In contrast to the higher endemism of Antarctic seaweeds (approx. 27%; Oliveira et al. 2020), few endemic species have been reported from the Arctic (Wiencke and Amsler 2012). These include the brown algae *Chukchia pedicellata, C. endophytica, Arcticophycus glacialis, Platysiphon verticillatus*, the red alga *Haemescharia*

Fig. 3.7 Seaweeds from the Canadian Arctic. **a**. *Ascophyllum nodosum* ("rockweed"); **b**. *Chondrus crispus*; **c**. Delesseriaceae; **d**. *Ulva* sp.; **e, f**. *Fucus vesiculosus* with bladders. (Images: F. Pellizzari 2024)

Fig. 3.8 Seaweeds from the Canadian Arctic. **a**. Sea ice during the winter; **b**. Fertile plant (receptacula) of F. vesiculosus; **c**. *A. nodosum* with red algae epiphytes; **d**. Underwater Kelp bed. (Image: R. Scheibling, retrieved from: https://canadiangeographic.ca/articles); **e**. *Fucus* sp. in artificial substrate; **f**. Filamentous red algae epiphytes on Ascophyllum. (Except **d**, images: F. Pellizzari 2024)

polygyna and the green alga *Acrosiphonia incurva*. Most recorded Arctic species have distributions that extend well into temperate regions, such as the red algae *Devaleraea ramentacea, Turnerella pennyi, Dilsea socialis, Pantoneura fabriciana* and *Laminaria solidungula* (Müller et al. 2009).

Overall, Arctic marine macroalgal species richness is between 150 and 210 species (Wilce 1990). The exact number of species is not certain and varies with the studied location (and country). Furthermore, there are estimation discrepancies in the literature, at the most extreme, ranging from 160 species for the entire Arctic (Bluhm et al. 2011) to 210 in only the Canadian Arctic (Archambault et al. 2010). Documented species richness, for both polar regions, is inevitably increasing primarily due to the availability of improved taxonomical tools and knowledge (including molecular approaches), as well as with new algal biodiversity survey efforts in remote areas (Küpper et al. 2016), although there is also the possibility of migration into new ice-free areas and distribution increase of some species (see synopses by Pellizzari et al. 2020; Goldsmith et al. 2021).

In the European Arctic, the best-studied macroalgal community composition and zonation patterns are those of Svalbard, between 74 and 81°N (Florczyk and Latala 1989; Weslawski et al. 1993; Hop et al. 2002, 2012; Wiencke et al. 2004). Fredriksen et al. (2019) report that a total of 197 species of macroalgae have been recorded in Svalbard, with 84 occurring in Kongsfjorden. Malavenda (2021), in a study in the mouth of a Spitsbergen fjord (Svalbard), listed 68 species, 17 of these being first records, suggesting that Arctic macroalgae richness is still underestimated. Among the algal groups, in the Arctic, green seaweeds have lower richness (and biomass) if compared to red and brown species. Besides large brown seaweeds, encrusting red algae are also abundant in the sublittoral zone.

Mathieson et al. (2010), in a comparative latitudinal study, documented the seaweed flora from James Bay (51°N, Canada) and three north-east Canadian Arctic areas. Between Ellesmere and Baffin Islands (65–80°N), 131 taxa were recorded, with 106 from Hudson Strait, 81 from Hudson Bay (60–62°N) and 44 in James Bay. The Arctic Ellesmere-Baffin zone had the highest richness and was characterised by an open rocky shore with higher salinities compared to James Bay, an area that is anthropogenically impacted by hydroelectric development and freshwater discharge. The total phycoflora from these areas consisted of 164 taxa (49 red, 65 brown, 49 green, 1 chrysophyte). Twenty-five taxa were exclusively recorded in the Ellesmere and Baffin Islands. Four new records were listed for James Bay: *Chaetomorpha minima, Elachista fucicola, Phyllophora pseudoceranoides* and *Spyridia filamentosa.* Küpper et al. (2016) assessed the biodiversity of seaweeds in northern Baffin Island (72°N), providing a baseline inventory. A total of 59 taxa were present (33 Phaeophyceae, 24 Rhodophyceae, 12 Ulvophyceae). Molecular barcoding of live isolates raised from incubated substratum samples revealed the presence of 20 species of brown seaweeds, including gametophytes of kelp and of a previously unsequenced *Desmarestia* species closely related to *D. viridis,* two species of *Pylaiella*, the kelp endophyte *Laminariocolax aecidioides* and 11 previously unsequenced species of the Ectocarpales, highlighting the necessity to apply algal culturing in combination with molecular techniques to fully unravel cryptic algal diversity. La Guardia et al. (2023) showed that kelp species composition around Southampton Island (Nunavut, Canada) was typical of ice-covered waters in the western Arctic and sub-Arctic, dominated by the perennial high-canopy species *Saccharina latissima, Alaria esculenta* and *Laminaria solidungula* and the low-canopy species *Agarum clathratum* (see also Filbee-Dexter et al. 2019, 2022; Bonsell and Dunton 2018; Filbee-Dexter et al. 2022).

In the Russian High Arctic, Mikhaylova et al. (2022) studied macroalgal assemblages of the Severnaya Zemlya Archipelago (82–84°N) in the AO north of Siberia. They listed 48 taxa, including 17 Chlorophyta, 18 Rhodophyta and 13 Phaeophyceae. Twenty-eight species were new records for the archipelago, and three were new records for the entire Russian Arctic. For the first time, a wide distribution of kelps was reported in the sub-tidal zone of this High Arctic part of the Siberian shelf, which is severely impacted by ice disturbance and glacial processes.

Similarly, records of macroalgal distribution in northern Greenland remain scarce. The assemblages are similar to, although with lower richness, those of the Canadian Arctic or even Svalbard (Bartsch et al. 2016; Borum et al. 2002). Rocks are often covered with encrustations of red coralline algae (*Phymatolithon* spp.). Species richness decreases in eastern Greenland from the south (North Atlantic) to the north (Arctic Circle). In Disko Island (west Greenland), there is a dense community of macroalgae. Here Bischoff and Wiencke (1993) studied the temperature requirements of 16 species, demonstrating that stenothermal members of this assemblage were better adapted to Arctic temperatures (e.g. *Acrosiphonia arcta, A. sonderi, Urospora pennicilliformis, Devaleraeae ramentaceae, Desmarestia aculeata, Pilayella littoralis*). *A. arcta, U. pennicilliformis* and *P. littoralis* also occur in Antarctica, being bipolar species (Pellizzari et al. 2017).

Only the northernmost point of Iceland is within the Arctic Circle, specifically Grimsey Island. Although Gunnarsson and Jonsson (2002) listed 269 seaweed species for Iceland, no specific data for Grimsey are available. The seas surrounding Iceland are at the boundary where warmer Atlantic waters meet cold Arctic waters. This highlights the importance of sea-ice dynamics from the Arctic Ocean, which can occasionally block access to the north and east coasts of Iceland and are associated with phycofloristic differences. The warmer Atlantic waters are a branch of the Gulf Stream, known in this area as the Irminger Current, that divides into two branches to the west of Iceland. As a mixing zone, about 220 species of seaweeds have been registered in this area (Iceland National Report on Biological Diversity 2001, https://www.cbd.int/doc/world/is/is-nr-01-en.pdf).

Lebrun et al. (2022) highlight the importance of coralline red algae as ecosystem engineers in the Arctic. These long-lived species (ca. 1200–1600 years for individuals found in the Labrador Sea) can form crusts or rhodolith beds that host key marine biodiversity. In the US Arctic, several studies have documented seaweed communities growing on boulders as well as smaller rock and hard-shell substrates in the Alaskan Beaufort Sea. These communities are dominated by the kelp *Laminaria solidungula* along with lesser numbers of the kelps *Saccharina latissima* and *Alaria esculenta*, six species of foliose Rhodophytes, and crustose coralline algae (Dunton 1992).

Arctic seaweed diversity has likely been driven by repeated isolation into disparate refugia during glaciation. This compilation of data highlights the need for ongoing monitoring of these unique Arctic assemblages. Human activities, amplification of warming and loss of perennial ice cover are set to dramatically alter available Arctic coastal habitats, with the potential loss of diversity, including seaweeds as the basis of the food web, and a decline in ecosystem resilience.

3.2.2 Seaweed Distribution and Depth Zonation in Antarctic and Arctic waters

Vertical zonation of macroalgae along depth gradients is driven by various environmental factors, and is species-dependent, inhabiting distinct depth zones, whose display different physiological performance and attributes. In some cases, morpho-functional traits have been used to assess the distribution of communities, giving clues regarding structure and ecosystem functioning along different environmental gradients (Gomez et al. 2019)). In polar zones, sea ice is a key physical driver, and to understand the ecological effects of sea ice and light demands for benthic algal assemblages, it is particularly important to focus on current and future climate change scenarios.

Antarctica

Due to their utility, algal functional groups (AFG) approaches have been used to detect changes in the ecosystem condition, distributional patterns, or to evaluate shifts of functional diversity in scenarios under global climate change (Gomez et al. 2019)). Studies revisiting the relationship between AFGs and the depth distributions of macroalgae in different geographical regions have reported contrasting results, supporting the hypothesis that there is not a uniform scheme of zonation, but that this varies locally depending on the species composition and physical and biological processes operating at each site. On the other hand, macroalgae can show a variety of morphologies within one life history and even during their ontogeny. However, most studies agree that gradients in irradiance define the position of seaweeds along the depth gradient and, thus, their morpho-functional attributes (Gómez et al. 2011). However, a morphology-based form-function concept could not explain the observed physiological and ecological patterns displayed by macroalgae in the context of stress tolerance, especially in those intertidal assemblages periodically submitted to extreme temperatures, salinity, desiccation, solar radiation, etc. (Gómez et al. 2011). For example, various metabolic adaptations developed to enhance resilience to enhanced solar radiation, such as photochemical down-regulation via photoinhibition, are strongly dependent on the position on the shore, but not on gross morphology (Aguilera et al. 2002). This can have important consequences for primary productivity as the life history strategy is a trade-off between morphology-related traits and environmental tolerance characteristics.

Antarctic macroalgae not only possess shade-adapted photosynthetic characteristics (Gómez et al. 2009, 2019) but also are metabolically well equipped to resist environmental stresses (e.g. temperature changes and UV radiation). For example, UV tolerance mechanisms can still operate in algae at depth, which are not normally exposed to UV radiation (Gómez and Huovinen 2015). Thus, it is essential to evaluate whether distinct coexisting species sharing similar functional traits in terms of gross morphology, taxonomic or biogeographical affinity show comparable photosynthetic and bio-optical properties in the Antarctic environment to estimate the ecological significance of functional diversity in community assembly.

Abundance, distribution, and species diversity have been associated with the prevalence of some dominant traits that ultimately determine different properties of the community, such as stability, niche partitioning and facilitative interactions. These features are strongly challenged when the community is subjected to multiple perturbations. In littoral systems exposed to strong disturbance by ice, species richness and functional diversity are important factors that allow the benthic community to withstand physical stress (Gómez et al. 2019). In King George Island, the structure of the benthic ecosystem is strongly influenced by species diversity, while mesoscale biomass patterns were accounted for mainly by the abundance of algal taxa with key functional traits, such as canopy-forming algae with leathery and coarsely-branched morphology (Valdivia et al. 2014). Under scenarios of future climate change in Antarctica, the importance of diversity and the distribution of macroalgal traits that confer tolerance to a broad spectrum of environmental stressors are likely to be important drivers of future community structure.

Studies of Antarctic algal zonation have highlighted the importance of stress tolerance as a relevant physiological trait defining broad vertical distribution patterns (Flores-Molina et al. 2016). For example, seaweed specimens between intertidal and 5 m depth showed up to 15% reduction in photosynthesis, and 75% increase in DNA damage when exposed for 2 h to summer UV radiation levels. In contrast, algae from greater depth (30–40 m) exhibited up to 35% reduction in photosynthesis and DNA damage increased by up to 150% (Gómez et al. 1997). There was also recovery from photodamage, correlated with the synthesis of photoprotective substances and antioxidant activity.

Species such as *Desmarestia menziesii* and *Ascoseira mirabilis*, frequently exposed to high UV radiation, demonstrate short-term tolerance due to efficient damage repair. These defence mechanisms are pronounced in Phaeophyceae and endemic species, which exhibit a rapid recovery of maximum PSII quantum yield under UV stress (Gómez et al. 1997, 2011). This resilience may be a key physiological trait in the face of global change scenarios, where efficient management of UV stress may facilitate adjustments in the vertical distribution of UV-sensitive macroalgae. However, recent studies with endemic Antarctic algae such as *D. menziensii* indicate that, although sporophytes and gametophytes can survive higher temperature stress, the species' northern distribution limit in sub-Antarctic South Georgia is defined by low temperature requirements for the reproduction of gametophytes (Matula et al. 2022). Global warming is, therefore, likely to impact the distribution of this and other Antarctic species through influencing their growth and reproduction.

Gómez et al. (2019) studied the photosynthetic and bio-optical properties of 29 Antarctic seaweeds from King George Island to address the question of whether different functional traits, grouped according to gross morphology, bathymetric distribution, taxonomic category and biogeographic affinity, could explain patterns of vertical distribution and relevant ecosystem services. The study was based on analysis of light requirements for photosynthesis, thallus absorptance and thickness, photosynthetic efficiency and tolerance to stress (UV and temperature). The authors reported that the studied algae showed high tolerance to stress from short-term exposure to UV radiation and enhanced temperature (7 °C). The stress tolerance was independent of gross morphology, with the greatest differences found along the depth gradient and between algae of different biogeographic affinity. The studied traits were highly interlinked in terms of photosynthetic and bio-optical properties. Overall, these results highlight the importance of morpho-functional traits of macroalgae in explaining their distribution across depth gradients and provide insights into their potential resilience to a changing polar environment.

Antarctic macroalgae also show distinct patterns of depth zonation according to their photosynthetic capabilities, where ambient PAR levels are sufficient to exceed the saturation and compensation points of photosynthesis (Gómez and Huovinen 2015). Some brown algae, such as *Desmarestia anceps* and *Himantothallus grandifolius*, require half the irradiance normally required for species from temperate zones (Runcie and Riddle 2006), i.e. they are strongly adapted to low light levels, providing a competitive advantage over other species.

Antarctic macroalgae are known to grow at depths of 40 m and potentially even deeper, taking advantage of their highly shade-adapted characteristics (Wiencke and Amsler 2012). At high polar latitudes, solar radiation is strongly seasonal; however, even during summer when there is 24 h daylight, sea-ice, phytoplankton blooms and glacial flour in runoff from the land can restrict light from reaching the benthos (Bischof et al. 2006; Vernet et al. 2019). *Palmaria decipiens* is a common red seaweed species in the sublittoral zone of Antarctica (Pellizzari et al. 2017, 2023). However, dredged material or photographs of this species have been obtained at depths up to 300 m, although this does not prove that macroalgae are viable at these depths. Recently, Robinson et al. (2022) reported living specimens of *P. decipiens* from 100 m depth, representing a new confirmed depth record. The specimens were sampled at Adelaide Island, WAP (67°34′S) on steep rocky slopes, using a ROV with a manipulative claw and thrusters. With this new record, the authors consider that estimates of macroalgal biomass, energy fluxes and potential energy fixation may need adjusting to consider deeper growing depths, particularly in the context of glacial retreat along the AP, revealing areas of rocky substrata for macroalgal colonisation, with wider implications for assessments of benthic productivity and food webs in Antarctica.

Macroalgae with fleshy and robust thalli dominate the hard bottom of shallow waters along the WAP. Amsler et al. (2023b) recently provided data on the depth distribution of conspicuous brown and red macroalgae in a latitudinal transect from the northern and less-known southern portion of the WAP, based on collections in the subtidal zone at 14 sites between Anvers Island (64.8°S) and Marguerite Bay (68.7°S). Although brown algae that can be common along the northern WAP (e.g. *Desmarestia anceps* and *Cystosphaera jacquinotii*) are mostly absent further south, the vertical distributions of *D. menziesii* and *Himantothallus grandifolius* are similar to those found in the northern WAP, but with much lower percentage cover. Likewise, the vertical distributions of the 14 most widespread red macroalgae mirrored those known from the northern WAP, even though macroalgal cover, biomass and total species richness declined markedly towards the south across this region due to increasing sea ice concentrations.

Finally, using a macroecological modelling approach, Sanches et al. (2016) investigated the role of the ACC as a biogeographic barrier limiting the distribution of macroalgal populations in the SSI, EAP, WAP and sub-Antarctic islands, and as an influence on ecological and evolutionary macroalgal processes. They analysed how ocean circulation and current and past patterns of continental drift could isolate or connect seaweed assemblages. These organisms were selected as models because they are benthic, reproduce using planktonic spores or propagules and spend part of their life cycle being transported by marine currents. These reasons also support the use of macroalgae as a sensitive indicator of changes in the biogeographic distribution patterns of marine ecosystems (Fraser et al. 2020). Sanches et al. (2016) focused on understanding the role of the ACC as a constraint on seaweed distribution patterns at different taxonomic levels in relation to the southern sub-tropical areas and to South America (1: north), influenced by the Belligshousen Sea (2: WAP) and by the Weddell Sea Gyre (3: EAP). The nMDS showed differences in the distributional patterns of species and genera in relation to these areas. The predicted latitudinal gradient of species richness formed two main clusters: 1. Sub-tropical and sub-Antarctic, and 2. Polar Front (Antarctic Peninsula) and Southern Antarctic Circumpolar Front (areas surrounding continental Antarctica). However, a higher similarity was observed between the second group (EAP—under the influence of the Weddell Sea Gyre, also including the maritime Antarctic South Orkney Islands and sub-Antarctic South Georgia and Crozet Archipelago) and the first group (Patagonia and the Cape Horn archipelago, the Falkland Islands and Antarctic Peninsula). When considering recently reported changes in diversity patterns of these locations, this study could support the existence of a clear species distributional flux, despite evidence of the ACC limiting dispersal. Such approaches provide a tool for monitoring further biogeographical changes in these connected regions, resulting either from natural dispersal associated with global meta-oceanographic changes or from biological invasions related to anthropogenic activities.

Arctic

Climate change is transforming marine ecosystems through expansions and contractions of species ranges. Sea ice loss and warming temperatures are expected to expand habitat availability for macroalgae along stretches of Arctic coastlines. Kelps are a dominant macrophyte group and primary producer in Arctic nearshore areas, providing significant ecosystem services. The quantification of these services in the Arctic is constrained by limited estimates of subtidal kelp depth extent, which creates uncertainties in the total coverage.

Kelp cover can be a proxy for biomass, individual size and productivity. Patterns of regional increases and decreases in abundance are consistent with trends at the global scale from a time series of kelp abundance over the past half-century (Krumhansl et al. 2016). Over that period, there was no detectable change in kelp abundance in 35% of the global ecoregions assessed and decreases in kelp abundance in 38% of ecoregions. It is possible that the combination of geographic and oceanographic complexity, with resulting gradients in nutrients, salinity, sea ice polynyas and varying seafloor substrata, creates a complex regional pattern of kelp abundance in the eastern Canadian Arctic that results in localised regions of gains and losses in abundance.

Goldsmith et al. (2021) studied the distribution of kelp forests in the Eastern Canadian Arctic, using identifications and percentage cover from sampling effort, supplemented with occurrence records from global biodiversity databases, literature and museum records. Environmental information and occurrence records were used to develop ensemble models for predicting habitat suitability and a Random Forest model to predict kelp cover for the dominant kelp species in the region—*Agarum clathratum*, *Alaria esculenta*, *Laminaria solidungula* and *Saccharina latissima*. Ice thickness, sea temperature and salinity explained the highest proportions of kelp distribution. Modelling showed that the current extent of Arctic kelps is potentially much greater than the available records suggest based on the most extreme emission scenario (RCP 8.5). The models agreed that the distribution of kelp in the Eastern Canadian Arctic is likely to expand to more northern locations under future warming scenarios, except for the endemic Arctic kelp *L. solidungula*, which is likely to lose a significant proportion of its suitable habitat, emphasising the importance of kelp forest monitoring and resolving the current underestimation of their potential distribution in Arctic ecosystems. Goldsmith et al. (2021) also reported inconsistent changes in kelp cover through time, increasing in some regions while decreasing in others. This suggests that, although overall suitable habitat for kelp is expanding, local or regional drivers of kelp abundance in the Eastern Canadian Arctic led to more mixed responses to climate change. Regions where all species are predicted to increase in cover are in the northern sections of Baffin Island, where the climate is expected to change less drastically than in other areas. For example, the decline in sea ice cover has been relatively slow in recent decades (Stammerjohn et al. 2008), meaning that this region could be a refugium for kelps, including Arctic endemic species.

La Guardia et al. (2023) studied the environmental drivers of the depth extent of Arctic kelp in Southampton Island (SI - Nunavut, Canada), as a model region after an initial survey found deep Arctic kelp (at least 50 m) with relatively low grazing pressure within diverse hydrographic conditions. Kelp cover increased with depth from the subsurface to 20 m and then decreased. It is known that Arctic kelp ecosystems can be abundant and widely distributed; however, uncertainties in estimating their depth extent are a key constraint to quantifying their total extent and estimating the ecosystem services they provide (Goldsmith et al. 2021; Filbee-Dexter et al. 2022). La Guardia et al. (2023) confirmed the presence of extensive kelp beds below 40 m in some areas of SI, supporting recent studies demonstrating kelp extending below 30 m in regions from the Arctic to the equator (Hop et al. 2012; Krause-Jensen et al. 2016). Kelps also follow a gradient of increasing cover moving offshore to around 20 m depth, which is consistent with zonation patterns reported from other Arctic coastlines where sea ice scours the seafloor in the nearshore zone, resulting in it supporting less biomass (Filbee-Dexter et al. 2022). The lack of relationship being apparent between kelp depth extent and water temperature (at their depth limit) could be influenced by the relatively low variation in temperature between stations and the overall low temperatures (−0.6 to 4 °C). These findings agree with recent observations in the Eastern Canadian Arctic that suggest that kelp distribution is driven mainly by sea ice rather than temperature (Filbee-Dexter et al. 2022). These authors also suggested that the marine flora of the Canadian Arctic is similar to that found on Svalbard, macroalgal zonation consists of a subtidal belt of *Fucus evanescens* and *Pylaiella littoralis* with a number of ephemeral green algae, such as *Urococcus foslieanus, Ulothrix flacca,* and *Acrosiphonia arcta* (the former two spp. are also found in the Antarctic Peninsula, even though there may be hitherto undetected cryptic diversity between the two polar regions).

La Guardia et al. (2023) also reported that the lower depth extent of kelps along western Greenland was correlated with the latitude, with kelps growing 10–15 m deeper at the southernmost station, where there was a high number of open water days with light (OWL, a proxy for the underwater light environment). The Arctic crustose calcareous algae (*Phymatolithon* spp.) also occur up to 50 m depth (Roberts et al. 2002). The alga *Saccharina latissima* in Greenland is adapted to low light, compensating for this through high pigment concentrations and greater photosynthetic efficiency (Borum et al. 2002).

Malavenda (2021) described algal distribution in a fiord in Spitsbergen (Svalbard), where most species are high-boreal and Arctic–boreal. *Devaleraea ramentacea* dominated the biomass, while the green algae *Ulvaria obscura* and *Acrosiphonia arcta* were widespread. A community comprising *Urospora penicilliformis*, *Ulothrix* spp. and *Ulva* spp. was present on the rocks at the mouth of the fjord, but large areas of the sublittoral zone had no macrophytes because of the predominance of clay sediments. The kelp belt, formed by *S. latissima, A. esculenta,* and *L. digitata*, began at the mouth of the fjord. On the western coast, at 0–3 m depth, a belt consisting of *Fucus distichus* and *F. vesiculosus* that included some epiphytic *Ectocarpus* spp. at the lower boundary is present. The algal flora of Svalbard generally comprises species also found in other regions of the North Atlantic (AN). The southern part of the archipelago may be the northern distribution limit for some boreal species where Arctic endemic species presently occur (Weslawski et al. 1997). Wiktor et al. (2022) propose that a warmer Arctic with less sea ice could favour the growth of macroalgae, in addition to their migration.

Lebrun et al. (2022) noted that, despite low annual growth rates in the Arctic, coralline algae have high potential for calcium carbonate accumulation, providing a three-dimensional habitat dominated by benthic invertebrates. Rhodolith beds are an important part of the coastal ecosystem, and biofilms growing on their surface represent a major food source for many grazing organisms. Thus, fucoid species, kelp forests, rhodolith beds and their associated biota provide hard substrates for benthic communities along Arctic coasts. Under the influence of global changes, expansions of both native and non-native kelp species have been documented in recent decades (Weslawski et al. 1997; Krause-Jensen et al. 2020). Warmer sea temperatures and longer ice-free periods may favour kelp growth, reproduction and survival (Krause-Jensen and Duarte 2014). Fucoids are also spreading in the Arctic, doubling in coverage and possibly tripling by the end of the current century (Lebrun et al. 2022).

Long-term research from Greenland, Russia and Norway suggests that a warmer Arctic with less sea ice may support higher kelp productivity and biomass by expanding the northern range and lower depth limits of these species (Filbee-Dexter et al. 2019; Krause-Jensen et al. 2020). However, the degree to which these changes will positively affect individual kelp species will likely vary regionally, and depend on the extent of glacial ice melt, permafrost collapse, turbidity increase and salinity reduction in coastal areas.

These changes in seaweed distribution limits (vertical and horizontal), as well as spatial and temporal datasets, suggest specific groups of seaweeds (e.g. kelp in the Arctic, green and some brown species in Antarctica) will play a larger role in the coastal marine ecosystem in a warmer future as the length of the open-water period increases. The expansion and/or migration of these communities will increase biomass as well as carbon flow and habitat diversity, exerting cascading effects on polar coastal ecosystems.

3.2.3 Integrative Taxonomy, Cryptic Species, and Hidden Diversity: Reassessing Polar Seaweed Facing Climate Change

There are many unresolved taxonomic questions relating to seaweed species currently recorded from Antarctica and the Arctic. Especially in Antarctica and parts of the Russian Arctic, many records are based on drifting or dredged specimens, while some species are known from very few records, and in some cases, the names of morphologically similar taxa from beyond the polar regions have been misapplied. Caution, therefore, is required when comparing lists of taxa recorded from distinct regions. For this reason, the integration of data sourced from classical taxonomy, such as morphological, reproductive and biochemical features, with a combination of molecular markers is essential to solve taxonomic issues at the species level, mainly for *hidden diversity*, and for the resolution of instances of cryptic speciation (please see also Sect. 3.4). Cryptic species are taxa that show apparently indistinguishable morpho-anatomical features and, often, reproductive characters, but that are distinct at the DNA sequence level. Algal genera belonging to at least three seaweed groups are known to have cryptic representatives.

For example, Guillemin et al. (2018) and Dubrasquet et al. (2018) performed molecular marker investigations to review the diversity and distribution of macroalgae along the WAP and SSI. Other studies have attempted to distinguish cryptic taxa, such as Tonicelli et al. (2021) in the Cystocloniaceae; Shilling et al. (2021)

with *Plocamium* spp. and Hughey et al. (2020) who repositioned the genus *Gigartina* to *Sarcopeltis* gen. Nov. (Gigartinaceae, Rhodophyta), recognising *S. skottsbergii* comb. nov. as a species that occurs in South America, and *S. antarctica* sp. nov. as a new species occurring in Antarctica. Ocarranza-Barrera et al. (2019) similarly reported molecular divergence between *Iridaea 'cordata'* (Rhodophyta) from the Antarctic Peninsula and representatives from the Magellan Region. The occurrence of cryptic speciation in some genera and families present in Antarctica remains uncertain, particularly relating to species currently thought to occur on both sides of the Drake Passage (e.g. *Cladophora* spp., *Phymatoliton* (Corallinaceae)) and complexes such as *Lithothamnion/Lithophyllum*, *Callophyllis/Kallymenia* and *Monostroma/Protomonostroma* (Fraser et al. 2020; Pellizzari et al. 2020).

The existence of cryptic taxa turns species identification a challenge, and it is challenging to address species introduced by anthropogenic activity or via natural routes, such as those carried by floating macroalgae rafts that can travel among ecoregions. These can also carry many smaller species (e.g. filamentous or sheet-like species), many of which are potentially cryptic, that can then recruit in new environments. Deception Island (SSI, Antarctica) hosts a high richness of cryptic species in addition to bipolar species (taxa that occur in both Antarctica and the Arctic with a latitudinal interruption of distribution in between). Bipolar seaweeds have disjunct amphi-equatorial distributions and, according to molecular studies, the biogeographic disjunctions between the respective species are recent and probably date back to the maximum of the Wurm/Wisconsin glaciation only 18,000 years ago (van Oppen et al. 1993). The occurrence of bipolar species may result from anthropogenic introductions or by dispersal of microscopic and encysted reproductive sexual or asexual cells, which are resistant to high temperatures and could have passed through the tropics during periods of lower water temperatures (Wiencke et al. 1994; Bischof and Wiencke 1995). As an example, the bipolar distribution of the rhodophyte genus *Palmaria* consists of a single species in the North Atlantic and parts of the Arctic (*P. palmata*), seven species present in the North Pacific, and the circumpolar Southern Ocean species, *P. decipiens*. Pellizzari et al. (2017, 2023) recorded the bipolar species, *Monostroma grevillei*, on Deception Island, which also has almost the longest history of human activity in the Antarctic. While this may be a potentially anthropogenically introduced organism, with the only record for the species in Antarctica being from Deception, further investigation is required to confirm this origin.

Considering the Arctic and possible biogeographic connections, Mols-Mortensen et al. (2012) suggested studying cryptic groups to reach a more complete understanding of species diversity and distribution. The authors used molecular markers (rbcL and cox1 sequences) to assess the diversity of foliose Bangiales species present in Iceland and the Faroe Islands (62°N). These analyses of material revealed a total of 13 species representing four genera. *Boreophyllum birdiae* (new to Iceland), *Porphyra dioica, P. linearis, P. purpurea, P. umbilicalis, Pyropia 'leucosticta'*, *Pyropia njordii* (sp. nov.), *Wildemania amplissima* and *W. miniata* were common to both areas, while *Pyropia thulaea* and *W. abyssicola* (Kjellman) A. Mols-Mortensen and J. Brodie, comb. nov. were reported from Iceland but not from the Faroe Islands.

P. njordii was also recorded from Greenland, New England, USA and Nova Scotia, Canada.

Integrated taxonomic studies are useful for inferring the origin of possible dispersal events, reassessing endemism data and introductions (passive or active), as well as for feeding back into monitoring databases. The data generated informs on the biogeographical distribution of seaweeds from the polar region, helping clarify whether these result from natural dispersal, biological invasions related to anthropogenic activities and/or meteorological and oceanographic changes (Hughes and Convey 2014; McCarthy et al. 2019; Macaya et al. 2020; Pellizzari et al. 2020).

Thus, a polar reference baseline of seaweed diversity is urgently required to integrate different inventories based on both morphology and life cycles, molecular data increasingly being generated from rapid, high-throughput and low-cost methods (Oliveira et al. 2018). This will enable researchers to effectively monitor environmental changes and help in the conservation of the unique polar environment. Moreover, laboratory culture techniques associated with morphological and molecular analyses will likely reveal much hidden diversity in the Antarctic and Arctic marine floras, especially with respect to smaller species such as epiphytes and endophytes, again including potential alien or cryptic species. Integrative taxonomical studies should similarly be applied to less known groups (e.g. crustose algae, Ulvales, Gigartinales), and multiple molecular approaches applied, including metagenomics, DNA barcoding and studies of environmental DNA (eDNA) techniques, to clarify the phylogenetic and biogeographic relationships of polar seaweeds.

3.3 Biogeographic Processes Influencing Antarctic and Arctic Seaweeds

3.3.1 Historical Isolation of Antarctic and Arctic Seaweeds

Biogeographic studies aim to understand species distributions and are becoming increasingly relevant to the establishment of baselines to monitor environmental changes. There are clear differences between the two polar regions, especially with respect to their origins and their biogeographic isolation histories. For instance, sea ice cover in the Arctic did not develop before 2 Ma (Clarke et al. 2012), whereas the Antarctic continent was already heavily glaciated 14 Ma (Zachos et al. 2001).

The Arctic Ocean is sometimes referred to as a 'Mediterranean' ocean today, being a relatively small ocean surrounded by continental landmasses, whereas Antarctica is an ancient continent that has long been biologically isolated from the rest of the world by the ACC atmospheric circulation (Sanches et al. 2016). Arctic coastlines and landmasses are generally connected to the temperate coasts and inland areas of America and Eurasia, whereas the Southern Ocean (SO) has no land bridges with South America and southern Australia since the late Mesozoic (Lüning et al. 1990; Convey et al. 2018). These differences are the main reasons for the different degrees of endemism between the two poles and are strongly reflected in the biogeographical relationships of their phycoflora.

Antarctica

The opening of the Drake Passage between South America and Antarctica, around 30–40 million years ago, and the subsequent onset of the ACC around 20–30 Ma (Convey et al. 2009; Fraser et al. 2020), geographically isolated Antarctica. In the Cenozoic, a combination of reduction in atmospheric CO_2 concentration and the onset of the ACC enhanced cooling and initiated the glaciation of Antarctica, one result of which was a marked reduction in diversity in some marine groups (DeConto and Pollard 2003; Convey et al. 2009). There are several well-recognised circumpolar oceanic fronts in the SO, including the Subantarctic Front and the Antarctic Polar Front (see Sect. 3.1.1). These oceanic features enhance the strong environmental distinctions between Antarctica, sub-Antarctic and temperate ecosystems.

The SO, with its strong westerly wind belt, great depths and challenging sea conditions, as well as the powerful eastward-flowing ACC and its fronts, provides a huge challenge for marine organisms to cross (Gonzalez-Wevar et al. 2012; Pellizzari et al. 2020). The well-documented high endemism of seaweeds in the SO, although perhaps now being eroded (Oliveira et al. 2020), results from the regional biota's long periods of evolution and adaptation in isolation and the region's extreme environmental conditions (Pellizzari et al. 2020). However, it is increasingly recognised that the SO and ACC are susceptible to dispersal events in Antarctica (Clarke et al. 2005; Barnes et al. 2006; Fraser et al. 2020) and at higher frequencies than previously thought (Pellizzari et al. 2020, 2023). Poulin et al. (2014) propose that the incongruence in molecular divergence between sub-Antarctic and Antarctic marine organisms based on geological analysis is due to the continuous connectivity of both continents due to larval dispersal or rafting.

Furthermore, the most important processes influencing movement of seaweeds in the SO are the mean northward drift of Ekman transport, transient north and south movement by eddies, and storm-driven Stokes drift, in addition to the predominantly zonal movement of the large-scale horizontal ocean currents (Fraser et al. 2020). the AP and Scotia Arc share a common geological history with southern South America. In southern Patagonia, at the beginning of the Andean orogeny (Jurassic extensional tectonics), a basin was created whose surviving extension towards the south-east formed the Weddell Sea. Later in the Cretaceous, part of the Weddell basin escaped Magellanic compressional tectonics and remained an oceanic basin. The SO has acted to separate Antarctica from the other southern continents since ca. 26 Ma (Turner et al. 2020).

Seaweed ecology, physiology and biogeography, including dispersal processes, in the Antarctic and sub-Antarctic regions are strongly driven by environmental parameters (Fraser et al. 2020; Pellizzari et al. 2020). The eastward movement of the ACC or the West Wind Drift (WWD) has strongly defined the oceanography of the Southern Hemisphere and, consequently, the diversity, biogeography, and ecology of seaweeds in all the adjacent coastal regions (Sanches et al. 2016). Poulin et al. (2014) describe how critical speciation seems to be part of a common evolutionary process in the Antarctic Ocean, being a possible explanation for the high degree of divergence present in subantarctic and Antarctic populations. A study carried out on the species *Iridaea cordata*, analysed using mt DNA on both continents, suggests that in Antarctica there is a single genetic unit along the peninsula (Ocaranza et al. 2019), which is possibly explained by the effect of historical disturbances in Antarctic populations. On the other hand, subantarctic *I. cordata* populations show genetic differentiation, possibly explained by the presence of microrefugia during the last glacial maxima (Ocaranza et al. 2019).

The genera *Devaleraea* and *Palmaria* could be an example of the corridor between Antarctica and the sub-Antarctic region of Magellan; these genera present low genetic divergence. Bustamante et al. (2022) describe a new species for the genus *Devaleraea*, the species *D. yagan* (Palmariales, Rhodophyta), restricted to the Magallanes region along the Beagle Channel. This species is adapted to estuarine conditions of the channels, being part of the three species reported for the southern hemisphere. Generally, these three species extend in environments of cold temperate zones and Antarctica (Bustamante et al. 2022), being a possible example of recent speciation of the genus Palmariacea in the subantarctic region (Breeman 1988). Thus, biogeographic barriers and ecological corridors are fundamental to understanding the macroecological and evolutionary processes that may isolate or connect groups of marine organisms, including seaweeds.

Arctic

The origin of the Arctic's seaweed flora has been widely debated. It was previously hypothesised that over 90% of seaweeds in the Arctic originated from Atlantic populations (Wilce 1990; Dunton 1992), but more recently several studies have reported more affinities to the Pacific and Indo-Pacific regions, such as indicated by the high proportion of macroalgae of Pacific origin in the Canadian Arctic (Lindstrom 2006). Molecular investigations by Adey et al. (2008) showed a strong molecular biogeographic link between Arctic and North Pacific macroalgae, although the Atlantic biogeographic connection remains particularly obvious in regions with a strong influx of Atlantic waters, such as Svalbard (Hop et al. 2012).

Adey et al. (2008) noted that the Arctic is geologically and biogeographically young (latest Tertiary and Pleistocene) and, following the opening of the Bering Strait about 3.5 Ma ago, the 'Great Trans-Arctic Biotic Interchange' populated the Arctic with organisms strongly dominated by species of North Pacific origin. The Thermogeographic Model (TM) demonstrates why climate and geography continued to support this pattern in the Pleistocene. Thus, Arctic and Atlantic sub-Arctic species of seaweeds are likely to have evolutionary origins in the North Pacific, sub-Arctic species are likely to be widespread in the warmer Arctic, and species of Atlantic boreal or warmer origin are unlikely to be present in the Arctic and sub-Arctic. Although Arctic seaweeds have been thought to have a greater affinity with the North Atlantic, Adey et al. (2008) reanalysed the Arctic endemic algal flora, using the TM and evolutionary trees based on molecular data, to demonstrate otherwise. Examining 35 congeneric species, including six abundant Arctic Rhodophyta,

32 species (91%) occurred in the North Pacific and two (6%) in the boreal or warmer Atlantic Ocean, with a single species being pan-oceanic but restricted to the sub-Arctic within the Arctic region. Thus, both the physical/time-based TM and the dominant biogeographic pattern of relatives of Arctic seaweeds suggest that the Arctic marine flora is largely of Pacific origin.

Wulff et al. (2009) stated that Arctic macroalgae are both of Atlantic and Pacific origin, with only a few cosmopolitan or endemic species, and many species with pan-Arctic distributions, still debating the earlier suggestions that Arctic macroalgae are mostly of Pacific origin (Adey et al. 2008). The algal flora in Svalbard is generally composed of species also found in other regions of the North Atlantic, with very few species having a strong distributional centre in the Arctic (e.g. *Laminaria solidungula, Devaleraea ramentacea*). Seaweed assemblages similar to those of Svalbard extend eastwards along the Russian Arctic coast. Overall, the macroalgal flora along the Russian Arctic coast includes an impoverished assemblage of the cold-temperate flora reported from Svalbard. West of the Canadian Arctic Archipelago, on the Alaskan Beaufort Sea shelf, macroalgae are restricted to small areas where hard substrata protected from ice scour are available, such as the Boulder Patch (Dunton 1992), and where few species occur in the intertidal zone.

According to Bolton (2010), the development of the Arctic kelp communities was initiated with colonisation from the North Atlantic, followed by the opening of the Bering Strait, giving access from the Pacific (ca. 5.5 Ma). Most genera of kelp are monospecific, with many confined to either the western or eastern temperate North Pacific, whereas the distribution of the most species-rich genera (*Alaria, Laminaria, Saccharina*) includes the Arctic, from where they spread to the North Atlantic. This rapid species-level evolution is hypothesised to have been promoted by the relatively recent invasion of the Atlantic by these taxa. Many kelp species are ecologically dominant in shallow coastal rocky habitats in temperate and Arctic regions, although, more recently, extensive kelp beds have been found in tropical regions in specific cooler water habitats (Graham et al. 2007; Fraser et al. 2009). Considering the richness of kelp genera among the 11 world marine ecoregions where they occur, in the Arctic, they are represented by four families and five genera in total: Chordaceae (1), Alariaceae (1), Costariaceae (1), Laminariaceae (2). The bio- and phylogeography of kelps in some regions of the world has still not been comprehensively studied, including in Arctic Russia, where systematic studies remain morphologically based and there has been little recent work (Graham et al. 2007; Bartsch et al. 2016).

Despite the spread and presence of kelp assemblages in the Atlantic Ocean, *Macrocystis* and *Lessonia* in the Southern Atlantic, and *Laminaria* in the north, the Laminariaceae and Alariaceae include some genera that occur only in temperate waters of the northern Pacific. Many of these are monospecific genera (e.g. *Lessoniopsis, Pleurophycus, Pterygophora, Undariella, Pelagophycus, Postelsia, Streptophyllopsis* and *Nereocystis*). It has been suggested that this genus-level evolution has occurred since the cooling of the Arctic Ocean and Bering Sea (Robinson 2009). It could thus be hypothesised that both these families have evolved warmer water forms, currently recognised at the genus level, separately on either side of the

Arctic (Bolton 2010). Recent evidence has shown that the prevailing SSTs in the Arctic after the opening of the Bering Strait were much warmer than today, as high as 18 °C in the Pliocene, 3.0–3.3 Ma ago (Robinson 2009).

Lüning (1990) suggested that polar cooling occurred in the Miocene, explaining why few species can survive both Arctic and temperate conditions (e.g. *Chorda* sp. in the Chordaceae, *A. esculenta* in the Alariaceae, *Costaria costata* and *Agarum clathratum* in the Costariaceae, *S. latissima* and *L. digitata* in the Laminariaceae). The Laminariales is thus a temperate rather than an Arctic group, in both likely origins and diversity. The Arctic endemic *L. solidungula* is the only true Arctic seaweed discussed in detail by Adey et al. (2008). Molecular genetics combined with temperature tolerance studies of this genus may help to elucidate the evolution of temperature tolerance in brown algae. It is also notable that all three of the most species-rich genera of kelps (*Laminaria, Alaria, Saccharina*) are widespread, yet monospecific, in the Arctic and have reached the North Atlantic. As kelps clearly originated in the Northern Hemisphere and post-date the time when southern and northern temperate regions started to be separated by permanent tropical waters, the ancestors of Southern Hemisphere kelps must have crossed the tropics. Kelps can generally be described as Arctic and temperate organisms, which rarely occur where prevailing minimum monthly mean SSTs are above 20 °C (Hop et al. 2012). Thus, much is still to be learnt regarding the historical isolation of Arctic seaweeds, as well as their origins. Following Garbary (2001), there seems to be a general pattern of initial radiation in the more diverse Pacific Ocean, followed by dispersal through the Arctic Ocean, and then into the less diverse North Atlantic, but it is not clear whether speciation has subsequently occurred there, or how and when transfer to the southern polar regions took place. It should be highlighted that, unlike the Arctic, the Antarctic is uniquely devoid of kelps (Moe and Silva 1977).

3.3.2 Polar Seaweed Biogeographical Patterns

Biogeography is concerned with the distribution of species and the underlying ecophysiological, evolutionary and anthropogenic drivers. Although many areas of the Arctic have accessible rocky shores, in Antarctica, the more spatially limited availability of consolidated substrate is in the form of rocky outcrops or clusters of pebbles randomly distributed in the shallower depths near the shore. In both polar regions, a significant proportion of these hard substrates is covered or scoured by ice for most of the year. Changes in ice cover pattern result in cascading shifts in seaweed assemblages.

Antarctica

Biogeographic barriers are known to be fundamental in macroecological and evolutionary processes. Ocean circulation, adding in the context of present and past patterns of continental drift, can isolate or connect groups of marine organisms, including seaweeds (Fraser et al. 2020). These benthic organisms release spores and propagules as planktonic stages that drift with currents and provide a useful indicator of changes in biogeographic distribution patterns (Sanches et al. 2016; Navarro et al. 2020).

The distribution of SO biota is the result of major geological, oceanographic and climate changes over the last 50 Ma, and there is a paradox apparent between the broad distributions of some species and their inherently poor dispersal capacity. However, marine biogeographic studies supported by molecular data are increasingly revealing examples of this paradox and indicating that long-distance dispersal of macroalgae is possible (Fraser et al. 2020; Pellizzari et al. 2020). According to Sanches et al. (2016) and Fraser et al. (2018), the ACC, previously, although not entirely accurately considered a biogeographic barrier, may provide a new pathway allowing biotic interconnectivity.

Temperature is the most important factor controlling the biogeographic distribution of macroalgae, with shifts in distributions being observed during periods of change or long periods of instability. These assemblages are, therefore, considered sensitive bioindicators of abiotic changes or abrupt oscillations. Mystikou et al. (2014) note that Antarctic seaweeds display plasticity and adaptability in response to extreme environmental conditions such as low temperatures and limited light availability. It is, therefore, relevant to examine how environmental changes, such as those caused by meteorological and oceanographical changes, will affect algal seasonality, richness, depth zonation and latitudinal distribution. Shifts in seaweed distribution have been important during periods of changing temperatures, as suggested in past glaciations. Similar changes may be expected during the current period of global warming, with a significant impact on marine communities. Additionally, polar macroalgal communities are expected to expand north and south of their current respective distributions (Pellizzari et al. 2017, 2023), due to reduced ice abrasion and improved light conditions, while new ice-free areas created due to glacier and sea-ice retreat are likely to be colonised by seaweeds (Quartino et al. 2013).

The first compilation of biogeographic and systematic relationships of Antarctic and sub-Antarctic Rhodophyta was that of Hommersand et al. (2009). For the brown algae, the SO is the only region in the world devoid of kelps, particularly Laminariales (Moe and Silva 1977). This order is ecologically replaced in the SO by Desmarestiales, including the endemics *Himantothallus* and *Desmarestia antarctica*. This order is considered to have originated in the Antarctic region and subsequently radiated (Peters et al. 1997).

Recent changes noted in the composition and structure of Antarctic algal assemblages feature new records, including opportunistic green algae and cyanobacteria, as well as the detection of cryptic species and an overall reduction in levels of endemism at least in part through improvements in taxonomic knowledge (Pellizzari et al. 2017, 2023; Oliveira et al. 2020). Pellizzari et al. (2020), reviewing the biogeography of Antarctic seaweeds in the context of climate change, note the need to reassess biogeographical hypotheses in the light of population shifts and the disruption of ecological isolation. They also note that the distribution of some species could be altered by thermohaline changes, which, in turn, would alter the dispersal patterns of macroalgae due to severe meteorological and oceanographic events.

Other factors must be considered in addition to abiotic drivers. For example, rafting of floating organisms and objects appears to have a critical role in facilitating long-distance dispersal of algae and for (re)structuring mainly the intertidal ecosystems in Antarctica (Fraser et al. 2018). Many species that lack obvious transoceanic dispersal ability (e.g. invertebrates, small non-buoyant macroalgae, and terrestrial vertebrates and invertebrates) can disperse long distances in association with floating objects or organisms, such as vessels, wood or kelp rafts (Fraser et al. 2020; see Sect. 3.3.3).

Some studies have suggested that, in certain areas, seaweed diversity could be decreasing due to climatic changes (Pibernat et al. 2007; Wiencke and Amsler 2012). However, this pattern has not been reported in Antarctica (Pellizzari et al. 2017, 2023), where, rather, seaweed communities seem to be in transition towards higher connectivity and distribution expansion. Further taxonomic studies supported by molecular analyses and biogeographical models are required to develop an improved understanding of the effects of global changes on seaweed diversity and changes in their biogeographic distributions.

Arctic

As in the Antarctic, in the Arctic, the main structuring abiotic factors or environmental drivers controlling the growth, life cycles and, hence, the latitudinal distribution of macroalgae are availability of substratum, light (UV), photoperiod, nutrients, salinity and temperature, as well as sea-ice cover, and wave exposure (Krause-Jensen et al. 2012). Ocean currents may also be an important vector of propagules/spore transport and hence facilitate migration and lead to changes in biogeographic distribution. Wegeberg (2023) compiled data regarding the biogeography of high latitude macroalgal flora, using macroalgal diversity on the coast of Greenland as a proxy for polar seaweeds. A total macroalgal richness of 176 species was identified, distributed along a latitudinal gradient from 60 to 83°N. Generally, and as expected, a linear decline in species number from south to north was observed, influenced in the north by the cold Arctic Ocean currents. A higher richness of Rhodophyta was noted, as discussed in Sect. 3.2, with trans-Arctic speciation occurring after the opening of the Bering Strait.

Bringloe et al. (2020) suggest that species may have persisted in northern refugia probably centred in southern Greenland, through cycles of glaciation, leading to 'unique assemblages in polar waters, rather than being entirely derived from more southern (temperate) areas following glaciation, a hypothesis that is also starting to be considered for some Arctic terrestrial biota'. However, many Arctic species' origins are still considered to be Atlantic/boreal (Krause-Jensen et al. 2020; Bringloe et al. 2020). Therefore, the northward distribution of macroalgal species may be limited by strong seasonality (Krause-Jensen et al. 2020; Lindstrom 2006), unless otherwise limited by suitable substratum or other ecological barriers resulting from, e.g. glacial run off leading to lowered salinity and light penetration, wave exposure, or nutrient depletion.

As noted, Bringloe et al. (2020) suggested that South Greenland is a probable macroalgal species refuge during the Last Glacial Maximum within the region, including the Bering Sea, the north-west Atlantic and parts of the European coastline. Additionally, following Wegeberg (2023), South Greenland is oceanographically connected with the north-east Atlantic through the North Atlantic drift and Irminger Current, which may promote transport and introduction of northern temperate organisms to higher latitudes such as Cape Farewell, acting as a potentially critical gateway.

3.3.3 *Bioinvasions, Opportunistic, Alien and Cryptogenic Seaweed Species: What These Could Mean for Polar Zones?*

Antarctica

The Antarctic is considered a pristine environment compared to other regions of the Earth, but it is increasingly vulnerable to invasions by marine, freshwater and terrestrial non-native species (Hughes et al. 2020). Convey and Peck (2019) reported that the establishment of non-native organisms in Antarctic ecosystems may present an even greater impact than climate change itself, and that the marine environment has much more limited response flexibility, meaning that even small levels of warming are threatening. Non-native marine species (NNMS) interact with other anthropogenic stressors, also affecting Antarctic ecosystems, with irreversible damage to biodiversity and ecosystem services. The main reasons that meteoceanographic change increases the risk of NNMS establishing around Antarctica are increasing water temperature and decreasing ice cover, followed by coastal freshening; these factors are expected to have combined impacts.

McCarthy et al. (2019) reported Antarctica as the final frontier for marine biological invasions. The authors reviewed the knowledge of non-native marine species in Antarctica, the physical and physiological factors that resist the establishment of NNMS, changes to resistance under climate change, the role of legislation in limiting marine introductions, the effect of increasing human activity on vectors and pathways of introduction, and potential bioinvasions. The authors reported that evidences of NNMS is limited in the area, and just four marine non-native and one cryptogenic species that were likely introduced anthropogenically have been reported freely living in Antarctic waters; no established populations have been reported (or bioinvasion); and additional six species have been observed in pathways to Antarctica that are potentially at risk of becoming invasive, surprisingly none seaweed species was still considered. However, the increasing intensity of ship activity across fishing, tourism and research sectors is a concern and, in addition to climate change, the main risk of bioinvasions in the SO.

Hughes and Convey (2014) warned that the introduction of non-native species in association with human activities is the major threat to the Antarctic indigenous biodiversity. However, existing legislation contained in the Protocol on Environmental Protection to the Antarctic Treaty does not specifically make the eradication of non-native species mandatory. In 2005, an additional Annex (VI), known as the 'liability annex', was first proposed, with a focus on prevention, contingency planning and reclaiming costs incurred when responding to environmental emergencies. However, even now, this annex has not yet been adopted. According to Hughes et al. (2020), the Antarctic Peninsula (AP) is the most potentially invaded region of the entire Antarctica, and the risk of introduction of invasive non-native species (NNS) is likely to increase due to human activities. Parties that are signatories to the Antarctic Treaty have called for regional assessments of NNS risk. In response, taxonomic and Antarctic experts undertook a scanning exercise to identify the species that are likely to present the highest risk to biodiversity and ecosystems within the AP over the next 10 years. One hundred and three species, currently absent in the AP, were identified as relevant, with 13 species identified as presenting a high risk of invading the AP. Marine invertebrates dominated the list of highest risk species, with flowering plants and terrestrial invertebrates also represented, but once more, seaweeds are not in the list. The authors also recommend the development and application of appropriate biosecurity measures by all stakeholders active in the AP, aiming to reduce and or avoid introductions and bioinvasions, resulting in negative consequences for the biodiversity of the entire continent, as introduced species establish and spread further due to climate change.

Cryptogenic species whose phylogeography is unknown, or that cannot be confidently defined as native or exotic, present a particular challenge. For instance, the green alga *Monostroma grevillei* (Chlorophyta) was reported by Pellizzari et al. (2017, 2023) from Deception Island (SSI) in consecutive years. They also reported a new record of an opportunistic green seaweed and six species of filamentous Cyanobacteria as epiphytes on large macroalgae, all present in the intertidal and shallow subtidal zone. Antarctic cyanobacteria are conspicuous and well-studied in lake assemblages (Taton et al. 2006); however, much more limited information on their presence in marine ecosystems is available. As some Ulvophyceae and filamentous Cyanobacteria are opportunistic organisms that can inhabit many ecosystems, including environments with high and rapid variability or extreme conditions, their monitoring is essential in polar regions. In contrast with other islands in the SSI, the Deception Island flora is dominated in richness and biomass by Rhodophyta and Chlorophyta rather than Phaeophyceae. Given its unusual combination of ameliorated environmental conditions, Deception Island provides a natural laboratory for monitoring opportunistic, alien and or cryptogenic species in the AP region (Pellizzari et al. 2020).

Another factor to be considered is the potential for new species to arrive by natural processes (Hughes et al. 2006), such as the transport of reproductive cells or macroalgal propagules by birds and migratory marine mammals, or by floating seaweed rafts that may also carry other species (Fraser et al. 2020; Macaya et al. 2020). Lopez et al. (2017) studied the potential for dispersal of other algae anchored or 'hitchhiking' on large floating macroalgal species such as *Durvillaea antarctica*, and the possibility of reaching Antarctica by drift. The natural distribution of *D.*

antarctica is north of latitude 55°S, with large beds concentrated in Chile and New Zealand, where it forms a traditional foodstuff. Although its distribution is restricted to cool temperate and sub-Antarctic zones, *D. antarctica* has been frequently recorded as drifting rafts or stranded on beaches around the AP. Fraser et al. (2020) report new evidence that buoyant seaweeds such as *D. antarctica* can cross the Drake Passage to reach Antarctic coasts. These macroalgal rafts carry diverse 'passengers', including marine invertebrates, other non-buoyant seaweeds and microorganisms (Barnes et al. 2010). Modelling indicates that algal rafts from the sub-Antarctic could reach Antarctic coasts monthly (Fraser et al. 2018; Macaya et al. 2020), providing a continuous influx of marine propagules and highlighting higher connectivity between Antarctica and lower latitudes.

To date, only *Ulva intestinalis* has been reported as an established introduced marine alien species in Half Moon (SSI), suggested to have arrived through fouling on the hull of a visiting vessel (Clayton et al. 1997; Frenot et al. 2005; McCarthy et al. 2019). The capacity of Antarctic macroalgae to withstand competition from invaders is not well understood. Antarctic macroalgae are highly adapted to the local stenothermal environment. Some studies have reported temperature optima for some species between −2 °C and 10 °C, but most Antarctic species show limitation or even failure in growth or gametogenesis at temperatures higher than 5 °C (e.g. *Georgiella confluens, Sarcopeltis antarctica, Plocamium cartilagineum, Desmarestia antarctica*) (Wiencke et al. 2014). In the context of rapid temperature increase, populations of sub-Antarctic or even cool temperate species may in future be able to colonise Antarctic coastlines (Pellizzari et al. 2020, 2023).

Grazing is likely the most important biotic factor controlling the composition of seaweed-dominated communities. In Antarctic benthic systems, the early successional stages dominated by small-sized seaweeds and periphyton represent excellent models to study how grazing modifies different ecological properties not only of native assemblages but also of alien species and or bioinvasions, whose arrival and establishment may be facilitated by climate change and or ecological niches transferring. Valdivia (2020) suggested that native Antarctic grazing amphipods are sensitive to climate change factors such as warming and acidification, which could hamper their ability to control native and mainly alien macroalgae. Alien macroalgae that modify the abiotic environment, such as *Ulva intestinalis* in intertidal pools, represent potentially superior competitors in Antarctic ecosystems. These authors used mathematical simulations to assess the potential impact of mesograzers in sub-Antarctic and Antarctic locations connected by dispersal. *Ulva* spp., being opportunistic species, are regarded as alien species in several sites around the globe, and are highly competitive in the Antarctic as well. The results obtained indicated that Antarctic mesograzers have a deterministic and clear effect on the biomass of the alien seaweeds. However, projected climate-change-driven shifts in temperature (or pH) could decrease the potential of amphipod grazers to control the development of these invaders (Gómez and Huovinen 2020a, b).

Mrowicki and Brodie (2023) provide the first report of a non-native seaweed established on sub-Antarctic South Georgia, *Ulva fenestrata* (Ulvophyceae, Chlorophyta), as well as confirming its widespread presence in the Falkland Islands (with the earliest record in a 2013 collection by Küpper, Asensi, Mystikou, van West and Marcou), supported by molecular data. These form the second Southern Hemisphere records for this species, which otherwise occurs in the Northern Hemisphere. It is important now to monitor this potentially invasive species and to determine its likely source and mode of arrival, while emphasising the importance of robust biosecurity measures. An additional process that has recently been recognised as extremely important for the dispersal of floating surface material is non-linear advection by surface waves, known as Stokes drift (Fraser et al. 2020).

In Antarctica, baseline datasets that would allow detection of local loss of native species or their replacement by alien assemblages are generally lacking. Clearly, there is an urgent need to extend the geographical range and number of surveys, develop inventories of phylogenetic markers and increase knowledge of less conspicuous algal groups such as crustose, cryptic and endophyte species. Also, mitigation measures that will substantially reduce the risk of introductions to Antarctica and the sub-Antarctic must focus on reducing propagule loads on humans, and their food, cargo, and transport vessels, including research, military or tourism.

Arctic

Climate changes in the Arctic in combination with anthropogenic activities often lead to eutrophication of marine systems and may result in sometimes massive growth of opportunistic macroalgae, which also may easily become detached and transported by coastal currents, forming large mats of drifting algae (Grall et al. 2006). Carbon flow through ecosystems is determined by the nature, availability, incorporation and passage of basal resources to the higher trophic levels of food webs. When pollution by organic material occurs, the development of abundant opportunistic algae often diversifies trophic niches within food webs. Green or brown tides of macroalgae have increased in frequency and intensity and are expected to continue to do so under global change, even in polar regions. Understanding the impacts of high biomass of opportunistic and or invasive macroalgae on the functioning of polar ecosystems is crucial.

Quillien et al. (2016) studied in temperate zones whether and how opportunistic green seaweeds modify macrotidal beach shore trophic structure and functioning. They investigated whether these effects propagate through the web, looking at different scales and trophic levels using over 1200 individual stable isotope and 180 biodiversity samples obtained at pristine and macroalgal-bloom-affected sites, together with biomass-weighted isotope signatures and isotopic functional indices. They found that the opportunistic species affected the food web across several trophic levels, by inducing shifts in trophic interactions and that, over time, opportunistic algae dominance simplified the entire food web.

In contrast, Ulaski et al. (2023) studied the beach-stranded and drifting seaweeds as a foraging source for marine and terrestrial macroinvertebrates at high latitudes. The drifting algal rafts were generally occupied by more marine-derived fauna, especially Amphipoda, Gastropoda, Mytilida and Polychaeta. They highlighted the importance of this ecologically important habitat to macroinvertebrates and its influence on macroinvertebrate community structure, higher trophic level consumers and key ecological processes on beaches.

Fucus serratus (Phaeophyceae) is currently considered invasive in Icelandic waters but is native to other ecoregions in the Arctic and sub-Arctic. Coyer et al. (2006) reported that this opportunistic species was probably anthropogenically introduced to Iceland and the Faroe Islands from Europe, as previous genetic studies have confirmed that life-history constraints preclude long-distance dispersal. They genotyped 19 populations throughout northern Europe, Iceland and the Faroes, suggesting that the Oslofjorden region (Norway) was the likely source of the Icelandic populations, and that the Hafnarfjordur region (Iceland) was the likely source of the single Faroese population. The introduction to Iceland probably took place during the nineteenth century, while that to the Faroes occurred during the late twentieth century. Their molecular data also confirmed hybridisation between the introduced *F. serratus* and the native *F. evanescens*.

Mathieson et al. (2010), in a comparative study of seaweed flora from James Bay (Canada) and three north-east Canadian Arctic locations, recorded the cryptic taxon *Spyridia filamentosa*, sourced possibly from a warm-water location in the north-west Atlantic, as the species was previously known only from one site in Nova Scotia, several in southern Maine/New Hampshire, south of Cape Cod, Massachusetts. A single introduced species (*Dumontia contorta* from Europe) was recorded from James Bay and the Ellesmere-Baffin Islands area.

Native boreal, temperate, non-native, and invasive species are likely to expand their ranges to higher-latitude regions, facilitated by global change. A recent horizon-scanning exercise for marine invasive species in Hudson Bay identified macroalgae as one of the taxonomic groups with the highest risk of invasion (Goldsmith et al. 2021). Invasive macroalgae have been found to induce ecological impacts even when they are outside of their thermal range of origin, and according to Diehl et al. (2020), the endemic Arctic kelp *L. solidungula* might be particularly impacted and become locally extinct from Arctic fjord systems, where temperature increase is most pronounced. Furthermore, increased temperatures cause glacier and sea ice melting and enhanced terrestrial run-off from snowfields, which can result in hyposaline conditions. Diehl et al. (2020) conducted a multiple-stressor experiment at four temperatures (0, 5, 10, 15 °C) and two salinities (25, 35 psu) to investigate the combined effects of increasing temperature and decreasing salinities on the physiological and biochemical status of young *L. solidungula* sporophytes. Both drivers had significant and interacting impacts, either in an additive or antagonistic way, depending on the response variable. Even though absolute pigment content was not affected, the deep oxidation state of the xanthophyll cycle increased with intensified stress. Higher temperatures affected the C:N ratio significantly, mainly due to reduced nitrogen uptake, while 25 psu supported nitrogen uptake, resulting in an attenuation of the effect. The concentration of mannitol decreased at 25 psu. In the control 35 psu, the mannitol level remained steady between 0 °C and 10 °C but significantly decreased at 15 °C. These data confirmed that this endemic species is very susceptible to both drivers of climate change, especially in combination.

These studies suggest that the macroalgal communities of the polar regions could face major challenges soon. Studies providing baseline and ongoing monitoring emphasise the need for the application of strict biosecurity measures and provide evidence to guide mitigation actions, aiming to minimise the risk of introducing non-native species into fragile polar marine ecosystems.

3.3.4 The Importance of Polar Thermohaline Patterns (and Other Abiotic Features) on Seaweed Distribution

Over the past three decades, strong trends in sea-ice extent and sea surface temperatures (SST) have been observed in both Arctic and Antarctic regions (Fig. 3.9). For decades, Arctic sea-ice extent has been characterised by a rapid decline, whereas Antarctic sea-ice extent and its decline are highly variable inter-annually. Purich and Doddridge (2023) reported the minimum Antarctic sea-ice record-breaking, being the third in seven years. Following the summer minimum, circumpolar Antarctic sea-ice coverage remained low during the autumn and winter, leading to the largest negative ice extent anomaly observed over the satellite era. The authors showed data of the confluence of SO subsurface warming and record minima, suggesting that ocean warming has played a role in pushing Antarctic sea-ice into a new low-extent state, and that the underlying processes controlling the coverage may also have been altered.

SST in the polar oceans can be at the freezing point (around −1.8 °C) during winter, and in higher latitudes where sea ice persists year-round, although it is slightly warmer at greater distances from the Antarctic continent. Some well-adapted macroalgal life-stages can survive beneath sea ice during the winter, but there is higher biomass and richness during the summer months in both polar regions.

Ocean fronts are characterised by sharp transitions in temperature, salinity and biogeochemical properties and align in the Southern Ocean with the eastward currents of the ACC. Traditionally, three primary fronts have been identified in the SO: the Subantarctic Front, the Polar Front and the Southern Antarctic Circumpolar Current Front (Orsi et al. 2002). A front appears as a near step-change from warmer, saltier waters in the north to colder, fresher waters on the southern side. The separation of warm and cold waters across the ACC fronts (until approx. 1000 m depth) has led to the widely-held concept that there is very little mixing or transport across the fronts.

Some hypotheses address how temperature may influence biological distributions: (1) by setting lethal limits that occur at higher or lower temperature tolerances corresponding to maximum and minimum temperature isotherms, (2) multiple ways not just driven by extremes, such as: sub-lethal stresses, rates of change, extreme events (e.g. marine heatwaves -MHWs-, rather than averages), and (3) by setting limits associated with constraints on completion of the life history (i.e. settlement, vegetative growth, and germination of reproductive structures). Notably, macroalgae can be grouped into temperature 'assemblages', whose distribution limits seem to be controlled by similar features, leading to the descriptions of warm temperate, cold temperate and polar assemblages.

Temperature is a major factor controlling seaweed growth and reproduction, as well as global biogeographic patterns. Hence, global warming is predicted to induce shifts in the latitudinal distribution of algal species, changes in overall temperature regimes and increases in the frequency and amplitude of extreme events such as MHWs. Such changes can often impair primary production or even survival of macroalgae, with consequences for standing stock biomass and even loss of important foundation communities (Smale et al. 2019, and references therein). SST has increased by 0.63 °C globally in the period 1986–2005 (IPCC 2022). However, at regional scales, the SST increase has been greater in some global regions. Under different climate change scenarios, global mean SST is likely to increase by 1.6 °C, and even up to more than 4 °C, by the end of this century and, additionally, MHWs are expected to become more severe (IPCC 2022). MHWs are defined as periods of SST increases of 3–5 °C above the long-term SST mean, which can last from a few days to several months (Diehl et al. 2021). Worldwide, the annual duration of MHWs has increased by 54% in recent decades (Oliver et al., 2018). Thomsen et al. (2019) recorded the local extinction of bull kelp (*Durvillaea* spp.) in New Zealand sub-Antarctic waters due to a strong MHW in the summer of 2017/18. They documented that, although gradual changes in algal communities are attributed to increases in global average temperatures, abrupt changes in local temperatures and anomalies associated with heat waves can cause more rapid and detrimental impacts and biological changes. They monitored temperature data from New Zealand's South Island and investigated whether MHW affected *D. antarctica,* in addition to *D. poha* and *D. willana*. *Durvillaea* spp. are large marine algae that inhabit the lower intertidal zone of exposed coasts, where they support the biodiversity and functioning of these southern ecosystems at the temperate-polar, or sub-Antarctic-Antarctic, interface. SST during the studied period included the strongest MHW recorded in 38 years of available ocean satellite data for the region. Air temperatures were also high and resulted in strong desiccation stress during daytime low tides. Analysis of drone images of four rock platforms showed that changes in cover and biomass losses varied between species and reefs, with the greatest (100%) loss of *D. poha* in Pile Bay, where SST exceeded 23 °C and air temperature 30 °C. Under these conditions, all bull kelp banks were eliminated. This was followed by the recruitment of the invasive alga *Undaria pinnatifida,* suggesting the initiation of another cascade effect.

Park et al. (1998) studied the thermohaline structure of the Antarctic Surface Water (ASW) and Winter Water (WW) in the Indian sector of the Southern Ocean (Fig. 3.9), illustrating horizontal distributions of the WW properties with the positions of the Polar Front, placed at the 1–2°C isotherm and salinity between 33 and 34 psu at the winter ice limit). Based on the most extensive cover of winter pack ice in the Weddell Sea, they concluded that the major cause of summer freshening of the surface layer in the Enderby Basin (EA) is the eastward advection of meltwater

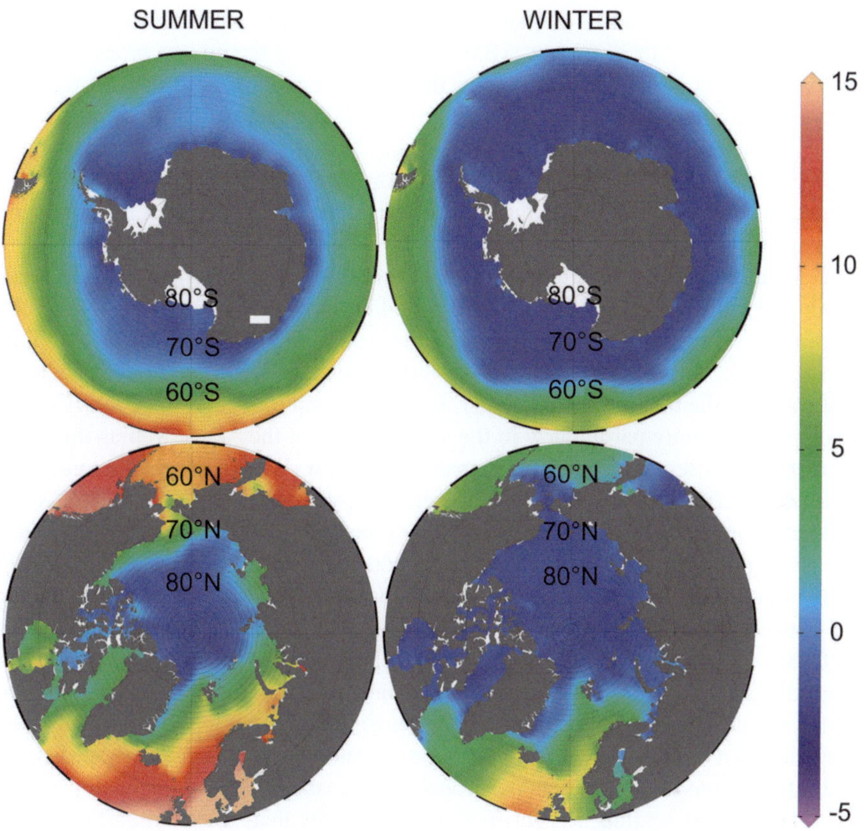

Fig. 3.9 Modelling map generated in Ocean Data View (ODV) using SST mean data during summer and winter months (January to March and July to September for both hemispheres) between 2015 and 2022. (Data retrieved from Locarnini et al. (2024); available in https://explore.webodv.awi.de/—WOA 2023 (World Ocean Atlas) —edited by NOAA atlas)

from the Weddell Basin (EAP). At higher latitudes, current glacier melting rates in WA substantially exceed those around the EA margin, leading to significant differences in thermohaline patterns between EA and WA. An exception is given by Wilkes Land in EA (66°15′ °S 110°31′ °E), where significant retreat took place between 2000 and 2012, underlining its sensitivity to climate change (Park et al. 1998). This process is strongly influenced by ocean dynamics, which in turn changes in accordance with changes in ice sheet dynamics. Bensi et al. (2022) studied temperature, salinity, and dissolved oxygen during the austral summer from the continental shelf break to approximately 3000 m depth, to investigate water mass distribution offshore from the Sabrina Coast (EA). The main water masses of the studied area, identified by their thermohaline properties, were (i) the ASW with potential temperature between θ and −1.5 °C and salinity 34.2 psu, (ii) the WW between −1.92 and −1.75 °C and 34.0 to 34.5 psu, (iii) the modified Circumpolar

Deep Water with 0 °C and 34.5 psu, and (iv) Antarctic Bottom Water between –0.50 and 0 °C and 34.63 to 34.67 psu. ABW are influenced by the mixing processes they undergo as they move westward along the Antarctic margin, also interacting with the warmer Circumpolar Deep Water. The spatial distribution of water masses offshore of the EA coast appears to be strongly linked with the complex morpho-bathymetry of the slope and rise area, supporting the hypothesis that downslope processes contribute to shaping the architecture of the distal portion of the continental margin.

Generally, in Antarctica, temperature and salinity are characterised by seasonal, depth and latitude-dependent gradients, in addition to the presence or absence of melting ice along the coast. Although generally does not experience rain (precipitation occurs as snow) or rivers that drain into the sea during the summer, as happens in the Arctic, the melting (increasingly accelerated and random) significantly alters near surface salinity particularly in the surroundings of the AP, as well as influencing pH and nutrient concentrations close to the coast (Wiencke and Clayton 2002; Wulff et al. 2009), where algal assemblages with greater biomass and richness are present. Where nutrient decreases have been reported around Antarctica, causes are suggested to be dilution with meltwater and uptake by strong phytoplankton blooms (Kang et al. 2002), but such nutrient reductions are not considered a limiting factor for seaweed growth throughout the year (Peters et al. 2005; Zacher et al. 2009).

Climate change impacts thermohaline circulation and, as a consequence, impacts marine biota. Li et al. (2023) suggested alarming signs that the overturning thermohaline circulation near Antarctica is also slowing because increasing volumes of meltwater are reducing the salinity of surface waters, hence reducing their density and rate of input to greater depths. Global change is accelerating this disruption of the thermohaline regime. Recent modelling suggests that, if global carbon emissions continue at the current rate, then the rate of Antarctic overturning will slow by more than 40% in the next 30 years (Li et al. 2023). If the deep ocean current collapses, the oceans globally below 4 km depth will stagnate, with nutrients trapped in the deep ocean rather than returned to surface waters, with likely major impacts on the entire marine biota and ecosystems.

Arctic

During the summer months, increasingly large areas of the AO are becoming exposed to direct solar radiation, with a direct impact of increasing SST. In turn, these warmed surface waters accelerate the melting of the surrounding sea ice, exposing more open water. Summer SST at the ice-free margins of the Arctic basin can be up to several degrees above 0 °C (Fig. 3.9), particularly in the vicinity of Pacific and Atlantic inflows (Timmermans and Marshall 2020). In both Canadian and Eurasian Basins, a layer of warm Atlantic-origin water is characterised by SST between 0–5 °C, and at 200–300 m depth, the seawater temperature can reach +2 °C. These also drive the formation of a halocline several meters below the surface, which is a defining feature of the AO. The stratification of fresher overlying more saline waters, separated by a strong halocline, is known as a β-ocean, where β refers to the saline contraction coefficient (Timmermans and Marshall 2020). In contrast, the subtropical α-oceans (where α refers to the thermal expansion coefficient) are characteristically stratified by temperature, with warmer waters overlying cooler.

AO salinity above 100 m depth is typically 30–32 psu (Timmermans and Marshall 2020). Marked gradients in temperature, salinity and density are confined to the top few hundred meters of the water column. River discharge, predominantly from the six main Arctic Siberian and North American rivers (Ob, Yenisey, Lena, Kolyma, Yukon, Mackenzie), is a major source of fresh water to the AO (McClelland et al. 2012) and, while the AO constitutes only 1% of the world's ocean by volume, it receives approximately 10% of global river discharge. The AO also receives fresh water through direct precipitation (in contrast with Antarctica) and lower salinity waters from the Pacific Ocean via the Bering Strait. Annually, the AO receives 50% of its freshwater input from rivers, 25% from Pacific water inflow and 25% by direct precipitation (Carmack et al. 2016). Surface freshwater from all these sources is drawn toward the centre of the Canadian Basin by the anticyclonic winds of the Beaufort High, preserving the strong halocline stratification. As AO ice extent grows in winter and moves, concentrated brine is rejected from the sea ice and there is a distillation of freshwater.

Duarte et al. (2017) proposed an empirical model to describe sea-ice thermohaline dynamics and biogeochemistry in the AO, aiming to analyse ice microalgal bloom dynamics in different types of sea-ice. They reported that snow precipitation varied between 0 and 9.2×10^{-2} kg m^{-2} s^{-1}, with incident short and longwave radiation in the ranges 0–500 and 170–310 W m^{-2}, respectively. SST was between −1.9 and 0.6 °C, and salinity varied between 33.6 and 34.3 psu at the surface, while nitrate and silicic acid varied between 0.4 and 10.4μM, and 2.4 and 4.2μM, respectively. Different ice algal bloom and net primary production (NPP) patterns were identified, suggesting that ice algal maximal growth rates will increase, while sea ice will decrease in the AO.

Malavenda (2021), using compiled data, reported that seawater temperature is increasing in Svalbard, and there is increased flow of Atlantic Water, with significant variability in thermohaline patterns being observed over the past decade. La Guardia et al. (2023) suggested that the lack of relationship between kelp depth extent and water temperature at the depth extent could be influenced by thermal variability driven by latitude (SST from −0.6 to 4 °C), suggesting that, in the Eastern Canadian Arctic, kelp distribution could be driven by sea ice (and open water days with light – OWL) rather than temperature (see also Filbee-Dexter et al. 2022). However, following Bello and Higuchi (2019), temperature in the AO may become a more important driver of regional kelp depth distribution in the future, with predicted ocean warming of 1 to 3 °C.

Some kelp species, such as *Saccharina latissima*, may respond positively to regional increases in ocean temperature (Bello and Higuchi 2019; Diehl et al. 2021), while endemic Arctic species such as *L. solidungula* may suffer negative impacts with even moderate increases in temperature (Goldsmith et al. 2021). Understanding the drivers of Arctic kelp depth extent will help to improve estimates of their current and future distribution, helping to better understand Arctic coastal ecology and the role of kelp forests facing climate change. La Guardia et al. (2023) also noted that the west coast of Greenland is characterised by cold (polar) waters, high water transparency, many OWL, low grazing pressure, abundant suitable substrata and deep and extensive kelp forests, and that OWL was the principal variable that could explain kelp depth extent amongst all these variables. However, when controlling for the latitudinal gradient and OWL, other environmental features, such as water transparency, also became important predictors.

Considering temperature tolerance limits, species with a wide latitudinal distribution are generally more tolerant than narrow-ranging species (Sunday et al. 2015). It is likely that most seaweeds with temperate to Arctic distributions will respond neutrally or even positively to predicted levels of warming, and that the same will apply to broadly distributed macroalgal species such as kelps. However, local adaptation and phenotypic plasticity can result in intraspecific differences in thermal tolerance, and population declines and losses might be expected both close to the 'trailing' (warmer) edges of distributions but also in other regions (King et al. 2018). Recently, a study of *L. digitata* revealed that, despite considerable plasticity at intermediate temperatures, plasticity at the upper survival limit was low (Franke et al. 2021). Thus, models assuming a uniform climatic envelope for wide-ranging species, particularly relating to the local temperature threshold for survival, may underestimate local adaptation and extinction (Filbee-Dexter 2020; Diehl et al. 2021).

Recent (2014–2020) rapid declines in summer AO ice cover have increased the OWL and annual PAR reaching the ocean surface (La Guardia et al. 2023). Additionally, climate warming combined with a longer ice-free season has increased coastal erosion, runoff and frequency of sediment resuspension from storm events, resulting in increased turbidity (Singh et al. 2022). Increasing turbidity could result in a shallowing of the kelp depth extent due to a reduction in light reaching the benthos, or a shift of distribution towards deeper, but clearer, waters offshore (Krause-Jensen et al. 2016). Other environmental conditions may also drive kelp depth redistribution. For instance, decreasing salinity close to the coast due to increasing river runoff can affect the physiology of some kelp species negatively (Krause-Jensen et al. 2020). Silt and sediment in the water column can have negative effects on the physiology of kelp when they deposit on kelp fronds or accumulate on the substratum, forming a thick sandy/silty layer unsuitable for kelp attachment and recruitment (Huovinen and Gomez 2020). Conversely, sediment deposition on kelp blades might provide a benefit by mitigating harmful ultraviolet radiation effects (Roleda and Dethleff 2011).

Li et al. (2023) report a series of sudden draining events, rapidly releasing very large quantities of freshwater into the AO. With a dramatic salinity drop, the low-density surface waters do not sink. If the thermohaline circulation reduces, less warm water could be moved from the equatorial Atlantic to higher latitudes, resulting in a cooler regional climate (Roleda and Dethleff 2011). The impacts of such a new and rapid wide-scale cooling event could be severe and widespread, leading to a wave of regional extinctions and substantial changes in distributions of biota, including key primary producer groups.

3.4 Phylogeography of Antarctic and Arctic Seaweeds

The unique environmental conditions of the Antarctic and Arctic present challenging environments, shaping the evolution and distribution of polar seaweeds. Understanding the origin, diversity and phylogeographic patterns of these organisms offers insights into their adaptation to polar environments and informs conservation efforts in the face of a changing environment. The phylogeny of Arctic and Antarctic seaweeds reveals a complex evolutionary history shaped by both historical and contemporary processes. Molecular phylogenetic analyses have elucidated the relationships among different taxa, including some with bipolar distribution, that also occur in the Arctic and Subarctic region, such as the green alga *Acrosiphonia arcta*. Internal transcribed spacer (ITS) studies between populations of both poles indicated a high degree of similarity, showing a recent separation of these populations (van Oppen et al. 1993), highlighting evolutionary patterns influenced by historical glaciations, oceanographic currents, and isolation. Studies employing molecular markers have provided valuable insights into the evolutionary relationships and diversification patterns of seaweeds from both polar regions (Bringloe and Saunders 2018, 2019a, b; Bringloe et al. 2017, 2020).

3.4.1 Molecular Tools for Understanding Seaweeds' Origin, Diversity and Distribution in Polar Regions

The taxonomy of species in many groups of seaweeds can be challenging due to their simple morphology and frequent occurrence of both morphological plasticity and cryptic diversity, requiring molecular-assisted taxonomy for reliable identification. In 1993, van Oppen et al. used the sequence of the ribosomal DNA internal transcribed spacer (ITS) regions to study biogeographic disjunctions in two bipolar species, *Acrosiphonia arcta* (Chlorophyta) and *Desmarestia viridis/willii* (Phaeophyceae), showing that these disjunctions are of recent origin (last glacial maximum around 18,000 years ago). In the last decades, this type of approach has proven to be useful in the distinction of cryptic and introduced species, as well as for phylogeny and for mapping polar seaweed distribution. Different molecular markers have been used depending on the seaweed group, including nuclear (e.g. ITS and ribosomal genes coding for the large rRNA subunit LSU and the small rRNA subunit SSU), plastid (e.g. rbcL gene coding for the large subunit of Rubisco, UPA universal plastid amplicon and tufa coding for an elongation factor) and mitochondrial (mostly the DNA barcode markers COI-5P, coding for the 5'end of the cytochrome oxidase subunit 1) markers (Coyer et al. 2011; Nelson et al. 2022; Oliveira et al. 2020; van Oppen et al. 1994). The use of molecular markers, combined with the description and identification of morphological, anatomical and reproductive characters, can fill taxonomic, phylogenetic, distributional and ecological gaps, besides being essential to quickly detect current biotic changes (Mrowicki and Brodie 2023).

3.4.2 Genetic Diversity, Surviving Glacial Cycles and Species Refugia

Both in Antarctica and the Arctic, seaweeds cope with extreme environmental conditions such as low temperatures, light and substrate availability, ice scours and intense seasonality. In Antarctica, these narrow environmental limits likely persisted over most of the last million years, impacting the evolutionary history of its biota (Barnes et al. 2006). But environmental stability in both polar regions is jeopardised by climate change, and shifts in the biota of both regions are expected to occur in a short time frame. Therefore, there is an urgent need to understand and describe the phylogeographic patterns in both polar regions, and these patterns can be affected by the ongoing changes.

Antarctica

According to Guillemin et al. (2020), rapid climatic fluctuations in the Southern Ocean during the Quaternary induced contractions and bottlenecks, drastically impacting marine biota. For seaweeds restricted to shallow waters, survival in deep-water refugia was not possible, and in this sense, the authors compared patterns of distribution of genetic diversity using sequences of mitochondrial and chloroplast markers in distinct species of green, brown and red macroalgae, aiming to detect common responses to the effect of these glacial cycles. The results suggested that all the Antarctic macroalgae were characterised by very low genetic diversity, absence of genetic structure and significant signatures of recent population expansion. The eight studied species seem to have barely survived glacial events in situ, in a unique refugium from which they recolonised their current distribution area, so the authors proposed that polynyas or areas showing long-term geothermal activity along Antarctic continental margins or peri-Antarctic islands should be good candidates as glacial refugia. Also, they verified that common haplotypes, scattered over hundreds or even thousands of kilometres of coastline, point to long-distance dispersal of fronds drifting on the strong oceanic currents in the region as the main mechanism of postglacial colonisation and expansion.

In Antarctica, comparatively with other ecoregions in the Southern Ocean, the higher seaweed diversity observed in the SSI (Pellizzari et al. 2017, 2020, 2023; Oliveira et al. 2020), a transitional area, bring the imminent need of long-term biological and abiotic monitoring to establish conservation guidelines across the Antarctic and sub-Antarctic zones, especially upon increasing tourism, global climate, and oceanographic changes. Pellizzari et al. (2017, 2023) reported new records of macroalgae for the SSI, and Oliveira et al. (2020) published a new complete list for Antarctica, a database for medium and long-term monitoring purposes, and reported a reduction in endemic species.

Arctic

Over the decades, the understanding of Arctic seaweed assemblages, mainly kelps, has gradually evolved from a one-vector ecosystem stemming from European refugia to many melding pathways, and from a biologically depauperate expanse into a complex genetic landscape. Arctic marine species are potentially born from selective pressures during Cenozoic global cooling and eventual ice conditions beginning in the Pleistocene. Arctic endemic diversity was likely additionally driven by repeated isolations into globally disparate refugia during glaciation. The Arctic is characterised by a turbulent climatic history and the prospect of further change. Repeated glaciations over the past 2.6 Ma had a lasting impact on biological communities, forcing populations to repeatedly contract and expand with the formation and retreat of ice sheets (Muller et al. 2009).

Laughinghouse et al. (2015) used phylogenetic and haplotype analyses to test the phylogenetic relationships and biogeography of *Fucus distichus*, suggesting that there is a distinct Arctic haplotype that is the source of subspecies in both the North Pacific and North Atlantic. *F. distichus* occupies a low tide zone habitat, and in Arctic/Subarctic regions, it is adapted to the severe stress of sea ice coverage. The Arctic is experiencing a rapid shift toward warmer regimes, calling for a need to understand levels of biodiversity and ecosystem responses to climate cycles. Currently, the warming in the Arctic is significantly exceeding the Northern Hemisphere average (Bringloe et al. 2020), and boreal and temperate regimes are expected to shift northwards (migrate) as a result, so many studies are trying to understand the responses of Arctic marine seaweed communities facing climate change (synopsis in Bringloe and Saunders 2019a, b). However, this is complex because it depends on a high level of marine biodiversity understanding, their recent history in Arctic ecosystems, and their potential for adaptation. Marine forests can provide insight into anticipated responses of Arctic marine communities to environmental changes. These ecosystems are structurally complex, ubiquitous worldwide, and provide valuable ecosystem and economic services.

Bringloe et al. (2020) aimed to work on the existing paradigm that marine Arctic seaweed ecosystems are depauperate extensions of the temperate assemblages established in the wake of recent glaciation, fundamentally changing how these systems should be viewed and interpreted. In an attempt to better understand, the authors worked on the hypotheses regarding the history of Arctic marine systems, particularly the endemism being an integral feature of Arctic biomes, and presented a firm framework for future evolutionary research in this system, typically viewed as ‘ecologically immature’. So, they presented genetic data (by DNA barcode surveys) for 109 Arctic seaweed species, revealing contiguous populations extending from the Bering Sea to the northwest Atlantic, with high levels of genetic diversity in the east Canadian Arctic. One-fifth of the species appeared to be restricted to Arctic waters. Further supported by hindcasted species distributions during the Last Glacial Maximum, the authors hypothesised that Arctic coastal systems were recolonised from many geographically disparate refugia, leading to enriched diversity levels in the east Canadian Arctic, with important contributions stemming from northerly refugia likely centred along southern Greenland. So, Bringloe et al. (2020) results suggested that the Arctic marine biomes persisted through cycles of glaciation, leading to unique assemblages in polar waters, rather than being entirely derived from southerly areas following glaciation. As such, Arctic marine species are potentially born from selective pressures during Cenozoic global cooling and eventual ice conditions beginning in the Pleistocene. Arctic endemic diversity, which is lower compared to Antarctica, was likely additionally driven by repeated isolations into globally disparate refugia during glaciation, highlighting the need to stock the Arctic marine biodiversity facing the current huge loss of perennial ice cover, and consequently the potential loss of diversity, and decline in ecosystem resilience.

Probably the marine Arctic environment was recolonised from numerous and globally distributed source populations, including unrecognised far northern refugia that additionally contribute to endemism in polar waters. Besides, probably the complex evolutionary processes are born out of the Arctic environment, particularly the interplay between incipient speciation and secondary contact (species-pump) (Bringloe and Saunders 2019a, b), and suggests Arctic marine benthic ecosystems are adapted to polar environments (evolution of Arctic assemblages throughout the Cenozoic and into the Pleistocene) rather than exclusively being cold-tolerant extensions of temperate flora/fauna. Amplification of warming trends (Muller et al. 2009) and the rapid decline of perennial ice cover has left an uncertain path forward for the Arctic biome. For the marine realm, the answer is not simple; while increased temperatures and the loss of multiyear sea ice will likely open new Arctic habitat, unstable coastlines due to permafrost melt and changes to water chemistry are likely to hamper potential gains in marine forests, with these impacts unevenly distributed throughout the Arctic Ocean (Filbee-Dexter et al. 2019). Net gains and losses of diversity in the Arctic, therefore, remains an area for further research. As such, taking stock of biodiversity remains among the first steps toward monitoring the resilience of Arctic marine ecosystems to ongoing environmental changes.

3.5 Threats to Antarctic and Arctic Seaweeds: Are We Looking Only for the Tip of the Iceberg?

3.5.1 Human Activities

Direct human activities, particularly over the last two centuries, as travel and trade have globalised and modified the distribution of biodiversity through pollution, habitat destruction and land-use change, over-exploitation and the accidental or intentional introduction of organisms. These environmental changes have almost universally negatively impacted native biodiversity. The study of pristine or non-pristine polar ecosystems as model systems now provides an opportunity for the examination of other indirect global environmental issues, particularly that of global climate change caused primarily by anthropogenic carbon emissions since the start of the Industrial Revolution, and the complex interplays of processes that contribute to or mitigate it. In this context, the oceans act as a key carbon sink and have removed ~20–30% of the CO_2 resulting from anthropogenic emissions, thereby buffering the consequential atmospheric warming (large as it still is), but at the same time leading to increased ocean acidification (Huovinen and Gómez 2020). Acidification can threaten marine organisms, causing reduced calcification rates in calcareous, skeleton- and or shell-forming species (IPCC 2022). Polar oceans are especially vulnerable to changes in temperature, salinity and pH due to several reasons described above, and that will be addressed in this final section.

The presence of numerous emerging organic contaminants (EOCs) and the remobilisation of legacy persistent organic pollutants (POPs) in polar regions have generated increasing concern amongst the scientific community, public groups, and stakeholders. Xie et al. (2022a, b) and Pedersen et al. (2022) review the occurrence of EOCs and POPs and their long-range environmental transport (LRET) processes via atmosphere and ocean currents from continental sources to the polar regions. Concentrations of various POPs have been monitored in the atmosphere at several Arctic stations and showed seasonal variations and declining trends. The same chemicals were also the major POPs reported in Antarctica, although their concentrations were lower than those in the Arctic, suggesting the importance of remoteness and the limited number of potential local sources in Antarctica. EOCs have been investigated in air, water, snow, ice and biota in the Arctic, but Antarctic data remain limited. Climate change may induce re-emission of legacy POPs and EOCs accumulated in glaciers, sea-ice and snow, altering available concentrations and amplifying their effects in polar regions. Globally, research is required to understand the various biogeochemical and geophysical processes involving anthropogenic pollutants under the combined drivers of climate change and changing anthropogenic pressures.

Antarctica

While human impacts are generally at a lower scale along Antarctica in comparison with the Arctic, there is growing evidence of anthropogenic activities, causing pollution, especially in the SSI and WAP regions (Tin et al. 2009; Convey 2020; Zakaria et al. 2021). Ivar do Sul et al. (2011) mentioned that we are looking only at the ‘tip of the iceberg’ in terms of marine conservation in SO, no matter how isolated Antarctica is.

In the SSI, Deception Island offers a rare opportunity to study the effects of environmental changes. As with other locations in the SSI, the island has one of the longest histories of human activity in Antarctica. This commenced in the 1820s with the initiation of sealing in Antarctica, and has since included exploration, whaling, second world war military station establishment and, since the mid-twentieth century, scientific research and now tourism (Pellizzari et al. 2020). The island was the location of one of only two whaling stations built on land in Antarctica (the other being on Signy Island in the South Orkney Islands) and, currently, hosts two national Antarctic summer-operating research stations (Argentina and Spain). The entire island is designated an 'Antarctic Specially Managed Area' (ASMA No. 4) in recognition of the multiple activities and operators that utilise it. Within the ASMA, 11 specific localities have been designated as sub-sites of Antarctic Specially Protected Area (ASPA No. 140). Given the intensity of human activity on and around the island, it is exceptionally vulnerable generally to human impacts and, specifically, to the impacts of introduced non-native species. The higher frequency of human visitation, mild climate (compared with other SSI islands), safe natural harbour, and the presence of geothermal activity may facilitate the establishment of alien marine and terrestrial species than any other Antarctic location and, indeed, the island already hosts the largest number of established terrestrial non-native species documented at any location in Antarctica (Hughes et al. 2015). In contrast, the island is the only location along SSI where CCA is not recorded, due to the natural acidification of the island associated with volcanic features (Pellizzari et al. 2017).

However, compared with other continents and given the overall scale of Antarctica, these impacts remain relatively low, even though some long-range transported airborne pollutants such as lead from fuel combustion, pesticides and microplastics have now been detected on the remote polar ice sheet close to the South Pole (Zakaria et al. 2021). Emerging contaminants, such as macro and microplastics (Barnes et al. 2010; Cincinelli et al. 2017), and even nanoplastics (Aves et al. 2022), have also been reported in studies from coastal regions in Antarctica. Given the concentration of both research and tourism operations in this part of Antarctica, the AP region is increasingly threatened by and subject to human impacts, which is evidenced by chemical contamination, including heavy metals (Chu et al. 2019) and polycyclic aromatic hydrocarbons (PAHs). Research stations and tourism are the main sources of locally sourced contamination in Antarctica. Fuel combustion, accidental oil spills, waste incineration and sewage disposal are amongst the primary sources of heavy metal contaminants in Antarctica, besides natural sources such as animal excrement, geological deposits, and volcanism.

Chu et al. (2019) considered the potential impacts of heavy metal pollution in Antarctica on algae, with a primary focus on the terrestrial environment. In both terrestrial and marine environments, adverse impacts on algae may affect organisms at the higher trophic levels, consequently disrupting food webs. Heavy metals can accumulate in algae and be biomagnified through the food chain, although robust toxicity testing protocols have yet to be applied to Antarctic seaweeds. More widely, macroalgae have been recognised as reliable metal-biomonitoring organisms due to their accumulation capacity and physiological responses. Lavergne et al. (2022) studied metal accumulation and photosynthetic parameters in 13 macroalgal species from areas with high human influence, close to national operator stations and at a control area on King George Island (SSI). Green seaweeds (mainly the foliaceous *Monostroma hariotti*) displayed higher metal contents (Al, Cu, Cr and Fe) close to stations. However, significant concentrations were not detected in brown macroalgae, although the reasons for this difference were not clear. The study did not detect a relationship between metal accumulation and photosynthetic performance, suggesting that, at least at present, there are low levels of anthropogenically-derived impacts associated with heavy metals in this area.

The fate of environmental contaminants can be drastically modified under changing climate and ozone depletion scenarios (Huovinen and Gómez 2020). Increased ice and snow melt can release chemical contaminants that have been stored for prolonged periods (Sulzberger et al. 2019). Even if PAH levels in the environment are still typically low in Antarctica, as reported on King George Island (Préndez et al. 2011), their potential for UV-induced or UV-enhanced toxicity under ozone depletion suggested an unaddressed risk may exist. UV radiation can also contribute to the degradation of plastic pollution, generating microplastics (Andrady et al. 2019), as well as interacting with other emerging contaminants such as antibiotics, hormones, and sunscreens. Even in the presence of Antarctica's natural barriers (the ACC and atmospheric circulation) and remoteness, the continent receives pollution such as aerosols and POPs from other regions through long-range atmospheric transport (LRAT) and oceanic transport (Ivar do Sul et al. 2011; Huovinen and Gómez 2020).

Arctic

As in the Antarctic, human activities in the Arctic are generally assumed to be modest, but it is now clear that they have greater influences on Arctic ecosystems than previously thought, further exacerbated by acting in synergy with climate change. Huntington et al. (2010) noted that the sources of most drivers impacting the Arctic lay beyond the region itself. In the case of persistent organic pollutants, Inuit and other Arctic indigenous populations pressed for global restrictions and were, in part, responsible for the negotiation and adoption of the Stockholm Convention (2001). Overfishing was the primary cause of the collapse of Atlantic cod (*Gadus morhua*) and other groundfish fisheries off eastern Canada in the early 1990s (Pedersen et al. 2017), a highly publicised crisis. In contrast, the collapse of the salmon (*Oncorhynchus* spp.) fishery in Alaska in the 1970s, which had, until then, harvested an average of more than 25 million salmon annually, became public only in the 1990s. Despite some conservation measures, in 1997, however, the salmon returns were far below the expected levels. Contrasting with the Antarctic, agricultural and human societal crises, oil exploration and several types of mining provide stark examples of major human impacts in the Arctic.

'Synergistic interactions', in which resource-uses amplified the biological (and human) consequences of climate variations, were apparent in the Arctic in the early twentieth century, for instance in Greenland's transition from a seal hunting to a cod fishing economy, or later, from cod to shrimp fishing, each time as alternatives to mitigate impacts (Huntington et al. 2010). A major obstacle to the development of Arctic resources is access, as most of the Arctic is too remote from supply sources and markets to justify expensive and challenging overland transportation. Access via sea or rivers is vital in different parts of the Arctic, while the presence of sea ice can both facilitate or provide a further barrier. The extent of summer sea ice, however, is currently decreasing rapidly. As sea ice declines, the Northern Sea Route along the Siberian coast, which is already being used, will become an increasingly viable ice-free route for ships to transit between the North Atlantic and North Pacific Oceans in the summer months (Humpert 2011). ACIA (2005) estimated that, between 2000 and 2050, the navigation season for the Arctic Ocean Route will increase from 70 days to 125 days of open water conditions. Increase in ship traffic will require investment in infrastructure, ports, and facilities for the transfer of cargo, supporting aviation, rescue and surveillance facilities. Sections of the North-West Passage, through the Canadian Archipelago, are also likely to become extensively ice-free, allowing greater access to most of the Canadian Arctic, a feature that is already being taken advantage of by the 'expedition cruise' industry. Development of economic resources in both the Siberian and Canadian Arctic, with the transition to ice-free summer seas, will become more attractive, raising prospects for population and economic growth, but with considerably increased threat of environmental impacts.

Permafrost degradation, itself a serious consequence of climate change, can also be further catalysed in synergy with local human activities such as road construction or an act as simple as disturbance of vegetation cover by driving a vehicle across the tundra. Most roads are built on gravel to insulate the permafrost beneath and prevent it from thawing, resulting in a physical barrier that can cause snow accumulation, water damming and flooding. Pooling water and increased absorption of solar radiation may then initiate thermokarst formation (subsidence of the surface following melting of buried ice) in the adjacent permafrost, changing the dominant vegetation and including impacts on the marine flora.

In various locations around the Arctic, edible seaweeds are traditionally exploited, including by native peoples such as the Inuits, both from natural beds and by direct cultivation. These activities primarily utilise *Porphyra* (nori), *Palmaria* (ribbon seaweed), *Nereocystis* (bull kelp), *Macrocystis* (giant kelp) and *Saccharina latissima* (sugar kelp). In Bristol Bay (Bering Sea) and elsewhere in sub-Arctic southern Alaska, *Fucus laden* eaten with herring eggs is a treasured spring food (https://aseaweed.com). *S. latissimi*, also found in Arctic waters around Greenland, Iceland, Alaska and Canada, is commonly used in culinary purposes. Diehl and Bischof (2021) studied mechanisms of acclimation in this seaweed, *S. latissima,* on Spitsbergen (Svalbard), suggesting that, in line with its broad latitudinal distribution range and adaptability, *S. latissima* was highly resilient to changing abiotic drivers and its growth will likely be promoted by warming in the future Arctic. *Alaria esculenta* ('winged kelp') is also widely distributed in Arctic waters, growing naturally in Svalbard and Iceland, as well as cold-temperate northern Europe and Atlantic North America, and being cultivated in Greenland, the Canadian Maritime provinces and Arctic Labrador (https://www.oceanrainforest.com/saccharina-latissima). In 2024, Iceland held the Arctic Algae Conference, focusing on 'Economics of micro- and macroalgae in the surroundings of North Pole'.

Greenland kelp species (e.g. *Agarum clathratum, A. esculenta, Hedohyllum nigripes, S. latissima*, and the fucoids *Ascophyllum nodosum, Fucus distichus* and *F. vesiculosus*) constitute an Arctic marine resource, as well as being important monitoring organisms in connection with arctic mining activities and other environmental changes. Wegeberg et al. (2023) studied the natural element content of kelps and their potential seasonal variation, aiming to design a monitoring program and plan harvests for the highest biomass quality and value. Their ICP-MS results showed dominance of potassium, sodium, calcium, magnesium and phosphorus in seaweed thalli, with the main trace elements being iron, aluminium and arsenic. Individual element concentrations were species-specific, with higher concentrations during winter. The contents of the compounds laminarin and mannitol were highest during late summer, the best season for optimal harvest yield of those compounds.

In a biomonitoring study, Obluchinskaya and Zakharova (2021) studied the metal concentrations in three species of *Fucus* on the coast of the Barents Sea, in locations with different salinity. The levels of Cu, Cd and Pb were much higher than the background levels at most stations for all algae, exceeding the background levels by 5–6, 1.5–2.5 and 3–5 times, respectively. They also noted that salinity level and interaction between site and season combined had the strongest influence on metal concentrations in the seaweeds. Metal concentrations in algae were also well correlated with those in water samples.

The High Arctic is one of the world's fastest-warming biomes, one consequence of which is allowing access to previously inaccessible land for resource extraction. Thus, scenarios of environmental change in the Arctic are susceptible to manipulation and mitigation attempts by humans (Huntington et al. 2010). This is particularly pertinent in the contexts of transitions and acceleration points in both environmental change and human activity.

Changes in thermohaline circulations and nutrient inputs, primarily from anthropogenic activities, can occur particularly rapidly in the Arctic. Reduced Arctic Sea ice cover is likely to increase solar radiation reaching the coastal seafloor, facilitating the expansion of kelp forests and seagrass meadows (Krause-Jensen and Duarte 2014). These new and productive ecosystems will support habitat development, food production, and contribute to enhanced CO_2 sequestration and protect the shoreline from erosion, thereby providing key ecosystem services. This train of events would provide a solid underpinning to guide forecasts on the spread of marine macrophytes into the Arctic driven by climate change and contribute to a balanced understanding of the consequences of anthropogenic impacts on polar ecosystems through a focus on the role of macroalgae as engineering species. Anticipating such changes in ecosystem structure and function is key to developing management and mitigation strategies to maximise their ecosystem services in a future warmer planet.

3.5.2 *Extreme Meteorological and Oceanographic Events*

Polar seaweed assemblages are clearly changing. Shifts in macroalgae richness and distribution have repeatedly occurred during previous geological periods of ocean temperature variation and, in that sense, have been a natural phenomenon over time (Bykova et al. 2020), although contemporary global changes of all types are argued to be exceptional both because of their scale over extremely short time periods, and the single species anthropogenic cause. Current studies emphasise temperature as a key determinant in the distribution of macroalgae, which inherently rely on oceanographic dispersal processes (Wiencke et al. 2014; Piñeiro-Corbeira et al. 2018; Pellizzari et al. 2020). As the ocean warms, the frequency, duration, intensity and range of marine heatwaves (MHW) increase (see also Sect. 3.3.4). The polar zones are now experiencing climatic and oceanographic extremes, including MHW, where high temperatures and large anomalies with respect to historical averages have been recorded, impacting marine fauna and phycoflora (Thomsen et al. 2019, 2021; Krause-Jensen et al. 2020; Siegert et al. 2023, Pellizzari et al. 2025). In Arctic and Antarctic communities, the combination of multiple environmental stressors such as sea surface temperature (SST), salinity, pH, nutrient availability, irradiance and sea ice dynamics can act antagonistically and/or synergistically, driving the composition of assemblages (Wiencke and Amsler 2012; Convey and Peck 2019) that have been profoundly affected by climate change in recent decades (Meredith et al. 2019).

There is increasing knowledge of 'extreme events' amongst the range of variables changing through climate change (Siegert et al. 2023). A pertinent example of this is given by a major event that impacted parts of continental Antarctica in 2022. According to the Climate Data Guide (https://climatedataguide.ucar.edu), Concordia Station (approx. 3000 m altitude) recorded a temperature of −12.2 °C in March 2022, while −28 °C is the long-running average for this season. No climatic event of this magnitude has previously been reported. At the same time, Vostok Station, the world's record of lower temperature, reported −17.7 °C, when the average would be −47.7 °C.

In the same period, in the Arctic, recent average monthly winter temperatures have reached 3.3 °C above normal and, in certain areas, close to the North Pole, instantaneous temperatures rose 30 °C above normal. It is unusual for heatwaves to be recorded at both poles simultaneously, even though these were separate events. At the same time, Svalbard recorded a new instantaneous maximum temperature for that date of 3.9 °C. Sea ice was also very low at this time, and on land, there was record rainfall across Greenland (https://nsidc.org/home). Thomsen et al. (2021) suggest that increasing frequency of heatwaves will possibly accelerate earlier sea-ice break-up, resulting in severe impacts on marine flora (including macroalgal) and faunal communities. It has also been suggested that unexpected MHWs may be related to the recent and strong decrease in sea-ice extent in Antarctica (González-Herrero et al. 2022). In the Arctic, modelling studies suggest that large-scale climate patterns are becoming more variable, suggesting that the unpredictable occurrence of heatwaves will provide one of the main challenges in the uncertain future ahead (https://www.ipcc.ch/ar6-syr/).

Approximately 22,000 years ago, during the Last Glacial Maximum, northern North America, Europe and Asia were covered by thick glaciers and experienced freezing temperatures for most or all of the year. Gradually, warming then occurred, and 5000 years ago temperatures were warmer than today. Over the same 17,000-year period, global temperature rose by an estimated 6 °C. In contrast, contemporary Antarctic warming is taking place about 25 times more rapidly (https://antarctic.org.au/).

Understanding how key marine species respond to short-term MHWs is urgent for predicting damage to coastal ecosystems. Nauer et al. (2022) studied *Hypnea pseudomusciformis* (Rhodophyta) under in vitro conditions, sampling in different floristic provinces along the South Atlantic coastline (tropical, transition and warm-temperate), evaluating the effects of simulated MHWs by measuring growth rates, pigment contents and chlorophyll fluorescence. Perturbation levels were assessed at average intensity heatwave (Δ +2 °C), maximum intensity HW (Δ +4 °C) and extreme intensity HW (Δ +6 °C). The populations exhibited distinct tolerance and physiological responses to the different HW scenarios, where the tropical and transition samples were affected by Δ + 4 °C and + 6 °C scenarios, while only the warm-temperate samples recovered from all temperature tested, indicating that tropical and transition species may be at greater risk of local extinction. Such data are essential for developing mitigation proposals under the predicted increased frequencies of MHW.

Gao et al. (2021) studied the impacts of MHWs on algal community structure and carbon sequestration in conjunction with ocean warming and acidification, providing evidence that MHWs cause a decline in large canopy macroalgae and increase the formation and abundance of algal turfs, with a consequential decrease in carbon sequestration by kelps. Furthermore, MHW increased grazing pressure on kelps due to tropicalisation, facilitating the expansion of turf formation that prevents kelp recovery through monopolising light and space. Other experiments performed using calcareous and non-calcareous algae indicate that responses to acidification depend on the taxonomic status, biogeographic location, prevailing metabolism of species, type of experimental approach and time of responses (Hofmann and Bischof 2014). Overall, interactions between ocean acidification (OA), climate change and UV radiation have been found to negatively impact calcifying seaweed species by reducing structural UV protection or affecting photosynthesis (Huovinen and Gómez 2020). These species possess lower buffering capacity, particularly in systems affected by freshening due to glacier melt, where acidification can be exacerbated, resulting in an increase in calcium carbonate corrosion (Jewett and Romanou 2017). Recently, Fernández et al. (2024) studied inorganic carbon (Ci) uptake strategies and CO_2 concentrating mechanisms (CCM), revealing species-specific responses and their implications for OA along the WAP. Among red seaweeds, the majority relied solely on CO_2 as an exogenous Ci source, with a high percentage of non-CCM species. Green seaweeds exhibited depth-dependent variations in CCM status, with the proportion of non-CCM species increasing at greater depths. In contrast, brown seaweeds exhibited a higher prevalence of CCM species, even in deep waters, indicating the use of CO_2 and HCO_{3-}. The authors concluded that OA may potentially increase the abundance of non-CCM species. Also, Tait et al. (2024) studied deep-living Antarctic seaweeds from depths >70 m (such as *H. grandifolius* and *P. decipiens*) and specimens of crustose coralline algae (CCA) as deep as 125 m, mentioning these assemblages as potentially important contributors to global carbon fixation and sequestration. The authors used global light modelling and published photosynthetic rates, and estimated values between 0.9% and 2.8% of global macroalgal C fixation, higher rates than previously reported.

Ross et al. (2023) considered the potential role of seaweeds in climate change mitigation, highlighting that macroalgal carbon accounting is yet to be fully elucidated. However, the group has the potential to contribute to climate change mitigation through (i) expanding sustainable nearshore seaweed cultivations, and (ii) offsetting industrial CO_2 emissions using seaweed products for emission abatement. Despite uncertainties about the quantification of the net impact of carbon export from seaweed restoration and seaweed farming on atmospheric CO_2, evidence suggests that nearshore farming contributes to carbon storage in sediments below the farm sites, but how scalable this is for sequestering significant amounts of carbon remains unclear.

3.5.3 *Reevaluating Connection, Eco-Regions, and Endemism in Antarctica*

With 90% of the globe's ice cover, Antarctica is considered the primary regulator of the planet's temperature. Therefore, recent thermal records, heatwaves and other severe events in the SO suggest that the rest of the planet, particularly South America and Oceania, will also face an increase in extreme events. The SO is experiencing increased warming, declining sea ice, acidification, increased UV, changes in upwelling and nutrient replenishment, lower salinity and the southwards migration of ocean fronts (Deppeler and Davidson 2017). The diverse environments within the vast SO will be impaired differentially by climate change, while the magnitude of abiotic factors driving biotic change will vary among bioregions. Predicting the net effect of multiple climate-induced stressors over a range of environments is complex, but it has already been reported that predicted changes to seaweed communities will affect SO biogeochemistry, carbon export and nutrient transfer to higher trophic levels.

The AP region has faced accelerated warming and multiple other facets of environmental change since the mid-twentieth century (Turner et al. 2009; Dubrasquet et al. 2018; Chown et al. 2022a, b). Some groups of macroalgae respond quickly to these changes in climatic and oceanographic patterns, altering the biogeography of ecologically important species (Wernberg et al. 2018). The impacts on Antarctic seaweed assemblages include (i) shifts in richness, diversity and distributions caused by changes in dispersal patterns, including species with invasive potential, (ii) changes in primary productivity and biogeochemical fluxes, (iii) changes in the gene pool with consequences for phenotypic plasticity and adaptive responses, and (iv) changes in the intensity and direction of biological interactions (Pellizzari et al. 2020).

With current global climatic and oceanographic changes, it is predicted that many organisms may have potential for poleward migration (and upward altitudinal migration on land (Walther et al. 2002; Fraser et al. 2020). Consequently, these existing connectivity pathways between southern regions may provide new entry routes for invasive species. Invasive alien (anthropogenically transported) or naturally dispersing non-native species represent a major threat to Antarctic biota. Understanding how permeable the SO 'barrier' is to dispersal of different organisms is a prerequisite for understanding how vulnerable Antarctic biodiversity might be to future colonisation events.

Considering the WAP sector, analyses of macroalgal assemblages over recent decades have revealed significant changes in cover, biomass, and species richness (Ducklow et al. 2013; Ko et al. 2020; Amsler et al. 2023b). The WAP and SSI are hotspots for glacial retreat, and the projected reduction in sea-ice extent and duration leads to the expectation of a substantial increase in richness and biomass in the central WAP, facilitating the expansion of these macroalgae to newly exposed areas (Deregibus et al. 2023). These authors studied the potential macroalgal expansion and blue carbon gains due to glacial retreat in Potter Cove, concluding that seaweeds are colonising the newly ice-free areas. They also estimated the potential for new carbon sinks and export at this location alone of ~0.2–0.4 tonne y^{-1}. Amsler et al. (2023b) reported southern algal communities in the WAP becoming similar to those present in northern areas.

Naughten et al. (2023) highlighted that ocean-driven melting of floating ice-shelves in the Amundsen Sea is currently the main process controlling Antarctica's contribution to sea level rise. Using a regional ocean model, they presented a projection of rapid ocean warming at around triple the historical rate (Pre-warming period). When internal climate variability was considered, no difference was observed between mid-range emissions scenarios and the most ambitious targets of the Paris Agreement, suggesting that the mitigation of greenhouse gas release has limited power to prevent ocean warming that could lead to the collapse of the West Antarctic Ice Sheet (WAIS). This would result in a sea level rise of ~5 m if lost completely. Previous studies (Roleda and Dethleff 2011) have suggested that the WAIS may be doomed to collapse over the course of centuries, but Naughten et al. (2023) show that even drastic emissions cuts in the coming decades will not slow the melting, and even if the world meets the most ambitious Paris Agreement target of keeping global heating below 1.5 °C above pre-industrial levels. In this scenario, billions of people living in coastal regions will be vulnerable to the impacts of SLR.

The future of Antarctic seaweed assemblages facing climate change in a complex reality, characterised by regional nuances, microhabitats, and species-specific responses to environmental stressors, is highly uncertain (Filbee-Dexter et al. 2018). However, understanding the influence of these changes on the biogeographical distributions of macroalgae and the impact they may have on marine life is essential. Macroalgae in Antarctica are sentinels for environmental changes, indicating abrupt shifts due to their reproductive cycles and physiology highly adapted to polar conditions. Furthermore, opportunistic macroalgae can disperse rapidly and establish in new regions under suitable conditions (Pellizzari et al. 2020, 2023). Robust and ongoing monitoring in priority impact areas is a crucial tool for polar conservation management, as well as to improve predictions of global climate change consequences and other anthropogenically-caused severe impacts.

Arctic

The Arctic has 35% of the global coastlines, supporting highly productive marine forests of macroalgae (Assis et al. 2022). Goldsmith et al. (2021) consider that the Arctic is the epicentre of the global climate crisis, facing ocean warming, an increase in precipitation and freshwater discharge, as well as lower salinity, sea-ice decline, changes in underwater light, amplification of ocean acidification and changing stratification and circulation dynamics that modify benthic habitats and their communities. Warming in the AO and surrounding regions is at least twice as rapid as the global average (Timmermans and Marshall 2020), and, without drastic and immediate action to reduce this, some areas are now almost certain to be ice-free in summer before 2050.

Recent climate change in the Arctic has been well documented in the Arctic Climate Impact Assessment (ACIA, see also Filbee-Dexter et al. 2019; Pedersen et al. 2022). However, there remain critical knowledge gaps regarding the current state of its biota. These gaps must be filled before substantial changes in assemblages occur. Kelp forests, comprising large brown macroalgae along rocky coastlines, play a crucial ecological role in the local marine ecosystem and are already experiencing the impacts of climate change (Filbee-Dexter et al. 2019). These areas can be particularly sensitive to changes in environmental conditions (e.g. temperature increase, pollution, sedimentation), suffering abrupt changes especially during early life stages (Monserrat et al. 2022). The Canadian Arctic has been highlighted as lacking historical baseline data for kelp assemblages (Filbee-Dexter et al. 2019; Bringloe et al. 2020). However, data from the Svalbard archipelago indicate significant changes in seaweed assemblages, including borealisation, with increasing SST being the main factor driving changes in species composition (Malavenda 2021).

Recently, Filbee-Dexter et al. (2019, 2022) compiled information on the distribution and diversity of Canadian Arctic kelp forests, demonstrating that in some regions their growth and extent are rapidly increasing due to reduced sea-ice cover, increased light and warming. The possible expansion of kelp beds should provide new habitats for fish and other marine organisms, increase carbon storage and alter a suite of other ecosystem services along Arctic coastlines. However, in other coastal areas, these ecosystems could be negatively impacted by increasing freshwater input and coastal erosion due to receding sea ice and melting permafrost. Krause-Jensen et al. (2020) studied the effects of climate change on distribution changes of pan-Arctic seagrasses and seaweeds beds, while Díaz et al. (2021) studied the effects of macroalgal beach bloom events from Arctic intertidal zones and Goldsmith et al. (2021) studied the global decline of kelp forests These studies highlighted the discrepancies in impacts identified depending on the studied area and, hence, in the need for appropriate management actions for these natural resources.

Changes in food-web structure and the carbon cycle are already occurring and are reshaping Arctic benthic ecosystems. However, many factors are not well understood, making it difficult to predict the future trajectory of the coastal Arctic ecosystem structure. Lebrun et al. (2022) review short- and long-term changes in environmental features and macroalgae assemblages from shallow hard-bottom

Arctic coasts, considering seasonality, functional diversity, distribution, life cycle and adaptations, as well as the responses of biological and ecological processes to climate change. The study also considered 'winners' and 'losers', and the effects on ecosystem functioning of fucoid species, kelps and coralline algae, which are key ecosystem engineers in these areas. Fucoid and several kelp species are expected to increase in dominance, leading to possible local extinctions of native species.

In temperate-polar regions, native kelps face challenges from the arrival of invasive species and from intense grazing by sea urchins (Wernberg et al. 2018), with reductions in their abundance attributed to the impacts of rising temperatures also documented (Filbee-Dexter et al. 2018). Gordillo et al. (2022) reported that the increase in temperature during the polar night may also restrict the occurrence of perennial species. However, paradoxically, a warmer Arctic with reduced ice cover may favour the expansion of macroalgae on a pan-Arctic scale. While some warmer populations found at distributional edges have recently been reported to decline, expansions of species have been recorded at their cold distribution edges. Reduced population resilience can contribute to kelp habitat loss; hence, understanding intra-specific variations in physiological responses across a species' latitudinal distribution is crucial for conservation.

Diehl et al. (2021) investigated the potential local physiological responses of the broadly distributed Northern Hemisphere kelp, *Saccharina latissima,* to marine heatwaves (MHW) in summer, collecting sporophytes in populations exposed to the coldest (N) and warmest (S) local temperature regimes. Growth was highest in northern populations and lowest in southern populations. Overall, northern populations were largely unaffected by increasing summer temperatures up to 1.6 °C. Conversely, sporophytes from the south were markedly stressed at 1.6 °C, experiencing tissue necrosis and a significantly elevated de-epoxidation state of the xanthophyll cycle. Some models suggest poleward migration in response to warming based on knowledge of temperature tolerance in combination with predicted future isotherm migration (Jueterbock et al. 2013), while scenarios for *Ascophyllum nodosus* distribution suggest retreat of some populations even along the species' northern distribution limit. However, Marbà et al. (2017) suggested that warming would lead to an overall expansion of the northern-most populations of *A. nodosum*, and that future distribution boundaries and the speed of northward migration will depend on the dispersal capacity of the species. The fate of kelp forests in this region remains a critical gap in our understanding.

Arctic calcified algal distributions will be constrained by the upper temperature limit in the south and by calcium carbonate saturation in the north of their distributional ranges (Chan et al. 2020). Monserrat et al. (2022) tested responses to the interactive effects of ocean warming (OW) and acidification (OA) on survival of the crustose coralline alga, *Neogoniolithon brassica-florida*, which is associated with the later successional species forest-forming *Cystoseira compressa*. The density of recruits was lower under a combination of OW and OA, while individual size was negatively affected by temperature increase but positively influenced by low pH, also suggesting that the interaction between climate change and the presence of crustose coralline algae may have a negative impact on the recruitment of *Cystoseira*.

Shifts in Arctic seaweed species distribution, abundance and performance are therefore predicted (Wiencke and Amsler 2012). Furthermore, warming and increased advection are facilitating poleward migration, bringing thermally tolerant boreal-sub-Arctic species that can compete with local species adapted to cold waters (Assis et al. 2022). These changes are expected to alter carbon cycling as well as the timing and magnitude of benthic primary production. At the same time, recently available biotechnologies have opened the door to marine forest restoration, highlighting that the interactions of multiple drivers and species interactions must be considered to achieve long-term population sustainability. The AO is considered a sentinel region for understanding the magnitude of global changes and their impacts on ecosystems. Prioritising research in areas of rapid environmental change will enable more effective management of these habitats, and better predictions of future changes in the coastal ecosystems they support, as well as the ecosystem services they provide.

3.6 Conclusion

Patterns of climate change and extreme events occurring at both poles are distinct, but in both, anomalies and severe climate events are becoming increasingly common and unpredictable. The polar regions are strategically vital elements in ocean-atmosphere exchange and the distribution of heat and cold across the globe.

Changes in benthic algal distributions, seasonality, functional diversity and their contribution to the carbon cycle, as well as shifts in polar food-web structure, are already occurring and are reshaping Antarctic and Arctic coastal ecosystems. Some opportunistic and or invasive species, such as some green seaweeds in Antarctica, and fucoid and kelp species in the Arctic, are expected to increase in dominance in some areas, with possible local extinctions of native, including endemic, species. Compared to the levels observed in Antarctica, few macroalgal taxa are considered endemic in the Arctic.

It is often assumed that biogeographic barriers such as the ACC protect Antarctica and the SO from colonisers originating from the north. However, it is now realised that this barrier is not, and never has been, as rigid as many have assumed. Current climatic and oceanographic changes are enabling parts of Antarctica and the SO to be more connected to other regions and susceptible to colonisation by or introduction of currently non-native species.

In the Arctic, kelps are foundation species in coastal ecosystems and are currently experiencing shifts in their distribution patterns caused by ocean warming and changes in ice distribution. The impacts of global changes are clear in richness shifts in coastal macroalgal beds, in seasonal development, and in interactions between stressors and across species, all of which can be positive or negative. Arctic borealisation is expected to increase biodiversity at least in the short term (Wiktor et al. 2022), including new species interactions and changes in ecosystem functioning.

Studies indicate that algal regulatory mechanisms acting against free radicals in polar zones are extremely efficient under controlled experimental conditions, considering increases in temperature and UV radiation as the main factors (Obluchinskaya et al. 2022). Although the seaweed assemblages are clearly being affected by climate change, this suggests that some polar species may act as sentinels or bioindicators of environmental changes and comparative models for studies aiming to enhance stress tolerance. The use of molecular tools, transcriptomic analyses and metabolomics will also be fundamental in elucidating adaptive strategies underlying acclimatisation. It will be crucial to consider the impact of multiple stressors, including increased temperature, UV radiation, ocean acidification and decreased salinity, resulting from glacial and sea-ice melt. Research into the synergistic effects of these variables is essential for predicting the future of polar ecosystems.

Commitment to long-term monitoring and multifactorial studies of polar seaweed communities that consider abiotic interactions between stressors and biotic interactions between species is vital. To confidently identify the presence and role of biogeographic connections, molecular investigations of cryptic groups are required. Functional diversity and ecosystem services could be impaired in the polar marine ecosystems, impacting the processing of land-derived nutrients and organic matter, thereby significantly altering trophic structure and 'bottom-up' energy flow. Various factors are still poorly understood, making it difficult to understand the current situation and predict the future of seaweeds in Antarctic and Arctic coastal ecosystems.

References

ACIA. Climate Impact Assessment (2005) Arctic. Cambridge University Press, Cambridge, p 1042

Adams WW, Zarter CR, Mueh KE, Amiard V, Demmig-Adams B. 2008 Energy dissipation and photoinhibition: a continuum of photoprotection. In: Demmig-Adams, B., Adams, W.W. III, Mattoo, A.K. (eds) Photoprotection, photoinhibition, gene regulation, and environment. Advances in photosynthesis and respiration. Springer, Dordrecht. 21, pp 49–64. https://doi.org/10.1007/1-4020-3579-9_5

Adey WH, Lindstrom SC, Hommersand MH, Müller KM (2008) The biogeographic origin of Arctic endemic seaweeds: a thermogeographic view. J Phycol 44(6):1384–1394. https://doi.org/10.1111/j.1529-8817.2008.00605.x. PMID: 27039853

Aguilera J, Bischof K, Karsten U, Hanelt D, Wiencke C (2002) Seasonal variation in ecophysiological patterns in macroalgae from an Arctic fjord. II. Pigment accumulation and biochemical defence systems against high light stress. Mar Biol 140:1087–1095. https://doi.org/10.1007/S00227-002-0792-Y

Aminina NM, Karaulova EP, Vishnevskaya TI, Yakush EV, Kim YK, Nam KH, Son KT (2020) Characteristics of polyphenolic content in brown algae of the Pacific Coast of Russia. Molecules 25:3909. https://doi.org/10.3390/molecules25173909

Amsler CD (2008) In: Amsler CD (ed) Algal chemical ecology. Springer, Berlin/Heidelberg

Amsler CD, Amsler MO, Klein AG, Galloway AWE, Iken K, McClintock JB, Heiser S, Lowe AT, Schram JB, Whippo R (2023a) Strong correlations of sea ice cover with macroalgal cover along the Antarctic Peninsula: ramifications for present and future benthic communities. Elem Sci Anth 11:00020

Amsler CD, Amsler MO, Heiser S, JB MC, Iken K, Galloway AWE, Klein (2023b) Vertical distribution of brown and red macroalgae along the central Western Antarctic Peninsula. Bot Mar. https://doi.org/10.1515/bot-2023-0085

Andrady AL, Pandey KK, Heikkilä AM (2019) Interactive effects of solar UV radiation and climate change on material damage. Photochem Photobiol Sci 18(3):804–825. https://doi.org/10.1039/c8pp90065e

Archambault P et al (2010) From sea to sea: Canada's three oceans of biodiversity. PLoS One 5(8):e12182. https://doi.org/10.1371/journal.pone.0012182

Assis J, Serrão EA, Duarte CM, Fragkopoulou E, Krause-Jensen D (2022) Major expansion of marine forests in a warmer Arctic. Front Mar Sci 9:850368. https://doi.org/10.3389/fmars.2022.850368

Aves AR, Revell LE, Gaw S, Ruffell H, Schuddeboom A, Wotherspoon NE, LaRue M, McDonald AJ (2022) First evidence of microplastics in Antarctic snow. Cryosphere 16:2127–2145. https://doi.org/10.5194/tc-16-2127-2022

Barnes DKA, Hodgson DA, Convey P, Allen C, Clarke A (2006) Incursion and excursion of Antarctic biota: past, present and future. Glob Ecol Biogeogr 15:121–142

Barnes DK, Walters A, Gonçalves L (2010) Macroplastics at sea around Antarctica. Mar Environ Res 70:250–252

Bartsch I, Paar M, Fredriksen S, Schwanitz M, Daniel C, Hop H, Wiencke C (2016) Changes in kelp forest biomass and depth distribution in Kongsfjorden, Svalbard, between 1996–1998 and 2012–2014 reflect Arctic warming. Polar Biol 50. https://doi.org/10.1007/s00300-015-1870-1

Becker S, Walter B, Bischof K (2009) Freezing tolerance and photosynthetic performance of polar seaweeds at low temperatures. Bot Mar 52:609–616. https://doi.org/10.1515/bot.2009.079

Bello R, Higuchi K (2019) Changing surface radiation and energy budgets of the Hudson Bay Complex using the North American regional reanalysis (NARR) model. Arctic Sci 5(4):218–239. https://doi.org/10.1139/as-2018-0034

Bensi M, Kovacevic V, Donda F, O'Brien FE, Armbrecht L, Armand LK (2022) Water masses distribution offshore the Sabrina Coast (East Antarctica). Earth Syst Sci Data 14(65–78). https://doi.org/10.5194/essd-14-65-2022

Bischof B, Wiencke C (1993) Temperature requirements for growth and survival of macroalgae from Disko Island (Greenland). Helgoländer Meeresuntersuchungen 47:167–191. https://doi.org/10.1007/BF02430357

Bischof B, Wiencke C (1995) Temperature ecotypes and biogeography of Acrosiphoniales (Chlorophyta) with Arctic-Antarctic disjunct and Arctic/cold-temperature distributions. Eur J Phycol 30:19–27. https://doi.org/10.1080/09670269500650771

Bischof K, Gómez I, Molis M, Hanelt D, Karsten U, Lüder U, Roleda MY, Zacher K, Wiencke C (2006) Ultraviolet radiation shapes seaweed communities. In: Life in extreme environments. Springer, Dordrecht, pp 187–212. https://doi.org/10.1007/978-1-4020-6285-8_12

Bluhm BA, Gebruk AV, Gradinger R, Hopcroft RR, Huettmann F et al (2011) Arctic marine biodiversity: an update of species richness and examples of biodiversity change. Oceanography 24:232–248. https://doi.org/10.5670/oceanog.2011.75

Bolton JJ (2010) The biogeography of kelps (Laminariales, Phaeophyceae): a global analysis with new insights from recent advances in molecular phylogenetics. Helgol Mar Res 2010(64):263–279. https://doi.org/10.1007/s10152-010-0211-6

Bonsell C, Dulton KH (2018) Long-term patterns of benthic irradiance and kelp production in the central Beaufort Sea reveal implications of warming for Arctic inner shelves. Prog Oceanogr 162. https://doi.org/10.1016/j.pocean.2018.02.016

Borum J, Pedersen M, Krause-Jensen D, Christensen P, Nielsen K (2002) Biomass, photosynthesis and growth of Laminaria saccharina in a high-arctic fjord, NE Greenland. Mar Biol 141:11–19. https://doi.org/10.1007/s00227-002-0806-9

Breeman A (1988) Relative importance of temperatures and other factors in determining geographic boundaries of seaweeds: experimental and phenological evidence. Helgolander Meeresunters 42:199241

Brey L (2009) Acclimation of kelp photosynthesis to seasonal changes in the underwater radiation regime of an Arctic fjord system. Dissertation, University of Bremen. http://nbn-resolving.de/urn:nbn:de46-diss000115247

Bringloe TT, Saunders GW (2018) Mitochondrial DNA sequence data reveal the origins of post-glacial marine macroalgal flora in the Northwest Atlantic. Mar Ecol Prog Ser 589:45–58. https://doi.org/10.3354/meps12496

Bringloe TT, Saunders GW (2019a) Trans-arctic speciation of Florideophyceae (Rhodophyta) since the opening of the Bering Strait, with consideration of the "species pump" hypothesis. J Biogeogr 46:694–705. https://doi.org/10.1111/jbi.13504

Bringloe TT, Saunders GW (2019b) DNA barcoding of the marine macroalgae from Nome, Alaska (Northern Bering Sea) reveals many trans-Arctic species. Polar Biol 42:851–864. https://doi.org/10.1007/s00300-019-02478-4

Bringloe TT, Dunton KH, Saunders GW (2017) Updates to the marine algal flora of the Boulder Patch in the Beaufort Sea off northern Alaska as revealed by DNA barcoding. Arctic 70:3–6. https://doi.org/10.14430/arctic4679

Bringloe TT, Verbruggen H, Saunders GW (2020) Unique biodiversity in Arctic marine forests is shaped by diverse recolonization pathways and far northern glacial refugia. Proc Natl Acad Sci USA 117:22590–22596. https://doi.org/10.1073/pnas.2002753117

Bustamante D, Calderón M, Mansilla A (2022) Molecular analyses reveal a new species of Palmariaceae from Subantarctic Chile: *Devaleraea yagan* sp. nov. (Palmariales, Rhodophyta). Phycologia 61:3. https://doi.org/10.1080/00318884.2022.2045080

Bykova N et al (2020) Seaweeds through time: morphological and ecological analysis of Proterozoic and early Paleozoic benthic macroalgae. Precambrian Res 350:105875

Campana GL, Zacher K, Fricke A, Molis M, Wulff A, Quartino LM, Wiencke C (2009) Drivers of colonization and succession in polar benthic macro- and microalgal communities. Bot Mar 52:655–667. https://doi.org/10.1515/BOT.2009.076

Carmack EC, Yamamoto-Kawai M, Haine TW, Bacon S, Bluhm BA et al (2016) Freshwater and its role in the Arctic Marine System: Sources, disposition, storage, export, and physical and biogeochemical consequences in the Arctic and global oceans. J Geophys Res Biogeo 121:675–717. https://doi.org/10.1002/2015JG003140

Chan PTW, Halfar J, Adey WH, Lebednik PA, Steneck R, Norley CJD, Holdsworth DW (2020) Recent density decline in wild-collected subarctic crustose coralline algae reveals climate change signature. Geology 48(3):226–230. https://doi.org/10.1130/G46804.1

Chown SL, Leihy RI, Naish TR, Brooks CM, Convey P, Henley BJ, Mackintosh AN, Phillips LM, Kennicutt MC, Grant SM (2022a) Antarctic climate change and the environment: a decadal synopsis and recommendations for action. Scientific Committee on Antarctic Research, Cambridge, 109 pp.

Chown SL, Leihy RI, Naish TR, Brooks CM, Convey P et al (2022b) Antarctic climate change and the environment: a decadal synopsis and recommendations for action. Scientific Committee on Antarctic Research, Cambridge, 109 pp.

Chu WL, Dang NL, Kok Y, Ivan Yap KS, Phang SM, Convey P (2019) Heavy metal pollution in Antarctica and its potential impacts on algae. Pol Sci 20:75–83. https://doi.org/10.1016/j.polar.2018.10.004

Cincinelli A, Scopetani C, Chelazzi D, Lombardini E, Martellini T, Katsoyiannis A, Fossi MC, Corsolini S (2017) Microplastic in the surface waters of the Ross Sea (Antarctica): occurrence, distribution and characterization by FTIR. Chemosphere 175:391–400

Clark GF, Marzinelli EM, Fogwill CJ, Turney CSM, Johnston EL (2015) Effects of sea-ice cover on marine benthic communities: a natural experiment in Commonwealth Bay, East Antarctica. Polar Biol. https://doi.org/10.1007/s00300-015-1688-x

Clarke LJ, Robinson SA, Hua Q, Ayre DJ, Fink D (2012) Radiocarbon bomb spike reveals biological effects of Antarctic climate change. Glob Chang Biol 18:301–310. https://doi.org/10.1111/j.1365-2486.2011.02560.x

Clayton MN, Wiencke C, Klöser H (1997) New records of temperate and sub-Antarctic marine benthic macroalgae from Antarctica. Polar Biol 17:141–149. https://doi.org/10.1007/s003000050116

Collins M et al (2013) Long-term climate change: projections, commitments and irreversibility. In: Stocker TF et al (eds) Climate change 2013: the physical science basis. Contribution of working group I to the fifth assessment report of the intergovernmental panel on climate change, Cambridge, pp 1031–1106

Convey P (2020) The price of cumulative human activities in the Antarctic. Antarct Sci 32:425. https://doi.org/10.1017/S0954102020000577

Convey P, Peck LS (2019) Antarctic environmental change and biological responses. Sci Adv 5(11). https://doi.org/10.1126/sciadv.aaz0888

Convey P, Stevens MI, Hodgson DA, Smellie JL, Hillenbrand CD, Barnes DKA, Clarke A, Pugh PJA, Linse K, Cary SC (2009) Exploring biological constraints on the glacial history of Antarctica. Quat Sci Rev 28:3035–3048

Convey P, Bowman V, Chown SL, Francis JE, Fraser C, Smellie J, Storey B, Terauds A (2018) Ice bound Antarctica: biotic consequences of the shift from a temperate to a polar climate. In: Hoorn, Perrigo, Antonelli (eds) Mountains, climate, and biodiversity. Wiley. ISBN: 978-1-119-15987-2

Convey P, Biersma EM, Casanova-Katny A, Maturana CS (2020) Chapter 10: refuges of Antarctic diversity. In: Oliva M, Ruiz-Fernández J (eds) Past Antarctica. Academic Press, Burlington, pp 181–200. https://doi.org/10.1016/B978-0-12-817925-3.00010-0

Cormaci M, Furnari G, Scammacca B (1992) The benthic algal flora of terra Nova Bay (Ross Sea, Antarctica). Environ Sci Biol. https://doi.org/10.1515/botm.1992.35.6.541

Coyer JA, Hoarau GG, Skage M, Stam WT, Olsen JL (2006) Origin of fucus serratus (Heterokontophyta; Fucaceae) populations in Iceland and the faroes: a microsatellite-based assessment. Eur J Phycol 41(2):235–246

Coyer JA, Hoarau G, Van Schaik J, Luijckx P, Olsen JL (2011) Trans-Pacific and trans-Arctic pathways of the intertidal macroalga Fucus distichus L. reveal multiple glacial refugia and colonizations from the North Pacific to the North Atlantic. J Biogeogr 38(4):756–771. https://doi.org/10.1111/j.1365-2699.2010.02437.x

DeConto R, Pollard D (2003) Rapid cenozoic glaciation of Antarctica induced by declining atmospheric CO2. Nature 421:245–249. https://doi.org/10.1038/nature01290

Deppeler SL, Davidson AT (2017) Southern ocean phytoplankton in a changing climate. Front Mar Sci 4. https://doi.org/10.3389/fmars.2017.00040

Deregibus D, Quartino ML, Campana GL, Momo FR, Wiencke C, Zacher K (2016) Photosynthetic light requirements and vertical distribution of macroalgae in newly ice-free areas in Potter Cove, South Shetland Islands, Antarctica. Polar Biol 39:153–166. https://doi.org/10.1007/s00300-015-1679-y

Deregibus D, Campana DL, Neder C, Barnes DKA, Zacher K, Piscicelli JM, Jerosch K, Quartino ML (2023) Potential macroalgal expansion and blue carbon gains with northern Antarctic Peninsula glacial retreat. Mar Environ Res 189. https://doi.org/10.1016/j.marenvres.2023.106056

Díaz MJ, Buschbaum C, Renaud PE, Molis M (2021) Effects of detached seaweeds on structure and function of Arctic intertidal soft-bottom communities. Front Mar Sci 8:575885. https://doi.org/10.3389/fmars.2021.575885

Diehl N, Bischof K (2021) Coping with a changing Arctic: mechanisms of acclimation in the brown seaweed *Saccharina latissima* from Spitsbergen. Mar Ecol Prog Ser 657:43–57. https://doi.org/10.3354/meps13532

Diehl N, Karsten W, Bischof K (2020) Impacts of combined temperature and salinity stress on the endemic Arctic brown seaweed *Laminaria solidungula*. J Agardh Polar Biol 43:647–656. https://doi.org/10.1007/s00300-020-02668-5

Diehl N, Roleda MY, Bartsch I, Karsten U, Bischof K (2021) Summer heatwave impacts on the European Kelp *Saccharina latissima* across its latitudinal distribution gradient. Front Mar Sci 8. https://doi.org/10.3389/fmars.2021.695821

Duarte P, Meyer M, Olsen LM, Kauko HM, Assmy P et al (2017) Sea ice thermohaline dynamics and biogeochemistry in the Arctic Ocean: empirical and model results. J Geophys Res Biogeosci 122:1632–1654. https://doi.org/10.1002/2016JG003660

Dubrasquet H, Reyes J, Sanchez RP, Valdivia N, Guillemin ML (2018) Molecular-assisted revision of red macroalgal diversity and distribution along the Western Antarctic Peninsula and South Shetland Islands. Cryptogam Algol 39. https://doi.org/10.7872/crya/v39.iss4.2018.409

Ducklow H, Fraser W, Meredith M, Stammerjohn S, Doney S, Martinson D, Sailley S, Schofield O, Steinberg D, Venables H, Amsler C (2013) West Antarctic Peninsula: an ice-dependent coastal marine ecosystem in transition. Oceanography 26:190–203. https://doi.org/10.5670/oceanog.2013.62

Dunton K (1992) Arctic biogeography: the paradox of the marine benthic fauna and flora. Trends Ecol Evol 7:183–189. https://doi.org/10.1016/0169-5347(92)90070-R

Fernández PA, Amsler CD, Hurd CL, Díaz PA, Gaitán-Espitia JD, Macaya EC, Schmider-Martínez A, Garrido I, Murúa P, Buschmann A (2024) Diverse inorganic carbon uptake strategies in Antarctic seaweeds: revealing species-specific responses and implications for ocean acidification. Sci Total Environ 945. https://doi.org/10.1016/j.scitotenv.2024.174006

Filbee-Dexter K (2020) Ocean forests hold unique solutions to our current environmental crisis. One Earth 2:398–401. https://doi.org/10.1016/j.oneear.2020.05.004

Filbee-Dexter K, Wernberg T, Ramirez-Llodra E et al (2018) Movement of pulsed resource subsidies from shallow kelp forests to deep fjords. Oecologia 187:291–304

Filbee-Dexter K, Wernberg T, Fredriksen S, Norderhaug KM, Pedersen MF (2019) Arctic kelp forests: diversity, resilience and future. Glob Planet Chang 172:1–14. https://doi.org/10.1016/j.gloplacha.2018.09.005. ISSN 0921-8181

Filbee-Dexter K, MacGregor KA, Lavoie C, Garrido I, Goldsmit J et al (2022) Sea ice and substratum shape extensive kelp forests in the Canadian Arctic. Front Mar Sci 9:754074. https://doi.org/10.3389/fmars.2022.754074

Florczyk L, Latala A (1989) The phytobenthos of the Hornsund fiord, SW Spitsbergen. Polar Res 7:29–41. https://doi.org/10.3402/polar.v7i1.6828

Flores-Molina MR, Rautenberger R, Muñoz P, Huovinen P, Gómez I (2016) Stress tolerance of the endemic antarctic brown alga *Desmarestia anceps* to UV radiation and temperature is mediated by high concentrations of phlorotannins. Photochem Photobiol 92:455–466. https://doi.org/10.1111/php.12580

Franke K, Liesner D, Heesch S, Bartsch I (2021) Looks can be deceiving: contrasting temperature characteristics of two morphologically similar kelp species co-occurring in the Arctic. Bot Mar 64:163–175. https://doi.org/10.1515/bot-2021-0014

Fraser CI, Nikula R, Spencer HG, Waters JM (2009) Kelp genes reveal effects of sub-Antarctic sea ice during the last glacial maximum. PNAS 106:3249–3253

Fraser CI, Morrison AK, Hogg AM, Macaya EC, van Sebille E, Ryan PG, Padovan A, Jack C, Valdivia N, Waters JM (2018) Antarctica's ecological isolation will be broken by storm-driven dispersal and warming. Nat Clim Chang 1. https://doi.org/10.1038/s41558-018-0209-7

Fraser CI, Morrison A, Olmedo Rojas P (2020) Biogeographic processes influencing Antarctic and Sub-Antarctic seaweeds. In: Gómez I, Huovinen P (eds) Antarctic seaweeds, 1st edn. Springer, Cham, pp 43–58. ISBN: 978-3-030-39448-6

Fredriksen S, Karsten U, Bartsch I, Woelfel J, Koblowsky M, Schumman R, Roang R, Steneck R, Wiktor J, Hop H, Wiencke C (2019) Biodiversity of benthic macro- and microalgae from Svalbard with special focus on Kongsfjorden. In: Hop H, Wiencke C (eds) The ecosystem of Kongsfjorden, Svalbard, Advances in polar ecology, vol 2. https://doi.org/10.1007/978-3-319-46425-1_9

Frenot Y, Chown SL, Whinam J, Selkirk P, Convey P, Skotnicki M, Bergstrom D (2005) Biological invasions in the Antarctic: extent, impacts and implications. Biol Rev 80:45–72

Gallardo T, Pérez-Ruzafa IM, Flores-Moya A, Conde F (1999) New collections of benthic marine algae from Livingston and Deception Islands (South Shetland Islands) and Trinity Island (Bransfield Strait) Antarctica. Bot Mar 42:61–69. https://doi.org/10.1515/bot.1999.009

Ganachaud A (2003) Large-scale mass transports, water mass formation, and diffusivities estimated from World Ocean Circulation Experiment (WOCE) hydrographic data. J Geophys Res 108:3213. https://doi.org/10.1029/2002JC001565

Ganachaud A, Wunsch C (2000) Improved estimates of global ocean circulation, heat transport and mixing from hydrographic data. Nature 408:453–457

Gao G, Zhao X, Jiang M, Gao L (2021) Impacts of marine heatwaves on algal structure and carbon sequestration in conjunction with ocean warming and acidification. Front Mar Sci 8:758651. https://doi.org/10.3389/fmars.2021.758651

Garbary DJ (2001) Biogeography of marine algae. In: Encyclopedia of life sciences. Wiley, Hoboken, pp 1–9

Goldsmith J, Schlegel RW, Filbee-Dexter K, MacGregor KA, Johnson LE, Mundy CJ, Savoie AM, McKindsey CW, Howland KL, Archambault P (2021) Kelp in the Eastern Canadian Arctic: current and future predictions of habitat suitability and cover. Front Mar Sci. https://doi.org/10.3389/fmars.2021.742209

Gómez I, Huovinen P (2015) Lack of physiological depth patterns in conspecifics of endemic Antarctic brown algae: a trade-off between UV stress tolerance and shade adaptation? PLoS One 10. https://doi.org/10.1371/journal.pone.0134440

Gómez I, Huovinen P (2020a) Antarctic Seaweeds, 1st edn. Springer, Cham

Gómez I, Huovinen P (2020b) Brown algal phlorotannins: an overview of their functional roles. Antarct Seaweeds:365–388. https://doi.org/10.1007/978-3-030-39448-6_18

Gómez I, Weykam G, Koser H, Wiencke C (1997) Photosynthetic light requirements, daily carbon balance and zonation of sublittoral macroalgae from King George Island (Antarctica). Mar Ecol Progr Ser 148:281–293

Gómez I, Wulff A, Roleda MY, Huovinen P, Karsten U, Quartino ML, Dunton K, Wiencke C (2009) Light and temperature demands of marine benthic microalgae and seaweeds in Polar Regions. Bot Mar 52:593–608

Gomez I, Wulff A, Roleda MY, Huovinen P, Karsten U, Quartino ML, Dunton K, Wiencke C (2011) Light and temperature demands of marine benthic microalgae and seaweeds in Polar Regions. In: Wiencke C (ed) Biology of polar benthic algae. De Gruyter, Berlin, pp 195–220

Gómez I, Navarro NP, Huovinen P (2019) Bio-optical and physiological patterns in Antarctic seaweeds: a functional trait based approach to characterize vertical zonation. Prog Oceanogr 174:17–27. https://doi.org/10.1016/j.pocean.2018.03.013

González-Herrero S, Barriopedro D, Trigo RM et al (2022) Climate warming amplified the 2020 record-breaking heatwave in the Antarctic Peninsula. Commun Earth Environ 3:122. https://doi.org/10.1038/s43247-022-00450-5

González-Wevar CA, Díaz A, Gerard K, Cañete JI, Poulin E (2012) Divergence time estimations and contrasting patterns of genetic diversity between Antarctic and southern South America benthic invertebrates. Rev Chil Hist Nat 85:445–456

Gordillo FJL, Carmona R, Jimenez C (2022) A warmer Arctic compromises winter survival of habitat-forming seaweeds. Front Mar Sci 8. https://doi.org/10.3389/fmars.2021.750209

Graham MH, Kinlan BP, Druehl LD, Garske LE, Banks S. 2007. Deep-water kelp refugia as potential hotspots of tropical marine diversity and productivity. Proc Natl Acad Sci 104:16576–16580

Grall J, Le Loh F, Guyonnet B, Riera P (2006) Community structure and food web based on stable isotopes (δ 15 N and δ 13 C) analysis of a Northeastern Atlantic maerl bed. J Exp Mar Bio Ecol. https://doi.org/10.1016/j.jembe.2006.06.013

Guillemin ML, Dubrasquet H, Reyes J, Valero M (2018) Comparative phylogeography of six red algae along the Antarctic Peninsula: extreme genetic depletion linked to historical bottlenecks and recent expansion. Polar Biol 41. https://doi.org/10.1007/s00300-017-2244-7

Guillemin ML, González-Wevar C, Cárdenas L, Dubrasquet H, Garrido I et al (2020) Comparative phylogeography of Antarctic seaweeds: genetic consequences of historical climatic variations.

In: Gómez I, Huovinen P (eds) Antarctic seaweeds. Diversity, adaptation and ecosystem services. Springer, pp 103–127. https://doi.org/10.1007/978-3-030-39448-6
Guiry MD, Guiry GM (2025) AlgaeBase. World-wide electronic publication. https://www.algaebase.org
Gunnarsson K, Jónsson S (2002) Benthic marine algae of Iceland: revised checklist. Cryptogam Algol 23(2):131–158
Haine TWN, Martin T (2017) The Arctic-Subarctic sea ice system is entering a seasonal regime: implications for future Arctic amplification. Sci Rep 7(1):4618. https://doi.org/10.1038/s41598-017-04573-0
Hofmann LC, Bischof K (2014) Ocean acidification effects on calcifying macroalgae. Mar Ecol Prog Ser 22:261–279
Hommersand MH, Moe RL, Amsler CD, Fredericq S (2009) Notes on the systematics and biogeographical relationships of Antarctic and sub-Antarctic Rhodophyta with descriptions of four new genera and five new species. Bot Mar 52:509–534. https://doi.org/10.1515/bot.2009.081
Hop H et al (2002) The marine ecosystem of Kongsfjorden, Svalbard. Polar Res 21:167–208. https://doi.org/10.1111/j.1751-8369.2002.tb00073.x
Hop H, Wiencke C, Vögele B, Kovaltchouk NA (2012) Species composition, zonation, and biomass of marine benthic macroalgae in Kongsfjorden, Svalbard. Bot Mar 55:399–414
Hughes KA, Convey P (2014) Alien invasions in Antarctica-is anyone liable? Polar Res 33:1–13
Hughes K, Ott S, Bölter M, Convey P (2006) Chapter 3: colonisation processes. In: Bergstrom DM, Convey P, Huiskes AHL (eds) Trends in Antarctic terrestrial and limnetic ecosystems: Antarctica as a global indicator. Springer, Dordrecht, pp 35–54
Hughes KA, Pertierra LR, Molina-Montenegro MA, Convey P (2015) Biological invasions in Antarctica: what is the current status and can we respond? Biodivers Conserv 24:1031–1055. https://doi.org/10.1007/s10531-015-0896-6
Hughes KA et al (2020) Invasive non-native species likely to threaten biodiversity and ecosystems in the Antarctic Peninsula region. Glob Chang Biol 26(4):2702–2716. https://doi.org/10.1111/gcb.14938
Hughey JR, Leister GL, Gabrielson PW, Hommersand MH (2020) *Sarcopeltis* gen. nov. (Gigartinaceae, Rhodophyta), with *S. skottsbergii* comb. nov. from southern South America and S. antarctica sp. nov. from the Antarctic Peninsula. Phytotaxa 468. https://doi.org/10.11646/phytotaxa.468.1.4
Humpert H (2011) The future of the Northern Sea route – a "golden waterway" or a niche trade route. The Arctic Center for Circumpolar security studies. https://www.thearcticinstitute.org/future-northern-sea-route-golden-waterway-niche/
Huntington HP, Boyle M, Flowers GE, Weatherly JW, Hamilton LC, Hinzman L, Gerlach C, Zulueta R, Nicolson C, Overpeck J (2010) The influence of human activity in the Arctic on climate and climate impacts climatic change. https://doi.org/10.1007/s10584-006-9162-y
Huovinen P, Gómez I (2013) Photosynthetic characteristics and UV stress tolerance of Antarctic seaweeds along the depth gradient. Polar Biol 36:1319–1332. https://doi.org/10.1007/s00300-013-1351-3
Huovinen P, Gómez I (2020) Underwater light environment of Antarctic seaweeds. In: Gómez I, Huovinen P (eds) Antarctic seaweeds. Springer, Cham, pp 131–153. https://doi.org/10.1007/978-3-030-39448-6_7
Huovinen P, Ramírez J, Gómez I (2016) Underwater optics in sub-Antarctic and Antarctic coastal ecosystems. PLoS One 11(5):e0154887. https://doi.org/10.1371/journal.pone.0154887
Iken K, Amsler CD, Gorman KB, Klein AG, Galloway AWE, Amsler MO, Heiser S, Whippo R, Lowe AT, Schram JB, Schneider ZX, McClintock JB (2023) Macroalgal input into the coastal food web along a gradient of seasonal sea ice cover along the Western Antarctic Peninsula. Mar Ecol Prog Ser 718:1–22. https://doi.org/10.3354/meps14388
IPCC (2022) Climate change 2022: mitigation of climate change. Contributo do Grupo de Trabalho III para o Sexto Relatório de Avaliação do Painel Intergovernamental das Alterações Climáticas. https://doi.org/10.1017/9781009157926

Irving A, Connell S, Johnston E, Pile A, Gillanders B (2005) The response of encrusting coralline algae to canopy loss: an independent test of predictions on an Antarctic coast. Mar Biol 147:1075–1083. https://doi.org/10.1007/s00227-005-0007-4

Ivar do Sul JA, Barnes DKA, Costa MF, Convey P, Costa ES, Campos L (2011) Plastics in the Antarctic environment: are we looking only at the tip of the iceberg? Oecologia Australis 15:150–170

Jewett L, Romanou A (2017) Ocean acidification and other ocean changes. In: Wuebbles DJ, Fahey DW, Hibbard KA, Dokken DJ, Stewart BC et al (eds) Climate science special report: fourth national climate assessment, vol I. U.S. Global Change Research Program, Washington, DC, pp 364–392

Johnston E, Connell S, Irving A, Pile A, Gillanders B (2007) Antarctic patterns of shallow subtidal habitat and inhabitants in Wilke's Land. Polar Biol 30:781–788

Jordan TA, Riley TR, Siddoway CS (2020) The geological history and evolution of West Antarctic. Nat Rev Earth Environ 1(2):117–133. https://doi.org/10.1038/s43017-019-0013-6. ISSN 2662-138X

Jueterbock A, Tyberghein L, Verbruggen H, Coyer JA, Olsen JL, Hoarau G (2013) Climate change impact on seaweed meadow distribution in the North Atlantic rocky intertidal. Ecol Evol 3(5):1356–1373. https://doi.org/10.1002/ece3.541

Kain JM (1989) The seasons in the subtidal. Br Phycol J 24:203–215. https://doi.org/10.1080/00071618900650221

Kang JS, Kang SH, Lee J, Lee S (2002) Seasonal variation of microalgal assemblages at a fixed station in King George Island, Antarctica, 1996. Mar Ecol Progr Ser 229:19–32. https://www.jstor.org/stable/24865049

Karsten U, Bischof K, Wiencke C (2001) Photosynthetic performance of Arctic macroalgae after transplantation from deep to shallow waters. Oecologia 127:11–20. https://doi.org/10.1007/s004420000553

Karsten U, Wulff A, Roleda MY, Müller R, Steinhoff FS, Fredersdorf J, Wiencke C (2009) Physiological responses of polar benthic algae to ultraviolet radiation. Bot Mar 52:639–654. https://doi.org/10.1515/BOT.2009.077

Kaur M, Saini KC, Ojah H, Sahoo R, Gupta K, Kumar A, Bast F (2022) Abiotic stress in algae: response, signaling and transgenic approaches. J Appl Phycol 34:1843–1869. https://doi.org/10.1007/s10811-022-02746-7

Kawai H, Hanyuda T, Yamagishi T, Kai A, Lane C, McDevit D, Küpper FC, Saunders GW (2015) Reproductive morphology and DNA sequences of the brown alga *Platysiphon verticillatus* support the new combination *P. glacialis*. J Phycol 51:910–917

King NG, McKeown NJ, Smale DA, Moore PJ (2018) The importance of phenotypic plasticity and local adaptation in driving intraspecific variability in thermal niches of marine macrophytes. Ecography 41:1469–1484. https://doi.org/10.1111/ecog.03186

Ko YW, Choi HG, Lee DS, Kim JH (2020) 30 years revisit survey for long-term changes in the Antarctic subtidal algal assemblage. Sci Rep 10(1):8481

Krause-Jensen D, Duarte CM (2014) Expansion of vegetated coastal ecosystems in the future Arctic. Front Mar Sci. https://doi.org/10.3389/fmars.2014.00077

Krause-Jensen D, Marbà N, Olesen B, Sejr MK, Christensen PB, Rodrigues J et al (2012) Seasonal sea ice cover as principal driver of spatial and temporal variation in the depth extension and annual production of kelp in Greenland. Glob Change Biol 18:2981–2994. https://doi.org/10.1111/j.1365-2486.2012.02765.x

Krause-Jensen D, Marbà N, Sanz-Martin M, Hendriks IE, Thyrring J, Carstensen J et al (2016) Long photoperiods sustain high pH in Arctic kelp forests. Sci Adv 2:e1501938. https://doi.org/10.1126/sciadv.1501938

Krause-Jensen D, Archambault P, Assis J, Bartsch I, Bischof K, Filbee-Dexter K et al (2020) Imprint of climate change on Pan-Arctic marine vegetation. Front Mar Sci 7:617324

Krumhansl K, Okamoto DK, Rassweiler A et al (2016) PNAS 113(48):13785–13790. https://doi.org/10.1073/pnas.1606102113

Küpper FC (2022) Obituary Robert Thayer Wilce, pioneer of arctic marine botany. 9 December 1924–26 February 2022. Bot Mar 65(3):153–158. https://doi.org/10.1515/bot-2022-0025

Küpper FC, Peters AF, Sayer MDJ, Mystikou A, Hugh Brown EA, Dargent O, Strittmatter M, Brennan D, Asensi AO, van West P, Wilce RT (2016) Arctic marine phytobenthos of Northern Baffin Island. J Phycol 52:532–549

Küpper FC, Amsler CD, Morley S, de Reviers B, Reichardt A, Peck LS, Peters AF (2019) Juvenile morphology of the large Antarctic canopy-forming brown alga, Desmarestia menziesii. J Agardh Polar Biol 42:2097–2103. https://doi.org/10.1007/s00300-019-02584-3

La Guardia LC, Filbee-Dexter K, Reimer J, MacGregor et al (2023) Increasing depth distribution of Arctic kelp with increasing number of open water days with light. Elem Sci Anth 11(1):00051. https://doi.org/10.1525/elementa.2022.00051

Lamb IM (1977) Benthic marine algae of the Antarctic Peninsula. Biology of the Antarctic Seas. Antarctic research series, Vol. 24, 5, paper 4, p 4

Laughinghouse HD, Müller KM, Adey WH, Lara Y, Young R, Johnson G (2015) Evolution of the Northern Rockweed, fucus distichus, in a regime of glacial cycling: implications for benthic algal phylogenetics. PLoS One 10(12):e0143795. https://doi.org/10.1371/journal.pone.0143795

Lavergne C et al (2022) Macroalgae metal-biomonitoring in Antarctica: addressing the consequences of human presence in the white continent. Environ Pollut 292. https://doi.org/10.1016/j.envpol.2021.118365

Lebrun A, Comeau S, Gazeau F, Gattuso JP (2022) Impact of climate change on Arctic macroalgal communities. Glob Planet Chang 219. https://doi.org/10.1016/j.gloplacha.2022.103980

Levring T (1945) Proceedings on the 10th International Seaweed Symposium. Walter de Gruyter, Berlin, pp 345–350

Li Q et al (2023) Abyssal Ocean overturning slowdown and warming driven by Antarctic meltwater. Nature 615:841–847. https://doi.org/10.1038/s41586-023-05762-w

Lindstrom SC (2006) Biogeography of Alaskan seaweeds. In: Anderson R, Brodie J, Onsøyen E, Critchley AT (eds) Eighteenth international seaweed symposium. Developments in applied phycology, vol 1. Springer, Dordrecht. https://doi.org/10.1007/978-1-4020-5670-3_49

Locarnini RA, Mishonov AV, Baranova OK, Reagan JR et al (2024) World Ocean Atlas 2023, Volume 1: temperature. A. Mishonov Technical Editor, NOAA Atlas NESDIS 89

López BA, Macaya EC, Tala F, Tellier F, Thiel M (2017) The variable routes of rafting: stranding dynamics of floating bull kelp *Durvillaea antarctica* (Fucales, Phaeophyceae) on beaches in the SE Pacific. J Phycol 53:70–84. https://doi.org/10.1111/jpy.12479

Lumpkin R, Speer K (2007) Global ocean meridional overturning. J Phys Oceanogr:2550–2562. https://doi.org/10.1175/JPO3130.1

Lüning K, Yarish C, Kirkman H (1990) In: Yarish C, Kirkman H (eds) Seaweeds: their environment, biogeography, and ecophysiology. Wiley. https://books.google.com.br/books?id=YNrSdP9ei40C

Macaya EC, Tala F, Hinojosa IA, Rothäusler E (2020) Detached seaweeds as important dispersal agents across the southern ocean. In: Gómez I, Huovinen P (eds) Antarctic seaweeds, 1st edn. Springer, Cham, pp 59–82. ISBN: 978-3-030-39448-6

Malavenda SV (2021) Species diversity of macroalgae in Grønfjorden, Spitsbergen, Svalbard. Polar Res. https://doi.org/10.33265/polar.v40.3682

Mansilla A, Palacios M, Navarro N, Riquelme V (2005) Estudios sobre las poblaciones de Macrocystis pyrifera (L) C, Agarth en Tierra del Fuego, Chile. In: International symposium & worshop, 19–24 octubre Ushuaia, Argentina

Mansilla A, Méndez F, Murcia S, Rodríguez J, Marambio J, Rosenfeld S, Yokoya N, Bischof K (2016) Adjustment of pigment composition in *Desmarestia* (Desmarestiaceae) species along a sub-Antarctic to Antarctic latitudinal gradient. Polar Res 35

Marambio J, Bischof J (2021) Differential acclimation responses to irradiance and temperature in two co-ocurring seaweed species in Arctic fjord. Polar Res 40:5702

Marambio J, Diehl N, Bischof K (2022a) High ecophysiological plasticity of *Desmarestia aculeata* (Phaeophyceae) present in an Arctic fjord under varying salinity and irradiance conditions. Biology 11:1499

Marambio J, Rosenfeld S, Bischof K (2022b) Hyposalinity affects diurnal photoacclimation patterns in the rhodophyte *Palmaria palmata*, under mimicked Arctic summer conditions. J Photochem Photobiol 11:100124

Marambio J, Rodriguez JP, Rosenfeld S, Méndez F, Ojeda J, Ocaranza P, Bischof K, Mansilla A (2023) New ecophysiological perspectives on the kelp *Macrocystis pyrifera*: generating a basis for sustainability in the sub-Antarctic region. Front Mar Sci 10:1222178

Marbà N, Krause-Jensen D, Olesen B et al (2017) Climate change stimulates the growth of the intertidal macroalgae Ascophyllum nodosum near the northern distribution limit. Ambio 46(Suppl 1):119–131. https://doi.org/10.1007/s13280-016-0873-7

Mathieson AC, Moore GE, Short FT (2010) A floristic comparison of seaweeds from James Bay and three contiguous Northeastern Canadian Arctic sites. Rhodora 112(952):396–434. http://www.bioone.org/doi/full/10.3119/09-12.1

Matula CV, Quartino ML, Nuñez JD, Zacher K, Bartsch I (2022) Effects of seawater temperature and seasonal irradiance on growth, reproduction, and survival of the endemic Antarctic brown alga *Desmarestia menziesii* (Phaeophyceae). Polar Biol 45:559–572. https://doi.org/10.1007/s00300-021-02991-5

McCarthy AH, Peck LS, Hughes KA, Aldridge DC (2019) Antarctica: the final frontier for marine biological invasions. Glob Changes Biol 25. https://doi.org/10.1111/gcb.14600

McClelland JW, Holmes R, Dunton K, Macdonald R (2012) The Arctic Ocean estuary. Estuar Coasts 35(2):353–368

Meredith M, Sommerkorn M, Cassotta S et al (2019) Polar regions. In: Pörtner HO et al (eds) IPCC special report on the ocean and cryosphere in a changing climate

Michel LN, Denis B, Dubois P, Eleaume M, Fournier J, Gallut C, Jane P, Lepoint G (2019) Increased sea ice cover alters food web structure in east Antarctica. Sci Rep 9:8062. https://doi.org/10.1038/s41598-019-44605-5

Mikhaylova TA, Spiridonov VA, Gavrilo MV, Ivanov SD (2022) Macroalgae of the high-Arctic Severnaya Zemlya Archipelago. Bot Mar. https://doi.org/10.1515/bot-2022-0031

Miller GH, Pearse. (1991) Ecological studies of seaweeds in McMurdo Sound, Antarctica. Am Zool 31:35–48

Moe RL, Silva PC (1977) Antarctic marine flora: uniquely devoid of kelps source. Science 196(4295):1206–1208

Mols-Mortensen A, Neefus CD, Nielsen R, Gunnarsson K, Ttir SE, Pedersen PM, Brodie J (2012) New insights into the biodiversity and generic relationships of foliose Bangiales (Rhodophyta) in Iceland and the Faroe Islands. Eur J Phycol 47(2):146–159

Monserrat M, Comeau S, Verdura J et al (2022) Climate change and species facilitation affect the recruitment of macroalgal marine forests. Sci Rep 12:18103. https://doi.org/10.1038/s41598-022-22845-2

Monteiro C, Li J, Diehl N, Collén J, Heinrich S, Bischof K, Bartsch I (2021) Modulation of physiological performance by temperature and salinity in the sugar kelp *Saccharina latissima*. Phycol Res. https://doi.org/10.1111/pre.12443

Mrowicki RJ, Brodie J (2023) The first record of a non-native seaweed from South Georgia and confirmation of its establishment in the Falkland Islands: *Ulva fenestrata* Postels & Ruprecht. Polar Biol 46:489–496. https://doi.org/10.1007/s00300-023-03136-6

Müller R, Laepple T, Bartsch I, Wiencke C (2009) Impact of oceanic warming on the distribution of seaweeds in polar and cold-temperate waters. Bot Mar 52:617–638. https://doi.org/10.1515/BOT.2009.080

Mystikou A, Peters AF, Asensi AO, Fletcher KI, Brickle P, West P, Convey P, Küpper FC (2014) Seaweed biodiversity in the south-western Antarctic Peninsula: surveying macroalgal community composition in the Adelaide Island/Marguerite Bay region over a 35-year time span. Polar Biol 37:1607–1619. https://doi.org/10.1007/s00300-014-1547-1

Mystikou A, Asensi AO, Declerck O, Müller DG, Peters AF, Tsiamis K et al (2016) New records and observations of macroalgae and associated pathogens from the Falkland Islands, Patagonia and Tierra del Fuego. Bot Mar 59. https://doi.org/10.1515/bot-2015-0071

Nauer F, Oliveira MC, Plastino EM, Yokoya NS, Fujii M (2022) Coping with heatwaves: How a key species of seaweed responds to heat stress along its latitudinal gradient. Mar Environ Res 177. https://doi.org/10.1016/j.marenvres.2022.105620

Naughten KA, Holland PR, de Rydt J (2023) Unavoidable future increase in West Antarctic ice-shelf melting over the twenty-first century. Nat Clim Changes 13:1222–1228. https://doi.org/10.1038/s41558-023-01818-x

Navarro N, Huovinen P, Gómez I (2020) Life history strategies, photosynthesis, and stress tolerance in propagules of Antarctic seaweeds. In: Gómez I, Huovinen P (eds) Antarctic seaweeds. Springer, Cham, pp 193–215. https://doi.org/10.1007/978-3-030-39448-6_10

Nelson WA, Neill KF, Archino R, Sutherland JE (2022) Marine macroalgae of the Balleny Islands and Ross Sea. Antarct Sci 34(4):298–312. https://doi.org/10.1017/S0954102022000220

Neushul M (1965) Diving observations of sub-tidal Antarctic marine vegetation. Bot Mar 8(2–4). https://doi.org/10.1515/botm.1965.8.2-4.234

Noctor G et al (2019) Physiological responses of saplings of Caesalpinia echinacea Lam., a Brazilian tree species, under ozone fumigation. Ecotoxicol Environ Saf 238:291–293. https://doi.org/10.1111/tpj.14292

Obluchinskaya E, Zakharova L (2021) Metal concentrations in three species of Fucus L. on the Murmansk coast of the Barents Sea. Pol Sci 28:100646. https://doi.org/10.1016/j.polar.2021.100646

Obluchinskaya ED, Pozharitskaya ON, Zakharov DV et al (2022) The biochemical composition and antioxidant properties of fucus vesiculosus from the Arctic Region. Mar Drugs 20. https://doi.org/10.3390/md20030193

Ocaranza-Barrera P, González-Wevar C, Guillemin ML, Rosenfeld S, Mansilla A (2019) Molecular divergence between Iridaea cordata (Turner) Bory de Saint-Vincent from the Antarctic Peninsula and the Magellan Region. J Appl Phycol 31. https://doi.org/10.1007/s10811-018-1656-2

Oliveira EC, Absher TM, Pellizzari FM, Oliveira MC (2009) The seaweed flora of Admiralty Bay, King George Island, Antarctic. Polar Biol 32:1639–1647. https://doi.org/10.1007/s00300-009-0663-9

Oliveira MC, Repetti SI, Iha C, Jackson CJ, Díaz-Tapia P, Lubiana KMF, Cassano V, Costa JF, Cremen MCM, Marcelino VR, Verbruggen H (2018) High-throughput sequencing for algal systematics. Eur J Phycol 53(3):256–272. https://doi.org/10.1080/09670262.2018.1441446

Oliveira MC, Pellizzari FM, Yokoya NS (2020) Diversity of Antarctic seaweeds. In: Gómez I, Huovinen P (eds) Antarctic seaweeds, 1st edn. Springer, Cham, pp 23–42. https://doi.org/10.1007/978-3-030-39448-6. ISBN: 978-3-030-39448-6

Oliver ECJJ, Donat MG, Burrows MT, Moore PJ, Smale DA et al (2018) Longer and more frequent marine heatwaves over the past century. Nat Commun 9:1324. https://doi.org/10.1038/s41467-018-03732-9

Orsi AH, Smethie WM Jr, Bullister JL (2002) On the total input of Antarctic waters to the deep ocean: A preliminary estimate from chlorofluorocarbon measurements. J Geophys Res 107:3122. https://doi.org/10.1029/2001JC000976

Overland J, Hanna E, Hanssen-Bauer I, Kim SJ, Walsh J, Wang M, Bhatt U (2019) The Arctic surface air temperature [in "State of the Climate in 2018"]. Bull Am Meteorol Soc 100(9):97–97

Park YH, Charriaud E, Fieux M (1998) Thermohaline structure of the Antarctic Surface Water and Winter Water in the Indian sector of the Southern Ocean. J Mar Syst 17:5–23

Pasqualetti CB, Carvalho MAM, Mansilla A, Avila M, Yokoya NS (2023) Variations on primary metabolites of the carrageenan-producing red algae Sarcopeltis skottsbergii from Chile and Sarcopeltis Antarctica from Antarctic Peninsula. Front Mar Sci 10:1151332. https://doi.org/10.3389/fmars.2023.1151332

Pavlov AK et al (2019) The underwater light climate in Kongsfjorden and its ecological implications. In: Hop H, Wiencke C (eds) The ecosystem of Kongsfjorden, Svalbard, Advances in polar ecology, vol 2

Pedersen EJ, Thompson PL, Ball RA, Fortin MJ, Gouhier TC, Link H, Moritz C, Nenzen H, Stanley RR, Taranu ZE, Gonzalez A, Guichard F, Pepin P (2017) Signatures of the collapse and incipient recovery of an overexploited marine ecosystem. R Soc Open Sci. https://doi.org/10.1098/rsos.170215

Pedersen ÅØ et al (2022) Five decades of terrestrial and freshwater research at Ny-Ålesund, Svalbard. Polar Res 41. https://doi.org/10.33265/polar.v41.6310

Pehlke H, Brey T, Konijnenberg R, Teschke K (2022) A tool to evaluate accessibility due to sea-ice cover: a case study of the Weddell Sea, Antarctica. Antarct Sci. https://doi.org/10.1017/S0954102021000523

Pellizzari F, Silva MCC, Silva EMM, Medeiros A, Oliveira MCC, Yokoya NSS, Pupo D, Rosa LHH (2017) Diversity and spatial distribution of seaweeds in the South Shetland Islands, Antarctica: an updated database for environmental monitoring under climate change scenarios. Polar Biol 40:1671–1685. https://doi.org/10.1007/s00300-017-2092-5

Pellizzari F, Rosa LH, Yokoya NS (2020) Biogeography of Antarctic seaweeds facing climate changes. In: Gómez I, Huovinen P (eds) Antarctic seaweeds. Springer, Cham, pp 83–102. https://doi.org/10.1007/978-3-030-39448-6_5. ISBN: 978-3-030-39448-6

Pellizzari F, Mello JPS, Osaki VS, Santos-Silva MC, Brandini FP, Convey P, Rosa LH (2023) New records and updated distributional patterns of macroalgae from the South Shetland Islands and northern Weddell Sea, Antarctica. Antarct Sci. https://doi.org/10.1017/S095410202300010X

Pellizzari FM, Santos KC, Osaki S, Rosa LH (2025) Marine heatwaves and changes in macroalgaerichness and composition from Antarctic Peninsula and South Shetland Islands: concise review. An Acad BrasCienc 97(1):e20240580. https://doi.org/10.1590/0001-3765202520240580

Perovich D, Meier W, Tschudi M, Farrell S, Hendricks S, Gerland S (2019) The Arctic sea ice cover [in "State of the Climatein 2018"]. Bull Am Meteorol Soc 100(9):146–150

Peters AF, van Oppen MJH, Wiencke C, Stam WT, Olsen JL (1997) Phylogeny and historical ecology of the Desmarestiaceae (Phaeophyceae) support a southern hemisphere origin. J Phycol 33:294–309

Peters KJ, Amsler CD, Amsler MO, McClintock JB, Dunbar RB, Baker BJ (2005) A comparative analysis of the nutritional and elemental composition of macroalgae from the western Antarctic Peninsula. Phycologia 44:453–463. https://doi.org/10.2216/0031-8884(2005)44[453:acaotn]2.0.co;2

Pibernat RA, Ellis-Evans C, Hinghofer-Szalkay HG (2007) Life in extreme environments. Springer, Dordrecht. https://doi.org/10.1007/978-1-4020-6285-8

Piñeiro-Corbeira C, Barreiro R, Cremades J, Arenas F (2018) Seaweed assemblages under a climate change scenario: functional responses to temperature of eight intertidal seaweeds match recent abundance shifts. Sci Rep 8(1):12978. https://doi.org/10.1038/s41598-018-31357-x

Plastino E, Mansilla A (2004) Light and photosynthesis. In: Werlinger C (ed) Marine biology and oceanography: concepts and processes. Trama Impresores S.A, Concepcion, pp 229–252

Porter SE, Parkinson CL, Thompson E (2016) Bellingshausen Sea ice extent recorded in an Antarctic Peninsula ice core. J Geophys Res Atmos. https://doi.org/10.1002/2016JD025626

Poulin E, González-Wevar C, Díaz A, Gerard K, Hüne M (2014) Divergence between Antarctic and South American marine invertebrates: what molecular biology tells us about Scotia Arc geodynamics and the intensification of the Antarctic Circumpolar Current. Glob Planet Chang 123:392–399

Purich A, Doddridge EW (2023) Record low Antarctic sea ice coverage indicates a new sea ice state. Commun Earth Environ 4:314. https://doi.org/10.1038/s43247-023-00961-9

Quartino M, Kloeser H, Schloss I, Wiencke C (2001) Biomass and associations of benthic marine macroalgae from the inner Potter Cove (King George Island, Antarctica) related to depth and substrate. Polar Biol 24:349355

Quartino ML, Zaixso HE, Boraso de Zaixso AL (2005) Biological and environmental characterization of marine macroalgal assemblages in Potter Cove, South Shetland Islands, Antarctica. Bot Mar 48. https://doi.org/10.1515/bot.2005.029

Quartino ML, Deregibus D, Campana GL, Latorre GJ, Momo FR (2013) Evidence of macroalgal colonization on newly ice-free areas following glacial retreat in Potter Cove (South Shetland Islands), Antarctica. PLoS One. https://doi.org/10.1371/journal.pone.0058223

Quillien N, Nordström MC, Schaal G, Bonsdorff E, Grall J (2016) Opportunistic basal resource simplifies food web structure and functioning of a highly dynamic marine environment. J Exp Mar Biol Ecol 477:92–102. https://doi.org/10.1016/j.jembe.2016.01.010

Rangel KC, Villela LZ et al (2020) Assessment of the photoprotective potential and toxicity of Antarctic red macroalgae extracts from *Curdiea racovitzae* and *Iridaea cordata* for cosmetic use. Algal Res 50:101984. https://doi.org/10.1016/j.algal.2020.101984

Rautenberger R, Wiencke C, Bischof K (2013) Acclimation to UV radiation and antioxidative defence in the endemic antarctic brown macroalga Desmarestia anceps along a depth gradient. Polar Biol 36:1779–1789. https://doi.org/10.1007/s00300-013-1397-2

Riaz A, Deng F, Chen G, Jiang W, Zheng Q, Riaz B, Mak M, Zeng F, Chen ZH (2022) Molecular regulation and evolution of redox homeostasis in photosynthetic machinery. Antioxidants 11. https://doi.org/10.3390/antiox11112085

Rintoul SR, Garabato AN (2013) Chapter 18 – dynamics of the southern ocean circulation. In: Siedler G, Griffies SM, Gould J, Church JA (eds) International geophysics, vol 103. Academic Press, pp 471–492. https://doi.org/10.1016/B978-0-12-391851-2.00018-0

Ríos C, Arntz W, Gerdes D, Mutschke E, Montiel A (2007) Spatial and temporal variability of the benthic assemblages associated to the holdfasts of the kelp *Macrocystis pyrifera* in the Straits of Magellan, Chile. Polar Biol 31:89–100

Roberts RD, Kuhl M, Glud RN, Rysgaard S (2002) Primary production of crustose coralline red algae in a high arctic fjord. J Phycol 38:273–283. https://doi.org/10.1046/j.1529-8817.2002.01104.x

Robinson MM (2009) New quantitative evidence of extreme warmth in the Pliocene Arctic. Stratigraphy 6:265–276

Robinson BJO, Morley SA, Rizouli A et al (2022) New confirmed depth limit of Antarctic macroalgae: Palmaria decipiens found at 100 m depth in the Southern Ocean. Polar Biol 45:1459–1463. https://doi.org/10.1007/s00300-022-03071-y

Roleda M, Dethleff D (2011) Storm-generated sediment deposition on rocky shores: Simulating burial effects on the physiology and morphology of Saccharina latissima sporophytes. Mar Biol Res 7(3):213–223. https://doi.org/10.1080/17451000.2010.497189

Ross FWR, Boyd PW, Filbee-Dexter K, Watanabe K et al (2023) Potential role of seaweeds in climate change mitigation. Sci Total Environ. https://doi.org/10.1016/j.scitotenv.2023.163699

Ruban A (2016) Nonphotochemical chlorophyll fluorescence quenching: mechanism and effectiveness in protecting plants from photodamage. Plant Physiol 170:1903–1916. https://doi.org/10.1104/pp.15.01935

Runcie JW, Riddle MJ (2006) Photosynthesis of marine macroalgae in ice-covered and ice-free environments in East Antarctica. Eur J Phycol 41:223–233. https://doi.org/10.1080/09670260600645824

Sáez CA, Troncoso M, Navarrete C, Rodríguez-Rojas F, Navarro N et al (2023) Photoprotective responses of three intertidal Antarctic macroalgae to short-term temperature stress. Front Mar Sci 10. https://doi.org/10.3389/fmars.2023.1223853

Saiz JI, García FJ, Manjón-Cabeza ME, Parapar J, Peña-Cantero A et al (2008) Community structure and spatial distribution of benthic fauna in the Bellingshausen Sea (West Antarctica). Polar Biol 31:735–743. https://doi.org/10.1007/s00300-008-0414-3

Sanches PF, Pellizzari FM, Horta PA (2016) Multivariate analyses of Antarctic and sub-Antarctic seaweed distribution patterns: an evaluation of the role of the Antarctic circumpolar current. J Sea Res 110:29–38. https://doi.org/10.1016/j.seares.2016.02.002

Sánchez-Bonet D, García-Oms S, Belda-Antolí M, Padrón-Sanz C, Lloris-Carsi JM, Cejalvo-Lapeña D (2021) RP-HPLC-DAD determination of the differences in the polyphenol content

of Fucus vesiculosus extracts with similar antioxidant activity. J Chromatogr B 1184:122978. https://doi.org/10.1016/j.jchromb.2021.122978
Schoenrock KM, Amsler CD, McClintock JB, Baker BJ (2015) Life history bias in endophyte infection of the Antarctic rhodophyte, Iridaea cordata. Bot Mar 58:1–8. https://doi.org/10.1515/bot-2014-0085
Schoenwaelder M, Wiencke C (2000) Phenolic compounds in the embryo development of several northern hemisphere fucoids. Plant Biol 2(1):24–33
Schwarz AM, Hawes I, Andrew N, Norkko A, Cummings V, Thrush S (2003) Macroalgal photosynthesis near the southern global limit for growth; Cape Evans, Ross Sea, Antarctica. Polar Biol 26:789–799. https://doi.org/10.1007/s00300-003-0556-2
Shilling AJ, Heiser S, Amsler CD, McClintock JB, Baker BJ (2021) Hidden diversity in an Antarctic algal forest: metabolomic profiling linked to patterns of genetic diversification in the Antarctic red alga *Plocamium* sp. Mar Drugs 19. https://doi.org/10.3390/md19110607
Siegert MJ, Bentley MiJ, Atkinson A, Bracegirdle TJ, Convey P, Davies B, Downie R, Hogg AE,Holmes C, Hughes KA, Meredith MP, Ross N, Rumble J, Wilkinson J (2023) Antarctic extreme events.Frontiers in Environmental Science (11). https://doi.org/10.3389/fenvs.2023.1229283
Singh RK, Vader A, Mundy CJ, Søreide JE, Iken K, Dunton KH, La Guardia LC, Sejr MK, Bélanger S (2022) Satellite-derived photosynthetically available radiation at the coastal Arctic seafloor. Remote Sens 14(20):5180. https://doi.org/10.3390/rs14205180
Skottsberg C (1953) On two collections of Antarctic marine algae. Arkiv Botanik 2:531–566
Sloyan BM, Rintoul SR (2001) The Southern Ocean limb of the global deep overturning circulation. J Phys Oceanogr 31:143–173
Smale DA, Wernberg T, Oliver ECJ, Thomsen M, Harvey BP, Straub SC et al (2019) Marine heatwaves threaten global biodiversity and the provision of ecosystem services. Nat Clim Chang 9:306–312. https://doi.org/10.1038/s41558-019-0412-1
Stammerjohn SE, Martinson DG, Smith RC, Yuan X, Rind D (2008) Trends in Antarctic annual sea ice retreat and advance and their relation to El Niño–Southern oscillation and Southern annular mode variability. J Geophys Res 113(C3). https://doi.org/10.1029/2007jc004269
Steneck RS, Dethier MN (1994) A functional group approach to the structure of algal-dominated communities. Oikos 69(3):476–498. https://doi.org/10.2307/3545860
Stévant P, Schmedes PS, Le Gall L, Wegeberg S, Dumay J, Rebours C (2023) Concise review of the red macroalga dulse, Palmaria palmata (L.) Weber & Mohr. J Appl Phycol 35:523–550. https://doi.org/10.1007/s10811-022-02899-5
Sulzberger B, Austin AT, Cory RM, Zepp RG, Paul ND (2019) Solar UV radiation in a changing world: roles of cryosphere-land-water-atmosphere interfaces in global biogeochemical cycles. Photochem Photobiol Sci 18:747–774. https://doi.org/10.1039/c8pp90063a
Summers N, Fragoso GM, Johnsen G (2023) Photophysiologically active green, red, and brown macroalgae living in the Arctic Polar Night. Sci Rep 13:17971. https://doi.org/10.1038/s41598-023-44026-5
Sunday JM, Pecl GT, Frusher S, Hobday AJ, Hill N, Holbrook NJ et al (2015) Species traits and climate velocity explain geographic range shifts in an ocean-warming hotspot. Ecol Lett 18:944–953. https://doi.org/10.1111/ele.12474
Tait LW, Chin C, Nelson W, George S, Marriott P, O'Driscoll RL, Lamare M, Mills VS, Cummings VJ (2024) Deep-living and diverse Antarctic seaweeds as potentially important contributors to global carbon fixation. Commun Earth Environ 5:205. https://doi.org/10.1038/s43247-024-01362-2
Taton A, Grubisic S, Balthasart P, Hodgson DA, Laybourn-Parry J, Wilmotte A (2006) Biogeographical distribution and ecological ranges of benthic cyanobacteria in East Antarctic lakes. FEMS Microbiol Ecol 57(2):272–289. https://doi.org/10.1111/j.1574-6941.2006.00110.x
Thomsen MS, Mondardini L, Alestra T, Gerrity S, Tait L, South PM, Lilley SA, Schiel DR (2019) Local extinction of Bull Kelp (*Durvillaea* spp.) due to a Marine heatwave. Front Mar Sci 6:84. https://doi.org/10.3389/fmars.2019.00084

Thomsen MS, Mondardini L, Thoral F, Gerber D et al (2021) Cascading impacts of earthquakes and extreme heatwaves have destroyed populations of an iconic marine foundation species. Divers Distrib. https://doi.org/10.1111/ddi.13407

Timmermans ML, Marshall J (2020) Understanding Arctic Ocean circulation: a review of ocean dynamics in a changing climate. J Geophys Res Oceans 125:e2018JC014378. https://doi.org/10.1029/2018JC014378

Tin T, Fleming Z, Hughes KA, Ainley D, Convey P et al (2009) Impacts of local human activities on the Antarctic environment: a review. Antarct Sci 21:3–33

Tonicelli GA, Croce ME, Díaz-Tapia P, Fredericq S, Freshwater DW, Gauna MC, Parodi ER, Verbruggen H, Hommersand MH (2021) Meridionella gen. nov., a new genus of Cystocloniaceae (Gigartinales, Rhodophyta) from the Southern Hemisphere, including M. obtusangula comb. nov. and M. antarctica sp. nov. J Phycol 57. https://doi.org/10.1111/jpy.13127

Turner J, Bindschadler R, Convey P, di Prisco G, Fahrbach E, Gutt J, Hodgson D, Mayewski P, Summerhayes C (eds) (2009) Antarctic climate change and the environment. Scientific Committee on Antarctic Research, Cambridge, xi + 526 pp.

Turner J, Guarino MV, Arnatt J, Jena B, Marshall GJ, Phillips T, Bajish CC, Clem K, Wang Z, Andersson T, Murphy EJ, Cavanagh R (2020) Recent decrease of summer sea ice in the Weddell Sea, Antarctica. Geophys Res Lett 47(11). https://doi.org/10.1029/2020GL087127

Ulaski BP, Sikes DS, Konar B (2023) Beach-cast and drifting seaweed wrack is an important resource for marine and terrestrial macroinvertebrates in high latitudes. Mar Environ Res 187. https://doi.org/10.1016/j.marenvres.2023.105970

Valdivia N (2020) Seaweed-herbivore interactions: grazing as biotic filtering in intertidal antarctic ecosystems. In: Gómez I, Huovinen P (eds) Antarctic seaweeds. Springer. https://doi.org/10.1007/978-3-030-39448-6_13

Valdivia N, Díaz MJ, Holtheuer J, Garrido I, Huovinen P, Gómez I et al (2014) Up, down, and all around: scale-dependent spatial variation in rocky-shore communities of Fildes Peninsula, King George Island, Antarctica. PLoS One 9:e100714. https://doi.org/10.1371/journal.pone.0100714

van Oppen MJH, Olsen JL, Stam WT, van den Hoek C, Wiencke C (1993) Arctic-Antarctic disjunctions in the benthic seaweeds *Acrosiphonia arcta* (Chlorophyta) and *Desmarestia viridis/willii* (Phaeophyta) are of recent origin. Mar Biol 115:381–386. https://doi.org/10.1007/BF00349835

van Oppen M, Diekmann OE, Wiencke C, Stam WT, Olsen JL (1994) Tracking dispersal routes: phylogeography of the Arctic-antarctic disjunct seaweed *Acrosiphonia arcta* (chlorophyta) 1. J Phycol 30(1):67–80

Vernet M, Geibert W, Hoppema M, Brown PJ, Haas C, Hellmer HH et al (2019) The Weddell Gyre, Southern Ocean: present knowledge and future challenges. Rev Geophys. https://doi.org/10.1029/2018rg000604

Walther GR, Post E, Convey P, Parmesan C, Menzel M, Beebee TJC, Fromentin JM, Hoegh-Guldberg O, Bairlein F (2002) Ecological responses to recent climate change. Nature 416:389–395

Wegeberg S (2023) Biogeography of high latitude macroalgal flora: Greenland diversity in a macroecological perspective. Res Square. https://doi.org/10.21203/rs.3.rs-3068259/v1

Wegeberg S, Søndergaard J, Geertz-Hansen O (2023) Elements and sugars in kelp and fucoid species in Greenland, correlation and seasonality. Algal Res 75. https://doi.org/10.1016/j.algal.2023.103240

Wernberg T, Krumhansl KA, Filbee-Dexter K, Pedersen MF (2018) Status and trends for the world's kelp forests. In: Sheppard C (ed) World seas: an environmental evaluation, Vol III: ecological issues and environmental impacts. Elsevier

Weslawski JM, Wiktor J, Zajaczkowski M, Swerpel S (1993) Intertidal zone of Svalbard: 1. Microorganism distribution and biomass. Polar Biol 13:73–79. https://doi.org/10.1007/s003000050157

Wesławski JM, Zajączkowski M, Wiktor J, Szymelfenig M (1997) Intertidal zone of Svalbard: 3. Littoral of a subarctic, oceanic island: Bjornoya. Polar Biol 18:45–52. https://doi.org/10.1007/s003000050157

Weykam G, Gómez I, Wiencke C, Iken K, Klöser H (1996a) Photosynthetic characteristics and C:N ratios of macroalgae from King George Island (Antarctica). J Exp Mar Biol Ecol 204:1–22. https://doi.org/10.1016/0022-0981(96)02576-2

Weykam G, Gómez I, Wiencke C, Iken K, Klöser H (1996b) Photosynthetic characteristics and C:N ratios of macroalgae from King George Island (Antarctica). J Exp Mar Biol Ecol 204(1–2):1–22. https://doi.org/10.1016/0022-0981(96)02576-2

Wiencke C, Amsler CD (2012) Seaweed biology, vol 219, pp 265–292. https://doi.org/10.1007/978-3-642-28451-9

Wiencke C, Clayton MN (2002) Antarctic seaweeds. In: Wagele JW (ed) Synopses of the Antarctic Benthos. Lichtensein, German, p 241

Wiencke C, Bartsch I, Bischoff B, Peters AF, Breeman AM (1994) Temperature requirements and biogeography of Antarctic, Arctic and amphiequatorial seaweeds. Bot Mar 37:247–260. https://doi.org/10.1515/botm.1994.37.3.247

Wiencke C, Gómez I, Pakker H, Flores-Moya A, Altamirano M, Hanelt D, Bischof K, Figueroa FL (2000) Impact of UV-radiation on viability, photosynthetic characteristics and DNA of brown algal zoospores: Implications for depth zonation. Mar Ecol Prog Ser 197:217–229. https://doi.org/10.3354/meps197217

Wiencke C, Vögele B, Kovaltchouk NA, Hop H (2004) Species composition and zonation of marine benthic macroalgae at Hansneset in Kongsfjorden, Svalbard. Ber Polarforsch Meeresforsch 492:55–62. http://epic.awi.de/11710/

Wiencke C, Gómez I, Dunton K (2009) Phenology and seasonal physiological performance of polar seaweeds. Bot Mar 52:585–592. https://doi.org/10.1515/bot.2009.078

Wiencke C, Amsler CD, Clayton MN (2014) Macroalgae. In: De Broyer C, Koubbi P, Griffiths HJ, Raymond B, Udekem d'Acoz C (eds) Biogeographic atlas of the southern ocean. Scientific Committee on Antarctic Research, Cambridge, pp 66–73

Wiktor JM, Tatarek A, Kruss A, Singh RK, Søreide JE (2022) Comparison of macroalgae meadows in warm Atlantic versus cold Arctic regimes in the high-Arctic Svalbard. Front Mar Sci. https://doi.org/10.3389/fmars.2022.1021675

Wilce RT (1990) Role of the Arctic Ocean as a bridge between Atlantic and Pacific Oceans: fact and hypothesis. In: Garbary DJ, South GR (eds) Evolutionary biogeography of the marine algae of North Atlantic. Springer, Berlin, pp 323–348

Wilce RT (2016) The "Arctic Stamp," its imprint on an endangered marine flora. The arctic benthic algal flora and its environment seen from 65 years of arctic research. Perspect Phycol 3:155–180

Wilson KL, Skinner MA, Lotze HK (2018) Projected 21st-century distribution of canopy-forming seaweeds in the Northwest Atlantic with climate change. Divers Distrib. https://doi.org/10.1111/ddi.12897

Wulff A, Iken K, Quartino ML, Al-Handal A, Wiencke C, Clayton MN (2009) Biodiversity, biogeography and zonation of marine benthic micro- and macroalgae in the Arctic and Antarctic. Bot Mar 52. https://doi.org/10.1515/BOT.2009.072

Xie Z et al (2022a) Legacy and emerging organic contaminants in the Polar regions. Sci Total Environ 835. https://doi.org/10.1016/j.scitotenv.2022.155376

Xie Z, Zhang P, Wu Z, Zhang S, Wei L, Mi L et al (2022b) Legacy and emerging organic contaminants in the Polar Regions. Sci Total Environ 835:155376. https://doi.org/10.1016/j.scitotenv.2022.155376

Zacher K, Hanelt D, Wiencke C, Wulff A (2007) Grazing and UV radiation effects on an Antarctic intertidal microalgal assemblage: a long-term field study. Polar Biol 30:1203–1212. https://doi.org/10.1007/s00300-007-0278-y

Zacher K, Rautenberger R, Hanelt D, Wulff A, Wiencke C (2009) The abiotic environment of polar marine benthic algae. Bot Mar 52. https://doi.org/10.1515/bot.2009.082

Zachos J, Pagani M, Sloan L, Thomas E, Billups K (2001) Trends, rhythms, and aberrations in global climate 65 Ma to present. Science 292:686–693. https://doi.org/10.1126/science.1059412

Zakaria NN, Convey P, Gomez-Fuentes C, Zulkharnain A, Sabri S, Shaharuddin NA, Ahmad SA (2021) Oil bioremediation in the marine environment of Antarctica: a review and bibliometric keyword cluster analysis. Microorganisms 9:419. https://doi.org/10.3390/microorganisms9020419

Zaneveld JS (1966a) *Lola irregularis* (Chlorophyta-Cladophoraceae): a new species from the Ross Sea, Antarctica. J Phycol 2:45–47

Zaneveld JS (1966b) The occurrence of benthic marine algae under shore fast-ice in the western Ross Sea, Antarctic. Proc Int Seaweed Symp 5:217–231

Zaneveld JS, Sanford RB (1980) Crustose corallinaceous algae (Rhodophyta) of the New Zealand and United States scientific expedition to the Ross Sea, Balleny Islands, and Macquarie Ridge, 1965. Blumea Biodivers Evol Biogeogr Plants 26(1):205–231

Zielinski K (1990) Bottom macroalgae of Admiralty Bay (King George Island, South Shetland Island, Antarctic). Polar Res 11:95–131

Chapter 4
Life in the Abyss: Exploring Microbial Diversity and Adaptation in the Deep Sea

Roshan Lal Sahu and Robin Anigo Minj

Abstract Microorganisms are the pioneers of life on Earth, adapting and thriving in diverse environments. The deep sea, also known as the abyss, remains one of the most enigmatic and least explored environments on Earth. This vast and dark underwater realm is characterized by extreme pressure, low temperatures, and limited nutrient availability. Despite these harsh conditions, an astonishing array of life thrives in the depths, including unique microbial communities that have adapted to survive under extreme pressure. This study delves into the fascinating world of piezophiles, microorganisms that have evolved to thrive under immense pressure in the deep sea. Simultaneously, this study helps us to understand the adaptation mechanism of microbes under high-pressure environments such as those found in the deep sea. Membrane structure adaptations, protein structure adaptations, DNA and RNA stability, pressure-responsive gene expression, compatible solutes, pressure-responsive enzymes, and enzyme stability are some common adaptation mechanisms in piezophiles microbes. By utilizing advanced sampling techniques and cutting-edge molecular analysis, scientists have uncovered a remarkable diversity of microbial life in the abyss. Additionally, it explores the potential biotechnological applications of these extremophiles, with their enzymes showing promise in various industrial processes and pharmaceutical development.

Keywords Piezophiles · Extremophiles · Microbial diversity · Adaptation

R. L. Sahu (✉)
Department of Botany, Government Naveen College, Kamleshwarpur, Chhattisgarh, India

R. A. Minj
Directorate of Forensic Science and Laboratory, Ranchi, Jharkhand, India

A. W. Fernandes Duarte et al. (eds.), *Polar Microbes and Climate Change*,
https://doi.org/10.1007/978-3-032-07096-8_4

4.1 Introduction

Most challenging habitats or extreme environments are characterized by extreme temperatures, pH levels, salinity, pressure, and radiation. Piezophiles, also known as barophiles, are organisms that thrive under high-pressure conditions, particularly in the deep sea. The deep sea is characterized by increasing hydrostatic pressure with depth, reaching several hundred atmospheres in the deepest parts of the ocean. Piezophiles are typically found at depths greater than 1000 m (3300 feet), where the pressure exceeds 100 times that at sea level. They have evolved unique adaptations to withstand the extreme pressure found in the depths of the ocean. These microorganisms have not only adapted to the physical challenges of their environment but have also developed intricate symbiotic relationships with the organisms around them, further enriching the complexity of this ecosystem (Dick 2019). The study showcases the significance of microbial communities in deep-sea ecosystems, acting as primary producers and playing vital roles in nutrient cycling and energy transfer (Thurber et al. 2014). Despite the remoteness of the deep sea, it is not immune to anthropogenic disturbances, such as deep-sea mining and climate change. Understanding the microbial diversity and ecological dynamics in this environment becomes crucial for developing effective conservation strategies to safeguard these unique habitats and their invaluable biodiversity. The findings of this book chapter contribute significantly to our understanding of deep-sea ecosystems and their potential applications in biotechnology, while also emphasizing the importance of responsible stewardship in preserving these delicate habitats for future generations.

4.2 Adaptations of Piezophiles

Piezophiles, also known as barophiles or piezotolerant microbes, are organisms that thrive in the high-pressure environments of the deep sea, where pressures can reach extreme levels. These organisms have evolved specific mechanisms of adaptation to cope with the immense pressure in the deep ocean.

4.2.1 Membrane Structure Adaptations

Membrane structure adaptations in deep-sea organisms play a crucial role in enabling their survival in the extreme pressure and cold temperatures of the abyss. These adaptations are essential for maintaining cellular integrity, regulating transport processes, and responding to changes in the environment. Several studies have shed light on the unique membrane adaptations found in deep-sea organisms. Deep-sea organisms experience tremendous hydrostatic pressure, which can disrupt cellular structures (Macdonald 2021). To counteract this pressure, they possess specialized membranes with unique lipid compositions. Studies have shown that deep-sea bacteria and certain eukaryotic organisms have higher proportions of unsaturated fatty acids in their membranes, which enhances membrane fluidity and flexibility, allowing them to withstand the extreme pressure (DeLong and Yayanos 1986). Deep-sea environments are characterized by low temperatures, which can lead to decreased membrane fluidity and hinder essential cellular processes. To adapt to cold temperatures, many deep-sea organisms incorporate more polyunsaturated fatty acids (PUFAs) into their membranes (Bartlett and Bidle 1999). These PUFAs prevent membrane rigidity and enhance the fluidity required for proper cellular function in cold conditions (Bartlett and Bidle 1999). Additionally, some deep-sea fish have evolved antifreeze proteins on their cell membranes to prevent ice crystal formation (Hazel and Williams 1990a, b). Membranes of deep-sea organisms must withstand rapid changes in pressure due to vertical migration. Some species have adapted by modifying the lipid composition of their membranes to maintain stability under varying pressure conditions (Bartlett et al. 2007). For example, certain deep-sea bacteria increase the proportion of phospholipids with branched chains, allowing them to maintain membrane integrity even when exposed to drastic pressure changes (Bartlett et al. 2007). Membrane adaptations in deep-sea organisms also play a role in optimizing nutrient uptake. Studies have found that certain deep-sea bacteria have specialized transport proteins in their membranes that facilitate the uptake of important nutrients, such as phosphorus and nitrogen, from the surrounding environment (Bartlett and Bidle 1999). These adaptations are crucial for the survival of microorganisms in the oligotrophic conditions of the deep sea (Lauro et al. 2004). Overall, these membrane structure adaptations in deep-sea organisms are a testament to the remarkable ability of life to adapt to extreme environments. The understanding of these adaptations not only expands our knowledge of the deep-sea ecosystem but also has potential biotechnological implications in areas such as industrial processes and pharmaceutical development. Phospholipids that incorporate unsaturated fatty acids exhibit decreased melting points and increased molecular flexibility in comparison to their saturated counterparts (Roach et al. 2004). As a result, numerous organisms, such as bacteria, plants, and poikilothermic animals, adjust the proportion of unsaturated fatty acids in their phospholipids to maintain optimal membrane fluidity and functionality in response to fluctuations in environmental temperature and pressure. This adaptive process is termed "homeoviscous" adaptation (DeLong and Yayanos 1985).

4.2.2 Protein Structure Adaptations

Piezophiles produce proteins that maintain their stability and function under high pressure (Sarma et al. 2023). These proteins often have increased structural compactness and stability, allowing them to resist denaturation and maintain their activity in the deep-sea environment (Gross and Jaenicke 1994; Sarma et al. 2023).

Deep-sea organisms often possess proteins with enhanced structural stability to withstand the high hydrostatic pressure. These proteins have evolved to resist denaturation and maintain their native conformations even under extreme pressure conditions. Researchers have found that deep-sea enzymes, such as proteases and lipases, have greater rigidity in their active sites and other regions critical for catalysis, which contributes to their stability in high-pressure environments (Bartlett 2002). Proteins from deep-sea organisms have adapted to function optimally in cold temperatures. These proteins often possess flexible regions and increased structural dynamics, allowing them to maintain activity at lower temperatures (Gross and Jaenicke 1994). Additionally, deep-sea enzymes have been found to have a higher proportion of charged and polar residues on their surfaces, which contributes to their solubility and functionality in the cold environment (Siddiqui and Cavicchioli 2006). In the nutrient-poor environment of the deep sea, proteins in deep-sea organisms have adapted to perform efficiently with limited resources. For example, transport proteins that facilitate nutrient uptake have evolved to have higher affinity for scarce nutrients, allowing organisms to effectively scavenge and utilize essential compounds (Kaur et al. 2022). Additionally, some proteins involved in nutrient metabolism may be more efficient and less dependent on cofactors that might be scarce in the deep sea (Campanaro et al. 2008). The deep-sea environment can be challenging for proper protein folding, leading to an increased risk of protein misfolding and aggregation. Deep-sea organisms have adapted by producing specialized chaperone proteins that assist in proper protein folding, stabilization, and refolding of denatured proteins under pressure and low-temperature conditions (Gross and Jaenicke 1990). These chaperones play a critical role in maintaining protein homeostasis and preventing proteotoxicity in the deep sea (Pruski and Dixon 2007). Overall, the protein structure adaptations in deep-sea organisms are fascinating examples of how life has evolved to thrive in extreme environments. The understanding of these adaptations not only enriches our knowledge of deep-sea biology but also has potential applications in biotechnology and medicine, inspiring the development of novel proteins and enzymes that can function efficiently under extreme conditions.

4.2.3 DNA and RNA Stability

High pressure in the deep sea can cause DNA and RNA damage (Chen et al. 2021). Piezophiles have evolved mechanisms to stabilize their nucleic acids and protect against pressure-induced damage. DNA and RNA stability adaptations in deep-sea organisms are critical for the maintenance and functionality of genetic material in the extreme conditions of the abyss. These adaptations are necessary to protect the integrity of the genetic code, ensure proper gene expression, and enable survival in the high-pressure, cold, and nutrient-limited environment. Deep-sea organisms have evolved efficient DNA repair mechanisms to counteract the increased damage caused by high levels of ultraviolet (UV) radiation and oxidative stress in the upper layers of the ocean. These repair systems include nucleotide excision repair (NER), base excision repair (BER), and photoreactivation repair, among others, to correct DNA lesions and maintain genomic stability (Shao et al. 2022). Similarly, RNA repair mechanisms, such as RNA editing and degradation pathways, help ensure the fidelity of RNA transcripts and minimize errors in gene expression (Cheng et al. 2019). Deep-sea organisms may possess specialized DNA and RNA polymerases with enhanced processivity and fidelity to cope with the high hydrostatic pressure and low temperatures (Carré et al. 2023). These adaptations help maintain accurate replication and transcription despite the challenging environmental conditions. Studies have shown that certain deep-sea bacteria and archaea have DNA polymerases with increased thermostability and resistance to denaturation under pressure (Jin et al. 2019). Deep-sea organisms may have specific RNA secondary structures that confer stability to RNA molecules under cold temperatures (Bernhardt and Tate 2012). Stable RNA secondary structures, such as hairpins and stem-loops, can prevent RNA degradation and maintain the integrity of functional RNA molecules, such as transfer RNA (tRNA) and ribosomal RNA (rRNA) (Deutscher 2006). Additionally, deep-sea organisms may produce unique RNA-binding proteins that protect and stabilize RNA transcripts in the harsh environment (DeLong and Yayanos 1986). In eukaryotic organisms, histones and chromatin structure play essential roles in DNA packaging, gene regulation, and DNA repair. Deep-sea organisms may have adapted histone proteins that provide increased protection to DNA and maintain stable chromatin structures under high pressure and cold conditions. These adaptations ensure that the genetic material is adequately protected and accessible for essential cellular processes (Shao et al. 2022). Overall, the DNA and RNA stability adaptations in deep-sea organisms are remarkable examples of how life has evolved to thrive in extreme environments. These adaptations ensure the accurate transmission of genetic information and proper gene expression, ultimately enabling survival and successful reproduction in the deep sea. The understanding of these mechanisms not only enhances our knowledge of deep-sea biology but also has implications in various scientific and biotechnological fields.

4.2.4 Pressure-Responsive Gene Expression

Piezophiles exhibit specific gene expression patterns in response to changes in pressure. They use pressure-responsive genes to adapt their cellular processes and maintain optimal functioning in high-pressure conditions (Michoud and Jebbar 2016). High pressure in the deep sea also affects osmotic balance within cells. Deep-sea organisms may employ various osmoregulation mechanisms to control the movement of water and ions across their cell membranes and maintain cellular homeostasis (Csonka and Hanson 1991). These adaptations help prevent osmotic stress and its detrimental effects on cellular function. In response to changes in pressure, deep-sea organisms activate specific gene expression patterns to adapt to their environment. Pressure-responsive genes are regulated by transcription factors or other regulatory elements that sense changes in pressure and trigger the appropriate gene expression responses. For example, certain genes involved in stress responses, DNA repair, and membrane stabilization are upregulated under high pressure to maintain cell integrity and functionality. Overall, pressure-responsive gene expression and adaptations in high pressure are crucial for the survival and functionality of deep-sea organisms. These adaptations allow them to thrive in an environment where pressure can fluctuate drastically with depth and oceanographic conditions. The study of pressure-responsive gene expression provides valuable insights into the molecular mechanisms underlying life's ability to adapt and thrive in extreme environments. Protein synthesis is the most pressure-sensitive process (Scheck and Landau 1982). Numerous research studies have aimed to pinpoint the step that acts as a bottleneck. In this process, initially, experiments involving ribosomes from E. coli conducted in a controlled lab environment showed that ribosomal subunits undergo separation when subjected to elevated pressure (Gross and Jaenicke 1990; Schulz et al. 1976). More recently, within living organisms, a shift towards dissociation in the balance of the ribosomal complex and its individual components has been observed, notably through differential scanning calorimetry (DSC) (Alpas et al. 2003; Niven et al. 1999). This phenomenon has been linked to the pressure's impact on cell viability. These findings suggest two key points: (i) ribosome dissociation is favored under high pressures due to a reduction in volume, and (ii) the stability of ribosomes is closely tied to cell viability under high-pressure conditions (Scheck and Landau 1982).

4.2.5 Compatible Solutes

Some piezophiles accumulate compatible solutes, such as proline, glycine, glutamate, and betaine, to protect cellular structures from pressure-induced damage and maintain osmotic balance (Simonato et al. 2006). Deep-sea organisms face a challenging environment characterized by extreme pressure, cold temperatures, and limited food availability. To survive in these conditions, they have evolved various adaptations, including the accumulation of certain compounds known as compatible solutes or osmolytes. Compatible solutes are small, organic molecules that help organisms maintain cellular function and stability under stressful conditions, such as high pressure, low temperature, and salinity. These solutes are typically highly soluble in water and can protect biological macromolecules from denaturation or damage caused by these extreme conditions (Yancey 2004). In the deep-sea environment, organisms accumulate compatible solutes to counteract the effects of high pressure. The increased pressure at greater depths can affect cell membranes, protein structures, and enzyme activities. By accumulating compatible solutes, these organisms can counterbalance the impact of pressure on their cellular processes and maintain the functionality of essential biomolecules (Yancey 2004). Compatible solutes found in deep-sea organisms include Trimethylamine N-oxide (TMAO): TMAO is a common compatible solute in marine animals, including deep-sea organisms (Yancey 2004). It helps to stabilize proteins and maintain the structural integrity of biological macromolecules under high-pressure conditions. Betaine is another osmolyte that accumulates in deep-sea organisms. It plays a role in protein stabilization and helps to protect enzymes from pressure-induced denaturation (Simonato et al. 2006). Ectoine and Hydroxyectoine are two compatible solutes found in various microorganisms, including deep-sea bacteria. They act as protectants against high-pressure environments, protecting cellular structures and enzymes from damage (Czech et al. 2018). Taurine is an amino acid derivative found in some deep-sea organisms. It helps regulate osmotic balance and counteracts the effects of high pressure on cell membranes (Yancey et al. 2002). Glycine Betaine: This osmolyte is involved in osmotic regulation and protein stabilization, making it beneficial for deep-sea organisms coping with high-pressure conditions. Accumulating compatible solutes is just one of several strategies that deep-sea organisms have evolved to survive in their extreme habitat. Additionally, they may have specialized enzymes, heat shock proteins, and other adaptations to thrive in the high-pressure, cold, and nutrient-poor conditions of the deep sea. These adaptations have fascinated scientists and provided valuable insights into the limits of life on Earth and the potential for life in extreme environments elsewhere in the universe.

4.3 Biotechnological Potential

Piezophiles and other deep-sea microbes have attracted attention for their unique enzymes and metabolic pathways with potential biotechnological applications. Enzymes from piezophiles have been studied for their industrial use in applications such as food processing, detergent production, and pharmaceuticals (Abe and Horikoshi 2001). High-pressure adaptive organisms produce enzymes that display increased stability and activity under high-pressure conditions. These enzymes, known as piezozymes, have garnered attention for their potential industrial applications. For instance, high-pressure-stable proteases and lipases have been used in the detergent industry, as they remain active and stable under the high-pressure conditions present in washing machines, leading to better cleaning efficiency (Abe and Horikoshi 2001). Furthermore, piezozymes hold promise in various biotechnological processes where high pressures are involved, such as biofuel production and pharmaceutical synthesis (González-González et al. 2017). On the other hand, high-pressure adaptive organisms produce a variety of biomolecules with unique properties that can be harnessed for biotechnological purposes. For example, lipids and membrane proteins from deep-sea organisms have been studied for their potential use in the development of stabilizers for pharmaceutical formulations and nanotechnology applications. Additionally, certain pigments and bioactive compounds produced by these organisms have shown promising pharmaceutical and nutraceutical properties (Kiuru et al. 2014). High-pressure adaptive microorganisms have the potential to contribute to bioremediation efforts in deep-sea oil spills and other high-pressure contaminated environments. These organisms possess specialized metabolic pathways that can degrade hydrocarbons and other pollutants even under extreme pressure conditions. Their use in bioremediation could provide more efficient and effective approaches for cleaning up oil spills and other environmental disasters in the deep ocean (Zhang et al. 2018a, b). Moreover, high-pressure processing (HPP) is an emerging food preservation technique that relies on the use of high pressures to inactivate spoilage microorganisms and pathogens while preserving the sensory and nutritional qualities of food products. High-pressure adaptive microorganisms play a role in HPP, as their enzymes and cellular components can withstand the pressure conditions involved in this process. Additionally, the discovery of pressure-tolerant yeasts and bacteria has opened new possibilities for fermentation-based food and beverage production under high-pressure conditions (Reineke et al. 2013). Furthermore, these organisms inhabit extreme environments with limited resources, leading to the evolution of unique secondary metabolites and bioactive compounds. Bioprospecting efforts focused on these organisms have revealed the presence of novel natural products with potential pharmaceutical applications. These compounds can serve as leads for the development of new drugs to treat various diseases, including cancer, bacterial infections, and neurological disorders (Raddadi et al. 2015). The biotechnological applications of high-pressure adaptive organisms have shown promising results across various industries. Enzymes with enhanced stability, novel biomolecules, and potential bioremediation applications are some of the exciting areas being explored. Additionally, the discovery of unique bioactive compounds from these organisms has opened new avenues for drug discovery and development.

4.4 Challenges of Studying Deep-Sea Microbes

The extreme depths of the ocean make it challenging to collect samples and study these organisms directly. Laboratory conditions: Replicating high-pressure conditions in the lab for studying piezophiles is technically demanding. Studying deep-sea microbes presents a lot of challenges due to the extreme and remote nature of their habitat. These challenges span from technological limitations to logistical complexities, making it a fascinating yet demanding field of research. The harsh conditions of the deep sea, characterized by high pressure, low temperatures, and limited light availability, make sampling and experimentation challenging. Traditional sampling methods often struggle to retrieve intact samples without disturbing delicate microbial communities. Additionally, maintaining the pressure and temperature conditions during sample retrieval and transportation to the surface is critical for accurate analysis. The lack of light in deep-sea environments has also limited the use of visual methods for studying these microbes. Unlike surface ecosystems where visual observation is common, researchers rely on alternative methods such as remotely operated vehicles (ROVs) and autonomous underwater vehicles (AUVs) equipped with advanced imaging and sensing technologies. These tools provide valuable insights but are costly to deploy and operate, restricting their use and coverage. Cultivating deep-sea microbes in the lab remains a significant challenge. The majority of deep-sea microorganisms have not been successfully cultured, limiting our understanding of their physiology, metabolism, and interactions. This "microbial dark matter" represents a substantial knowledge gap that hinders comprehensive ecosystem studies. Recent advances in culturing techniques, such as high-pressure bioreactors and microfluidic systems, offer promising avenues to address this challenge (Zhang et al. 2018a, b). Furthermore, the vastness and complexity of deep-sea environments contribute to logistical hurdles. Conducting research in remote, deep-sea locations demands extensive planning, sophisticated equipment, and skilled personnel. Limited research vessel availability, unpredictable weather conditions, and high costs associated with deep-sea expeditions can hinder consistent and widespread data collection (Danovaro et al. 2014). Molecular analysis of deep-sea microbes is another challenge. Extracting genetic material and analyzing complex microbial communities from environmental samples require specialized methods. However, these techniques can be prone to biases, making it challenging to accurately assess the diversity and function of deep-sea microbial populations. Developing robust and standardized molecular protocols is crucial for reliable comparative studies. Collaboration among researchers from various disciplines is vital for overcoming these challenges. Oceanographers, microbiologists, engineers, and data scientists must work together to develop innovative approaches and technologies. Integrating multi-omics analyses, which combine data from genomics, metagenomics, transcriptomics, and other fields, can provide a more comprehensive understanding of deep-sea microbial ecosystems (Danovaro et al. 2014). The inference is that studying deep-sea microbes is a rewarding endeavor that unveils the hidden secrets of one of Earth's most enigmatic ecosystems. However, the challenges presented by extreme conditions, technological limitations, and logistical complexities are considerable. Overcoming these challenges demands collaborative efforts, innovative technologies, and interdisciplinary approaches to unravel the intricate web of life in the deep sea.

4.5 Importance of Deep-Sea Microbes in Climate Change

Deep-sea microbes play a crucial role in regulating the global carbon cycle, influencing the sequestration and release of carbon dioxide in the oceans. Deep-sea microbes contribute significantly to carbon sequestration, playing a crucial role in the biological pump. Phytoplankton in surface waters photosynthesize and produce organic matter, which sinks to the deep sea as marine snow. Microbes in the deep sea then break down this organic matter, facilitating its storage in the deep ocean, thus removing carbon dioxide from the atmosphere (Thurber et al. 2014). Moreover, deep-sea microbes help to control methane emissions, which are a potent greenhouse gas. They participate in anaerobic processes that break down methane, preventing its release into the atmosphere. Cold seeps and methane hydrate environments are home to methane-consuming microbes, which play a crucial role in maintaining methane balance (Niemann et al. 2006). Deep-sea ecosystems can act as refuges from ocean acidification impacts due to their relatively stable pH levels compared to surface waters (Thurber et al. 2014). Microbes in these environments can have unique metabolic pathways and tolerance mechanisms, potentially offering insights into how microbial communities could adapt to changing ocean conditions (Glover and Smith 2003). Climate change can alter deep-sea currents, nutrient transport, and temperature gradients. These changes impact microbial communities and their functions, potentially leading to feedback loops that further influence climate processes. Understanding the intricate interactions between deep-sea microbes and the changing climate is crucial for predicting and managing these feedbacks (Levin and Le Bris 2015). Deep-sea microbes play pivotal roles in climate change dynamics, participating in carbon sequestration, methane regulation, and ecosystem resilience. Recognizing their contributions and vulnerabilities is essential for holistic climate change mitigation and adaptation strategies.

4.6 Conclusion

The adaptations of piezophiles, or deep-sea organisms, to extreme pressure environments showcase the remarkable resilience and ingenuity of life. Membrane structure adaptations, including unique lipid compositions and antifreeze proteins, enable these organisms to thrive in the abyss. Protein structure adaptations, such as increased rigidity and charged residues, ensure stability and functionality in high-pressure and cold conditions. DNA and RNA stability mechanisms, including repair systems and specialized polymerases, safeguard genetic material in the challenging deep-sea environment. On the other hand, pressure-responsive gene expression allows piezophiles to dynamically regulate their cellular processes, while the accumulation of compatible solutes, like TMAO and betaine, provides a defence against pressure-induced damage. Deep-sea organisms play a pivotal role in nutrient cycling, contributing to carbon, nitrogen, sulfur, and phosphorus cycles, highlighting their importance in maintaining the delicate balance of the ocean ecosystem. These adaptations not only deepen our understanding of deep-sea biology but also hold promise for biotechnological and medical applications. As we unravel the mysteries of life in extreme environments, we gain insights into the potential for life beyond Earth and the intricacies of sustaining ecosystems in the harshest corners of our planet. Ultimately, the study of piezophiles broadens our perspective on the adaptability and resilience of life in the face of seemingly insurmountable challenges and opens up the scope towards the biotechnological potential.

References

Abe F, Horikoshi K (2001) The biotechnological potential of piezophiles. Trends Biotechnol 19(3):102–108

Alpas H, Lee J, Bozoglu F, Kaletunç G (2003) Evaluation of high hydrostatic pressure sensitivity of Staphylococcus aureus and Escherichia coli O157: H7 by differential scanning calorimetry. Int J Food Microbiol 87(3):229–237

Bartlett DH (2002) Pressure effects on in vivo microbial processes. Biochim Biophys Acta 1595(1–2):367–381
Bartlett DH, Bidle KA (1999) Membrane-based adaptations of deep-sea piezophiles. In: Enigmatic microorganisms and life in extreme environments. Springer, Dordrecht, pp 501–512
Bartlett DH, Lauro FM, Eloe EA (2007) Microbial adaptation to high pressure. In: Physiology and biochemistry of extremophiles. Wiley, pp 331–348
Bernhardt HS, Tate WP (2012) Primordial soup or vinaigrette: did the RNA world evolve at acidic pH? Biol Direct 7(1):1–12
Campanaro S, Treu L, Valle G (2008) Protein evolution in deep sea bacteria: an analysis of amino acids substitution rates. BMC Evol Biol 8:1–15
Carré L, Henneke G, Henry E, Flament D, Girard É, Bruno F (2023) DNA polymerization in icy moon abyssal pressure conditions. Astrobiology
Chen H, Wang M, Li M, Lian C, Zhou L, Zhang X et al (2021) A glimpse of deep-sea adaptation in chemosynthetic holobionts: Depressurization causes DNA fragmentation and cell death of methanotrophic endosymbionts rather than their deep-sea Bathymodiolinae host. Mol Ecol 30(10):2298–2312
Cheng J, Hui M, Sha Z (2019) Transcriptomic analysis reveals insights into deep-sea adaptations of the dominant species, Shinkaia crosnieri (Crustacea: Decapoda: Anomura), inhabiting both hydrothermal vents and cold seeps. BMC Genomics 20:1–16
Csonka LN, Hanson AD (1991) Prokaryotic osmoregulation: genetics and physiology. Ann Rev Microbiol 45(1):569–606
Czech L, Hermann L, Stöveken N, Richter AA, Höppner A, Smits SH et al (2018) Role of the extremolytes ectoine and hydroxyectoine as stress protectants and nutrients: genetics, phylogenomics, biochemistry, and structural analysis. Genes 9(4):177
Danovaro R, Snelgrove PV, Tyler P (2014) Challenging the paradigms of deep-sea ecology. Trends Ecol Evol 29(8):465–475
DeLong EF, Yayanos AA (1985) Adaptation of the membrane lipids of a deep-sea bacterium to changes in hydrostatic pressure. Science 228(4703):1101–1103
DeLong EF, Yayanos AA (1986) Biochemical function and ecological significance of novel bacterial lipids in deep-sea procaryotes. Appl Environ Microbiol 51(4):730–737
Deutscher MP (2006) Degradation of RNA in bacteria: comparison of mRNA and stable RNA. Nucleic Acids Res 34(2):659–666
Dick GJ (2019) The microbiomes of deep-sea hydrothermal vents: distributed globally, shaped locally. Nat Rev Microbiol 17(5):271–283
Glover AG, Smith CR (2003) The deep-sea floor ecosystem: current status and prospects of anthropogenic change by the year 2025. Environ Conserv 30(3):219–241
González-González R, Fucinos P, Rúa ML (2017) An overview on extremophilic esterases. In: Extremophilic enzymatic processing of lignocellulosic feedstocks to bioenergy. Springer, Cham, pp 181–204
Groß M, Jaenicke R (1990) Pressure-induced dissociation of tight couple ribosomes. FEBS Lett 267(2):239–241
Gross M, Jaenicke R (1994) Proteins under pressure: the influence of high hydrostatic pressure on structure, function and assembly of proteins and protein complexes. Eur J Biochem 221(2):617–630
Hazel JR, Williams EE (1990a) The role of alterations in membrane lipid composition in enabling physiological adaptation of organisms to their physical environment. Progress in Lipid Research 29(3):167–227
Hazel JR, Williams EE (1990b) The role of alterations in membrane lipid composition in enabling physiological adaptation of organisms to their physical environment. Prog Lipid Res 29(3):167–227
Jin M, Gai Y, Guo X, Hou Y, Zeng R (2019) Properties and applications of extremozymes from deep-sea extremophilic microorganisms: a mini review. Mar Drugs 17(12):656
Kaur J, Vishnu AL, Khipla N, Kaur J (2022) Microbial life in cold regions of the deep sea. In: Survival strategies in cold-adapted microorganisms. Springer, Singapore, pp 63–86

Kiuru P, D'Auria MV, Muller CD, Tammela P, Vuorela H, Yli-Kauhaluoma J (2014) Exploring marine resources for bioactive compounds. Planta Med 80(14):1234–1246

Lauro FM, Bertoloni G, Obraztsova A, Kato C, Tebo BM, Bartlett DH (2004) Pressure effects on Clostridium strains isolated from a cold deep-sea environment. Extremophiles 8:169–173

Levin LA, Le Bris N (2015) The deep ocean under climate change. Science 350(6262):766–768

Macdonald A (2021) Life at high pressure. Springer, Cham

Michoud G, Jebbar M (2016) High hydrostatic pressure adaptive strategies in an obligate piezophile Pyrococcus yayanosii. Sci Rep 6(1):27289

Niemann H, Lösekann T, De Beer D, Elvert M, Nadalig T, Knittel K et al (2006) Novel microbial communities of the Haakon Mosby mud volcano and their role as a methane sink. Nature 443(7113):854–858

Niven GW, Miles CA, Mackey BM (1999) The effects of hydrostatic pressure on ribosome conformation in Escherichia coli: an in vivo study using differential scanning calorimetry. Microbiology 145(2):419–425

Pruski AM, Dixon DR (2007) Heat shock protein expression pattern (HSP70) in the hydrothermal vent mussel Bathymodiolus azoricus. Mar Environ Res 64(2):209–224

Raddadi N, Cherif A, Daffonchio D, Neifar M, Fava F (2015) Biotechnological applications of extremophiles, extremozymes and extremolytes. Appl Microbiol Biotechnol 99:7907–7913

Reineke K, Schlumbach K, Baier D, Mathys A, Knorr D (2013) The release of dipicolinic acid – the rate-limiting step of Bacillus endospore inactivation during the high pressure thermal sterilization process. Int J Food Microbiol 162(1):55–63

Roach C, Feller SE, Ward JA, Shaikh SR, Zerouga M, Stillwell W (2004) Comparison of cis and trans fatty acid containing phosphatidylcholines on membrane properties. Biochemistry 43(20):6344–6351

Sarma J, Sengupta A, Laskár MK, Sengupta S, Tenguria S, Kumar A (2023) Microbial adaptations in extreme environmental conditions. In: Bacterial survival in the hostile environment. Academic Press, pp 193–206

Scheck AC, Landau JV (1982) The effect of high hydrostatic pressure on eukaryotic protein synthesis. Biochim Biophys Acta 698(2):149–157

Schulz E, Jaenicke R, Knoche W (1976) Pressure-jump relaxation studies of the association-dissociation reaction of E. coli ribosomes. Biophys Chem 4(3):253–257

Shao G, He T, Mu Y, Mu P, Ao J, Lin X et al (2022) The genome of a hadal sea cucumber reveals novel adaptive strategies to deep-sea environments. Iscience 25(12)

Siddiqui KS, Cavicchioli R (2006) Cold-adapted enzymes. Annu Rev Biochem 75:403–433

Simonato F, Campanaro S, Lauro FM, Vezzi A, D'Angelo M, Vitulo N et al (2006) Piezophilic adaptation: a genomic point of view. J Biotechnol 126(1):11–25

Thurber AR, Sweetman AK, Narayanaswamy BE, Jones DO, Ingels J, Hansman RL (2014) Ecosystem function and services provided by the deep sea. Biogeosciences 11(14):3941–3963

Yancey PH (2004) Compatible and counteracting solutes: protecting cells from the Dead Sea to the deep sea. Sci Prog 87(1):1–24

Yancey PH, Blake WR, Conley J (2002) Unusual organic osmolytes in deep-sea animals: adaptations to hydrostatic pressure and other perturbants. Comp Biochem Physiol A Mol Integr Physiol 133(3):667–676

Zhang S, Hu Z, Wang H (2018a) A retrospective review of microbiological methods applied in studies following the deepwater horizon oil spill. Front Microbiol 9:520

Zhang Z, Wu Y, Zhang XH (2018b) Cultivation of microbes from the deep-sea environments. Deep-Sea Res II Top Stud Oceanogr 155:34–43

Chapter 5
The Biocatalytic Potential of Extremozyme and Efficient Downstream Processing

Mohit Gururani, Ajay Kumar, and Rajpal Srivastav

Abstract Extremophilic microbes thrive in extreme environmental conditions and are an important resource for industries, especially in enzyme, metabolite production, etc. Extreme conditions of temperature, salinity, pressure, etc., led to the emergence of extremophiles with enzymes designed to survive in these hostile conditions. Extremophiles have evolved enzymes with remarkable structural stability and catalytic activity in extreme environments through adaptive evolution, which presents potential opportunities for biocatalysis and enzyme mechanistic studies. There are various significant aspects associated with extremozyme extraction, fermentation condition optimization, isolation, and purification methods, such as ultrafiltration and chromatography for producing high-purity enzyme preparations, which are prerequisites for efficient industrial use. The current article discusses the potential of enzymes and downstream processing targeted at improving functionality from extremophilic microorganisms.

Keywords Biocatalysis · Extremophiles · Extremozymes · Stress tolerance · Downstream processing

5.1 Introduction

Extremophiles are microorganisms capable of flourishing in exceedingly harsh settings, such as Arctic and Antarctic regions with very low temperatures, acidic lakes, high-pressure environments, and highly saline lakes. These organisms are categorized based on the extreme conditions they can tolerate. The microbes can survive in extreme polar conditions. Thermophiles and hyperthermophiles can grow at temperatures above 80 °C, with some even surviving at temperatures exceeding 110 °C. Psychrophiles can withstand temperatures as low as 2–20 °C. Piezophiles

M. Gururani · A. Kumar · R. Srivastav (✉)
Amity Institute of Biotechnology, Amity University Uttar Pradesh, Noida, India
e-mail: rsrivastav2@amity.edu

A. W. Fernandes Duarte et al. (eds.), *Polar Microbes and Climate Change*,
https://doi.org/10.1007/978-3-032-07096-8_5

Table 5.1 Techniques for the analysis and chemical characterization of pigments produced by fungi

Technique	Information	Advantage	Limitation
UV-Vis	Absorption spectrum in the ultraviolet and visible	Low-cost instrumentation; useful for exploring optical properties	Low diagnostic importance
FT-IR (mid)	Vibrational spectrum of chemical bonds	Speed; high sensitivity; non-destructive analysis	Complex spectra containing broad bands with overlapping bands
NMR	Differentiation of hydrogens and carbons in molecular structure	High diagnostic significance in the structural elucidation of organic compounds; non-destructive analysis	Low sensitivity, requiring considerable amounts of analyte
MS	Molecular weight	High sensitivity; allows determining the molecular formula of a compound; analyzes compounds of different molecular dimensions and polarities	Requires specific ionization methods for certain classes of organic compounds
LC-MS	Molecular weight	High sensitivity; analysis of complex samples	Longer analysis time
GC-MS	Molecular weight	High sensitivity; analysis of complex samples; reproducible spectra	Restricted to the analysis of volatile compounds

can survive under pressures greater than 500 atmospheres. Acidophiles thrive in environments with a pH below 4, while alkaliphiles prefer environments with a pH above 9. Halophiles can survive in high salinity conditions, such as environments with 30% salt content. Extremophiles generate extremozymes capable of catalyzing reactions in environments unsuitable for mesophiles (Table 5.1).

Enzymes derived from mesophilic bacteria exhibit optimal activity at moderate temperatures but fail to catalyze chemical reactions under harsh conditions. The use and development of novel extremozymes is being done to meet industry requirements. The global industrial enzyme market has seen consistent growth over the years. Since their introduction into business operations during the 1960s, the market has witnessed steady expansion. It is expected that by 2024, the market will further expand to reach $10.5 billion, indicating a compound annual growth rate of 5.7%. Extremozymes have revolutionized biotechnology and industrial processes, particularly those operating in harsh conditions. These enzymes have evolved unique processes and molecular mechanisms to ensure the stability of their cellular components in challenging environments (Kakkar and Wadhwa 2022). The extremophiles are classified based on their ability to survive in various extreme conditions. The extremophilic microbes can survive in various kinds of extreme environments such as extreme heat, extreme cold, extreme pH, high salinity, and high pressure (Fig. 5.1). To facilitate their survival in such extreme environments, microbes have developed several adaptations, as discussed in the following sections.

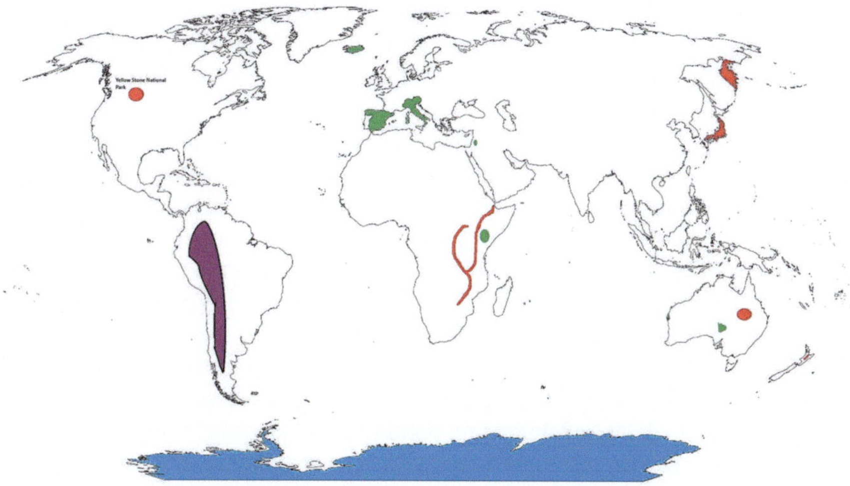

Fig. 5.1 The representation of geographical distributions of extremophiles across the globe. Extremophiles, organisms that thrive in extreme conditions, have been found on every continent across the globe

5.2 Extremophilic Adaptations

Thermophilic organisms are known for their ability to thrive in elevated temperatures, a trait attributed to the stability of their membrane lipids. These lipids form a permeability barrier regulating the passage of nutrients across the cell, alongside crucial components such as rRNA and tRNA molecules. Additionally, these organisms exhibit distinct features, including higher G + C concentrations, a prevalence of amino acids with higher charges on their surfaces, and the presence of intramolecular salt bridges. These characteristics collectively support their adaptations to high-temperature environments (Pucci and Rooman 2017).

Psychrophiles are organisms capable of thriving in temperatures ranging from 0 to 20 °C, with their optimal proliferation occurring at 5 °C. These organisms possess unique adaptations that enable them to thrive in cold environments. One such adaptation is the presence of unsaturated fatty acids in their cell membranes, which remain in a liquid state even at low temperatures, facilitating the movement of substances across the membrane. Additionally, psychrophiles produce cold shock proteins and cryoprotectants, which aid in the synthesis of cold-adapted enzymes. Various alterations in membrane fluidity, reduced levels of transcription and translation, and modifications in ribosomal structure contribute to the adjustments in cellular machinery necessary for survival (Srivastav et al. 2014). To prevent freezing, psychrophiles employ different strategies, such as accumulating compatible solutes like glycine betaine, mannitol, and sucrose, or producing antifreeze proteins that inhibit the formation of ice crystals. Furthermore, psychrophiles utilize the formation of exopolysaccharides as a defensive mechanism against freezing conditions

(De Maayer et al. 2014). These organisms exhibit complex metabolic adjustments, including altered nutrient transportation methods, protein denaturation in cold conditions, internal ice formation, and modifications in protein folding (Grover et al. 2019).

Alkaliphiles, organisms thriving in high pH environments, exhibit adaptive strategies to maintain cytoplasmic pH balance and counteract alkalinity. They achieve this through electrogenic, secondary cation/proton antiporters, facilitating the absorption of H+ ions. Additionally, these microorganisms employ protective mechanisms such as the secretion of acidic compounds like acidic amino acids, teichuronopeptide, and teichuronic acid outside the cell to mitigate the effects of high pH levels. Some alkaliphiles produce cytochrome C on their outer membrane, which plays a role in regulating proton transfer across the membrane. Moreover, alkaliphiles are known to produce phosphoserine aminotransferase, enhancing hydrophobic interactions and increasing stability in alkaline environments by incorporating negatively charged amino acids at the interface (Somayaji et al. 2022).

Acid-loving microorganisms thrive in environments characterized by extreme acidity, for example, Acidobacterium, *Leptospirillum, Picrophilus,* and *Ferroplasma.* These microorganisms typically require a neutral intracellular pH; the mechanisms underlying pH regulation in acidophiles remain poorly understood. Membrane impermeability plays a crucial role in controlling proton influx into the cell, influencing intracellular pH balance. Archaea possess a natural resilience to low pH due to the composition of their cell membrane, particularly the tetraether lipids, which limit proton permeability (Johnson and Quatrini 2020).

Halophiles, a group of organisms capable of thriving in high salt environments, exhibit remarkable adaptations to sustain themselves amidst elevated salt concentrations. These microorganisms can endure salt concentrations as high as 15% or 2.5 M. Their survival hinges on intricate adjustments within their cellular and enzymatic systems to maintain osmotic balance. To cope with high salt levels, halophiles employ various strategies. One such approach involves the production of osmoprotectants, which are organic solutes aiding in osmotic equilibrium. Additionally, halophiles employ techniques like the low salt, organic solute-in, and high salt-in strategies to maintain a balance between cellular and ambient salt concentrations. Furthermore, halophiles have evolved enzymes that are optimized for their saline habitats. These enzymes are characterized by an increased proportion of negatively charged amino acids, which facilitate ion exchange and contribute to protein solubility. Overall, the adaptive mechanisms of halophiles enable them to thrive in hostile saline environments (Yin et al. 2015).

Organisms that thrive under intense hydrostatic pressure are commonly referred to as piezophiles. These unique organisms exhibit adaptations to withstand such pressures through various mechanisms. For instance, studies on *S. violaceae* have revealed that piezophiles counter high-pressure environments by upregulating genes that operate effectively under such conditions. Additionally, they may regulate the expression of genes involved in transcription, translation, or alter proteome structure to enhance stability under pressure. In a specific case study involving the piezophilic *Pyrococcus yayanosii* CH1, researchers have identified several pressure-responsive genes. These genes play crucial roles in pathways such as chemotaxis, formate and hydrogenases metabolism for energy production, translation processes, and the identification of palindromic sequences linked to the degradation of vulnerable proteins (Michoud and Jebbar 2016).

5.3 Extremophile Plant Interactions

The microbial counterpart and the plant have important interactions; bacteria help plants become more resilient to environmental stressors and increase their fitness, while plants can influence the dynamics of the microbial community to foster beneficial relationships, including nutritional and niche requirements (Tiwari et al. 2021). Additionally, the associated microbial communities affect how plants respond to changing climate conditions. For example, plant-growth-promoting rhizobacteria (PGPR), fungal endophytes, mycorrhizal associations, nitrogen fixation symbiosis, and others all affect how flexible and resilient plants are under adverse circumstances (Uzilday and Ganie 2023).

Microbes known as acidophiles are those that can survive in extremely acidic conditions—pH less than three. Acidophiles inhabit a variety of ecological niches, such as animal stomachs, deep-sea vents, volcanic zones, and hydrothermal regions. Excessive acidity has been shown to negatively impact plant growth, resulting in alterations to the pH and nutrient availability of the soil. The plant maintains its internal pH to counteract highly acidic environments. Acidic environments are home to a variety of rhizospheric microorganisms. Acid-tolerant plant-growth-promoting bacteria (PGPB), such as *Acidithiobacillus*, *Flavobacterium*, *Lysinibacillus, Pseudomonas,* and *Methylobacterium*, have been identified and studied. Bacterial domains and Archaea contain the best-characterized acidophiles. It has been discovered that the microbial communities connected to the agricultural plants are crucial for the health of the soil, crop productivity, and the generation of siderophores, which reduce abiotic stress (Mukhtar et al. 2019; Srivastav et al. 2019). Furthermore, in an acidic environment, these siderophores mediate iron uptake and convert Fe3+ to Fe2+ ions. Acidic soils have low pH levels, which encourage the use of PGP bacteria for plant growth (Tiwari et al. 2021). Alkaliphiles are microorganisms from alkaline environments that can withstand high pH levels (>9). These extremophiles have physiological and genetic defenses against hostile environments. Many types of bacteria thriving in environments abundant in methane and hydrogen are commonly discovered in the rhizosphere of plants that thrive in alkaline conditions. Examples of such plants include *Smallanthus sonchifolius, Chrysanthemum morifolium*, and *Dichanthium annulatum* (Mukhtar et al. 2019). In these alkaline conditions, microbes maintain cytoplasmic pH through protein and enzyme activity. There seems to be a complex yet sensitive relationship at the cellular and metabolic level (Tiwari et al. 2021; Radhakrishnan et al. 2024).

5.4 Temperature and Its Role

Elevated temperatures have a negative impact on plant growth by changing the permeability of membranes, germination of seeds, and photosynthesis rate. The bacteria found in the rhizosphere of plants that thrive in high temperatures, such as *Cupressus dupreziana, Triticum aestivum,* and *Sporobolus indicus*, aid in the growth of plants under hot circumstances. Through a number of processes, including nitrogen fixation, phosphorus and zinc solubilization, phytohormone synthesis, HCN, and siderophore synthesis, PGPB promotes plant growth in *Dichanthium annulatum*. Many bacterial species, including *Arthrobacter*, *Streptomyces*, *Pseudomonas*, and *Staphylococcus* sp., are employed as bioformulations for plant growth in high-temperature environments because of the potential of plant-growth-promoting bacteria (PGPB) to promote plant growth. Extremophile metabolism and physiology are negatively impacted by high temperatures; therefore, microbial enzymes boost the expression of heat-tolerant proteins to aid in high-temperature acclimatization and the preservation of cell integrity and structure. A number of fungal species, including *Myceliophthora fergusii, Aspergillus terreus, T. byssochlamydoides, Scytalidium thermophilum, Myceliophthora thermophila,* and *Talaromyces thermophilus*, have been isolated from hot environments (Yang et al. 2020). Psychrotrophic microorganisms are cold-adapted bacteria with potential applications in the biotech and pharmaceutical industries. Cold-adapted microorganisms can be found everywhere: on the summits of mountains, in glaciers, frozen lakes, and snow, often coexisting with plants that are adapted to such environments. Numerous microorganisms were found and classified into categories like *Bacteroidetes, Euryarchaeota, Ascomycota, Basidiomycota, Chlamydiae, Actinobacteria, Cyanobacteria,* and *Chloroflexi* using both culture-dependent and independent approaches. Microorganisms that inhabit cold climates are significant from an ecological perspective because they have an impact on many aquatic and terrestrial ecosystems, or both. Moreover, Antarctica and other incredibly cold regions of the planet have reported the existence of cold-region extremophiles. Among the extremophiles with a wide range of characteristics are the unusual psychrophilic bacteria *Octadecabacter arcticus, Sphingomonas glacialis, Halobacterium lacusprofundi,* and *Cellulophaga algicola* (Kanekar and Kanekar 2022).

Further, an important factor is salt concentration in the environment. Salinity threatens most agricultural land globally because it causes osmotic stress and ion-induced toxicity, which is detrimental to plant growth and affects microbiological functioning. Moreover, salinity in the soil affects plants negatively in a variety of ways, such as seed germination, nutrient and water uptake, crop productivity, and ecological balance (Bacilio et al. 2016). Numerous studies have demonstrated the potential benefits of endophytic bacteria and PGP in reducing the negative effects of salinity in soil. Salt-tolerant bacteria use a range of direct and indirect tactics, including the production of phytohormones, resource mobilization, nitrogen fixation, and siderophore synthesis, to fight salt stress. These microbial methods increase root length, number of roots, and surface area through nutrient absorption. Additionally, the principal indirect strategy of salt-tolerant bacteria gives resistance against pathogen invasion by reducing the frequency of pathogen infection (Shrivastava and Kumar 2015).

5.5 Extremozymes from Extremophiles

5.5.1 Thermophilic Proteins

Over the past four decades, extensive research has been conducted on thermophilic bacteria, particularly focusing on their extremophilic nature. These organisms exhibit the remarkable ability to thrive in temperatures ranging from 41 to 122 °C. A diverse array of enzymes derived from thermophilic bacteria has been extensively studied and characterized. Among these enzymes are lipases, pectinases, phytases, amylases, cellulases, proteases, mannanases, esterases, xylanases, pullulanases, and chitinases. Enzymes sourced from thermophilic bacteria demonstrate robustness under harsh conditions, including exposure to denaturing agents, organic solvents, and high salinity. Their versatility enables them to withstand proteolysis and various adverse environments. Applications of these enzymes encompass the reduction of contamination risk, maintenance of substrate low adhesiveness, and enhancement of substrate solubility (Raddadi et al. 2015).

Thermozymes exhibit unique physical attributes and electrostatic interactions that enable them to maintain their activity even under extreme temperatures. These enzymes demonstrate various adaptations, including the capacity to uphold their structural integrity and functionality when exposed to extreme heat. Additionally, they can enhance the accumulation of hydrophobic residues, facilitating bisulfide bonding between oppositely charged ions. In both biotechnological and industrial settings, there is a demand for thermostable enzymes like lipases, which find applications in diverse processes such as grease hydrolysis, esterification, interesterification, transesterification, and organic biosynthesis. Furthermore, thermozymes play a crucial role in the production of optical nanosensors and analytes. Lipase finds extensive use across multiple sectors, including the paper industry, milk processing, pigment production, leather processing, and healthcare. Among the thermozymes, proteases are utilized for synthesizing dipeptides and in starch processing, while cellulase, hemicellulases, and xylanases are integral to the paper bleaching process and environmental remediation efforts. Modern biodetergents incorporate enzymes such as amylase, protease, cellulase, and lipase, which exhibit resilience to harsh conditions (Nadaroglu and Polat 2022). Thermophilic enzymes like amylase sourced from *Pyrococcus furiosus* have demonstrated utility in genetic studies. Mutations in pancreatic fistula amylase have resulted in increased production of maltoheptaose from β-cycloamyloses. Maltoheptaose serves as a carrier in various industries, including food, cosmetics, and pharmaceuticals. Despite their high-temperature performance, thermophilic enzymes may not exhibit catalytic activity at lower temperatures. Nonetheless, their potential for biotechnological applications remains significant, particularly in high-temperature environments in present industrial settings (Radhakrishnan et al. 2023; Siddiqui 2015).

5.5.2 *Piezophilic Proteins and Molecules*

Piezophiles, organisms thriving in high hydrostatic pressures prevalent in deep-sea environments like deep-sea and volcanic zones, such as *Pyrococcus abyssi*, have garnered attention. One example is the piezophilic protein, like the stable peptidase sourced from *Pyrococcus horikoshii*, which maintains its integrity even under extreme pressure conditions. Notably, piezophilic bacteria lack specialized saline channels for stability, unlike thermopiezophiles that adapt to both low temperature and high pressure. The potential of piezophilic enzymes in various industrial applications is substantial. However, these enzymes from piezophilic microbes remain limited. Studies have revealed intriguing findings, such as the α-amylase from piezophilic proteins producing trisaccharides and tetrasaccharides from maltooligosaccharides under high-pressure and low-energy conditions. This enzymatic reaction holds promise for industrial and biotechnological applications, particularly in the food industry. Piezophilic proteins have demonstrated remarkable effectiveness in diverse sectors, including detergents, food production, and chemical manufacturing (Bhatt et al. 2020).

5.5.3 *Acidophilic Proteins*

Organisms known as acidophiles thrive in environments with a pH below 3–4. Despite this, there is still much to uncover about the adaptability of proteins within acidophiles. For instance, research has shown that the endo-β-glucanase enzyme found in *Sulfolobus solfataricus* remains stable at its optimal pH of 1.8. However, it loses stability when subjected to pH levels below 2. α-glucosidase from *Ferroplasma acidiphilum* shows a preference for a pH of 3 over the internal pH of *Ferroplasma acidiphilum*, which is 5.6. Enzymes derived from acidophilic bacteria have significant potential in various biotech and pharma industries, including the production of biofuels and ethanol. Specifically, cellulolytic and xylanolytic enzymes function effectively in acidic environments, even at high temperatures, facilitating the hydrolysis of cellulose-based materials for better control and utilization (Kour et al. 2019).

5.5.4 Halophilic Proteins

Halophilic microorganisms exhibit a remarkable ability to thrive in environments with high salt concentrations, typically at least 1 M NaCl. They have developed various chemical, structural, and physiological adaptations to support the selection and stability of proteins possessing specific physicochemical properties. These halophilic microbes can be categorized into three groups based on their preferred salt concentrations: extreme halophiles, which thrive in environments with 2500–5200 mM NaCl; moderate halophiles, which prefer concentrations ranging from 500 to 2500 mM NaCl; and slightly halophilic organisms, which can survive in environments with 200–500 mM NaCl. Enzymes derived from halophiles employ distinct adaptation mechanisms, enabling them to remain stable even under conditions of low water activity and in the presence of organic solvents (Elleuche et al. 2014). Extremozymes sourced from halophiles, including xylanases, amylases, proteases, and lipases, are produced by a variety of organisms such as *Acinetobacter, Haloferax, Halobacterium, Halorhabdus, Marinococcus, Micrococcus, Natronococcus, Bacillus, Halobacillus,* and *Halothermothrix*. Among these, lipases and esterases hold significant promise for industrial applications, particularly in the realms of polyunsaturated fatty acids, food processing, and biodiesel production. Notably, the enzymatic activity of ligase N from *Haloferax volcanii* is notably enhanced in the presence of KCl, whereas its activity is markedly reduced in the presence of NaCl (Schreck and Grunden 2014). As characteristics, halophilic enzymes have very little solubility in aqueous/organic and non-aqueous solutions (Dumorné et al. 2017).

5.6 Significance of Extremozymes

Proteins and enzymes are highly sensitive to modifications in chemical or physical properties in vivo and in vitro (Srivastav et al. 2014; Hu et al. 2023). Efforts to bolster the diminishing reservoir of fossil fuels have led to a focused search for manufacturing alternative fuels from biomass sources like corn, beets, wheat, and sugar cane. Biofuels are categorized into first and second generation, the former being derived from readily hydrolyzed sugars, starches, and oils of easily accessible crops, while the latter is sourced from lignocellulosic material, which is less prone to hydrolysis. Biofuels are further distinguished by their resulting end products: butanol, ethanol, hydrogen, methane, and biodiesel. Recent developments have seen the emergence of larger-scale systems utilizing microorganisms like *Caldicellulosiruptor saccharolyticus* and *Thermotoga elfii*. Strains of *Caldicellulosiruptor, Thermoanaerobacterium, Pyrococcus,* and *Aeropyrum* hold significant potential. Although still in its nascent stages, recent breakthroughs such as the engineering of hyperthermophiles show promise. Among the compounds produced by microorganisms, biodiesel and butanol stand out for their notable economic success (Dadwal

Table 5.2 The example of different types of extremozymes produced by extremophiles and their industrial applications

SN	Extremophile(s)	Enzyme (s)	Industrial use(s)	Reference(s)
1	*Pyrolobus fumarii*	DNA polymerases	Genetic engineering	Vieille and Zeikus (2001)
2	*Psychromonas ingrahamii*	B-galactosidase	Lactose hydrolysis	Bhaskar et al. (2023)
3	*Pircrophilus oshimae*	Sulfur oxidation	Desulfurization of coal	Blohs et al. (2019)
4	*Natronococcus occultus*	Pectinases	Pulp bleaching	Chinnathambi (2015)
5	*Halobacterium*	Rheological polymers	Oil recovery	DasSarma and DasSarma (2015)
6	*Thermococcus barophilus*	Whole microorganism	Formation of starch granules	Gupta et al. (2014)

et al. 2021). Extremophiles and their enzymes have drawn interest in the mining sector in addition to their usage in biofuels. This process, known as bioleaching, uses microorganisms to dissolve metal oxides or sulfides that are intractable. Compared to traditional heap leaching techniques, which rely on different chemicals, such as cyanide, to separate some minerals/metals from others, bioleaching offers a safer and more environmentally friendly option for metal recovery (Table 5.2).

Proteases, lipases, and glycosyl hydrolases collectively constitute a significant portion of the enzyme supply, exceeding 70%. Among these, proteases stand out as the most widely employed enzymes. Their versatility finds applications across various industries. Notably, in laundry detergents, proteases have been integral components. They play a crucial role in breaking down and removing stains of protein origin (Vojcic et al. 2015). Proteases find significant application in various industries such as cheese production, brewing, and baking. Typically, these industries employ mesophilic proteases sourced from Bacillus microbes. However, there have been attempts to utilize psychrophilic proteases for cold water washing, although their effectiveness is limited by instability at room temperature. Through guided evolution, a hybrid psychrophilic/mesophilic protease was developed, enhancing performance during cold water washing.

Lipases constitute a lucrative industry, valued at billions, due to their versatility in industrial settings, characterized by a wide substrate range, high specificity, and stability. Despite the demand for lipases in harsh conditions like low temperatures in laundry detergents and low water activity in organic synthesis, the predominant lipases utilized are mesophilic, often derived from organisms like *Bacillus* and *Aspergillus* species. Extremophilic lipases are typically overlooked, although lipases from thermophilic Bacillus species have demonstrated superior efficacy. Glycosyl hydrolases play a crucial role in breaking glycosidic bonds between carbohydrates and other molecules, classified into numerous families exceeding 100. The hydrolysis process usually involves only two amino acids—an acid proton donor and a nucleophile/base—resulting in retention or inversion of the anomeric structure of the resultant carbohydrate (Coker 2016).

5.7 Enzymes Synthesis and Isolation

Enzymes can be derived from various sources, including animals (such as pepsin and trypsin), plants (like papain and bromelain), and microorganisms (including pectinase, glucoamylase, and α-amylase). Microbes offer a viable alternative source of enzymes, alongside plants and animals. They are particularly advantageous due to their ease of handling and high growth capacity. Notably, a single strain of microorganism can produce multiple types of enzymes, facilitating the operation of a versatile fermentation facility capable of producing a range of enzymes or focusing on specific ones as needed. This approach enables producers to adapt to fluctuations in market demand for particular enzymes. Additionally, utilizing microorganisms for enzyme production allows for comprehensive control over all stages, from sourcing to the final product. Advances in genetic modification techniques, such as DNA recombinant and editing methods, offer the potential to enhance enzyme yield and diversify the range of enzymes available for commercial use. Ultimately, the widespread adoption of microbial enzymes in industry is driven by the abundance of suppliers and the availability of standardized formulations (Vitolo 2019).

5.8 Downstream Processing of Extremozymes

The cell's protective envelopes, comprising the cytoplasmic membrane and cell wall, pose natural barriers hindering the release of macromolecules. Particularly in the creation of high-value products from microbial, plant, or animal sources, the increased cost may occur. Moreover, costs of production may be offset by simultaneously isolating various products post-cell disruption and/or by directly dissolving them in the fermentation broth (Md Nadzir et al. 2023). In downstream processing, a series of unit operations is utilized to concentrate and purify an enzyme commonly found in a raw extract (Fig. 5.2). The procedures involve a series of steps, including cell disruption, ultrafiltration, crystallization, coagulation, precipitation, chromatography, drying, sedimentation, standardization, packaging, evaporation, flocculation, centrifugation, and filtration. There may be significant interactions of these contaminants at the molecular level, which can be cancerous (Radhakrishnan et al. 2024). Therefore, ensuring the safety of the enzyme product is paramount, necessitating its freedom from allergens, hazardous compounds, and pathogenic bacteria.

5.8.1 Phase Separation

A range of centrifuge types, such as the tubular bowl, multi-chamber, disc bowl, solid-bowl scroll, and perforate bowl basket centrifuges, is available to meet the needs of phase separation. Among the industrial filtration methods, the plate and

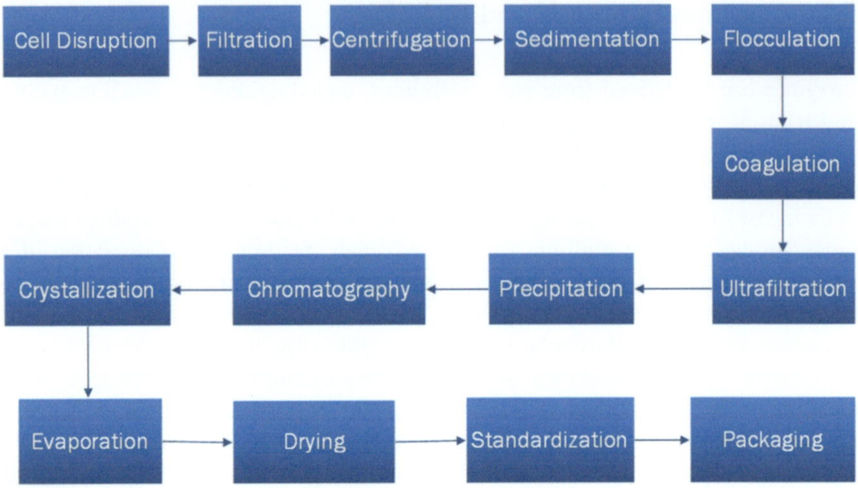

Fig. 5.2 The representation of the workflow of the downstream processing of extremozymes

frame press and the rotary drum filter are commonly employed. The plate and frame press comprises corrugated plates with filter cloths sandwiched for further separation through filtration (Chisti and Moo-Young 2020). These unit operations take advantage of differences in the density of the particles in relation to the surrounding fluid. Gravity-driven sedimentation is a procedure that helps separate solids from liquids; it usually takes place in rectangular or circular flow tanks. In the meantime, centrifugation uses mechanical force to separate solids from liquids and has become a popular technique for removing solids, particularly when filtration is insufficient, including when removing colloidal or gelatinous materials. An important phenomenon is ultracentrifugation, which happens when centrifugal force exceeds gravity by a factor of 500,000 or more. The force experienced by colloidal solutions in the ultracentrifuge rotor cells is determined by the angular velocity of the rotor and the distance of the particles from the axis of rotation. This force, along with the features of the particles and the suspending liquid, determines the speeds at which the particles travel. Ultracentrifugation is used to separate nuclear, microsomal, and mitochondrial fractions from damaged tissue cells and to purify viral proteins in addition to tissue proteins (Williamson et al. 2015).

5.8.2 *Extraction*

When the charges of minute particles in a liquid are neutralized, coagulation takes place, causing the particles to stick to one another directly. By adding polyvalent ions with opposing charges, such as inorganic salts or organic polyelectrolytes, the particles are neutralized and allowed to combine. Conversely, flocculation entails the creation of aggregates with a loose structure. In this case, the flocculating agent serves as a link between the particles and can take the form of materials like gelatin or charged/uncharged synthetic polymers. Although they can help with charge neutralization and thereby promote flocculation, inorganic ions are unable to cause flocculation. On the other hand, coagulation and flocculation can be induced simultaneously by organic polyelectrolytes. Extraction refers to the process of isolating enzymes from a raw cell-free extract. This can be achieved through various methods such as ultrafiltration, solvent extraction, or aqueous two-phase systems. Ultrafiltration, a direct separation method, involves using a membrane with a specific cut-off to retain the desired enzyme while removing undesired by-products. Solvent extraction, particularly using water-immiscible organic solvents, is a widely utilized technique in the chemical industry for separating lipophilic compounds, often on a large scale. In addition to its applications in the chemical industry, extraction plays a crucial role in isolating antibiotics (Nadar et al. 2017). Further, there exist several chromatographic techniques for separating or purifying biological substances: (a) Ion-exchange chromatography relies on the electric charge of molecules for separation. Its efficacy is particularly notable in initial downstream stages when handling large quantities. The resolution, retention capabilities, and speed depend on the matrix used, such as divinyl-styrene copolymer, crystalline cellulose, or dextran. Gel filtration is efficient for desalting and buffer exchanging (Ahluwalia 2023).

One of four liquid chromatography techniques can be used to separate proteins based on their distinct characteristics. The first technique, affinity chromatography, depends on specific interactions between the protein and the resin. The second technique, known as ionic exchange chromatography, takes advantage of the proteins' surface charge and their interactions with the resin. The third is reversed-phase chromatography, which depends on the hydrophobicity of the enzyme. The fourth and last technique is size exclusion chromatography, which separates the proteins based on their molecular weight. Following purification, the target protein is evaluated using a variety of biochemical characteristics, including ideal pH and temperature, heat resistance, inhibitor sensitivity, maximal catalytic velocity (Vm), and Michaelis-Menten constant (Km), among others. The comprehensive characterization procedure is necessary to determine the catalytic effectiveness of the enzyme (Fenwick et al. 2011; Srivastav et al. 2019). At this point, achieving a high level of purification is crucial to get a clean and complete amino acid sequence of the protein, which is necessary for the identification of future genes within the microorganism's genome. The discovery of this gene, which is discussed in more detail in the next section, may make it possible to produce the recombinant form of the protein. It is important to remember that enzyme characterization is crucial since it offers essential details regarding the enzyme's activity for industrial uses. Based on its functioning, this information allows evaluation of the many markets and processes in which the enzyme may find application. Numerous examples of extremozymes have been discovered to date, successfully purified, and characterized for use in both industrial and research contexts thanks to the conventional functional screening procedure (Radhakrishnan et al. 2023). Examples of such enzymes that have been isolated from various Antarctic microbes are lipases, nitrilase, glutamate dehydrogenase, and superoxide dismutase (Cáceres-Moreno et al. 2019).

5.8.3 *Formulations*

The process involves developing an enzyme formulation suitable for sale in either powder or solution form, tailored to meet the specific requirements of the industry in terms of type and purity. For microbial extracellular catalysts, a lyophilized cell-free supernatant processed to eliminate colored components is deemed acceptable. Industrial enzymes necessitate a baseline level of purification, with downstream protocols focused on enhancing catalytic efficiency per unit weight rather than purification. Given the industry's emphasis on cost reduction, enzyme purity levels should be minimized. In scenarios such as brewing broth and textile sizing, or in the removal of glucose traces from certain food products like dried eggs, mixed enzyme preparations may be necessary, while in others, their combination may be undesirable for example, fungal amylase used as a flour additive should be devoid of protease to prevent excessive foaming and browning resulting from the release of soluble amino acids from proteins. Conversely, enzymes used in clinical assays must not interfere with test results, and pharmaceutical enzymes such as hyaluronidase and asparaginase must be free from pyrogens. The overall percentage recovery holds significance, particularly considering the relatively low prices of raw materials and fermentation (Vitolo 2019). The industrial production of extremozymes has been powered by advancements in fermentation optimization, extraction techniques, and purification technologies (Nadar et al. 2017; Radhakrishnan et al. 2023). Further, the use of metabolic engineering and synthetic biology offers the opportunity to customize extremozymes for commercial uses, increasing their adaptability and efficiency to meet the industry demand (Suneja and Srivastav 2021).

5.9 Conclusion

The extremophilic microbes are an important part of the extreme ecosystem, including polar microbes. The extremozymes obtained from these microbes find significant applications in the pharmaceutical and biotechnology industries. Understanding their extremozymes will help us to understand biocatalysis strategies, biofuel generation, bioremediation, pharmaceuticals, etc. Furthermore, for efficient use of extremozymes, effective extraction, purification methods, and downstream processing efficiency need to be optimized.

Acknowledgments The authors acknowledge Amity University Uttar Pradesh for infrastructural support.

References

Ahluwalia VK (2023) Instrumental methods of chemical analysis

Bacilio M, Moreno M, Bashan Y (2016) Mitigation of negative effects of progressive soil salinity gradients by application of humic acids and inoculation with Pseudomonas stutzeri in a salt-tolerant and a salt-susceptible pepper. Appl Soil Ecol 107:394–404

Bhaskar P, Thoudam J, Tripathi R, Gupta V, Bisht S, Sharma A (2023) Psychrophiles as a novel and promising source of cold-adapted industrial enzymes. Appl Biol Chem J 4(2)

Bhatt BM, Trivedi UB, Patel KC (2020) Extremophilic amylases: microbial production and applications, pp 185–205

Blohs M, Moissl-Eichinger C, Mahnert A, Spang A, Dombrowski N, Krupovic M, Klingl A (2019) Archaea-an introduction. In: Encyclopedia of microbiology, pp 243–252

Cáceres-Moreno P, Muñoz-Ibacache SA, Monsalves MT, Amenabar MJ, Blamey JM (2019) Functional approach for the development and production of novel extreme biocatalysts. In: Next generation biomanufacturing, ACS symposium series, vol 1329. American Chemical Society, pp 1–22

Chinnathambi A (2015) Industrial important enzymes from alkaliphiles-an overview. Biosci Biotechnol Res Asia 12(3):2007–2016

Chisti Y, Moo-Young M (2020) Fermentation technology, bioprocessing, scale-up and manufacture. In: Biotechnology. CRC Press, pp 177–222

Coker JA (2016) Extremophiles and biotechnology: current uses and prospects. In: F1000Research, vol 5. Faculty of 1000

Dadwal A, Sharma S, Satyanarayana T (2021) Thermostable cellulose saccharifying microbial enzymes: characteristics, recent advances, and biotechnological applications. Int J Biol Macromol 188:226–244

DasSarma S, DasSarma P (2015) Halophiles and their enzymes: negativity put to good use. Curr Opin Microbiol 25:120

De Maayer P, Anderson D, Cary C, Cowan DA (2014) Some like it cold: understanding the survival strategies of psychrophiles. EMBO Rep 15(5):508–517

Dumorné K, Córdova DC, Astorga-Eló M, Renganathan P (2017) Extremozymes: a potential source for industrial applications. J Microbiol Biotechnol 27(4):649–659

Elleuche S, Schröder C, Sahm K, Antranikian G (2014) Extremozymes – biocatalysts with unique properties from extremophilic microorganisms. Curr Opin Biotechnol 29(1):116–123

Fenwick RB, Esteban-Martín S, Salvatella X (2011) Understanding biomolecular motion, recognition, and allostery by use of conformational ensembles. Eur Biophys J 40:1339–1355. https://doi.org/10.1007/s00249-011-0754-8

Grover R, Burse SA, Shankrit S, Aggarwal A, Kirty K, Narta K, Srivastav R, Ray AK, Malik G, Vats A, Motiani RK, Thukral L, Roy SS, Bhattacharya S, Sharma R, Natarajan K, Mukerji M, Pandey R, Gokhale RS, Natarajan VT (2019) Myg1 exonuclease couples the nuclear and mitochondrial translational programs through RNA processing. Nucleic Acids Res 47(11):5852–5866. https://doi.org/10.1093/nar/gkz371

Gupta GN, Srivastava S, Khare SK, Prakash V (2014) Extremophiles: an overview of microorganism from extreme environment. Int J Agric Environ Biotechnol 7(2):371

Hu Y-T, Hong X-Z, Li H-M, Yang J-K, Wei S, Wang Y-W, Liu Y-H (2023) Modifying the amino acids in conformational motion pathway of the α-amylase of *Geobacillus stearothermophilus* improved its activity and stability. Front Microbiol 14:2023. https://doi.org/10.3389/fmicb.2023.1261245

Johnson DB, Quatrini R (2020) Acidophile microbiology in space and time. Curr Issues Mol Biol 39(1):63–76

Kakkar P, Wadhwa N (2022) Extremozymes used in textile industry. J Text Inst 113(9):2007–2015

Kanekar PP, Kanekar SP (2022) Psychrophilic, psychrotrophic, and psychrotolerant microorganisms, pp 215–249

Kour D, Rana KL, Kaur T, Singh B, Chauhan VS, Kumar A, Rastegari AA, Yadav N, Yadav AN, Gupta VK (2019) Extremophiles for hydrolytic enzymes productions: biodiversity and potential biotechnological applications. In: Bioprocessing for biomolecules production. Wiley, pp 321–372

Md Nadzir S, Yusof N, Nordin N, Kamari A, Yusoff MZM (2023) A review of microalgal cell wall composition and degradation to enhance the recovery of biomolecules for biofuel production. Biofuels 14(9):979–997

Michoud G, Jebbar M (2016) High hydrostatic pressure adaptive strategies in an obligate piezophile Pyrococcus yayanosii. Sci Rep 6(1):1–10

Mukhtar S, Mehnaz S, Malik KA (2019) Microbial diversity in the rhizosphere of plants growing under extreme environments and its impact on crop improvement. Environ Sustain 2(3):329–338

Nadar SS, Pawar RG, Rathod VK (2017) Recent advances in enzyme extraction strategies: a comprehensive review. Int J Biol Macromol 101:931–957

Nadaroglu H, Polat MS (2022) Microbial extremozymes: novel sources and industrial applications. Academic Press, London, pp 67–88

Pucci F, Rooman M (2017) Physical and molecular bases of protein thermal stability and cold adaptation. Curr Opin Struct Biol 42:117–128

Raddadi N, Cherif A, Daffonchio D, Neifar M, Fava F (2015) Biotechnological applications of extremophiles, extremozymes and extremolytes. Appl Microbiol Biotechnol 99(19):7907–7913

Radhakrishnan A, Balaganesh P, Vasudevan M, Natarajan N, Chauhan A et al (2023) Bioremediation of hydrocarbon pollutants: recent promising sustainable approaches, scope, and challenges. Sustainability 15(7):5847. https://doi.org/10.3390/su15075847

Radhakrishnan A, Gangopadhyay R, Sharma C, Kapardar RK, Sharma NK, Srivastav R (2024) Unwinding helicase MCM functionality for diagnosis and therapeutics of replication abnormalities associated with cancer. Mol Diagn Ther. https://doi.org/10.1007/s40291-024-00701-5

Schreck SD, Grunden AM (2014) Biotechnological applications of halophilic lipases and thioesterases. Appl Microbiol Biotechnol 98(3):1011–1021

Shrivastava P, Kumar R (2015) Soil salinity: a serious environmental issue and plant growth promoting bacteria as one of the tools for its alleviation. Saudi J Biol Sci 22(2):123–131

Siddiqui KS (2015) Some like it hot, some like it cold: temperature dependent biotechnological applications and improvements in extremophilic enzymes. Biotechnol Adv 33(8):1912–1922

Somayaji A, Dhanjal CR, Lingamsetty R, Vinayagam R, Selvaraj R, Varadavenkatesan T, Govarthanan M (2022) An insight into the mechanisms of homeostasis in extremophiles. Microbiol Res 263:127115

Srivastav R, Kumar D, Grover A, Singh A, Manjasetty B, Sharma R, Taneja B (2014) unique subunit packing in mycobacterial nanoRNase leads to alternate substrate recognitions in DHH phosphodiesterases. Nucleic Acids Res 42(12):7894–7910

Srivastav R, Sharma R, Tandon S, Tandon C (2019) Role of DHH superfamily proteins in nucleic acids metabolism and stress tolerance in prokaryotes and eukaryotes. Int J Biol Macromol 127:66–75. https://doi.org/10.1016/j.ijbiomac.2018.12.123

Suneja G, Srivastav R (2021) Impact of microbial genome sequencing advancements in understanding extremophiles. In: Pandey A, Sharma A (eds) Extreme environments, unique ecosystems – amazing microbes, 1st edn. CRC Press. ISBN 9780367350161, p 13

Tiwari P, Srivastava Y, Bae H (2021) Trends of pharmaceutical design of endophytes as anti-infective. Curr Top Med Chem 21(17):1572–1586

Uzilday RO, Ganie SA (2023) Editorial: extremophiles: tolerance mechanisms and use in crop improvement. Front Plant Sci 14:1233202

Vieille C, Zeikus GJ (2001) Hyperthermophilic enzymes: sources, uses, and molecular mechanisms for thermostability. Microbiol Mol Biol Rev 65(1):1

Vitolo M (2019) Extraction of invertase from Saccharomyces cerevisiae: effects of rheological parameters. World J Pharm Res 8(6):12

Vojcic L, Pitzler C, Körfer G, Jakob F, Martinez R, Maurer KH, Schwaneberg U (2015) Advances in protease engineering for laundry detergents. New Biotechnol 32(6):629–634

Williamson CD, Wong DS, Bozidis P, Zhang A, Colberg-Poley AM (2015) Isolation of endoplasmic reticulum, mitochondria, and mitochondria-associated membrane and detergent resistant membrane fractions from transfected cells and from human cytomegalovirus-infected primary fibroblasts. Curr Protoc Cell Biol 68(1):3.27.1-3.27.33

Yang XY, Zhang JX, Ding QY, He ZC, Zhu CY, Zhang KQ, Niu XM (2020) Metabolites from two dominant thermophilic fungal species thermomyces lanuginosus and Scytalidium thermophilum. Chem Biodivers 17(5):e2000137

Yin J, Chen JC, Wu Q, Chen GQ (2015) Halophiles, coming stars for industrial biotechnology. Biotechnol Adv 33(7):1433–1442

Chapter 6
Insights into the Diversity, Adaptations, and Ecological Function of Antarctic Microorganisms Provided by Genomics

Mayanne Karla da Silva, Leonardo da Silva Santos, Crísthian Phellipe Santos Lima, Averlane Vieira da Silva, João Pedro Ferreira Nunes, Luiz Henrique Rosa, and Alysson Wagner Fernandes Duarte

Abstract Antarctica is one of the most remote and unknown areas on the planet, which hosts a significant adapted microbiota, represented by archaea, bacteria, fungi, viruses, and other microeukaryotic communities. Genomics is a great ally in generating information and understanding the taxonomic diversity of Antarctic microorganisms. Approaches involving ribosomal gene metabarcoding, metagenomes, shotgun, and single-cells are genomic techniques that have made it possible to access, infer, and understand the prevalent groups and their ecological, adaptive, and biotechnological significance within this polar environment. Thus, this review aimed to gather recent studies that, using genomic methods, revealed updates in the taxonomy, ecology, adaptation, and applications of prokaryotes and eukaryotes in the Antarctic ecosystem.

Keywords Environmental biodiversity · Polar · Molecular techniques · Genome · Bioinformatics

M. K. da Silva · L. da Silva Santos · C. P. S. Lima · A. V. da Silva · J. P. F. Nunes · A. W. F. Duarte (✉)
Laboratório de Microbiologia, Imunologia e Parasitologia, Complexo de Ciências Médicas e Enfermagem, Campus Arapiraca, Universidade Federal de Alagoas, Arapiraca, AL, Brazil
e-mail: alysson.duarte@arapiraca.ufal.br

L. H. Rosa
Universidade Federal de Minas Gerais, Instituto de Ciências Biológicas, Belo Horizonte, MG, Brazil

A. W. Fernandes Duarte et al. (eds.), *Polar Microbes and Climate Change*,
https://doi.org/10.1007/978-3-032-07096-8_6

6.1 Introduction

Genomics has been a great ally in the taxonomic and functional understanding of prokaryotes and eukaryotes that make up the polar biota (Peck et al. 2004; Hilário et al. 2019), with a particular focus on those that are endemic and have high adaptive power and cryotolerance in these environments, which are strictly governed by perennial low temperatures, high incidence of solar radiation, and nutritional-water oligotrophy (Hilário et al. 2019). This large area of genetics allows the application of varied approaches, providing surveys of Antarctic terrestrial and marine biodiversity, which is informative for readings of taxa and the composition of Antarctic communities in the most varied habitats, from ice, snow, rocks, lichens, plants, animals, soils, lakes, sediments, planktonic water columns, deep sea to Antarctic volcanic zones (Duarte et al. 2019; Hilário et al. 2019; Rosa et al. 2019; Bendia et al. 2021).

Antarctic biological communities are studied using several methodologies that involve both in situ and in silico surveys, based on dependence or independence from cultivation techniques, especially with regard to microbial communities (Duarte et al. 2019; Rosa et al. 2019). Advances in molecular biology, genetic material analytical tools and sequencing platforms, especially next-generation sequencing (NGS), have made it easier to explore the different existing biological groups from a molecular point of view (Dimov et al. 2024), allowing greater robustness in the generation and interpretation of community nucleic acids from different domains of life, including all major microbial groups of Bacteria, Archaea, Fungi, Viruses, Protists and other microeukaryotes (Hilário et al. 2019; Jackson et al. 2021; Kinasz et al. 2022; Passarini et al. 2024), including the biological stages of these groups that are not normally detected in standard morphological investigations (spores, microscopic fragments and single cells) (Câmara et al. 2022).

The genomics field specializes in examining the entire or partial genome content of an organism, including all the genetic load contained in the coding and non-coding regions. In microbiology, genomics contributes directly to studies based on or not on microbial cultivation (Fig. 6.1). In common culture-based techniques, the genetic content of individual microorganisms in Antarctica recovered by cultivation is accessed via genomic DNA extraction, followed by amplification, sequencing (usually by the Sanger method) and bioinformatic analysis of standard biomarker regions (*barcoding*), corresponding to conserved sequences of ribosomal genes, against a reference database (Villanueva et al. 2021). Here, at first, genomics facilitates the molecular taxonomic definition of cultivable species isolated from Antarctic habitats together with the polyphasic treatment of identification or draft of the whole genome of the cultivated isolate (*single taxon*) (Fig. 6.1b, d) (Jo et al. 2020; Liu et al. 2022; da Silva et al. 2024).

On the other hand, in studies directed by independent cultivation in environmental substrates, genomic initiatives to provide community data on Antarctic species are conceived through metagenomics (Hilário et al. 2019), which can branch into two approaches (Fig. 6.1c, d). The first is targeted or amplicon-based

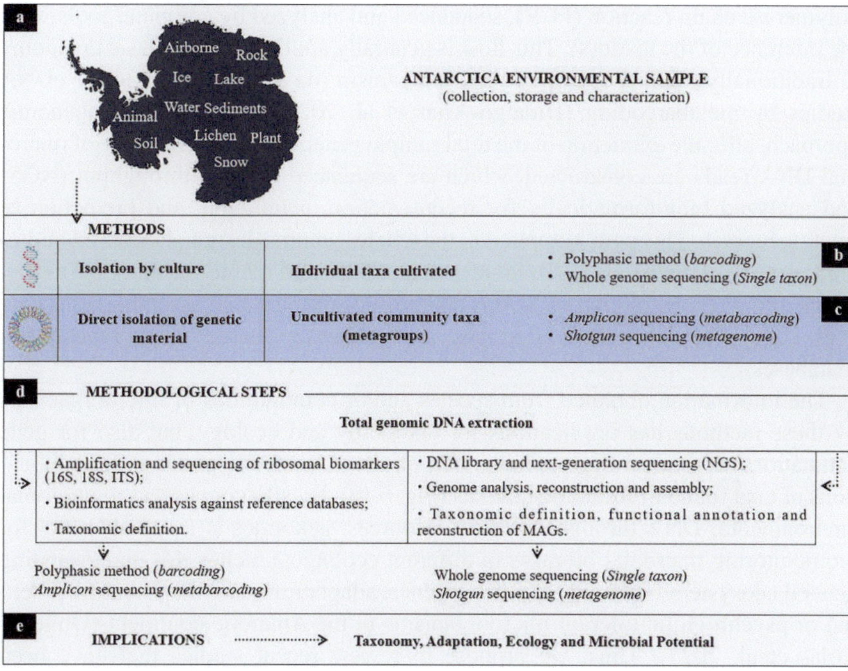

Fig. 6.1 Genomic experimental procedures for studying microbial taxa in Antarctic prokaryotic and eukaryotic communities

metagenomics, also known as metabarcoding, which involves high-throughput sequencing of DNA amplicons, that is, specific analyses of conserved gene regions of the genome, normally the 16S and 18S rRNA genes (prokaryotes and eukaryotes) and the ITS region (fungi) to characterize microorganisms, at least at the taxonomic level of phylum or other higher taxonomic level (family, order, or division) (Bendia et al. 2023; Guglielmin et al. 2023; Hidalgo-Arias et al. 2023). The second is through shotgun metagenomics, which is broader and has the principle of nonspecifically sequencing all DNA, allowing the reconstruction of entire genomes through metagenome assembly (MAGs), simultaneously providing taxonomic information with ecological functional validation of the set of microorganisms in the analyzed environment (Fig. 6.1c, d), with a greater chance of inferring individuals up to the taxonomic level of species (Bendia et al. 2021; Dithugoe et al. 2023; Greco et al. 2024).

To determine the Antarctic environmental microbial biomass, a methodological scheme is generally followed, as simplified in Fig. 6.1. When defining the study method, regardless of isolation or not from pure cultures, the entire flow begins with the selection of the substrate to be investigated, a collection, storage and morpho-physical-chemical characterization (abiotic aspects) procedure is used, followed by processing using a total genomic DNA extraction method. After this, in the case of amplicon = (meta) barcoding purposes, the informative region directed to prokaryotes and/or eukaryotes (ribosomal genes) is defined, which is amplified by polymerase chain reaction (PCR), sequenced and analyzed by computer tools, with the inference of the taxon(s). This flow is generally applied in polyphasic taxonomy to traditionally identify a cultured microorganism (da Silva et al. 2024) or eDNA studies by metabarcoding (Hidalgo-Arias et al. 2023). Under the metagenomic approach, after the extraction of the total sample genetic material, libraries of microbial DNA reads are constructed, which are sequenced by high-throughput (NGS) and analyzed bioinformatically for reconstruction, annotation, and prediction of (meta)genomes. This path is applied in the whole-genome design of cultured microorganisms (Coleine et al. 2020; Shaffer et al. 2023) and mainly in shotgun studies, which allow reconstruction of MAGs and profiles of microbial assemblages (Dutta et al. 2023). The data generated in these approaches are shared in data repositories (databases).

The information obtained from species and/or communities of microorganisms by these methods has implications for taxonomy and ecology, but also for gene annotation, survival strategies, mitigation of stressors, and hypotheses for applications of taxa found in Antarctic habitats (Fig. 6.1e). By allowing access to microbial environmental DNA through its many resources, genomics is a valuable tool for biomonitoring microbial biomass in different ecological niches, for understanding general ecosystem functions, exploring genes, adaptations, and the scientific potential of psychrophilic-tolerant microorganisms in the Antarctic continent (Olmedo-Rojas et al. 2023). Thus, we propose to review recent studies that have been dedicated to applying molecular genomic methods, providing insights into the diversity, ecological significance, and adaptive role of microbial communities in the Antarctic ecosystem.

6.2 Genomics in Microbial Diversity and Ecological Functionality in Antarctica

By applying genomic methods, we will see that the different microorganisms distributed across the biological domains that occur in Antarctica were accessed within the various contexts of this polar continent, exhibiting distinct trophic guilds, such as autotrophs, symbionts, saprotrophs, and pathotrophs, as will be seen in the following topics and summarized in Fig. 6.2.

6.2.1 Prokaryotes: Archaea Domain

Genomic techniques have contributed to surveys on the diversity and metabolic-ecological potential of extremophilic archaea in Antarctica. By combining 16S rRNA amplicon-gene and metagenomic methods, Li et al. (2019) recorded archaeal biomass in soils and rocks in mountains near the Union Glacier, even in low abundance when

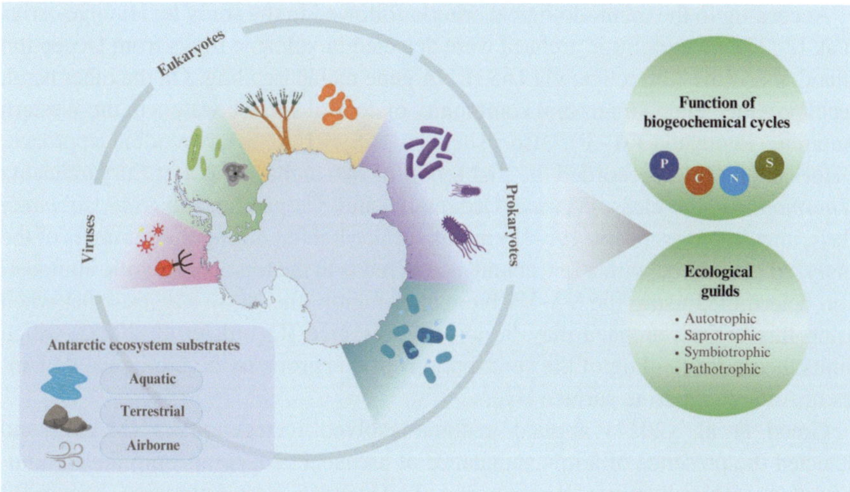

Fig. 6.2 Overview of microbial groups reported in the various terrestrial, aquatic, and airborne substrates of the Antarctic ecosystem through the application of genomic technologies and their environmental functions by taxa reported trophically as autotrophic, saprotrophic, symbiotrophic, or even pathotrophic

compared to bacteria. DNA readings were attributed to members of Euryarchaeota, with *Methanosarcina* (13.3%), *Methanocella* and *Methanothrix* (<5%), along with representatives of Crenarchaeota and Thaumarchaeota. Functional prediction of metagenomic archaeal gene reads revealed involvement in carbon, nitrogen, and sulfur metabolism, in the processes of methanogenesis (e.g., *Methanosphaerula* and *Methanococcus*), N fixation (*Methanococcus*—Euryarchaeota, diazotrophic), and S mineralization, demonstrating the possible involvement of these microorganisms with biogeochemical cycles in soils and rocks (Li et al. 2019).

In brines areas of lake systems in Boulder Clay (Northern Victoria Land, Antarctica), archaeal genomic diversity was accessed through 16S rRNA gene amplicon sequencing (Lo Giudice et al. 2021), the total and metabolically active community was composed of OTUs (operational taxonomic units) mainly attributed to Euryarchaeota (*Methanopyrus*, *Methanothermus*, *Methanoplanus*, *Methanosalsum*, among others), Crenarchaeota (*Aeropyrum*, *Desulfurococcus*, *Sulfolobus*, and *Sulfurisphaer*) and Korarchaeota, the authors highlight that the taxa found, such as methanogens and halophilics, can be active under the pressures of this extreme microenvironment (Lo Giudice et al. 2021). Also in an endoglacial hypersaline brine system, in this same Antarctic sampling area, through the detection of amplicon sequencing variants (ASVs) of the 16S rDNA region (V3–V4) via the Illumina platform (NGS sequencing), Guglielmin et al. (2023) reported archaea affiliated mainly with the methanogenic genus *Methanoculleus* (87%, Methanomicrobia) as components of the bacterial community, raising the hypothesis that they are active in the brines analyzed due to their ability to perform hydrogenotrophic methanogenesis.

According to the methodological criteria followed in the study by Hidalgo-Arias et al. (2023), no endolithic archaea were detected in volcanic rocks from Deception Island, maritime Antarctica, via 16S rRNA gene metabarcoding. On the other hand, accessing the pelagic microbial community of coastal surface waters in the Western Antarctic Peninsula (WAP), also using the 16S rRNA metabarcoding approach, Ozturk et al. (2022) recorded the first report of two archaeal taxa of Euryarchaeota (*Thermoplasmata* class, Marine Groups-II) and Crenarchaeota (*Bathyarchaeia* class, uncultured archaeon 19b-17) in this surface habitat during the summers of the analyzed area. Even with a low abundance of reads in the total prokaryotic composition determined using the V3–V4 biomarker region, the study suggests that when using this region, archaea may be underrepresented (Ozturk et al. 2022), which limits the understanding of the contributions of this group to the biological dynamics of Antarctic marine surface layers.

Greco et al. (2024) applied genome-resolved metagenomics (MAGs) and detected the presence of a low abundance of archaeal taxa composing the community of microbial benthic structures in Lake Untersee, East Antarctica, known for being perennially covered by ice, suggesting that this group is less specialized to colonize this lacustrine environment; representatives were not mentioned. Similarly, using the MAG approach, the study by Dithugoe et al. (2023) evidenced the genomes of archaeal groups linked to suspended and/or sinking marine particles in the Antarctic Southern Ocean. The metagenomic reads of the OTUs revealed main matches for *Nitrososphaeria*, *Thermoplasmata*, and *Methanosarcina*, together with *Archaeoglobi*, *Methanomicrobia*, *Poseidoniia*, *Thermoprotei*, and others with lower abundance in the detected community. Functional annotation revealed among the archaea copies of genes related to the degradation of organic nitrogen by the prediction of peptidases (amino-, carboxy-, endo-, exo-, and iso), in the degradation of high molecular weight compounds, such as carbohydrates (by CAZymes), polyphenolics, catechol and trans-cinnamate, as well as genes involved in denitrification and nitrification (ammonium monooxygenase; nitrite reductase), in CO2 fixation and production of recalcitrant dissolved organic carbon (RDOC) suggesting chemoautotrophic metabolism and containing an ABC-2 type transport system. Thus, the detected archaea exhibited genomic potential to contribute to the oceanic mobilization of nitrogen and carbon from particulate matter (Dithugoe et al. 2023).

By whole shotgun metagenomics, archaeal taxonomic signatures (0.14%) were found forming a heterogeneous microbial community in ice cores from Collins Glacier (King George Island) containing tephra layers from the Deception Island volcano, exhibiting annotated functional genes involved in glycolytic metabolism (GAPDH—glyceraldehyde-3-phosphate dehydrogenase) and nitrification (ammonium-oxidizing) (Kinasz et al. 2022). Shotgun metagenomes from the marine ecosystem of the Western Antarctic Peninsula (wAP) were evaluated by Dutta et al. (2023), with MAGs attributed to archaea of the phylum Thermoplasmatota (n = 9) and Thermoproteota (n = 1), whose clusters revealed evidence of genes linked to biotransformation processes of key elements such as carbon, in the autotrophic pathway through the Arnon-Buchanan cycle, and nitrogen, through the oxidation of ammonia, and correspondences to enzymes involved in the conversion of nitrite to nitric oxide (denitrification). In addition, ammonia-oxidizing archaea (AOA), notably *Nitrosopumilus* and *Nitrosopelagicus*, were reported by Alcamán-Arias et al. (2022) as participants in the Antarctic marine system of surface waters of Greenwich Island during the austral summer, after metagenomic analysis of marker genes linked to the N cycle, such as *amoA* and *amoC*.

In Antarctic fumarole samples (98 °C) on the volcanic Deception Island, shotgun reads were dominantly assigned to Archaea, especially the classes *Thermoprotei*, *Thermococci*, *Methanococci*, *Archaeoglobi*, *Methanobacteria*, *Methanopyri*, and *Methanomicrobia* (Bendia et al. 2021), with twelve annotated MAGs belonging to *Nitrososphaerales* (*Candidatus Nitrosocaldus*), *Nitrosoarchaeum*, *Nitrosotenius*, *Nitrospumilus*, *Desulfurococcales* (*Aeropyrum*), *Acidilobaceae*, *Pyrodictiaceae*, and *Woesearchaeia* (Nanoarchaeota), being related to the homologous gene annotation for carbon fixation by the phosphate acetyltransferase acetate kinase pathway (DI_MAG_00022—*Woesearchaeia*/archaeon GW2011_AR20), in the nitrogen cycle in nitrification (*amoABC*) and sulfate (*sat*, *cysH*, and *sir*), highlighting the annotated MAG DI_MAG_00049—*Nitrososphaerales*/*C. Nitrosocaldus* (Bendia et al. 2021). In the study by Garber et al. (2021), a single high-quality Archean MAG (MAG 48) was described in surface sediments collected in the Ross Sea, West Antarctica, phylogenomically related to the taxon *Nitrosopumilus* (Thaumarchaeota), which presented an operon of three genes, encoding hypothetical proteins, with remote homology to *amoABC*, which is directly linked to the conversion of ammonia to nitrite.

Exploring a 16S rRNA sequence dataset from different Antarctic habitats, composing the core microbiome, Bendia et al. (2023) recorded the correspondence of archaeal AVSs, mainly in fumarole and marine sediments sampled from Antarctic islands, with relative abundance (> 1%) for lineages within Crenarchaeota (*Acidilobaceae*, *Desulfurococcaceae*, *Nitrososphaeraceae*, and *Pyrodictiaceae*) and Nanoarchaeota (*Woesearchaeales*). In addition to other habitats such as snow (*Nitrosopumilales —Crenarchaeota*) and seawater (e.g., archaeal Marine Group II). The main archaeal lineages found in the study corresponded to groups with a thermophilic and hyperthermophilic profile potentially adapted to the thermally variable microenvironments of Antarctica (Bendia et al. 2023). Interestingly, Happel et al. (2022) detected archaeal genetic markers associated with microbiome stability in the Antarctic benthic sponge *Mycale* (*Oxymycale*) *acerata*.

Combined metagenomic and single-cell genome (SAG) technologies have also revealed genomic data from archaeal taxa in Antarctica (Llorenç-Vicedo et al. 2024; Myeong et al. 2024). For example, in the soil microbiome of high-altitude geothermal sites in northern Victoria Land (Mount Melbourne and Mount Rittmann), archaea affiliated with the phylum Thermoproteota (*Nitrososphaeria*) were detected by both SAGs and MAGs generated by PacBio HiFi (long-read) genome sequencing (Myeong et al. 2024). Similarly, using the same technologies mentioned, Llorenç-Vicedo et al. (2024) reported abundance for Thaumarchaeota (*Nitrosopumilus*) DNA signatures in the preserved (≈300 years) microbiome of the Filchner-Ronne Antarctic Sea Ice Core (B15), with possible genes implicated in chemolithoautotrophy by ammonia oxidation (*amoA, amoC*, and *amoX*).

6.2.2 Prokaryotes: Bacteria Domain

Recent studies estimating the ecological diversity of microbial groups in the Antarctic environment have shown the predominance of the Bacteria Domain in a wide range of substrates, providing information about its main representatives and the potential roles they play in this extreme environment. For example, using shotgun genomic sequencing of DNA extracted from rock samples from eight different locations, it was possible to identify bacterial genomes that belonged mainly to the phyla Actinobacteriota and Proteobacteria (Larsen et al. 2024). Using the same genomic sequencing technique, Waschulin et al. (2022) found the same phyla in a surface soil sample from the southeast coast of Alexander Island, in maritime Antarctica, with emphasis on non-cultivable taxonomic classes and orders, such as Acidimicrobiia (Actinobacteriota) and Burkholderiales (Proteobacteria).

Similarly, but using a MAG approach to identify bacterial representatives present in red (formed under glacier) and green (formed on meltwater) biofilms, on the volcanic Deception Island, in the South Shetlands archipelago, the same phyla mentioned above were found, including Bacteroidetes (Centurion et al. 2022). The functional annotations of the MAGs obtained indicated abundance in metabolisms associated with carbon sources, as well as associated with nitrogen metabolism (nitrite-oxidizing and denitrifying bacteria) and even with MAGs indicating the presence of photosynthetic systems (Cyanobacteria—*Pseudonabaena*). Also in Deception, bacteria present in sediments were identified from amplicon sequencing of the 16S rRNA gene as belonging to the genera *Geobacillus*, *Anoxybacillus*, and *Brevibacillus*, with functional annotations indicating the presence of aromatic compound metabolism (Schultz et al. 2022a).

In water samples collected in the western and eastern regions of Lake Bonney, in the McMurdo Dry Valleys, Lee et al. (2024) found a variety of genes and pathways for carbon, nitrogen, and sulfur cycling in the functional annotations of bacterial MAGs, indicating potential participation in biogeochemical cycles. The authors also identified MAGs with potential in the degradation of complex organic molecules, such as glycoside hydrolases and peptidases, and others with potential for photosynthesis (Cyanobacteria—*Pseudanabaena*), iron oxidation (Verrucomicrobiota, Chloroflexota, and Pseudomonadota), and iron reduction (*Geopsychrobacter*). Likewise, cyanobacteria were predominantly identified among the mat-forming bacterial groups covering soft, laminated structures of fine clay sediments in Lake Untersee, located in the Otto-von-Gruber-Gebirge of Queen Maud Land (East Antarctica), with MAGs linked to C and N cycling (Cyanobacteria, Bacteroidota and Verrucomicrobiota), pathways related to sulfur and phosphorus (Cyanobacteria, Actinobacteria, Bacteroidota, and Proteobacteria), as well as genes linked to photosynthetic processes (Cyanobacteria, Myxococcota, Gemmatimonadota, and Proteobacteria) being highlights among the functional annotations of the identified groups (Greco et al. 2024).

On the other hand, with the Shotgun sequencing approach in soil samples impacted and not impacted by marine animals (pinnipeds and birds) from two distinct locations, King George Island and Livingston Island, the taxonomic composition based on OTUs of the 16S rRNA gene consisted of the presence of phyla Proteobacteria, Bacteroidetes and Actinobacteria, with MAGs with relative abundance of genes linked to the nitrogen cycle (denitrification and nitrification), with emphasis on the nitrite reductase (*nirK*) and nitrous oxide reductase (*nosZ*) genes, being more abundant in soil samples impacted by birds than without animal impaction (Ramírez-Fernández et al. 2021). While exploring the complete genome sequence of a strain belonging to the genus *Rhodococcus* (Actinobacteria) isolated from floating ice near Zhongshan Research Station of Antarctica, using Single Molecule, Real-Time (SMRT) technology, it was possible to obtain functional annotations of genes linked to the degradation of dimethylsulfoniopropionate into dimethylsulfide, this being a more common characteristic in members of Proteobacteria, thus providing more information for understanding the sulfur cycle in Antarctica (Qu et al. 2019).

Using 16S rRNA gene metabarcoding combined with shotgun metagenomics to estimate the ecological diversity, composition, and functional potential of algal blooms, in snow samples obtained from the Fildes Peninsula, Doumer Island, and Isabel Riquelme Islet, Soto et al. (2022) found 84% of bacterial genes, of which the ASVs obtained were mostly from the genera *Hymenobacter*, *Pedobacter*, *Flavobacterium*, *Ferruginibacter*, and *Polaromonas*, with functional differences related to molecular responses to environmental factors (genes linked to osmoregulation, biofilm formation, proteins involved in biodegradation, and xenobiotic metabolism). The authors highlight that the metabolic models observed by the genome indicate that bacteria within algae, instead of forming cooperative communities, probably compete for resources (Soto et al. 2022). In contrast to this, also using 16S rRNA gene sequencing and shotgun metagenomics, in addition to cultivation and phylogenetic assays, Möller et al. (2024) investigated the diversity of the intestinal microbiota of an adult Antarctic krill (*Euphausia superba*) collected in the Bransfield Strait (West Antarctica), finding the genus *Pseudoalteromonas* that assists the holobiont in the hydrolysis of ingested polysaccharides, contributing favorably to the degradation of biopolymers.

Furthermore, new bacterial species and strains have been identified using specific primers and distinct bioinformatic approaches to explore their genomes and the functional role of these organisms. He et al. (2022), when isolating bacteria from intertidal sediment samples from King George Island (West Antarctic), assigned two new species (*Aureibaculum luteum* sp. nov. and *A. flavum* sp. nov.) to the genus *Aureibaculum*, of which genomic analysis identified genes linked to the degradation of sulfated polysaccharides by glycosyl hydrolases and sulfatases. In sediment samples collected in geothermal environments, from Whalers Bay and Fumarole Bay (Deception Island), 7 new strains of the genus *Anoxybacillus* were found, with genes linked to carbohydrate, potassium, and sulfur metabolism, as well as iron and aromatic compound metabolism (Schultz et al. 2022b). While from seawater samples collected in Biological Bay—adjacent to Fildes Peninsula, Zhang et al. (2023) identified a new species suggested as belonging to the genus *Maribacter*, with the proposed name *Maribacter aquimaris* sp. nov., with the functional annotations of its genome highlighting genes associated with the nitrogen cycle and cellulose degradation.

6.2.3 Eukaryotic Domain: Fungi

Surveys using molecular techniques demonstrated the predominance of fungal phyla dispersed in the Antarctic ecosystem, mainly associated with *Ascomycota* and *Basidiomycota*, which are the most numerous in taxonomic richness, together with other relevant ones of more intermediate/rare occurrence, including *Basidiobolomycota, Chytridiomycota, Glomeromycota, Mortierellomycota, Monoblepharomycota, Mucoromycota, Rozellomycota,* and *Zoopagomycota*, beyond the fungus-like taxa (*Oomycota*—Stramenopila and *Mycetozoa*) (Rosa et al. 2019; da Silva et al. 2022a; Gonçalves et al. 2024; Passarini et al. 2024). Recent studies have described genomic drafts (whole genome sequencing) of endemic Antarctic fungi, for example, Coleine et al. (2020) described the genome of the cryptoendolytic fungus *Friedmanniomyces endolithicus*, while Batista et al. (2020) characterized the blue fungus *Antarctomyces pellizariae*. And, the new species *Knufia victoria*, recovered from soil in the Antarctic region of North Victoria Land, was proposed through amplicon analysis of two ribosomal biomarkers (ITS-LSU multi-locus sequences) by Isola et al. (2022). These are relevant taxa that highlight the importance of preservation and mycological genomic studies in Antarctica to detect groups with limited endemic occurrence in this area of the planet.

A prevalence of ascomycete fungi in Antarctica has been confirmed in studies using metagenomic and DNA metabarcoding technologies. For example, among the metagenomic eukaryotic signatures found in microbial mats in Lake Untersee (East Antarctica), fungi composed the benthic community of this lacustrine habitat, revealing ascomycete taxa with a low dominance, possibly acting as decomposers (Greco et al. 2024). Likewise, after employing metagenomic shotgun, Kinasz et al. (2022) reported DNA reads (0.68%) matching the phylum *Ascomycota* in an ice-tephra sample from Collins Glacier (King George Island). In meltwater soils at Whalers Bay, Deception Island, ascomycetes were also abundant among fungal metagenomic OTUs detected in this substrate (Passarini et al. 2024), with top matches to genera *Pseudogymnoascus*, *Aspergillus*, and *Trichoderma*. Additionally, there were DNA readings for the phyla *Mucoromycota* and *Chytridiomycota* (Passarini et al. 2024). As components of the microbiotic community of Antarctic sponges, a significant proportion of fungal sequences were observed among the eukaryotic metagenomic reads, most exhibiting annotations with unknown functions, with dominance of *Saccharomycetales* (*Ascomycota*) and *Mucorales* (Moreno-Pino et al. 2024). These studies discussed are examples of metagenomic studies that confirm fungi as important participants in the different ecological contexts of Antarctica.

Fungal diversity in the most diverse Antarctic habitats was also efficiently accessed through environmental sequencing via metabarcoding application, from the marine/aquatic system, with fungal ASVs found in the deep-sea (Ogaki et al. 2020; Barone et al. 2022; da Silva et al. 2022a), lakes (Ogaki et al. 2021; de Souza et al. 2021; Gonçalves et al. 2022; Rosa et al. 2022; Marchetta et al. 2023) and lagoons (Doytchinov et al. 2024), to the terrestrial context, in soils (Rosa et al.

2020a; da Silva et al. 2022b; Gonçalves et al. 2023, 2024), rocks (de Menezes et al. 2021; Rabelo et al. 2024), moss carpets (Rosa et al. 2021a) and glaciers (Guglielmin et al. 2023), as well as in environmental samples of snow, airborne (Rosa et al. 2020b, 2021b; de Menezes et al. 2024), ice (de Menezes et al. 2022), fragments of historical Antarctic wooden structures (de Souza et al. 2023) and other ecological niches analyzed simultaneously to infer fungal assemblages in Antarctic areas such as lichens, sediments, macroalgae, mosses and plants (Zhang et al. 2022; Jiya et al. 2023).

In the study of Guglielmin et al. (2023), amplicon-based metagenomic sequencing of the ITS2 ribosomal region showed asci, basidia and mucoromycete fungi as constituents in the structured microbial community of an Antarctic endoglacial hypersaline brine system, including ASV reads from both dominant taxa with filamentous patterns, represented by *Cladosporium*, *Phoma*, *Penicillium*, *Phaeoisaria*, *Aspergillus*, *Cyphellophora*, *Periconia*, *Parengyodontium*, and *Mucor*, as well as yeast-like (*Glaciozyma*) and black yeast-like patterns, which evidences possible colonization of these eukaryotes in this type of ecosystem so peculiar to Northern Victoria Land. On the Byers Peninsula, in Livingston, Almela et al. (2019) showed the natural biomass of fungi interacting in the microecosystemic trophic web of microbial mats, after analysis of ribotypes via 18S rRNA metabarcoding and isotope signatures, with OTUs assigned especially to *Ascomycota* and *Basidiomycota*, which could be, from an ecological point of view, decomposing particulate organic matter to be metabolized in this evaluated community.

These distinct genetic investigations indicated functional predictions for fungi (Nguyen et al. 2016; Põlme et al. 2020), appointed as important actors in the dynamics of the polar ecosystem, possibly playing several trophic modes and ecological guilds with saprotrophic, pathotrophic, and symbiotrophic taxa (Gonçalves et al. 2023; Marchetta et al. 2023). These eukaryotes exhibit metabolic capabilities to contribute to local biogeochemical cycling as they are excellent bioconverters of complex polymers, releasing products with elements such as carbon, nitrogen and other chemical elements that are made bioavailable to the Antarctic food web (Gonçalves et al. 2023), therefore, they can colonize Antarctic habitats where there is availability of organic matter and which provides them with a certain stability so that they can deal with low temperatures (da Silva et al. 2022a; de Souza et al. 2023).

Typical taxa with a decomposer profile among some of the cited studies above included assignments to *Penicillium, Mrakia*, *Tetracladium*, and *Thelebolus* (da Silva et al. 2022a; de Menezes et al. 2021; de Souza et al. 2021), as well as mutualists/symbiotics such as *Cadophora*, *Microdochium*, *Mortierella*, and *Pezoloma* (de Souza et al. 2023; Marchetta et al. 2023). Readings of the genetic material of fungi recognized for their medical and agricultural importance, attributed as pathogens/parasites of animals and plants, were also recorded in Antarctica, exemplified by taxa of the genera *Acremonium*, *Alternaria, Aspergillus*, *Betamyces*, *Candida*, *Cladophialophora*, *Cryptococcus*, *Exophiala*, *Malassezia, Microdochium*, *Nigrospora, Parengyodontium*, *Phialemonium*, *Pseudogymnoascus*, *Rhodotorula*, and *Trichosporon* (da Silva et al. 2022a; Zhang et al. 2022; de Souza et al. 2023; Marchetta et al. 2023), these are data that demonstrate the power that fungi have in spreading across the most varied ecosystems, also reaching the extreme poles, but which are also sensitive sources that serve as a warning for environmental conservation issues in this unique polar area that is Antarctica, even if it is not known for sure whether these detected fungi are in cellular viability. Otherwise, interestingly, a significant percentage of fungal AVSs/OTUs from genomic studies conducted in Antarctica are not assigned to known taxa or are rare occurrence groups in this global polar area, reflecting the importance of studying the fungi present there in a more in-depth and timely manner (Guglielmin et al. 2023; Doytchinov et al. 2024).

6.2.4 Eukaryotic Domain: Non-fungal Microorganisms and Viruses

Microscopic beings such as viruses and microeukaryotes (microalgae, phytoplankton, and other protists) also live at the limits of life in the Antarctic icy continent and are explored in genomic studies. In this polar ecosystem, microscopic green algae (Viridiplantae) are mentioned in the sediment communities of three Antarctic lakes, on Vega Island, in the Weddell Sea (Antarctic Peninsula), being represented mainly by the Phylum Chlorophyta, whose readings via metabarcoding method revealed DNA sequences for chlorophytes of the classes Chlorophyceae, Trebouxiophyceae, and Ulvophyceae, with the order Chlamydomonadales (Chlorophyceae) by species *Chlorococcum microstigmatum* having the largest contribution in the number of readings recorded (Fonseca et al. 2022). In this study, 78 total taxa were assigned to this phylum, of which 32% had no previous records from Antarctica, including *Chlamydocapsa*, *Paulschulzia*, *Auxenochlorella*, *Didymogenes*, *Jaagichlorella*, *Desmococcus*, *Lobosphaera*, *Ploeotila*, and *Desmochloris* (Fonseca et al. 2022). Ecologically, algal taxa presented trophic status as planktonic (e.g., *Chlamydomonas* and *Desmodesmus*), lichenized (e.g., *Chlorococcum* and *Trebouxia*), and components of Antarctic snow algae communities (e.g., *Scotiella*, *Chloromonas*, and also *Chlamydomonas*) (Fonseca et al. 2022).

Some chlorophyte taxa of the orders Chlamydomonadales (*Chlamydomonas nivalis* and *Chloromonas pichinchae*) and Trebouxiales (*Dictyochloropsis splendida*, *Trebouxia asymmetrica*, and *T. usneae*) were also detected by metabarcoding in rocks from the Ellsworth Mountains, continental Antarctica, which trophically are also groups of snow algae, free-living photobionts, or symbiotics of lichens occurring in this analyzed Antarctic area (Câmara et al. 2022). Also composing the cryptoendolithic community in volcanic rocks at Deception, using the 18S biomarker gene via metabarcoding, Hidalgo et al. (2023) reported, among the eukaryotic readings, abundances for algae attributed to Chlorophyta, *Raphidonema, Trebouxia* (Chloroplastida), and *Ancylonema* (Streptophyta), indicating the ability of these phototrophic organisms to endolithically colonize this type of substrate rich in metals (Hidalgo et al. 2023).

The study by Pushkareva et al. (2024) aimed to describe the composition of microbial life present in a wide variety of environments (diverse biological crusts and soils, ephemeral wetland sands, biofilms, endolithic and hypolithic communities) in East Antarctica. Sequencing of 18S rRNA showed the prevalence of Chloroplastida, and the studied samples exhibited the presence and dominance of Chlorophyceae, Trebouxiophyceae, Ulvophyceae, Bacillariophyceae, and Zygnematophyceae within the algal community. Eukaryotic algae are frequently found in the terrestrial environments of Antarctica. Furthermore, algae from the order Prasiolales (Trebouxiophyceae) were present in all samples collected, on the surfaces of seal mummies and penguin feathers. It was also possible to note the positive correlation in the co-occurrence of four algal taxa (Chlorodendrophyceae, Coleochaetophyceae, Bacillariophyceae, and Zygnematophyceae) with other eukaryotic members belonging to Alvelota, Amoebozoa, Stramenopiles, and Rhizaria (Pushkareva et al. 2024).

On the McMurdo Ice Shelf, in Vitoria do Sul Land, where microbial mats are abundant and contribute a large part of biomass and biodiversity, microbial diversity was studied and analyzed also through amplicon of the 18S rRNA gene (Jackson et al. 2021), of which one of the communities dominant algae was led by Chlorophyta, in addition to taxa such as Bacillariophyta (6.63%) and Chlorophyceae (6.61%), while other microalgae included Trebouxiophyceae, Chrysophyceae, Xanthophyceae, and Eustigmatophyceae. Other microeukaryotes were also found, such as ciliate members of Litostomatea, Nassophorea, Oligohymenophorea, and Phyllopharyngea, as well as amoebas belonging to Tubulinea and Lobosa, as well as other protists such as Heterolobosea, Endomyxa, Oomycota, and Pirsonia (Jackson et al. 2021). The soils of the Antarctic region, McMurdo Dry Valleys (MDV), also harbor a unique diversity of protists with interesting ecological functions revealed by shotgun metagenome (Thompson et al. 2020), with good representation of phagotrophic protists, mainly abundances for the phyla Ciliophora and Cercozoa, phototrophs, such as Stramenopiles (diatoms of the genera *Hantzschia* and *Nitzschia*) and Chlorophyta (*Trebouxia* and *Chloroidium*), as well as parasites (*Dermocystidium* and *Pythium*) and mixotrophs (*Amphitrema* and *Hatena*). The diverse OTUs of these groups of protists reveal their importance in the microbiota composition in this type of substrate in Antarctica (Thompson et al. 2020).

Given the many abiotic pressures of Antarctica, viruses can associate with other microorganisms and exhibit ecological-adaptive importance, having infectious effects with lytic and/or (pseudo) lysogenic cycles, affecting the growth and physiological characteristics of their hosts. In the study of Demina et al. (2022), the genetic diversity of four bacteriophages that infect bacteria of the genera *Paraglaciecola* and *Octadecabacter* that are common in Antarctic sea ice was investigated, namely the viruses *Paraglaciecola* Antarctic GD 1 (PANV1), *Paraglaciecola* Antarctic JLT 2 (PANV2), *Octadecabacter* Antarctic BD 1 (OANV1), and *Octadecabacter* Antarctic DB 2 (OANV2). These double-stranded DNA (dsDNA) genome phages presented unique gene products, without assigned functions, with a high percentage of exclusive ORFs (protein-coding open reading frames) without homology in the analysis database (NCBI non-redundant proteins), the nucleotide sequence derived from their metagenomes showed similarities to scaffolds of viruses recorded from aquatic environments in Antarctica and also from other distant geographic points. Furthermore, from their genomes, no information about potential additional hosts was found, demonstrating the phage-specific nature of sea ice bacterial strains, which are important in this ecosystem (Demina et al. 2022).

From another perspective, exploring the virioplankton community on the Antarctic Ross Ice Shelf (RIS) by integrating single-cell genomic, transcriptomic, and metagenomic data, López-Simón et al. (2023) recorded a high abundance of assembled viral contigs (~90%) belonging to Caudoviricetes (Duplodnaviria; dsDNA viruses), abundant ones such as Monodnaviria (ssDNA viruses), Riboviria (RNA viruses, Chrymotiviricetes), and Varidnaviria (large nucleocytoplasmic DNA viruses and virophages). A significant fraction of the annotated ORFs had uncertain classifications, indicating that they could correspond to novel families not yet described, with many of the detected viruses representing singletons and/or outliers without any connection to other known ones, evidencing the high viral genomic diversity endemic and unique to RIS (López-Simón et al. 2023). Biogeographically, similarities with other viral genomes from the Southern Ocean have shown that these viruses are adapted to polar environments and may be capable of infecting microbial hosts important in biogeochemical cycles and primary production, especially active chemolithoautotrophic prokaryotes such as *Nitrosopumilus* (ammonia oxidation) and *Thioglobus* (sulfur oxidation), and slow-growing ones such as *Pelagibacter*. Genomic analyses of these viruses have highlighted auxiliary metabolic genes (AMG) that may potentially contribute to viral adaptation, acquisition, and metabolism of microelements (nitrogen, sulfur, and phosphorus) by their hosts and other microbial maintenance processes in the Antarctic RIS (López-Simón et al. 2023).

6.3 Genomics Revealing Adaptations and Potential of Antarctic Microorganisms

Genomic methodologies have revealed an arsenal of adaptive strategies of Antarctic biological communities, by predicting genes, metabolic functions, significant biosynthetic pathways, and potential applications (Cho and Lee 2024), as well as measurements of strains adapted to heterogeneous thermal variations in Antarctica by their thermo-tolerance/philia profiles (Bendia et al. 2023) and psychrophilia (Batista et al. 2020; Isola et al. 2022). Biosynthetic genes encoding enzymes (e.g., β-carotene ketolase, 15-cis-phytoene desaturase, and dioxygenase) linked to the production of photoprotective pigments such as carotenoids (carotenogenic) were detected in prokaryotes by metagenomics applied to samples of waters from the Southern Ocean (Cho and Lee 2024) and in ice-tephra layers of the Collins Glacier (Kinasz et al. 2022). These molecules detected from psychrophilic sources are interesting for industrial purposes as future alternatives in the cosmetic, food, and pharmaceutical areas (Kinasz et al. 2022).

Putative biosynthetic genes (BGCs) for the pigments violacein and prodigiosin, with antimicrobial and sunscreen effects, were annotated from the genome of the Antarctic strain DJMP01 *Massilia frigida* from microbial mats in the Don Juan Lagoon (DJP) basin, Victoria Land (Shaffer et al. 2023), experimentally after cultivation, when exposed to variations in UV radiation, the accumulation of prodigiosin was altered in this bacterium, evidence that provides a preliminary indication that microorganisms can respond differentially in the production of secondary metabolites to future climate disturbances in the Antarctic environment, in this case, more intense solar incidence (Shaffer et al. 2023).

From the assembled genome of the Antarctic endemic blue fungus *Antarctomyces pellizariae*, a putative IBP (*ice-binding-like*) gene encoding a 236 amino acid ice-binding protein was inferred, identified as APELI_8493-RA, which appears to be an important element in the response to low temperature, acting in the formation of cellular ice crystals (Batista et al. 2020). And some gene clusters for secondary metabolites were noted (SMGC: *secondary metabolite gene clusters*), but with most of them having no similarity to described clusters (Batista et al. 2020). The draft of this psychrophilic fungus from Antarctica also opened the potential for phylogenomic analysis of the taxonomic relationships of the class *Leotiomycetes*, with a focus on the order *Thelebolales*, in which the fungus is currently positioned (Batista et al. 2020). This highlights the importance of genomic techniques for biological systematics in Antarctica (Batista et al. 2020; Demina et al. 2022). In reconstructed bacterial metagenomes from Antarctic Lake microbial mats, stress adaptation genes were detected encoding chaperones and cold shock proteins, also including about ninety putative IBP genes, mainly in *Bacteroidota* (Greco et al. 2024). In addition to genes linked to cryoprotection in the metabolism of exopolymeric substances and biofilm formation, such as *Wza* (Greco et al. 2024).

Metagenomic data from bacteria from the Union Glacier in Antarctica revealed genes functionally involved in DNA repair mechanisms (e.g., base excision, nucleotide excision, mismatch excision), antioxidant activity (superoxide dismutase, catalase, and catalase peroxidase), cellular modulation (biosynthesis of unsaturated fatty acids), antifreeze proteins, and catalysis by cold-adapted enzymes (Li et al. 2019). In a shotgun approach to the genome of a new bacterial species (*Pedobacter polysacchareus* sp. nov) isolated from soil, Wang et al. (2023) identified the presence of genes encoding sigma factor (*rpoE*, *rpoD*, and *serB*), associated responses to environmental stresses such as cold shock (*cap*A, *nus*A, *pnp*, *rbf*A, *inf*A, and *inf*B), osmotic stress (*glg*ABC and *pro*VXW), and radiation exposure (*uvr*ABC and *umu*CD). Similar characteristics were observed in different strains of the genus *Anoxybacillus,* isolated from fumarole sediments in Deception, of which the potential functions of their genes indicated the capacity for DNA stabilization and repair (formamidopyrimidine–DNA glycosylase, uracil–DNA glycosylase, *UvrA*, *UvrB,* and *UvrC*), adaptation to thermal stress (chaperones DnaK and DnaJ for heat and genes *CspB* and *CsoD* for cold), and even for adaptation to alkaline pH (Schultz et al. 2022b).

Using pangenomic analyses from the shotgun strategy of the complete genome of *Acidithiobacillus ferrooxidans* isolated from surface sediments on King George

Island, genes related to cold tolerance were also found encoding the cold shock protein *CspE*, trehalose biosynthesis and fatty acid desaturase, as well as genes related to oxidative stress that include the encoding of enzymes such as superoxide dismutase—SOD (Fe/Mn-SOD), catalase, thiolperoxidase and alkyl hydroperoxide reductase involved in the elimination of reactive oxygen species—ROS, indicating adaptations to different stress factors (Muñoz-Villagrán et al. 2022). *A. ferrooxidans* also showed the ability to oxidize ferrous iron, which suggests potential applicability in the bioleaching of metals, in addition to having the presence of transporters for metals (Cu, Zn and Pb) and a set of operon genes *mer* (*mer*RTPCDAB) which confers mercury resistance (Muñoz-Villagrán et al. 2022). A similar feature was also observed in the functional annotations of MAGs obtained from soil from the Antarctic Peninsula, which detected metal resistance genes in 292 bacterial genera (mainly *Pseudomonas*, *Polaromonas*, *Streptomyces*, *Variovorax*, and *Burkholderia*), like mercury (*mer*RTP), chromium (*rec*G), zinc (*mnt*H), among other biocidal metals that can be leached by the microbial action of Antarctic strains (Marcoleta et al. 2022).

There are other records of studies using genomics to investigate the potential for environmental bioremediation in microorganisms derived from the Antarctic ecosystem. Many enzymes linked to the degradation of aromatic compounds (e.g., toluene, xylene, methylnaphthalene, and styrene) were annotated from the bacterial metagenome of glacial ice with volcanic sediments, in Deception (Kinasz et al. 2022). This reveals possible abilities to bioconvert these toxic compounds through microbial metabolism via genome scanning. Likewise, on this island, different thermophilic bacteria isolated from fumaroles ranging from 50 to 100 °C were identified, distributed among the genera *Geobacillus*—capable of degrading crude oil in vitro, and *Anoxybacillus*—with biosurfactant and bioemulsifying action, proven in in vitro tests and confirmed in functional annotations that indicated metabolism of aromatic compounds with benzoate and xylene degradation pathways, in addition to coding genes for the production of biosurfactants (*srf*A, *pps*A, and *aln*B) and linked to hydrocarbon degradation (*nah*C, *phe*A, and *bup*G), which gives these Antarctic microbial genera potential applications both in petroleum bioremediation and in the industry for the production of biosurfactants/emulsifiers (Schultz et al. 2022a).

Furthermore, results from the metagenomic approach of bacteria isolated from the intestine of adult krill indicated potential application of hydrolytic enzymes from the presence of hydrolases related to the degradation of lipids and proteins (Flavobacteriaceae and Pseudoalteromonadaceae), in addition to other potential plastic degraders (Moraxellaceae) such as polyethylene terephthalate (PET linked to carboxylic ester), polylactide (PLA), polybutylene adipate terephthalate (PBAT) and polyhydroxyalkanoate (PHA) (Möller et al. 2024). From the point of view of agricultural applications, metagenomics of microorganisms present in meltwater soils in Deception (Whalers Bay, Antarctica) highlighted groups of agronomic importance with genes linked to the capacity for phosphate solubilization in *Pseudomonas*, biological nitrogen fixation in the genera *Bradyrhizobium*, *Burkholderia*, *Cupriavidus*, *Mesorhizobium*, and *Streptomyces*, promoting plant growth with *Achromobacter*, *Azospirillum*, *Bacillus*, and *Paenibacillus*, in addition to acting in biological control in *Saccharopolyspora, Beauveria, Trichoderma,*

Metarhizium, Isaria, and *Pochonia*, with strong prospects in the formulation of bio-inoculants with biostimulant action for more sustainable plant production from Antarctic strains (Passarini et al. 2024). Readings for fungi known to have phyto-pathogenic activities were also evidenced by genomics, mainly *Aspergillus*, *Colletotrichum*, *Fusarium*, *Pythium*, *Rhizoctonia*, and *Sclerotium* (Passarini et al. 2024).

6.4 Conclusions and Future Perspectives of the Polar Genomic Approach: Optimization and Challenges

The omics sciences, in their breadth, will continue to contribute to the most diverse perspectives of scientific studies, with singular attention to microorganisms highly adapted to cryospheric contexts, being one of the main frontiers to advance in Antarctica. Genomic technologies will certainly continue to aid in the taxonomic identification of species in Antarctic habitats to propose new endemic lineages (along with the polyphasic method) and/or detect uncultivable taxa not yet described (polar unknown dark biomass), explore non-typical microecosystems (e.g., endoglacial brines, cryptoendoliths) or more remote continental areas of Antarctica, provide pipelines for genome assembly, gene prediction, metabolic pathways and functional networks useful to the biotechnological and environmental fields, as well as provide significant systematic inferences from this important polar region that is underrepresented in reference databases (Batista et al. 2020; Villanueva et al. 2021; Isola et al. 2022; Liu et al. 2022; Demina et al. 2022; Guglielmin et al. 2023).

A mix of different genomic approaches such as single-cell, metagenomics, amplicons, metatranscriptomics, and associated technologies including microcosm experiments, simulation chambers, in situ analysis, microscopic techniques (e.g., *Scanning Electron Microscopy, Epifluorescence*), use of stable isotopes, microbial cell count by flow cytometry, computational predictions of environmental phenomena at the molecular level (Almela et al. 2019; Duarte et al. 2019; Thompson et al. 2020; de Souza et al. 2021; Lavergne et al. 2021; Trefault et al. 2021; Hidalgo-Arias et al. 2023; Moreno-Pino et al. 2024; Pushkareva et al. 2024), may facilitate the understanding of Antarctic biological communities in terms of aspects of diversity, dynamics, metabolism, ecology, adaptations and their potential applications (Li et al. 2019; Lavergne et al. 2021; Llorenç-Vicedo et al. 2024; Myeong et al. 2024).

Genomics is also relevant in processes of assessing the effects of progressive climate changes on the biosensing of microbial genetic signatures, both ancient (fossil record) and recent, in this polar environment, aiming to anticipate strategies to deal with future climate disorders that may have a direct impact on Antarctica (Lavergne et al. 2021; Shaffer et al. 2023; Llorenç-Vicedo et al. 2024), as well as in seasonality studies, surveys of primary microbial succession in different niches, relationships between biological domains (Dimov et al. 2024), co-occurrence network (Pushkareva et al. 2024) and be a useful tool in microbial understanding within adaptive contexts in this extreme ecosystem, having implications in the field of ecophysiology and astrobiology (Bendia et al. 2021; Coleine et al. 2021; Hidalgo-Arias et al. 2023).

Some limitations that challenge genomic research in Antarctica primarily involve the logistics of accessing this remote continent due to its intrinsic global geoclimatic isolation (Peck et al. 2004), together with the requirements related to expertise for the application of systems biology methods, interpretation of biological data, development, and validation of technologies—also considering the high costs related to the use of tools, especially regarding the demands of high-throughput genetic sequencing (Hilário et al. 2019). Even in the face of challenges, the wide applicability of genomic methodologies remains very relevant in generating information about Antarctic species, stimulating local and community biological understanding of this unique polar ecosystem that is still little investigated in its entirety (Hilário et al. 2019), with a central focus on the contributions of (micro) prokaryotes-eukaryotes adapted to the Antarctic biota which, as seen in this chapter, are highly diverse, well distributed across different terrestrial and aquatic substrates, adapted and appear to perform crucial ecological activities within this environment, expanding their potential for targeted biotechnological studies.

References

Alcamán-Arias ME et al (2022) Surface ammonia-oxidizer abundance during the late summer in the west Antarctic coastal system. Front Microbiol 13:821902. https://doi.org/10.3389/fmicb.2022.821902

Almela P et al (2019) Carbon pathways through the food web of a microbial mat from Byers Peninsula, Antarctica. Front Microbiol 10:628. https://doi.org/10.3389/fmicb.2019.00628

Barone G et al (2022) Local environmental conditions promote high turnover diversity of benthic Deep-Sea Fungi in the Ross Sea (Antarctica). J Fungi 8(1):65. https://doi.org/10.3390/jof8010065

Batista TM et al (2020) Whole-genome sequencing of the endemic Antarctic fungus Antarctomyces pellizariae reveals an ice-binding protein, a scarce set of secondary metabolites gene clusters and provides insights on Thelebolales phylogeny. Genomics 112(5):2915–2921. https://doi.org/10.1016/j.ygeno.2020.05.004

Bendia AG et al (2021) Metabolic potential and survival strategies of microbial communities across extreme temperature gradients on Deception Island volcano. Antarct Environ Microbiol 23(7):4054–4073. https://doi.org/10.1111/1462-2920.15649

Bendia AG et al (2023) Insights into Antarctic microbiomes: diversity patterns for terrestrial and marine habitats. An Acad Bras Cienc 95(suppl 3):e20211442. https://doi.org/10.1590/0001-3765202320211442

Câmara PEAS et al (2022) Diversity of Viridiplantae DNA present on rock surfaces in the Ellsworth Mountains, continental Antarctica. Polar Biol 45(4):637–646. https://doi.org/10.1007/s00300-022-03021-8

Centurion VB et al (2022) Microbiome structure in biofilms from a volcanic island in Maritime Antarctica investigated by genome-centric metagenomics and metatranscriptomics. Microbiol Res 265:127197. https://doi.org/10.1016/j.micres.2022.127197

Cho WY, Lee PC (2024) Metagenomic analysis of Antarctic Ocean near the King Sejong Station reveals the diversity of carotenoid biosynthetic genes. Microorganisms 12(2):390. https://doi.org/10.3390/microorganisms12020390
Coleine C et al (2020) Peculiar genomic traits in the stress-adapted cryptoendolithic Antarctic fungus Friedmanniomyces endolithicus. Fungal Biol 124(5):458–467. https://doi.org/10.1016/j.funbio.2020.01.005
Coleine C et al (2021) Endolithic microbial composition in Helliwell Hills, a newly investigated Mars-like area in Antarctica. Environ Microbiol 23(7):4002–4016. https://doi.org/10.1111/1462-2920.15419
da Silva MK, de Souza LMD, Vieira R et al (2022a) Fungal and fungal-like diversity in marine sediments from the maritime Antarctic assessed using DNA metabarcoding. Sci Rep 12:21044. https://doi.org/10.1038/s41598-022-25310-2
da Silva TH, Câmara PEAS, Pinto OHB et al (2022b) Diversity of Fungi Present in Permafrost in the South Shetland Islands. Maritime Antarct Microb Ecol 83:58–67. https://doi.org/10.1007/s00248-021-01735-6
da Silva MK, Barreto DLC, Vieira R et al (2024) Diversity and enzymatic, biosurfactant and phytotoxic activities of culturable Ascomycota fungi present in marine sediments obtained near the South Shetland Islands, maritime Antarctica. Extremophiles 28:20. https://doi.org/10.1007/s00792-024-01336-4
de Menezes GCA, Câmara PEAS, Pinto OHB et al (2021) Fungal diversity present on rocks from a polar desert in continental Antarctica assessed using DNA metabarcoding. Extremophiles 25:193–202. https://doi.org/10.1007/s00792-021-01221-4
de Menezes GCA, Câmara PEAS, Pinto OHB et al (2022) Fungi in the Antarctic cryosphere: using DNA metabarcoding to reveal fungal diversity in glacial ice from the Antarctic Peninsula Region. Microb Ecol 83:647–657. https://doi.org/10.1007/s00248-021-01792-x
de Menezes GCA, Lopes FAC, Santos KCR et al (2024) Fungal diversity present in snow sampled in summer in the north-west Antarctic Peninsula and the South Shetland Islands, Maritime Antarctica, assessed using metabarcoding. Extremophiles 28:23. https://doi.org/10.1007/s00792-024-01338-2
de Souza LMD, Ogaki MB, Câmara PEAS et al (2021) Assessment of fungal diversity present in lakes of Maritime Antarctica using DNA metabarcoding: a temporal microcosm experiment. Extremophiles 25:77–84. https://doi.org/10.1007/s00792-020-01212-x
de Souza LMD, Teixeira EAA, da Costa Coelho L et al (2023) Cryptic fungal diversity revealed by DNA metabarcoding in historic wooden structures at Whalers Bay, deception Island, maritime Antarctic. Braz J Microbiol 54:213–222. https://doi.org/10.1007/s42770-022-00869-0
Demina TA et al (2022) Virus-host interactions and genetic diversity of Antarctic sea ice bacteriophages. mBio 13:e00651–e00622. https://doi.org/10.1128/mbio.00651-22
Dimov SG et al (2024) Observations on the antagonistic relationships between fungi, archaea, and bacteria in Livingston Island (Maritime Antarctica) with the use of amplicon-based metagenomics. Diversity 16(1):56. https://doi.org/10.3390/d16010056
Dithugoe CD et al (2023) Bacteria and archaea regulate particulate organic matter export in suspended and sinking marine particle fractions. Msphere 8(3):e00420–e00422. https://doi.org/10.1128/msphere.00420-22
Doytchinov VV, Peykov S, Dimov SG (2024) Study of the bacterial, fungal, and archaeal communities structures near the Bulgarian Antarctic research base "St. Kliment Ohridski" on Livingston Island, Antarctica. Life 14(2):278. https://doi.org/10.3390/life14020278
Duarte AWF, Centurion VB, Oliveira VM (2019) Uncultivated fungi from Antarctica. In: Rosa L (ed) Fungi of Antarctica. Springer, Cham. https://doi.org/10.1007/978-3-030-18367-7_2
Dutta A et al (2023) Depth drives the distribution of microbial ecological functions in the coastal western Antarctic Peninsula. Front Microbiol 14:1168507. https://doi.org/10.3389/fmicb.2023.1168507

Fonseca BM et al (2022) Green algae (Viridiplantae) in sediments from three lakes on Vega Island, Antarctica, assessed using DNA metabarcoding. Mol Biol Rep:1–10. https://doi.org/10.1007/s11033-021-06857-1

Garber AI et al (2021) Metagenomics of Antarctic marine sediment reveals potential for diverse chemolithoautotrophy. Msphere 6(6):e00770–e00721. https://doi.org/10.1128/mSphere.00770-21

Gonçalves VN et al (2022) Diversity and ecology of fungal assemblages present in lake sediments at Clearwater Mesa, James Ross Island, Antarctica, assessed using metabarcoding of environmental DNA. Fungal Biol 126(10):640–647. https://doi.org/10.1016/j.funbio.2022.08.002

Gonçalves VN et al (2023) Soil fungal diversity and ecology assessed using DNA metabarcoding along a deglaciated chronosequence at Clearwater Mesa, James Ross Island. Antarct Peninsula Biol 12(2):275. https://doi.org/10.3390/biology12020275

Gonçalves VN et al (2024) Fungal and fungal-like diversity present in ornithogenically influenced maritime Antarctic soils assessed using metabarcoding. J Basic Microbiol. https://doi.org/10.1002/jobm.202300601

Greco C et al (2024) Genome-resolved metagenomics reveals diverse taxa and metabolic complexity in Antarctic lake microbial structures. Environ Microbiol 26(6):e16663. https://doi.org/10.1111/1462-2920.16663

Guglielmin M, Azzaro M, Buzzini P et al (2023) A possible unique ecosystem in the endoglacial hypersaline brines in Antarctica. Sci Rep 13:177. https://doi.org/10.1038/s41598-022-27219-2

Happel L et al (2022) Stability of the microbiome of the sponge Mycale (Oxymycale) acerata in the Western Antarctic Peninsula. Front Microbiol 13:827863. https://doi.org/10.3389/fmicb.2022.827863

He X-y et al (2022) Description of Aureibaculum luteum sp. nov. and Aureibaculum flavum sp. nov. isolated from Antarctic intertidal sediments. Antonie Van Leeuwenhoek:1–15. https://doi.org/10.1007/s10482-021-01702-8

Hidalgo-Arias A et al (2023) Adaptation of the endolithic biome in Antarctic volcanic rocks. Int J Mol Sci 24(18):13824. https://doi.org/10.3390/ijms241813824

Hilário HO, Batista TM, Franco GR (2019) Genomics of Antarctic fungi: a new frontier. In: Rosa L (ed) Fungi of Antarctica. Springer, Cham. https://doi.org/10.1007/978-3-030-18367-7_15

Isola D et al (2022) Knufia obscura sp. nov. and Knufia victoriae sp. nov., two new species from extreme environments. Int J Syst Evol Microbiol 72(10):005530. https://doi.org/10.1099/ijsem.0.005530

Jackson EE, Hawes I, Jungblut AD (2021) 16S rRNA gene and 18S rRNA gene diversity in microbial mat communities in meltwater ponds on the McMurdo Ice Shelf, Antarctica. Polar Biol 44(4):823–836. https://doi.org/10.1007/s00300-021-02843-2

Jiya N, Prafulla S, Avinash S (2023) Diversity and composition of fungal communities across diverse environmental niches in Antarctica. Pol Sci 38:100973. https://doi.org/10.1016/j.polar.2023.100973

Jo S-W et al (2020) Isolation, identification, and biochemical characteristics of a cold-tolerant Chlorella vulgaris KNUA007 isolated from King George Island, Antarctica. J Mar Sci Eng 8(11):935. https://doi.org/10.3390/jmse8110935

Kinasz CT et al (2022) Taxonomic and functional diversity from Antarctic ice-tephra microbial community: ecological insights and potential for bioprospection. An Acad Bras Cienc 94(suppl 1):e20210621. https://doi.org/10.1590/0001-3765202220210621

Larsen S et al (2024) Geology and elevation shape bacterial assembly in Antarctic endolithic communities. Sci Total Environ 907:168050. https://doi.org/10.1016/j.scitotenv.2023.168050

Lavergne C et al (2021) Temperature differently affected methanogenic pathways and microbial communities in sub-Antarctic freshwater ecosystems. Environ Int 154:106575. https://doi.org/10.1016/j.envint.2021.106575

Lee H et al (2024) Microbial assemblages and associated biogeochemical processes in Lake Bonney, a permanently ice-covered lake in the McMurdo Dry Valleys. Antarct Environ Microbiome 19(1):60. https://doi.org/10.1186/s40793-024-00605-1

Li Y et al (2019) Reconstruction of the functional ecosystem in the high light, low temperature union glacier region, Antarctica. Front Microbiol 10:2408. https://doi.org/10.3389/fmicb.2019.02408

Liu Z et al (2022) Lysobacter antarcticus sp. nov., an SUF-system-containing bacterium from Antarctic coastal sediment. Int J Syst Evol Microbiol 72(2):005250. https://doi.org/10.1099/ijsem.0.005250

Llorenç-Vicedo A et al (2024) New avenues for potentially seeking microbial responses to climate change beneath Antarctic ice shelves. Msphere:e00073–e00024. https://doi.org/10.1128/msphere.00073-24

Lo Giudice A et al (2021) Prokaryotic diversity and metabolically active communities in brines from two perennially ice-covered Antarctic lakes. Astrobiology 21(5):551–565. https://doi.org/10.1089/ast.2020.2238

Lopez-Simon J, Vila-Nistal M, Rosenova A et al (2023) Viruses under the Antarctic ice shelf are active and potentially involved in global nutrient cycles. Nat Commun 14:8295. https://doi.org/10.1038/s41467-023-44028-x

Marchetta A et al (2023) A deep insight into the diversity of microfungal communities in arctic and Antarctic lakes. J Fungi 9(11):1095. https://doi.org/10.3390/jof9111095

Marcoleta AE et al (2022) The highly diverse Antarctic Peninsula soil microbiota as a source of novel resistance genes. Sci Total Environ 810:152003. https://doi.org/10.1016/j.scitotenv.2021.152003

Möller L et al (2024) Rich microbial and depolymerising diversity in Antarctic krill gut. Microbiol Spectrum 12(4):e04035–e04023. https://doi.org/10.1128/spectrum.04035-23

Moreno-Pino M et al (2024) Unveiling microbial guilds and symbiotic relationships in Antarctic sponge microbiomes. Sci Rep 14(1):6371. https://doi.org/10.1038/s41598-024-56480-w

Muñoz-Villagrán C et al (2022) Characterization and genomic analysis of two novel psychrotolerant Acidithiobacillus ferrooxidans strains from polar and subpolar environments. Front Microbiol 13:960324. https://doi.org/10.3389/fmicb.2022.960324

Myeong NR, Choe YH, Shin SC et al (2024) Genomic profiling of Antarctic geothermal microbiomes using long-read, Hi-C, and single-cell techniques. Sci Data 11:1023. https://doi.org/10.1038/s41597-024-03875-z

Nguyen NH, Song Z, Bates ST, Branco S, Tedersoo L, Menke J, Schilling JS, Kennedy PG (2016) FUNGuild: an open annotation tool for parsing fungal community datasets by ecological guild. Fungal Ecol 20:241–248. https://doi.org/10.1016/j.funeco.2015.06.006

Ogaki MB, Coelho LC, Vieira R et al (2020) Cultivable fungi present in deep-sea sediments of Antarctica: taxonomy, diversity, and bioprospecting of bioactive compounds. Extremophiles 24:227–238. https://doi.org/10.1007/s00792-019-01148-x

Ogaki MB, Câmara PEAS, Pinto OHB et al (2021) Diversity of fungal DNA in lake sediments on Vega Island, north-east Antarctic Peninsula assessed using DNA metabarcoding. Extremophiles 25:257–265. https://doi.org/10.1007/s00792-021-01226-z

Olmedo-Rojas P et al (2023) Soil environmental DNA metabarcoding in low-biomass regions requires protocol optimization: a case study in Antarctica. Antarct Sci 35(1):15–30. https://doi.org/10.1017/S0954102022000384

Ozturk RC, Feyzioglu AM, Altinok I (2022) Prokaryotic community and diversity in coastal surface waters along the Western Antarctic Peninsula. Pol Sci 31:100764. https://doi.org/10.1016/j.polar.2021.100764

Passarini MRZ, Robayo MIG, Ottoni JR et al (2024) Biotechnological potential in agriculture of soil Antarctic microorganisms revealed by omics approach. World J Microbiol Biotechnol 40:345. https://doi.org/10.1007/s11274-024-04114-8

Peck LS et al (2004) Genomics: applications to Antarctic ecosystems. Polar Biol 28:351–365. https://doi.org/10.1007/s00300-004-0671-8

Põlme S et al (2020) FungalTraits: a user-friendly traits database of fungi and fungus-like stramenopiles. Fungal Divers 105:1–16. https://doi.org/10.1007/s13225-020-00466-2

Pushkareva E et al (2024) Microbial community composition of terrestrial habitats in East Antarctica with a focus on microphototrophs. Front Microbiol 14:1323148. https://doi.org/10.3389/fmicb.2023.1323148
Qu C et al (2019) Complete genome sequence of Rhodococcus sp NJ-530, a DMSP-degrading actinobacterium isolated from Antarctic sea ice. 3 Biotech 9:1–8. https://doi.org/10.1007/s13205-019-1889-z
Rabelo NG et al (2024) Endolithic fungal diversity in antarctic oligocene rock samples explored using DNA metabarcoding. Biology 13(6):414. https://doi.org/10.3390/biology13060414
Ramírez-Fernández L et al (2021) Diversity of microbial communities and genes involved in nitrous oxide emissions in Antarctic soils impacted by marine animals as revealed by metagenomics and 100 metagenome-assembled genomes. Sci Total Environ 788:147693. https://doi.org/10.1016/j.scitotenv.2021.147693
Rosa LH et al (2019) Fungi in Antarctica: diversity, ecology, effects of climate change, and bioprospection for bioactive compounds. In: Rosa L (ed) Fungi of Antarctica. Springer, Cham. https://doi.org/10.1007/978-3-030-18367-7_1
Rosa LH, da Silva TH, Ogaki MB et al (2020a) DNA metabarcoding uncovers fungal diversity in soils of protected and non-protected areas on Deception Island, Antarctica. Sci Rep 10:21986. https://doi.org/10.1038/s41598-020-78934-7
Rosa LH, Pinto OHB, Šantl-Temkiv T et al (2020b) DNA metabarcoding of fungal diversity in air and snow of Livingston Island, South Shetland Islands, Antarctica. Sci Rep 10:21793. https://doi.org/10.1038/s41598-020-78630-6
Rosa LH, da Costa Coelho L, Pinto OHB et al (2021a) Ecological succession of fungal and bacterial communities in Antarctic mosses affected by a fairy ring disease. Extremophiles 25:471–481. https://doi.org/10.1007/s00792-021-01240-1
Rosa LH, Pinto OHB, Convey P et al (2021b) DNA metabarcoding to assess the diversity of airborne fungi present over Keller Peninsula, King George Island, Antarctica. Microb Ecol 82:165–172. https://doi.org/10.1007/s00248-020-01627-1
Rosa LH, Ogaki MB, Lirio JM et al (2022) Fungal diversity in a sediment core from climate change impacted Boeckella Lake, Hope Bay, north-eastern Antarctic Peninsula assessed using metabarcoding. Extremophiles 26:16. https://doi.org/10.1007/s00792-022-01264-1
Schultz J et al (2022a) Polyphasic analysis reveals potential petroleum hydrocarbon degradation and biosurfactant production by rare biosphere thermophilic Bacteria from Deception Island, an active Antarctic volcano. Front Microbiol 13:885557. https://doi.org/10.3389/fmicb.2022.885557
Schultz J et al (2022b) Unraveling the genomic potential of the thermophilic bacterium Anoxybacillus flavithermus from an Antarctic geothermal environment. Microorganisms 10(8):1673. https://doi.org/10.3390/microorganisms10081673
Shaffer JMC et al (2023) Genomic and phenotypic characterization of a red-pigmented strain of Massilia frigida isolated from an Antarctic microbial mat. Front Microbiol 14:1156033. https://doi.org/10.3389/fmicb.2023.1156033
Soto DF et al (2022) Functional filtering and random processes affect the assembly of microbial communities of snow algae blooms at Maritime Antarctic. Sci Total Environ 805:150305. https://doi.org/10.1016/j.scitotenv.2021.150305
Thompson AR, Geisen S, Adams BJ (2020) Shotgun metagenomics reveal a diverse assemblage of protists in a model Antarctic soil ecosystem. Environ Microbiol 22:4620–4632. https://doi.org/10.1111/1462-2920.15198
Trefault N, De la Iglesia R, Moreno-Pino M et al (2021) Annual phytoplankton dynamics in coastal waters from Fildes Bay, Western Antarctic Peninsula. Sci Rep 11:1368. https://doi.org/10.1038/s41598-020-80568-8
Villanueva P et al (2021) Description of the first four species of the genus Pseudogymnoascus from Antarctica. Front Microbiol 12:713189. https://doi.org/10.3389/fmicb.2021.713189

Wang W et al (2023) Genome-based classification of Pedobacter polysacchareus sp. nov., isolated from Antarctic soil producing exopolysaccharide. FEMS Microbiol Lett 370:fnad031. https://doi.org/10.1093/femsle/fnad031

Waschulin V et al (2022) Biosynthetic potential of uncultured Antarctic soil bacteria revealed through long-read metagenomic sequencing. ISME J 16(1):101–111. https://doi.org/10.1038/s41396-021-01052-3

Zhang T, Yan D, Ji Z et al (2022) A comprehensive assessment of fungal communities in various habitats from an ice-free area of maritime Antarctica: diversity, distribution, and ecological trait. Environ Microbiome 17:54. https://doi.org/10.1186/s40793-022-00450-0

Zhang Y, Zhai Y, Mu L et al (2023) Maribacter aquimaris sp. nov., isolated from seawater adjacent to Fildes Peninsula, Antarctica. Antonie Van Leeuwenhoek 116:753–761. https://doi.org/10.1007/s10482-023-01844-x

Chapter 7
Agricultural Applications of Antarctic Bacteria and Fungi

Averlane Vieira da Silva, Mayanne Karla da Silva, Adriely Vital de Souza Silva, Rafaela Ferreira da Silva Costa, Adeildo Júnior de Oliveira, Maria Nicolle Pereira da Silva, Aline Cavalcanti de Queiroz, Magna Suzana Alexandre Moreira, Andressa Letícia Lopes da Silva, Júlia Ronzella Ottoni, Michel Rodrigo Zambrano Passarini, Lara Durães Sette, Valéria Maia de Oliveira, Paulo Eduardo Camara, Luiz Henrique Rosa, and Alysson Wagner Fernandes Duarte

Abstract Antarctica is a natural laboratory, an environment with limiting characteristics for developing all life forms. Microorganisms are the life forms most adapted to the different Antarctic ecological niches and directly influence the development of mosses and the two angiosperms that occur at the South Pole. The aim of this review is to present the main biotechnological applications for use in the agricultural sector that fungi and bacteria isolated from Antarctica have been prospected in recent literature. In this sense, bacteria and fungi isolated from Antarctica have been reported to have biological activities of agricultural interest, including the pro-

A. V. da Silva · M. K. da Silva · A. V. de Souza Silva · R. F. da Silva Costa · A. J. de Oliveira · M. N. P. da Silva · A. C. de Queiroz · A. W. F. Duarte (✉)
Universidade Federal de Alagoas, Campus Arapiraca, Arapiraca, Brazil
e-mail: alysson.duarte@arapiraca.ufal.br

M. S. A. Moreira · A. L. L. da Silva
Universidade Federal de Alagoas, Campus AC Simões, Maceió, Brazil

J. R. Ottoni · M. R. Z. Passarini
Universidade Federal da Integração Latino Americana, Foz do Iguaçu, Brazil

L. D. Sette
Departamento de Biologia Geral e Aplicada, Instituto de Biociências, Universidade Estadual Paulista Júlio de Mesquita Filho (UNESP), Rio Claro, Brazil

V. M. de Oliveira
Centro Pluridisciplinar de Pesquisas Químicas, Biológicas e Agrícolas Universidade Estadual de Campinas, Paulínia, Brazil

P. E. Camara
Departamento de Botânica, Campus Darcy Ribeiro, Universidade de Brasília, Asa Norte, Brasília, Brazil

L. H. Rosa
Universidade Federal de Minas Gerais, Belo Horizonte, Brazil

A. W. Fernandes Duarte et al. (eds.), *Polar Microbes and Climate Change*,
https://doi.org/10.1007/978-3-032-07096-8_7

duction of metabolites related to phosphate solubilization, nitrogen fixation, phytohormones such as indole-3-acetic acid (IAA), substances with herbicidal and phytopathogen-inhibiting activities. Furthermore, as they are adapted to the different adverse conditions of Antarctica, understanding the possible applications of Antarctic fungi and bacteria could help in the near future to boost the intensification of agriculture sustainably, with the production of bioinoculants, for example, especially in a scenario of global climate change.

Keywords Bioactive metabolites · Endophyte · Herbicidal activity · Phytopathogens · Phosphate-solubilizing microorganisms

7.1 Introduction

Antarctica, located in the southern hemisphere, covers an area of approximately 14 million square kilometers, of which 98% is covered with ice and snow, and the few ice-free areas of this continent allow biological dispersion. It is surrounded by the Antarctic Ocean, with an average altitude of about 2.500 m, with large mountain ranges, valleys, and permanent glaciers (Mink et al. 2014; Lemonnier et al. 2020). The Antarctic territory comprises three regions: maritime, sub-Antarctic, and continental, each with its climate and unique biological and geological patterns (Varliero et al. 2024).

Extreme conditions are more severe in continental regions, with lower temperatures throughout the year and vegetation limited to two vascular plants, bryophytes, and lichens (Singh et al. 2015, 2018). Maritime Antarctica, the most coastal zone, has milder climate characteristics and species richness in biotic communities, including algae, bacteria, fungi, lichens, vascular plants, invertebrates, birds, and mammals. Comprising the Antarctic Peninsula and nearby islands, while the sub-Antarctic zone extends over a large area by the ocean, it is warmer throughout the year and includes islands such as South Georgia, Bouvet, Heard, and McDonald (Convey 2011).

The Antarctic environment presents arid and extreme conditions, such as low temperatures ranging from 0 to 10 °C in summer and an average of −30 °C during winter in the coastal regions, with lower temperatures in the innermost portion of the continent, where values of around −80 °C are observed (Alberdi et al. 2002). Strong winds in Antarctica and the low atmospheric humidity make the air dry and thin, given the little annual precipitation (ranged 3–50 mm water equivalent in the McMurdo Dry Valleys at roughly 78° south latitude) that occurs mainly in the form of ice and snow crystals (Wright et al. 2024). Furthermore, significant changes in the prevalence of photosynthetically active and UV radiation take place during different seasons (Ramírez et al. 2024). Due to these extreme factors and the difficulty of access, Antarctica is considered the most pristine continent with the least anthropogenic action (Zwoliński and Mercier 2025).

Another typically stressful feature in this environment is the low nutritional index of soils, such as N:P, due to the vast territorial range marked by the abundance of ice and the scarcity of raw organic matter. The most significant nutrient availability is found in coastal soils subject to the influence of the sea and in places where there are vertebrate droppings and bird nesting spaces, especially in soils and rocks (Chown and Convey 2006; Convey et al. 2014). This scarcity is one of the variants that restrict the diversity and biotic interaction within this environment, which colonizing organisms must tolerate. In Antarctica, alkaline soils are predominant, supplied with carbon, nitrogen, potassium, and phosphorus in moderately low quantities, influenced by wind and water movements, as well as the activities of animals (primarily birds). In the Antarctic, soils are classified according to the bioclimatic region where they are formed. Zonal soils are characteristic of drier areas, ornithogenic soils result from phosphate deposits of bird manure, and organic and lithosol soils (Bockheim 2015; Schmitz et al. 2024).

Amidst challenging conditions in Antarctica, bacteria and fungi play a critical role in the growth of vegetation in this region, and the endophytic organisms associated with two native flowering plants, i.e., *Colobanthus quitensis* (Kunth) Bartl. (*Caryophyllaceae*) and *Deschampsia antarctica* (Desv.) (Fig. 7.1) have been studied (Rosa et al. 2009; Santiago et al. 2017; Acuña-Rodriguez et al. 2024). Upson et al. (2009a) related the benefits of endophytic fungi on the growth and survival of *D. antarctica* in the natural environment, and according to their findings, fungal endophytes can mineralize peptides in the rhizosphere, which leads to a greater availability of nitrogen for the roots. More recently, Acuña-Rodriguez et al. (2024) reported that the two angiosperms that occur in Antarctica show low genetic diversity, although they express considerable phenotypic plasticity. The authors reinforce that the endophytic fungi associated with *C. quitensis* and *D. antarctica* possibly play a crucial role in the adaptation strategies and phenotypic patterns of these plants at the South Pole and that understanding the symbiont communities of these plants is essential for a broader understanding of how these plants develop in cold conditions and how they can respond to global climate change scenarios.

A) B)

Fig. 7.1 Vascular plants growing in Antarctica. (**a**) *Colobanthus quitensis* (Kunth) Bartl and (**b**) the Antarctic grass *Deschampsia antarctica* (Desv)

In this perspective, bacteria and fungi isolated from the Antarctic environment have been reported as promising organisms for application in agricultural activities, and the purpose of this review is to discuss the biotechnological potential of these microorganisms.

7.2 Microbial Communities from Antarctica

A great diversity of microbial life is found in Antarctica, including bacteria (Barros et al. 2022) and fungi (Zucconi et al. 2024), associated with native plants, soils, and lichens, for example (Duarte et al. 2016; Santiago et al. 2017; Godinho et al. 2015; Silva et al. 2021). Such microorganisms withstand the stressful conditions of the Antarctic environment through cellular adaptations and interaction with other organisms, such as vascular plants (Santiago et al. 2017) and macroalgae (de Freitas et al. 2024). Maintaining the cellular integrity of microorganisms that inhabit cold environments is essential for their survival. Therefore, the composition of their cell membrane is modified to contain a higher proportion of unsaturated fatty acids, which increases the plasticity against the effects of low temperature (Liu et al. 2023, Akulava et al. 2024). Additionally, many of them produce antifreeze proteins (PACs), enzymes, and glycerol to prevent the formation of ice crystals (Shivaji and Prasad 2009; Lopes et al. 2024). These microorganisms' psychrophiles are usually differentiated in their ability to grow in certain temperature ranges, especially between −20 and 10 °C (Cavicchioli 2016).

Fungi stand out among the known Antarctic microorganisms. This group of microorganisms is widely distributed in polar regions and can be found associated with various substrates. In Antarctica, for instance, such associations occur in plants, lichens, and mosses (Tosi et al. 2002; Santiago et al. 2015), as well as in the marine environment, where they associate with macroalgae (de Freitas et al. 2024), invertebrates (Godinho et al. 2019), and sediments (Wentzel et al. 2019). Rosa et al. (2021) reported the existence of fungi in moss carpets from King George Island, mainly belonging to the following phyla *Ascomycota*, *Basidiomycota*, *Rozellomycota*, *Chytridiomycota*, *Mortierellomycota*, and *Monoblepharomycota*. Furthermore, the genera *Pseudogymnoascus*, *Penicillium*, *Debaryomyces*, *Aspergillus*, and *Goffeauzyma* were evident in the soils of this continent (Godinho et al. 2015; Gomes et al. 2018). The presence of yeasts in soil samples, marine sediments, and Collins glacier soil also from King George Island, Antarctica, such as *Candida atlantica*, *Camptobasidium* sp., *Goffeauzyma gilvescens*, *Goffeauzyma gastrica*, and *Naganishia* sp., was described by Farias et al. (2022).

Antarctic lichen stalks such as *Usnea antarctica* and *Usnea aurantiacoatra* provide shelter for a variety of non-lichenized fungi, such as *Pseudogymnoascus* sp., *Fusarium* sp., *Antarctomyces psychrotrophicus*, *Vishniacozyma victoriae*, *Cystobasidium laryngis* (formerly *Rhodotorula laryngis*), and *Bensingtonia yamatoana* (Santiago et al. 2015). Duarte et al. (2016) found other lichen yeasts in the South Shetland Islands (Deception, Livingston, Half Moon, King George, Penguin, and Nelson Islands) including *Mrakia* sp., *Trichosporon* sp., and *Bannozyma arctica* (formerly *Rhodotorula arctica*), as well as yeasts associated with macroalgae, belonging to the genera *Metschnikowia*, *Rhodotorula*, and *Glaciozyma*.

Endophytic fungi were found in Antarctic angiosperms *D. antarctica* and *C. quitensis*, dispersed in Admiralty Bay on King George Island. According to the work of Santiago et al. (2017), the species *V. victoriae* is abundant in both plants, while *C. laryngis* was more frequent in *D. antarctica*, followed by *Rhodotorula mucilaginosa*, *Sporidiobolus ruineniae*, and *Leucosporidium golubevii*. In this context, these fungal species exhibit strong adaptation to the plant tissue, thus indicating their endophytic nature. In a similar study, Rosa et al. (2009) also indicated the presence of endophytic filamentous fungi in the leaves of *D. antarctica*, the *Ascomycetes*, *Alternaria* sp., and *Phaeosphaeria* sp.

Several studies described the presence of Antarctic bacteria distributed in different environmental samples in the region, such as native plants, macroalgae, lichens, and mosses. They reported the dominance of the phyla *Actinobacteria*, *Proteobacteria*, *Acidobacteria*, and *Firmicutes*. The prokaryotes found in the rhizosphere of the two angiosperms plants belong mainly to the genera *Bifidobacterium*, *Arcobacter*, and *Faecalibacterium* (Teixeira et al. 2010).

In the study performed by Park et al. (2016), the authors describe the importance of bacterial phyla by analyzing Antarctic lichens, including *Usnea*, *Buellia*, *Umbilicaria*, and *Cladonia*. Among them, the predominant bacterial class was *Alphaproteobacteria* (*Acetobacteraceae* family).

This microbial diversity found throughout Antarctica associated with varying environmental substrates, including, lichens, mosses, macroalgae, and particularly plants, performs numerous ecological functions within the polar ecosystem, whether in the production of metabolites in biogeochemical cycling, for example (Park et al. 2016; Alvarado et al. 2018; Graham et al. 2017). In many cases, because they are associated with other living organisms, as reported in this topic, fungi and bacteria demonstrate high adaptability to the demanding conditions of the Antarctic environment due to their association with other living organisms (Vero et al. 2019; Giudice and Arazzo 2019). Antarctic microorganisms have drawn significant global interest in agriculture, which are described in the following topics in greater detail.

7.3 Agricultural Application of Bacteria and Fungi

The biotechnological applications of Antarctic plant growth-promoting microbes (PGPM) in agriculture are outlined in Fig. 7.2. Tables 7.1 and 7.2 summarize the main studies with bacterial and fungal isolates, respectively.

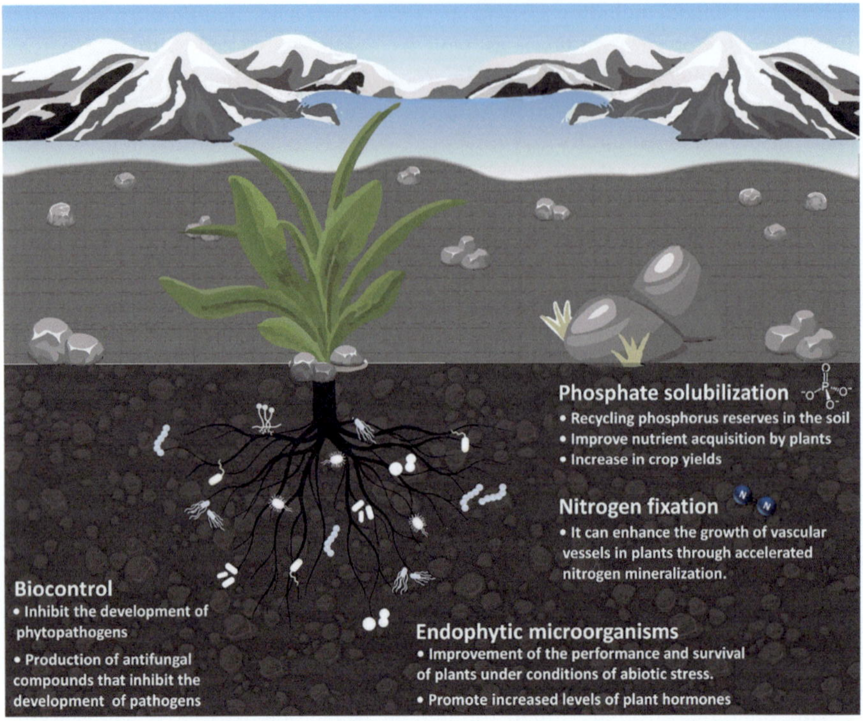

Fig. 7.2 Agricultural applications of Antarctic bacteria and fungi including phosphate solubilization, nitrogen fixation, endophyte microorganism and phytopathogen biocontrol

7.4 Nitrogen Fixation

Nitrogen fertilizers have been widely used in agriculture worldwide (Bodirsky et al. 2014). Nitrogen (N) is an essential component for the biosynthesis of cellular nitrogenous compounds and is considered one of the nutrients essential for plant development (Yuan et al. 2012). Proper nitrogen source supplementation, especially NH_4^+ and $NO3^-$, positively influenced the development of *Capsicum annuum* L., including nutrient accumulation and an increase in fruit quality, as studied by Zhang et al. (2019). However, the excessive use of fertilizer impacts the environment. It presents a high cost, depicting the need to develop alternative technologies that partially or fully replace the use of nitrogen fertilizers. In this context, nitrogen-fixing bacteria, such as *Azospirillum* sp., *Bradyrhizobium* sp., *Rhizobium* sp., and some species of *Pseudomonas* spp., are some examples of bacteria that present potential as an alternative to chemical fertilizers, acting as biofertilizers, which can enable efficient soil fertilization and plant development (Vassilev et al. 2015; Mahanty et al. 2017; Chauhan et al. 2023).

Nitrogen fixation, also known as diazotrophy, is the main conversion of nitrogen in the nitrogen cycle. Essentially inert atmospheric nitrogen gas is reduced in

Table 7.1 Agricultural applications of bacteria isolated from different Antarctic environments

Agricultural applications	Species	Phylum	Antarctic sample	Origin of sample	References
Phosphate solubilization and indole-acetic acid (IAA) production	*Pseudomonas* sp.	*Proteobacteria*	Rhizosphere soil samples of plant *Deschampsia antarctica* Desv. (Poaceae)	Not cited	Berríos et al. (2013)
Phosphate solubilization and indole-acetic acid (IAA) production	*Pseudomonas* sp.	*Proteobacteria*	Soil	Greenwich Island	Yarzábal et al. (2018)
Phosphate solubilization	*Pseudomonas* sp. (11.LB15 and 1.LB34)	*Proteobacteria*	Lichens (11.LB15 from *Caloplaca regalis* and 1.LB34 from *Usnea auratiacoatra*)	Barrientos and King George Island	Silva et al. (2021)
Uptake and fixation of nitrogen and indole-acetic acid (IAA) phytohormone	*Pseudomonas frederiksbergensis*, *Pseudomonas poae*, *Pseudomonas brassicacearum*, *Pseudomonas mandelii*, *Pseudomonas lini*, and *Janthinobacterium lividum*	*Proteobacteria*	Root rhizosphere of plant *D. antarctica* plant	South Shetland Islands: Gemelos Hill, Collins Glacier, Ardley Peninsula, and Hanna Point	Cid et al. (2018)
Antibacterial activity in vitro and in vivo against agents of bacterial spot of pepper	*Streptomyces* sp.	*Actinobacteria*	Soils	Not cited	Encheva-Malinova et al. (2014a)
Antifungal activity against phytopathogens of agricultural interest	*Arthrobacter* spp., *Streptomyces* spp., and *Rhodococcus* spp.	*Actinobacteria*	Soils, marine sediment and seawater	Not cited	Santos et al. (2019)

(continued)

Table 7.1 (continued)

Agricultural applications	Species	Phylum	Antarctic sample	Origin of sample	References
Antifungal activity against Fusarium *oxysporum* f. sp. *cubense*	*Streptomyces* sp. K6 and *Streptomyces fildesensis* E7	*Actinobacteria*	Soils	King George Island	Perez et al. (2024)
Antibacterial potential of streptomycetes	*Streptomyces* spp.	*Actinobacteria*	Soils	Livingston Island, Antarctica	Encheva-Malinova et al. (2014b)

ammonia (NH_3) or ammonium (NH_4^+) in soils with neutral pH, and biological nitrogen fixation (BNF) is the main mechanism for cycling N, molecules essential to life (Mahanty et al. 2017). Some cyanobacteria can fix nitrogen, contributing significantly to the nitrogen cycle and directly increasing yield in agriculture (Latysheva et al. 2012). However, the use of polar microorganisms as growth promoters in commercial crops has not yet been fully explored (Latysheva et al. 2012; Yarzábal et al. 2018), although a recent study using an omics approach with Antarctic soils pointed out that among the various agricultural applications prospected in the study with Deception soils, the main activity detected in bacteria-related OTU was nitrogen fixation (Passarini et al. 2024).

The nitrogen levels in Antarctic soils are significantly lower than those found in terrestrial soils in different environments, like forests and cultivated soils, for example (Cowan and Tow 2004). These soils are also poor in organic matter, with most of the nutrient-rich soils found in Antarctica being in coastal areas and in the McMurdo Dry Valleys (MDV) (Tahon et al. 2018), where fixed nitrogen can be derived from the contribution of diazotrophic cyanobacteria for microbial communities occurring in MDV streams (Kohler et al. 2018). However, in hypolithic MDV communities, all putative diazotrophs were found to be *Proteobacteria* (Lacap-Bugler et al. 2017).

In soils and marine sediments of Antarctica, the release of nitrogen from the decomposition of organic material is slow and limited due to low temperatures, imposing severe restrictions on plant life (Godinho et al. 2019; Oses-Pedraza et al. 2020). However, access to nitrogen can occur through a mutualistic symbiosis between plant and fungal endophytes and bacteria (Upson et al. 2009b; Nesme et al. 2018). Oses-Pedraza et al. (2020) studied two endophytic fungi from *D. antarctica* and *C. quitensis*, *Penicillium chrysogenum* and *Penicillium brevicompactum*, and both exhibit exolytic enzymatic activities that may be involved in cell wall degradation and nutrient absorption from carbon, nitrogen, and phosphorus groups. Furthermore, endophytic Antarctic fungi enhanced the growth of vascular vessels in Antarctic plants through accelerated nitrogen mineralization.

Table 7.2 Agricultural applications of fungi isolated from different Antarctic environments

Agricultural applications	Species	Phylum	Antarctic sample	Origin of sample	References
Phosphate solubilizing, nitrogen mineralization, and growth enhancement of their host plants	*Penicillium chrysogenum* and *P. brevicompactum*	*Ascomycota*	Rhizospheric soil of *Colobanthus quitensis* Kunth (Caryophyllaceae) and *Deschampsia antarctica* Desv. (Poaceae)	Antarctic Polish Base "Henryk Arctowski" on King George Island, South Shetland Islands	Oses-Pedraza et al. (2020)
Phosphate solubilizing	*Humicola marvinii*	*Ascomycota*	Soils	Signy Island, maritime Antarctic	Weinstein et al. (1997)
Phosphate solubilizing	*Vishniacozyma victoriae* 2.L15	*Basidiomycota*	*Polycauliona regalis* lichen	Barrientos	Silva et al. (2022)
Activity of secondary metabolites against *Xanthomonas* species	*Pseudogymnoascus*, *Penicillium*, *Cadophora*, *Paraconiothyrium*, and *Toxicocladosporium*	*Ascomycota*	Soil and marine sediment	Deception Island and Admiralty Bay, King George Island, South Shetland Islands	Vieira et al. (2018)
Activity of secondary metabolites against *Xanthomonas euvesicatoria* and *Xanthomonas axonopodis* pv. *passiflorae*	*Pseudogymnoascus* sp., *Penicillium*, *Paraconiothyrium*, *Cadophora*, *Toxicocladosporium*	*Ascomycota*	Soil and marine sediment	Admiralty Bay and Deception Island	Purić et al. (2018)
Herbicidal activities of extract fungal against *Allium schoenoprasum* and *Lactuca sativa*	*Pseudogymnoascus destructans*, *Mortierella amoeboidea*, *Mortierella* sp., *Penicillium tardochrysogenum*	*Ascomycota* *Mucoromycota*	Soil	Robert, Nelson, King George, and Penguin Islands at South Shetland archipelago	Gomes et al. (2018)
Herbicidal activity against *Lactuca sativa* and *Agrostis stolonifera*	*Aspergillus sydowii*, *Penicillium allii-sativi*, *P. brevicompactum*, *P. chrysogenum*, and *P. rubens*	*Ascomycota*	Soils	Ellsworth Mountains	Godinho et al. (2015)

(continued)

Table 7.2 (continued)

Agricultural applications	Species	Phylum	Antarctic sample	Origin of sample	References
Herbicidal activities of extract fungal against *Lactuca sativa* and *Allium schoenoprasum*	*Penicillium* cf. *palitans*	*Ascomycota*	Marine sediments	Livingston Island (Walker Bay), Deception Island (Whalers Bay), and English Strait	Da Silva et al. (2024)
Herbicidal activities of extract fungal against *Lactuca sativa* and *Allium schoenoprasum*	Strains of *Penicillium* sp.	*Ascomycota*	Macroalgae		De Freitas et al. (2024)
Antifungal activity against phytopathogens	*Leucosporidium scottii*	*Basidiomycota*	Soils	King George Island	Cary et al. (2010)
Control of postharvest pathogens of apples	*Candida sake*	*Ascomycota*	Soil and water samples	King George Island, South Shetland Islands	Arrarte et al. (2017)
Filamentous fungi against *X. citri* and *X. albilineans*	*Pseudogymnoascus* sp. and *Cladosporium* sp.	Ascomycota	Soils	Island (South Shetland Islands, Antarctica) and King George Island (South Shetland Islands, Maritime Antarctica)	Ferrarezi et al. (2019)
Plant growth (*Deschampsia antarctica*) and nutrient balance	*Mollisia* sp., *Tapesia* sp., *Oculimacula yallundae*	Ascomycota	Roots of the *Deschampsia antarctica*	South Georgia on the Subantarctic Islands and Signy, Coronation, King George, and Anchorage in Antarctic Maritime	Upson et al. (2009a)

7.5 Phosphate Solubilization by Antarctic Microorganisms

Phosphorus (P) is considered an essential and irreplaceable element in nature, as it plays a vital role in living organisms. The importance of this element lies in its participation in several processes, such as nucleic acid and protein synthesis, as well as in various metabolic activities (Nesme et al. 2018). Phosphorus has always played an essential role in agriculture and horticulture, being a limiting factor in agricultural production and in maintaining food production for population growth, with an estimated world population of 9.7 billion by 2050 (Sattari et al. 2012; Mogollón et al. 2018). Despite this importance, the average phosphorus content in the soil is approximately 0.05% (w/w). However, due to its poor solubility and fixation in the soil, a small amount of phosphorus (0.1%) is available to plants. Such a challenge of phosphorus deficiency in the soil is resolved through applying phosphate fertilizers (Maharajan et al. 2017).

The search for phosphate-solubilizing microorganisms for agricultural purposes involves the isolation of microorganisms, screening for desirable characteristics such as growth and solubilization in inorganic phosphate sources such as calcium phosphate, aluminum phosphate, and phosphate rock powder and at different temperatures, pH, salinity, and stability when applied in the field for possible formulation of inoculants in suitable formulations (Berríos et al. 2013; Janati et al. 2023). These important steps are part of the quest for sustainable agricultural development. Some species of microorganisms, especially those associated with plant roots (Çakmakçi et al. 2006) or lichens attached to rocks (saxicolous) (Sigurbjörnsdóttir et al. 2014), can solubilize phosphate from the soil, making it readily available for plant absorption and nutrition, thus are considered as plant growth promoters (Yarzábal et al. 2018).

Various mechanisms have been identified in phosphate-solubilizing microorganisms that help facilitate the P cycle in soil-plant systems. These mechanisms include the production of extracellular polysaccharides, siderophores, organic acids, phosphatases, and specific phosphate transport systems (Silva et al. 2021; Cheng et al. 2023; Silva et al. 2022). One widely reported indirect mechanism for solubilizing inorganic phosphate that is used by researchers today is the production of organic acids. This mechanism of release of organic acids occurs with the participation of hydroxyl and carboxyl groups, which chelate aluminum (Al), iron (Fe), and calcium (Ca) cations and decrease the pH and increase the bioavailability of phosphorus (Sagoe et al. 1998; Jha et al. 2013). *Pseudomonas* species isolated from cold regions have been commonly reported as producers of organic acids (Chauhan et al. 2023).

In 1997, Weinstein et al. conducted one of the first studies on the fungus *Humicola marvinii*. The fungus was isolated from fellfield soils at Signy Island in the maritime Antarctic. The study found that after 27 days of incubation at 15 °C, the fungus showed colony formation with a diameter of 22.0 mm. Additionally, there was a clear zone of phosphate solubilization with a diameter of 26.0 mm.

In 2008, Barrientos-Díaz et al. worked with rhizospheric soil samples from the plant *D. antarctica* and identified *Pseudomonas*, *Flavobacterium*, and *Arthrobacter*. Later, in 2013, Berríos et al. found that one of these isolates (*Pseudomonas* Da-bac TI-8) solubilized P when grown in a PVK agar medium, and the phosphate sources used were tricalcium phosphate, monobasic calcium phosphate, and rock phosphate. The capacity to solubilize inorganic P by the bacteria generated halos that varied, with the largest halo when the P source was monobasic calcium phosphate ($Ca(HPO_4)_2H_2O$) under conditions of 15 °C for 5 days. Subsequently, the same strain solubilized 34.3 mg/L P from phosphate rock by lowering the pH under conditions of 4 °C for 5 days. The authors also reported the production of organic acids (gluconic acid). This strain also showed superior performance in root elongation of the native plant *D. antarctica*, in vitro compared to the control treatment.

In 2018, Yarzábal et al. worked with 25 isolates of *Pseudomonas* spp., obtained from Antarctic soils on the Greenwich Island. They found that some isolates could solubilize phosphate when glucose, galactose, and sucrose were supplied as sole carbon sources. In general, using glucose as a carbon source resulted in greater P solubilization, with solubilization halos between 5.0 and 10.0 mm at 8 °C, in the NBRIP solid medium. In the same study, the authors demonstrated that some strains of *Pseudomonas* promoted the growth of wheat seedlings (*Triticum aestivum*) of their roots, when kept at 14 °C for 15 days, under controlled conditions, which may be helpful for future applications in agricultural systems. Other examples of Antarctic bacterial potential in agricultural applications are shown in Table 7.1.

Oses-Pedraza et al. (2020) tested endophytic fungi isolated from the roots of *C. quitensis* and *D. antarctica* for P solubilization in PVK agar medium. The fungi, identified as *P. chrysogenum* AFE001 and *P. brevicompactum* AFE002 solubilized P at 4 and 15 °C. It was also noted, in the experiment with native plants of Antarctica, that the accumulation of total plant biomass was greater in *C. quitensis* and *D. antarctica* with root endophytes compared to control trials of plant models without endophytes (Fig. 7.3). Silva et al. (2021) reported a high number of phosphate-solubilizing bacteria from Antarctic lichens at 15 °C (66 isolates), and *Pseudomonas* sp. (strains 11.LB15 and 1.LB34) showed the best activities at 25 and 30 °C, with the release of fumaric and tartaric acids during the solubilization process.

Yeasts isolated from Antarctic lichens have also exhibited phosphate-solubilizing activity. According to Silva et al. (2022), 26% of the yeasts tested demonstrated positive results at 15 °C. *Vishniacozyma victoriae* 2.L15 showed the best activity, with P solubilized at 25 and 30 °C and the production of tartaric and acetic acids after 72 h of incubation (Table 7.2).

Finally, we highlight in this topic what has been studied about the solubilization of P by microorganisms from Antarctica, with the purpose of application in agriculture. We emphasize the importance of future research with tests in a wide temperature range and the interaction between bacteria and fungi, being a promising approach to improve food production through the formulation of possible bioinoculants (Alori et al. 2017; Sarikhani et al. 2019).

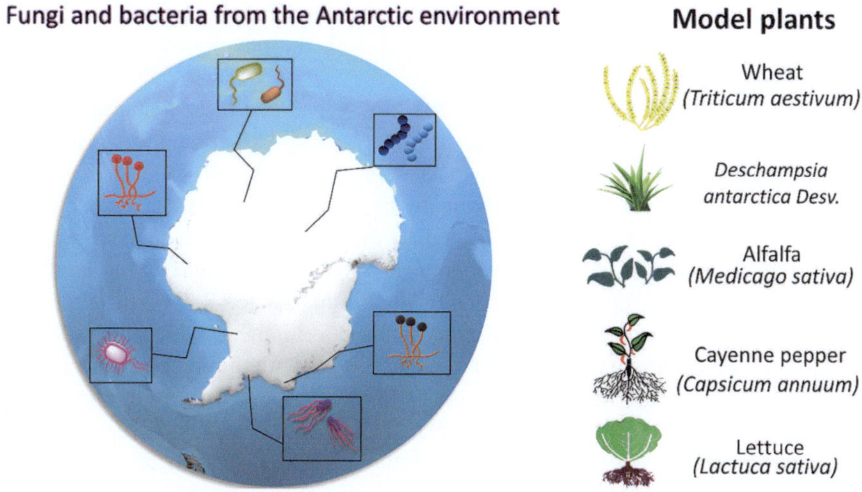

Fig. 7.3 Model plants used in studies with Antarctic fungi and bacteria that promote plant growth

7.6 Phytohormones Produced by Antarctic Bacteria

Plant growth-promoting microbes (PGPM) can produce compounds capable of stimulating plant development through different mechanisms, such as the production of phytohormones (Gray and Smith 2005). Rhizobacteria are considered plant growth promoters (PGPR) as they stimulate growth directly or indirectly (Grobelak et al. 2015), and among the microorganisms belonging to the PGPR group are the bacterial genera *Pseudomonas*, *Azospirillum*, *Burkholderia*, *Bacillus*, *Enterobacter*, *Rhizobium*, *Erwinia*, *Serratia*, *Alcaligenes*, *Arthrobacter*, *Acinetobacter*, and *Flavobacterium* (Grobelak et al. 2015; Rodríguez and Fraga 1999; Rosas et al. 2006).

Indole-acetic acid (IAA) may act as a reciprocal signaling metabolite in microbe-plant interactions (Spaepen and Vanderleyden 2011). The production of phytohormone such as IAA is one of the important characteristics related to plant growth promotion (Yarzábal et al. 2018), with a relevant physiological effect on the plant by direct stimulation of seed and tuber germination, triggering and proliferation of cell-plant division, and vegetative growth control, among other activities (Spaepen and Vanderleyden 2011; Tsavkelova et al. 2006).

One of the most studied bacterial groups with the potential to solubilize phosphate is the *Pseudomonas* genus, which has also shown the ability to produce phytohormone such as indole-acetic acid, a statement that was confirmed in the study by Yarzábal et al. (2018). According to the authors, an Antarctic isolate of *Pseudomonas* sp. showed positive activity for the production of IAA in vitro, which was also observed by Berríos and co-workers in 2013, who verified the activity of *Pseudomonas* sp. Da-bac TI-8, which could solubilize phosphate and produce IAA, showed greater production of this phytohormone at 20 °C. Above the ground, through a study on the genome sequencing of phyllosphere bacteria isolated from

D. antarctica, the presence of five strains of *Pseudomonas* and one *Janthinobacterium* was revealed, all containing genes associated with the phytohormone of indole-acetic acid (Cid et al. 2018).

In practical terms, Styczynsk et al. (2022) observed that three Antarctic strains (*Pseudomonas* sp., *Psychrobacter* sp., and *Bacillus* sp.) promoted the growth of alfalfa (*Medicago sativa*), grown in pots containing 2 kg of soil. Such strains increased shoot length and fresh biomass by up to 26 and 46%, respectively. Furthermore, these strains were also tested for their ability to produce plant growth-promoting compounds such as siderophores and indole-acetic acid, for example.

Therefore, Berríos et al. (2013) demonstrated that bacterial species inhabiting the Antarctic soil can promote plant growth and development, which may play an important role in the survival of plants residing on the Antarctic continent. In this topic, we elucidate the beneficial effects of producing IAA from microorganisms (especially bacteria) from Antarctica for plant health. These microorganisms are effective in plant development with the combination of P solubilization and IAA production (Yarzábal et al. 2018).

7.7 Indirect Mechanisms of Plant Growth Promotion by Antarctic Bacteria and Fungi

In addition to the direct mechanisms described in the previous topics, there are also indirect mechanisms that are related to plant growth promotion. Reports show that the ability of Antarctic microorganisms to tolerate high concentrations of salinity, pH, UV radiation, temperature, and other physical-chemical parameters can help plant development (Gallardo-Cerda et al. 2018). The potential of these microorganisms has been used to improve crop yield and resistance under stressful conditions in different contexts (Table 7.3).

Microorganisms present in the rhizosphere soil of *D. antarctica* and *C. quitensis* have been identified as important collaborators regarding the improvement of the survival of these plants and other plant cultures under conditions of abiotic stress, especially by establishing consortia or symbiotic relationships with the plant. This was verified in the work performed by Gallardo-Cerda et al. (2018), in which rhizobacteria obtained from native Antarctic plants proved tolerant to high salt concentrations and could promote plant growth. The authors found that both *D. antarctica* and *C. quitensis* had significantly higher survival rates when grown with rhizobacteria and/or irrigated solely with water (control). Additionally, those plants in the salt condition with rhizobacteria also had improved survival rates. These findings demonstrate the positive effects of rhizobacteria in stressful conditions.

A consortium consisting of *Arthrobacter* sp. and *Planococcus* sp. and endophytic fungi such as *P. chrysogenum* and *P. brevicompactum*, also obtained from the roots of Antarctic vascular plants, increased the chances of survival in three of four crops subjected to high salinity conditions, including cayenne pepper (*Capsicum*

Table 7.3 Resistance mechanisms promoted by bacteria and fungi isolated from Antarctic environments

Activity of agricultural interest (resistance promotion)	Species	Antarctic sample	Origin of sample	References
Bacteria				
Effect of temperature, NaCl, and UV radiation	*Pseudomonas* sp., *Sphingobacterium anhuiense*, *Sporosarcina aquimarina*, *Stenotrophomonas rhizophila*, *Dermacoccus profundi*, *Microbacterium foliorum*, *Brevundimonas vesicularis*, *Rothia* sp., *Frondihabitans* sp.	Green moss, black crustose lichen, orange lichen, and *D. antarctica*	The Ukrainian Antarctic Station "Akad. Vernadsky" and the biogeographical polligon on Galindez Island, Argentine Islands, maritime Antarctica	Vasileva-Tonkova et al. (2014)
Tolerance to salt stress	*Rhodococcus* sp., *Enterobacter* sp., *Devosia* sp., B*urkholderia* sp., *Arthrobacter* sp. and *Planococcus* sp.	Rhizospheric soil of plants *Colobanthus quitensis* and *D. antarctica*	Livingston Island	Gallardo-Cerda et al. (2018)
Tolerance to salt stress	Microbial consorce (*Arthrobacter* sp. and *Planococcus* sp. and two fungal, *Penicillium chrysogenum* and *Penicillium brevicompactum*)	Rhizosphere of Antarctic plants	Not cited	Acuña-Rodríguez et al. (2019)
Fungi				
Effect of temperature, NaCl, and UV radiation	*Rhodosporidium kratochvilovae*, *Cryptococcus terricola*, *C. victoriae*, *Candida davisiana*, *Exophiala xenobiotica*, *Leucosporidiella muscorum*	Green moss, black crustose lichen, orange lichen, and *D. antarctica*	The Ukrainian Antarctic Station "Akad. Vernadsky" and the biogeographical polligon on Galindez Island, Argentine Islands, maritime Antarctica	Vasileva-Tonkova et al. (2014)
Drought tolerance	*Penicillium brevicompactum* *P. chrysogenum*	Plant roots rhizosphere *C. quitensis* and *D. antarctica*	King George Island	Molina-Montenegro et al. (2016)
Salt stress				Molina-Montenegro et al. (2020)

(continued)

Table 7.3 (continued)

Activity of agricultural interest (resistance promotion)	Species	Antarctic sample	Origin of sample	References
Water stress	*Penicillium chrysogenum* *Penicillium brevicompactum* *Alternaria* sp. *Phaeosphaeria* sp. *Eupenicillium osmophilum*	*C. quitensis* plants	Punta Arenas (sub-antarctic zone), King George Island (South Shetland Islands) and Lagotellerie Island (Antarctic Peninsula)	Hereme et al. (2020)
Fungal endophytes on host plant performance under high level of UV-B radiation	Not cited	*C. quitensis* plants	Antarctic Station "Henryk Arctowski," South Shetland Islands	Ramos et al. (2018) Barrera et al. (2020)

annuum), onion (*Allium cepa*), lettuce (*Lactuca sativa*), and tomatoes (*Solanum lycopersicum*) (Acuña-Rodríguez et al. 2019). Similarly, Molina-Montenegro et al. (2020) evaluated the influence of root fungi (*P. brevicompactum* and *P. chrysogenum*) on tomatoes and lettuce crops under saline stress conditions. In general, the presence of endophytes significantly increased the probabilities of crop survival and productivity, and in the liquid photosynthetic rates, high foliar Na^+ content, and better uptake of nitrogen and potassium from the soil, effects were observed both under stress treatment and in control conditions.

Endophytic fungi can also influence the impact of UV-B radiation on *C. quitensis* when they are associated. These plants exhibited a greater number of flowers, total biomass, and lower lipid peroxidation in the presence of these microorganisms when compared to plants without these endophytes (Ramos et al. 2018). In addition, these fungi contributed to the increase in plant hormone levels such as salicylic acid, abscisic, indole-3-acetate, and jasmonate in the tissues of this Antarctic plant under high radiation due to the modulation of the expression of genes related to the adaptive response to UV light, improving the plant's tolerance to this limiting factor (Ramos et al. 2018; Barrera et al. 2020).

Another highly relevant and necessary abiotic factor for plants is humidity, which, in low availability, has serious implications for the proper growth of crops. The role of root microorganisms in mitigating the effects of drought on plants has been demonstrated (Molina-Montenegro et al. 2016; Hereme et al. 2020). These authors found that the presence of species of *Penicillium*, *Alternaria*, and *Phaeosphaeria* promoted increased drought tolerance in the plants with which they were associated (lettuce and *C. quitensis*). This effect was the result of differential gene expression, increased liquid photosynthesis, and greater efficiency in water use due to osmoprotective mechanisms such as the accumulation of sugars and proline, which prevent damage to cell membranes due to water deficit. The use of microorganisms tolerant to physical-chemical stresses represents, therefore, an alternative strategy to increase the survival of plant species under unfavorable conditions (Vasileva-Tonkova et al. 2014).

In the following paragraphs, we evidenced that the health of plants can be compromised by several agents, and it is of great interest to find new organic products and ecological sources to help control phytopathogens (Arrarte et al. 2017), and the use of metabolites produced by psychrophilic microorganisms may represent an alternative for sustainable development (Son et al. 2008; Purić et al. 2018). Plant diseases caused by the microbial infection of vegetable tissues (affecting seeds, leaves, roots, and other structures) are responsible for economic losses that reach billions of dollars annually, and biological control techniques need to be developed and/or improved (Shuping and Eloff 2017; Lahlali et al. 2022). Therefore, it is believed that the Antarctic microorganisms represent an untapped source of compounds with the capacity to control phytopathogens, thus collaborating with the supply of an essential need in agriculture (Santos et al. 2019).

Some strains of *Pseudogymnoascus* and *Cladosporium*, which were isolated from Antarctic soil, were found to inhibit the growth of two types of phytopathogenic bacteria, *Xanthomonas citri* and *Xanthomonas albilineans* (Ferrarezi et al. 2019). Vieira et al. (2018) evaluated extracts produced by Antarctic fungi belonging to the genera *Pseudogymnoascus*, *Penicillium*, *Cadophora*, *Paraconiothyrium*, and *Toxicocladosporium* and found that they exhibited antimicrobial activity against *X. citri* subsp. *citri* (Vieira et al. 2018). In another study, intracellular and extracellular extracts from fungi found in Antarctic marine sediments were found to have activity against the phytopathogenic isolates known to cause disease in peppers, tomatoes, and passion fruit such as *Xanthomonas axonopodis* pv. *passiflorae* and *Xanthomonas euvesicatoria* (Purić et al. 2018).

Godinho et al. (2015) conducted a study on the diversity and production capacity of bioactive compounds by fungi isolated from Antarctic oligotrophic soil. They found that extracts from different isolates of *Aspergillus sydowii*, *Penicillium allii-sativi*, *P. brevicompactum*, *P. chrysogenum*, and *Penicillium rubens*, particularly *P. brevicompactum*, exhibited antifungal activity against the phytopathogenic fungus *Colletotrichum gloeosporioides*. Vero et al. (2012) identified a strain of *Leucosporidium scottii* At17, isolated from soil samples from Antarctica, as a promising biocontrol agent for the blue and grey mold of two apple cultivars. This isolate produced soluble and volatile antifungal compounds that inhibited the development of pathogens *Penicillium expansum* and *Botrytis cinerea*. In another study, using a metagenomic approach with soils from Deception Island, one of the main agricultural applications of taxonomic groups related to Antarctic fungi has been biological control (Passarini et al. 2024).

Among the bacteria that produce secondary metabolites such as herbicides and antimicrobials, the *Actinobacteria* phylum is one of the most reported in soil samples from different regions of Antarctica (Santos et al. 2019; Silva et al. 2020; De França et al. 2023, Perez et al. 2024). Santos et al. (2019) tested the antagonistic effect of 28 *Actinobacteria* strains from Antarctic soils against *Botrytis* sp., *Fusarium oxysporum*, *Phytophthora infestans*, and *Rhizoctonia solani.* All strains presented antifungal activity by at least one of the studied fungi, most belonging to the genera *Arthrobacter*, *Streptomyces*, and *Rhodococcus*, and the strain *Arthrobacter polychromogenes* was able to inhibit the growth of all tested fungi.

Encheva-Malinova et al. (2014a, b) reported the antibacterial potential of 23 strains of *Actinomycetes* and *Streptomyces*, originating from Antarctic soils. Four of these strains exhibited excellent potential to produce different biologically active substances against phytopathogenic *Xanthomonas euvesicatoria*, *Xanthomonas gardneri*, *Xanthomonas perforans*, and *Xanthomonas vesicatoria*. Antarctic isolates of *Streptomyces* sp. K6 and *Streptomyces fildesensis* E7 grown in differential culture medium (casein-starch) showed activity against the phytopathogen *Fusarium oxysporum* f. sp. *cubense* (Table 7.1), a pathogen of the banana crop (Perez et al. 2024).

De Freitas et al. (2024) evaluated extracts of fungi isolated from macroalgae, and it was observed that 17.5% of the isolates evaluated ($n = 177$) showed activity in inhibiting the growth of monocotyledonous (*Allium schoenoprasum*) and dicotyledonous (*Lactuca sativa*) models (Fig. 7.3). In particular, the strain *Penicillium* sp. 9, isolated from *Caepidium antarcticum*, in which the extract with the lowest concentrations (250 μg mL^{-1} and 62.5 μg mL^{-1}) were effective for the *L. sativa* and *Allium schoenoprasum*, respectively. In another study, using the same plant models mentioned above, da Silva et al. (2024) observed that extracts of isolates of *Penicillium* cf. *palitans* isolated from Antarctic marine sediments showed a high phytotoxic profile at concentrations of 500 μg mL^{-1} and 1000 μg mL^{-1}. The search for natural herbicides is a demand of sustainable agriculture, given the losses from weeds and the replacement of synthetic herbicides due to their toxicity (Ofosu et al. 2023).

A major gap in the technological scenario for the use of Antarctic microorganisms for agricultural applications is the limitation of recent studies in the literature, since they were essentially on a laboratory scale and there were no field studies. However, a deeper understanding of the different agricultural application mechanisms already reported by these organisms could be strategically relevant in the search for more sustainable agriculture through the use of microorganisms adapted to different adverse conditions.

In view of the expectation of using Antarctic microorganisms in field experiments, a recent study by Rodríguez et al. (2023) carried out in a greenhouse, points to interesting results from transplanting microbiota from Antarctic soils into tomato-growing soils. These authors used soils from different Antarctic islands (Deception, Doumer, King George, Robert) and transferred them to agricultural soils (20% w/w soil inoculation) and monitored the development of tomatoes. Some taxa specific to Antarctic soils persisted (up to 96 weeks) in the rhizosphere of tomato plants, most notably *Bacillus* spp. and *Candidatus Nitrosocosmicus*. Vegetable crops showed better resistance to water stress when cultivated with the microbiota from transplanted Antarctic soils, reinforcing the possibility of using Antarctic microbiota for contemporary agriculture subjected to global climate change.

7.8 Conclusion and Perspective

Several microorganisms from Antarctic environments are endowed with characteristics of interest to the agricultural area, such as the potential to solubilize phosphate, fix nitrogen, and produce phytohormones and secondary metabolites, thus contributing to plant health. Some genera, such as *Pseudomonas* stand out in phosphate solubilization, while the consortium between the fungus *Penicillium* and the bacteria *Arthrobacter* and *Planococcus* promote plant development and tolerance to saline stress. Technological approaches made it possible to discover specific genes from strains isolated from Antarctica capable of fixing nitrogen. Finally, bacteria and fungi have an agricultural potential of environmental and commercial interest. It is worth mentioning that despite the efforts made so far, further research must be carried out using molecular biology techniques, and physical-chemical and physiological parameters will contribute to plant development and the production of low-cost commercial bioinoculants.FundingThis research received financial support from Fundação de Amparo à Pesquisa do Estado de Alagoas—FAPEAL (60030–0000000161/2022 and 60030 000542/2018), CNPq (433388/2018–8 and 314775/2023) and PROANTAR/CNPq (442258/2018–6).

References

Acuña-Rodríguez IS, Hansen H, Gallardo-Cerda J, Atala C, Molina-Montenegro MA (2019) Antarctic extremophiles: biotechnological alternative to crop productivity in saline soils. Front Bioeng Biotechnol 7:22. https://doi.org/10.3389/fbioe.2019.00022

Acuña-Rodríguez IS, Ballesteros GI, Gundel PE, Castro-Nallar E, Barrera A, Carrasco-Urra F, Molina-Montenego M (2024) Fungal endophyte symbionts enhance plant adaptation in Antarctic habitats. Physiol Plant 176(6):e14589. https://doi.org/10.1111/ppl.14589

Akulava V, Smirnova M, Byrtusova D, Zimmermann B, Ekeberg D, Kohler A, Blazhko U, Miamin U, Valentovich L, Shapaval V (2024) Explorative characterization and taxonomy-aligned comparison of alterations in lipids and other biomolecules in Antarctic bacteria grown at different temperatures. Environ Microbiol 16:e13232. https://doi.org/10.1111/1758-2229.13232

Alberdi M, Bravo LA, Gutiérrez A, Gidekel M, Corcuera LJ (2002) Ecophysiology of Antarctic vascular plants. Physiol Plant 115:479–486. https://doi.org/10.1034/j.1399-3054.2002.1150401.x

Alori ET, Glick BR, Babalola OO (2017) Microbial phosphorus solubilization and its potential for use in sustainable agriculture. Front Microbiol 8:971. https://doi.org/10.3389/fmicb.2017.00971

Alvarado P, Huang Y, Wang J, Garrido I, Leiva S (2018) Phylogeny and bioactivity of epiphytic Gram-positive bacteria isolated from three co-occurring Antarctic macroalgae. Antonie Van Leeuwenhoek 111:1543–1555. https://doi.org/10.1007/s10482-018-1044-6

Arrarte E, Garmendia G, Rossini C, Wisniewski M, Vero S (2017) Volatile organic compounds produced by Antarctic strains of *Candida sake* play a role in the control of postharvest pathogens of apples. Biol Control 109:14–20. https://doi.org/10.1016/j.biocontrol.2017.03.002

Barrera A, Hereme R, Ruiz-Lara S, Larrondo LF, Gundel PE, Pollmann S, Molina-Montenegro MA, Ramos P (2020) Fungal endophytes enhance the photoprotective mechanisms and photochemical efficiency in the Antarctic *Colobanthus quitensis* (Kunth) Bartl. exposed to UV-B radiation. Front Ecol Evol 8:122. https://doi.org/10.3389/fevo.2020.00122

Barrientos-Díaz L, Gidekel M, Gutiérrez A (2008) Characterization of rhizospheric bacteria isolated from *Deschampsia antarctica* Desv. World J Microbiol Biotechnol 24:2289–2296. https://doi.org/10.1007/s11274-008-9743-1

Barros CR, Baethgen LF, Mastrobuono G, Peixoto RS, Miranda KR, Carvalho ACS, Dias RCS, Rosado AS, Duarte RS (2022) Mycobacterial diversity in soil samples from King George Island, Antarctica. Pol Sci 34:100890. https://doi.org/10.1016/j.polar.2022.100890

Berríos G, Cabrera G, Gidekel M, Gutiérrez-Moraga A (2013) Characterization of a novel Antarctic plant growth-promoting bacterial strain and its interaction with Antarctic hair grass (*Deschampsia antarctica* Desv). Polar Biol 36:349–362. https://doi.org/10.1007/s00300-012-1264-6

Bockheim JG (2015) Soil-forming factors in Antarctica. In: Bockheim JG (ed) The soils of Antarctica, World soils book series. Springer, Cham, pp 5–20

Bodirsky BL, Popp A, Lotze-Campen H, Dietrich JP, Rolinski S, Weindl I, Schmitz C, Müller C, Bonsch M, Humpenöder F, Biewald A, Stevanovic M (2014) Reactive nitrogen requirements to feed the world in 2050 and potential to mitigate nitrogen pollution. Nat Commun 5:3858. https://doi.org/10.1038/ncomms4858

Çakmakçi R, Donmez F, Aydin A, Sahin F (2006) Growth promotion of plants by plant growth-promoting rhizobacteria under greenhouse and two different field soil conditions. Soil Biol Biochem 38:1482–1487. https://doi.org/10.1016/j.soilbio.2005.09.019

Cary SC, Mcdonald IR, Barretti JE, Cowan DA (2010) On the rocks: the microbiology of Antarctic Dry Valley soils. Nat Rev Microbiol 8:129–138. https://doi.org/10.1038/nrmicro2281

Cavicchioli R (2016) On the concept of a psychrophile. ISME J 10:793–795. https://doi.org/10.1038/ismej.2015.160

Chauhan M, Kimothi A, Sharma A Pandey A (2023) Cold adapted *Pseudomonas*: ecology to biotechnology. Front Microbiol 14:1218708. https://doi.org/10.3389/fmicb.2023.1218708

Cheng Y, Narayanan M, Shi X, Chen X, Li Z, Ma Y (2023) Phosphate-solubilizing bacteria: their agroecological function and optimistic application for enhancing agro-productivity. Sci Total Environ 901:166468. https://doi.org/10.1016/j.scitotenv.2023.166468

Chown SL, Convey P (2006) Biogeography. In: Bergstrom DM, Convey P, Huiskes AHL (eds) Trends in Antarctic terrestrial and limnetic ecosystems: Antarctica as a global indicator. Springer, Dordrecht, pp 55–69

Cid P, Maruyama F, Murase K, Graether SP, Larama G, Bravo LA, Jorquera MA (2018) Draft genome sequences of bacteria isolated from the Deschampsia antarctica phyllosphere. Extremophiles 22:537–552. https://doi.org/10.1007/s00792-018-1015-x

Convey P (2011) Antarctic terrestrial biodiversity in a changing world. Polar Biol 34:1629–1641. https://doi.org/10.1007/s00300-011-1068-0

Convey P, Chown SL, Clarke A et al (2014) The spatial structure of Antarctic biodiversity. Ecol Monogr 84:203–244. https://doi.org/10.1890/12-2216.1

Cowan DA, Tow LA (2004) Endangered Antarctic environments. Ann Rev Microbiol 58:649–690. https://doi.org/10.1146/annurev.micro.57.030502.090811

Da Silva MK, Barreto DLC, Vieira R, Neto AA, Oliveira FS, Convey P, Rosa CA, Duarte AWF, Rosa LH (2024) Diversity and enzymatic, biosurfactant and phytotoxic activities of culturable *Ascomycota* fungi present in marine sediments obtained near the South Shetland Islands, maritime Antarctica. Extremophiles 28:20. https://doi.org/10.1007/s00792-024-01336-4

De França P, Costa JH, Fill TP, Lancellotti M, Ruiz ALTG, Fantinatti-Garboggini F (2023) Genome mining reveals secondary metabolites of Antarctic bacterium *Streptomyces albidoflavus* related to antimicrobial and antiproliferative activities. Arch Microbiol 205:354. https://doi.org/10.1007/s00203-023-03691-w

De Freitas MS, De Carvalho CR, Pellizzari FM, Mansilla AO, Marambio J, Alves TMA, Murta SMF, Zani CL, Convey P, Rosa LH (2024) Diversity, distribution and phytotoxic and anti-*Trypanosoma* activities of cultivable fungi associated with Magellan sub-Antarctic strait and Maritime Antarctic macroalgae. Extremophiles 28:46. https://doi.org/10.1007/s00792-024-01363-1

Duarte AW, Passarini MRZ, Delforno TP et al (2016) Yeasts from macroalgae and lichens that inhabit the South Shetland Islands, Antarctica. Environ Microbiol Rep 8:874–885. https://doi.org/10.1111/1758-2229.12452

Encheva-Malinova M, Vancheva T, Badzhinerov N, Koleva V, Tishkov S, Bogatzevska N, Moncheva P (2014a) Antimicrobial activity of Antarctic streptomycetes against pepper bacterial spot causing agentes. In: First National Conference of Biotechnology, vol. 100, pp 216–222

Encheva-Malinova M, Stoyanova M, Avramova H, Pavlova Y, Gocheva B, Ivanova I, Moncheva P (2014b) Antibacterial potential of streptomycete strains from Antarctic soils. Biotechnol Biotechnol Equip 28:721–727. https://doi.org/10.1080/13102818.2014.947066

Farias GS, Santos JA, Giovanella P, Sette L (2022) Antarctic-derived yeasts: taxonomic identification and resistance to adverse conditions. An Acad Bras Cienc 94:e20210592. https://doi.org/10.1590/0001-3765202220210592

Ferrarezi JH, Santos JA, Sette LD, Ferreira H, Sass DC (2019) Anti-Xanthomonas activity of Antarctic fungi crude extracts. Afr J Biotechnol 28:713–718. https://doi.org/10.5897/AJB2019.16886

Gallardo-Cerda J, Levihuan J, Lavín P, Oses R, Atala C, Torres-Diaz C, Cuba-Diaz M, Barrera A, Molina-Montenegro MA (2018) Antarctic rhizobacteria improve salt tolerance and physiological performance of the Antarctic vascular plants. Polar Biol 41:1973–1982. https://doi.org/10.1007/s00300-018-2336-z

Giudice AL, Arazzo M (2019) Diversity and ecological roles of prokaryotes in the changing Antarctic marine environment. In: Sowinski SC (ed) The ecological role of microorganisms in the Antarctic environment. Springer, Dordrecht, pp 109–131

Godinho VM, Gonçalves VN, Santiago IF et al (2015) Diversity and bioprospection of fungal community present in oligotrophic soil of continental Antarctica. Extremophiles 19:585–596. https://doi.org/10.1007/s00792-015-0741-6

Godinho VM, De Paula MTR, Silva DAS et al (2019) Diversity and distribution of hidden cultivable fungi associated with marine animals of Antarctica. Fungal Biol 123:507–516. https://doi.org/10.1016/j.funbio.2019.05.001

Gomes ECQ, Godinho VM, Silva DAS et al (2018) Cultivable fungi present in Antarctic soils: taxonomy, phylogeny, diversity, and bioprospecting of antiparasitic and herbicidal metabolites. Extremophiles 22:381–393. https://doi.org/10.1007/s00792-018-1003-1

Graham LE, Graham JM, Knack JJ, Trest MT, Piotrowski MJ, Arancibia-Avila P (2017) A Sub-Antarctic peat moss metagenome indicates microbiome resilience to stress and biogeochemical functions of early Paleozoic terrestrial ecosystems. Int J Plant Sci 178:618–628. https://doi.org/10.1086/693019

Gray EJ, Smith DL (2005) Intracellular and extracellular PGPR: commonalities and distinctions in the plant–bacterium signaling processes. Soil Biol Biochem 37:395–412. https://doi.org/10.1016/j.soilbio.2004.08.030

Grobelak A, Napora A, Kacprzak M (2015) Using plant growth-promoting rhizobacteria (PGPR) to improve plant growth. Ecol Eng 84:22–28. https://doi.org/10.1016/j.ecoleng.2015.07.019

Hereme R, Morales-Navarro S, Ballesteros G et al (2020) Fungal endophytes exert positive effects on Colobanthus quitensis under water stress but neutral under a projected climate change scenario in Antarctica. Front Microbiol 11:264. https://doi.org/10.3389/fmicb.2020.00264

Janati W, Bouabid R, Mikou K, Ghadraoui LE, Errachidi F (2023) Phosphate solubilizing bacteria from soils with varying environmental conditions: occurrence and function. PLoS One 18(12):e0289127. https://doi.org/10.1371/journal.pone.0289127

Jha A, Saxena J, Sharma V (2013) An investigation on phosphate solubilization potential of agricultural soil bacteria as affected by different phosphorus sources, temperature, salt and pH. Commun Soil Sci Plant Anal 44:2443–2458. https://doi.org/10.1080/00103624.2013.803557

Kohler TJ, Stanish LF, Liptzin D, Barrett JE, Mcknight DM (2018) Catch and release: hyporheic retention and mineralization of N-fixing Nostoc sustains downstream microbial mat biomass in two polar desert streams. Limnol Oceanogr 3:357–364. https://doi.org/10.1002/lol2.10087

Lacap-Bugler DC, Lee KK, Archer S et al (2017) Global diversity of desert hypolithic cyanobacteria. Front Microbiol 8:867. https://doi.org/10.3389/fmicb.2017.00867

Lahlali R, Ezrari S, Radouane N, Kenfaoui J, Esmaeel Q, Hamss HE, Belabess Z, Barka EA (2022) Biological control of plant pathogens: a global perspective. Microorganisms 10:596. https://doi.org/10.3390/microorganisms10030596

Latysheva N, Junker VL, Palmer WJ, Codd GA, Barker D (2012) The evolution of nitrogen fixation in cyanobacteria. Bioinformatics 28:603–606. https://doi.org/10.1093/bioinformatics/bts008

Lemonnier F, Madeleine JB, Claud C, Palerme C, Genthon C, L'ecuyer T, Wood NB (2020) CloudSat-inferred vertical structure of snowfall over the Antarctic continent. J Geophys Res Atmos 125:e2019JD031399. https://doi.org/10.1029/2019JD031399

Liu H, Zheng G, Chen Z, Ding X, Wu J, Zhang H, Jia S (2023) Psychrophilic yeasts: insights into their adaptability to extremely cold environments. Genes 14(1):158. https://doi.org/10.3390/genes14010158

Lopes JC, Veiga VP, Seminiuk B, Santos LOF, Luiz AMC, Fernandes CA, Kinasz CT, Pellizari VH, Duarte RTD (2024) Freezing and thawing in Antarctica: characterization of antifreeze protein (AFP) producing microorganisms isolated from King George Island, Antarctica. Braz J Microbiol 55:1451–1463. https://doi.org/10.1007/s42770-024-01345-7

Mahanty T, Bhattacharjee S, Goswami M et al (2017) Biofertilizers: a potential approach for sustainable agriculture development. Environ Sci Pollut Res 24:3315–3335. https://doi.org/10.1007/s11356-016-8104-0

Maharajan T, Ceasar SA, Ajeesh Krishna TP, Ramakrishnan M, Duraipandiyan V, Abdulla AN, Ignacimuthu S (2017) Utilization of molecular markers for improving the phosphorus efficiency in crop plants. Plant Breed 137:10–26. https://doi.org/10.1111/pbr.12537

Mink S, López-Martínez J, Maestro A, Garrote J, Ortega JA, Serrano E, Durán JK, Schmid T (2014) Insights into deglaciation of the largest ice-free area in the South Shetland Islands (Antarctica) from quantitative analysis of the drainage system. Geomorphology 225:4–24. https://doi.org/10.1016/j.geomorph.2014.03.028

Mogollón JM, Beusen AHW, Van Grinsven HJM, Westhoek H, Bouwman AF (2018) Future agricultural phosphorus demand according to the shared socioeconomic pathways. Glob Environ Change 50:149–163. https://doi.org/10.1016/j.gloenvcha.2018.03.007

Molina-Montenegro MA, Oses R, Torres-Díaz C, Atala C, Zurita-Silva A, Ruiz-Lara S (2016) Root-endophytes improve the ecophysiological performance and production of an agricultural species under drought condition. AoB Plants 8:plw062. https://doi.org/10.1093/aobpla/plw062

Molina-Montenegro MA, Acuña-Rodríguez IS, Torres-Díaz C, Gundel PE, Dreyer I (2020) Antarctic root endophytes improve physiological performance and yield in crops under salt stress by enhanced energy production and Na+ sequestration. Sci Rep 10:5819. https://doi.org/10.1038/s41598-020-62544-4

Nesme T, Metson GS, Bennett EM (2018) Global P flows through agricultural trade. Glob Environ Change 50:133–141. https://doi.org/10.1016/j.gloenvcha.2018.04.004

Ofosu R, Agyemang ED, Márton A, Pásztor G, Taller J, Kazinczi G (2023) Herbicide resistance: managing weeds in a changing world. Agronomy 13:1595. https://doi.org/10.3390/agronomy13061595

Oses-Pedraza R, Torres-Díaz C, Lavín P, Retamales-Molina P, Atala C, Gallardo-Cerda J, Acuña-Rodríguez IS, Molina-Montenegro MA (2020) Root endophytic Penicillium promotes growth of Antarctic vascular plants by enhancing nitrogen mineralization. Extremophiles 24:721–732. https://doi.org/10.1007/s00792-020-01189-7

Park C, Kim K, Kim O, Jeong G, Hong S (2016) Bacterial communities in Antarctic lichens. Antarct Sci 28:455–461. https://doi.org/10.1017/S0954102016000286

Passarini MRZ, Robayo MIG, Ottoni JR, Duarte AWF, Rosa LH (2024) Biotechnological potential in agriculture of soil Antarctic microorganisms revealed by omics approach. World J Microbiol Biotechnol 40:345. https://doi.org/10.1007/s11274-024-04114-8

Perez JV, Serrano L, Viteri R, Sosa D, Romero CA, Diez N (2024) Antarctic streptomyces: promising biocontrol agents for combating *Fusarium oxysporum* f. sp. *cubense*. Biotechnol Rep 43:e00852. https://doi.org/10.1016/j.btre.2024.e00852

Purić J, Vieira G, Cavalca L, Sette L, Ferreira H, Vieira M, Sass D (2018) Activity of Antarctic fungi extracts against phytopathogenic bacteria. Lett Appl Microbiol 66:530–536. https://doi.org/10.1111/lam.12875

Ramírez CF, Cavieres LA, Sanhueza C, Vallejos V, Gómez-Espinoza O, Bravo LA, Sáez PL (2024) Ecophysiology of Antarctic vascular plants: an update on the extreme environment resistance mechanisms and their importance in facing climate change. Plants 13(3):449. https://doi.org/10.3390/plants13030449

Ramos P, Rivas N, Pollmann S, Casati P, Molina-Montenegro MA (2018) Hormonal and physiological changes driven by fungal endophytes increase Antarctic plant performance under UV-B radiation. Fungal Ecol 34:76–82. https://doi.org/10.1016/j.funeco.2018.05.006

Rodríguez H, Fraga R (1999) Phosphate solubilizing bacteria and their role in plant growth promotion. Biotechnol Adv 17:319–339. https://doi.org/10.1016/S0734-9750(99)00014-2

Rodríguez R, Barra PJ, Larama G, Carrion VJ, Mora ML, Hale L, Durán P (2023) Microbiome engineering optimized by Antarctic microbiota to support a plant host under water deficit. Front Plant Sci 14:1241612. https://doi.org/10.3389/fpls.2023.1241612

Rosa LH, Vaz ABM, Caligiorne RB, Campolina S, Rosa CA (2009) Endophytic fungi associated with the Antarctic grass Deschampsia antarctica Desv. (Poaceae). Polar Biol 32:161–167. https://doi.org/10.1007/s00300-008-0515-z

Rosa LH, Coelho CL, Pinto OHB, Carvalho-Silva M, Convey P, Rosa CA, Câmara PEAS (2021) Ecological succession of fungal and bacterial communities in Antarctic mosses affected by a fairy ring disease. Extremophiles 25:471–481. https://doi.org/10.1007/s00792-021-01240-1

Rosas SB, Andrés JA, Rovera M, Correa NS (2006) Phosphate-solubilizing Pseudomonas putida can influence the rhizobia–legume symbiosis. Soil Biol Biochem 38:3502–3505. https://doi.org/10.1016/j.soilbio.2006.05.008

Sagoe CI, Ando T, Kouno K, Nagaoka T (1998) Relative importance of prótons and solution calcium concentration in phosphate rock dissolution by organic acids. Soil Sci Plant Nutr 44:617–625. https://doi.org/10.1080/00380768.1998.10414485

Santiago IF, Soares MA, Rosa CA, Rosa LH (2015) Lichensphere: a protected natural microhabitat of the non-lichenised fungal communities living in extreme environments of Antarctica. Extremophiles 19:1087–1097. https://doi.org/10.1007/s00792-015-0781-y

Santiago IF, Rosa CA, Rosa LH (2017) Endophytic symbiont yeasts associated with the Antarctic angiosperms Deschampsia antarctica and Colobanthus quitensis. Polar Biol 40:177–183. https://doi.org/10.1007/s00300-016-1940-z

Santos A, Núñez-Montero K, Lamilla C, Pavez M, Quezada-Solís D, Barrientos L (2019) Antifungal activity screening of Antarctic actinobacteria against phytopathogenic fungi. Acta Biol Colomb 25:353–358. https://doi.org/10.15446/abc.v25n2.76405

Sarikhani MR, Khoshru B, Greiner R (2019) Isolation and identification of temperature tolerant phosphate solubilizing bacteria as a potential microbial fertilizer. World J Microbiol Biotechnol 35:126. https://doi.org/10.1007/s11274-019-2702-1

Sattari SZ, Bouwman AF, Giller KE, Van Ittersum MK (2012) Residual soil phosphorus as the missing piece in the global phosphorus crisis puzzle. PNAS 16:6348–6353. https://doi.org/10.1073/pnas.1113675109

Schmitz D, Siqueira RG, Michel RFM, Pereira AB, Putzke J, Francelino MR, Schaefer CEGR (2024) Pedodiversity and ornithogenesis of a tiny Antarctic Island (Half Moon): landformgeology-vegetation interrelationships. An Acad Bras Cienc 96(Suppl. 2):e20240581. https://doi.org/10.1590/0001-3765202420240581

Shivaji S, Prasad GS (2009) Antarctic yeasts: biodiversity and potential applications. In: Kunze G, Satyanarayana T (eds) Yeast biotechnology: diversity and applications. Springer, Dordrecht, pp 3–18

Shuping DSS, Eloff JN (2017) The use of plants to protect plants and food against fungal pathogens: a review. Afr J Tradit Complement Altern Med 14:120–127. https://doi.org/10.21010/ajtcam.v14i4.14

Sigurbjörnsdóttir MA, Heiðmarsson S, Jónsdóttir AR, Vilhelmsson O (2014) Novel bacteria associated with Arctic seashore lichens have potential roles in nutrient scavenging. Anadian J Microbiol 60:307–317. https://doi.org/10.1139/cjm-2013-0888

Silva LJ, Crevelin EJ, Souza DT et al (2020) Actinobacteria from Antarctica as a source for anticancer discovery. Sci Rep 10:1–15. https://doi.org/10.1038/s41598-020-69786-2

Silva AV, Oliveira AJ, Tanabe ISB et al (2021) Antarctic lichens as a source of phosphate-solubilizing bacteria. Extremophiles 25:181–191. https://doi.org/10.1007/s00792-021-01220-5

Silva AV, Silva MK, Oliveira AJ et al (2022) Phosphate solubilization by Antarctic yeasts isolated from lichens. Arch Microbiol 204:698. https://doi.org/10.1007/s00203-022-03310-0

Singh SM, Olech M, Cannone N, Convey P (2015) Contrasting patterns in lichen diversity in the continental and maritime Antarctic. Pol Sci 9:311–318. https://doi.org/10.1016/j.polar.2015.07.001

Singh J, Singh RP, Khare R (2018) Influence of climate change on Antarctic flora. Pol Sci 18:94–101. https://doi.org/10.1016/j.polar.2018.05.006

Son SW, Kim HY, Choi GJ, Lim HK, Jang KS, Lee SO, Lee S, Sung ND, Kim JC (2008) Bikaverin and fusaric acid from Fusarium oxysporum show antioomycete activity against Phytophthora infestans. J Appl Microbiol 104:692–698. https://doi.org/10.1111/j.1365-2672.2007.03581.x

Spaepen S, Vanderleyden J (2011) Auxin and plant-microbe interactions. Cold Spring Harb Perspect Biol 3:a001438. https://doi.org/10.1101/cshperspect.a001438

Styczynski M, Biegniewski G, Decewicz P, Rewerski B, Debiec-Andrzejewska K, Dziewit L (2022) Application of psychrotolerant Antarctic bacteria and their metabolites as efficient plant growth promoting agents. Front Bioeng Biotechnol 10:772891. https://doi.org/10.3389/fbioe.2022.772891

Tahon G, Tytgat B, Willems A (2018) Diversity of key genes for carbon and nitrogen fixation in soils from the Sør Rondane Mountains, East Antarctica. Polar Biol 41:2181–2198. https://doi.org/10.1007/s00300-018-2353-y

Teixeira LCRS, Peixoto RS, Cury JC et al (2010) Bacterial diversity in rhizosphere soil from Antarctic vascular plants of Admiralty Bay, maritime Antarctica. ISME J 4:989–1001. https://doi.org/10.1038/ismej.2010.35

Tosi S, Casado B, Gerdol R, Caretta G (2002) Fungi isolated from Antarctic mosses. Polar Biol 25:262–268. https://doi.org/10.1007/s00300-001-0337-

Tsavkelova EA, Klimova SY, Cherduntseva TA, Netrusoy AI (2006) Microbial producers of plant growth stimulators and their practical use: a review. Appl Biochem Microbiol 42:117–126. https://doi.org/10.1134/S0003683806020013

Upson R, Read DJ, Newsham KK (2009a) Nitrogen form influences the response of Deschampsia Antarctica to dark septate root endophytes. Mycorrhiza 20:1–11. https://doi.org/10.1007/s00572-009-0260-3

Upson R, Newsham KK, Bridge PD, Pearce DA, Read DJ (2009b) Taxonomic affinities of dark septate root endophytes of Colobanthus quitensis and Deschampsia Antarctica, the two native Antarctic vascular plant species. Fungal Ecol 2:184–196. https://doi.org/10.1016/j.funeco.2009.02.004

Varliero G, Lebre PH, Adams B, Chown SL, Convey P, Dennis PG, Fan D, Ferrari B, Frey B, Hogg ID, Hopkins DW, Kong W et al (2024) Biogeographic survey of soil bacterial communities across Antarctica. Microbiome 12:9. https://doi.org/10.1186/s40168-023-01719-3

Vasileva-Tonkova E, Romanovskaya V, Gladka G, Gouliamova D, Tomova I, Stoilova-Disheva M, Tashyrev O (2014) Ecophysiological properties of cultivable heterotrophic bacteria and yeasts dominating in phytocenoses of Galindez Island, maritime Antarctica. World J Microbiol Biotechnol 30:1387–1398. https://doi.org/10.1007/s11274-013-1555-2

Vassilev N, Vassileva M, Lopez A et al (2015) Unexploited potential of some biotechnological techniques for biofertilizer production and formulation. Appl Microbiol Biotechnol 99:4983–4996. https://doi.org/10.1007/s00253-015-6656-4

Vero S, Garmendia G, González MN, Bentancur O, Wisniewski M (2012) Evaluation of yeasts obtained from Antarctic soil samples as biocontrol agents for the management of postharvest diseases of apple (Malus x domestica). FEMS Yeast Res 13:189–199. https://doi.org/10.1111/1567-1364.12021

Vero S, Garmendia G, Martínez-Silveira A, Cavello I, Wisniewski M (2019) Yeast activities involved in carbon and nitrogen cycles in Antarctica. In: Sowinski SC (ed) The ecological role of microorganisms in the Antarctic environment. Springer, Dordrecht, pp 45–64

Vieira G, Purić J, Morão L, Santos J, Inforsato F, Sette L, Ferreira H, Sass D (2018) Terrestrial and marine Antarctic fungi extracts active against Xanthomonas citri subsp. citri. Lett Appl Microbiol 67:64–71. https://doi.org/10.1111/lam.12890

Weinstein RN, Palm ME, Johnstone K, Wynn-Williams DD (1997) Ecological and physiological characterization of Humicola marvinii, a new psychrophilic fungus from fellfield soils in the maritime Antarctic. Mycologia 89:706–711. https://doi.org/10.1080/00275514.1997.12026836

Wentzel LCP, Inforsato FJ, Montoya QV et al (2019) Fungi from Admiralty Bay (King George Island, Antarctica) soils and marine sediments. Microb Ecol 77:12–24. https://doi.org/10.1007/s00248-018-1217-x

Wright A, Gooseff M, Bergstrom A, Welch K (2024) The hydrologic and geochemical contributions from snow to streamflow in the McMurdo Dry Valleys of Antarctica. Hydrol Process 39:e15195. https://doi.org/10.1002/hyp.15195

Yarzábal LA, Monserrate L, Buela L, Chica E (2018) Antarctic Pseudomonas spp. promote wheat germination and growth at low temperatures. Polar Biol 41:2343–2354. https://doi.org/10.1007/s00300-018-2374-6

Yuan L, Yuan Y, Du J, Sun J, Guo S (2012) Effects of 24-epibrassinolide on nitrogen metabolism in cucumber seedlings under Ca(NO3)2 stress. Plant Physiol Biochem 61:29–35. https://doi.org/10.1016/j.plaphy.2012.09.00

Zhang J, Lv J, Dawuda MM et al (2019) Appropriate ammonium-nitrate ratio improves nutrient accumulation and fruit quality in pepper (Capsicum annuum L.). Agronomy 9:683. https://doi.org/10.3390/agronomy9110683

Zucconi L, Cavallini G, Canini F (2024) Trends in Antarctic soil fungal research in the context of environmental changes. Braz J Microbiol 55:1625–1634. https://doi.org/10.1007/s42770-024-01333-x

Zwolinski Z, Mercier D (2025) Management strategies for the Earth's surface processes in (sub) polar areas. In: Beylich AA, Tarrío DV, Li D, Oliva M, Marteles MM (eds) Climate and anthropogenic impacts on earth surface processes in the Anthropocene. Elsevier, pp 241–247

Chapter 8
Microbes-Mediated Remediation of Petroleum and Hydrocarbons in Antarctic Environments

Gabriela Sosa Benegas, Marahia Isabel Guevara Robayo, Dayana Lunkes Colaço, Camila Kessler, Valéria Maia de Oliveira, Luiz Henrique Rosa, and Michel Rodrigo Zambrano Passarini

Abstract The microbial-mediated remediation of petroleum and hydrocarbons in Antarctic environments involves the utilization of indigenous microorganisms to degrade harmful pollutants. Increased human activity has increased the risk of hydrocarbon spills in Antarctica. Microbial communities, mainly bacteria and fungi, play crucial roles in breaking down complex hydrocarbons through various metabolic processes and synergistic interactions between bacterial communities enhance the overall effectiveness. Hydrocarbon-degrading bacteria and fungi have been identified as key players in the degradation of petroleum compounds. Bacterial communities generally present higher degradation efficiencies at cold temperatures than single bacteria. Potential threats of polycyclic aromatic hydrocarbons (PAHs have aroused the interest of polar scientists and member nations of the Antarctic Treaty System, which has led to the growth of research into bioremediation approaches. Ongoing research is critical to understanding microbial interactions and optimizing bioremediation strategies tailored to extreme weather conditions

---

G. S. Benegas
Laboratório de Biotecnologia Ambiental, Universidade Federal da Integração Latino-Americana (UNILA), Foz do Iguaçu, Paraná, Brazil

Laboratorio de Agua y Sedimentos MARR.CE, Itaipu Binacional, Hernandarias, Paraguay

M. I. G. Robayo · D. L. Colaço · C. Kessler · M. R. Z. Passarini (✉)
Laboratório de Biotecnologia Ambiental, Universidade Federal da Integração Latino-Americana (UNILA), Foz do Iguaçu, Paraná, Brazil
e-mail: michel.passarini@unila.edu.br

V. M. de Oliveira
Centro Pluridisciplinar de Pesquisas Químicas, Biológicas e Agrícolas Universidade Estadual de Campinas, Paulínia, Brazil

L. H. Rosa
Departamento de Microbiologia, Universidade Federal de Minas Gerais, Belo Horizonte, MG, Brazil

A. W. Fernandes Duarte et al. (eds.), *Polar Microbes and Climate Change*,
https://doi.org/10.1007/978-3-032-07096-8_8

and remoteness. This chapter will provide an overview of the Antarctic oil spills and current information on the use of cold-environment microorganisms as bioremediation strains for petroleum-derived compounds.

Keywords Cold environment · Degradation · Psychrophiles · Environmental pollution · Bioremediation

8.1 Introduction

Earth's climate and ocean systems profoundly depend on the Antarctic continent, the world's most important natural laboratory. Antarctic research plays a key role in revealing the unique record of our planet's climate over the past one million years, predicting future environmental trends (Charman et al. 2018), and, simultaneously, being an exceptional source of microorganisms for biotechnology (Correa and Abreu 2020).

Antarctica is administrated under the Antarctic Treaty, a major international treaty with 54 national members that came into force in 1961. Over the treaty's existence, several environmental regulations have been adopted. Among them, in 1998 (Secretariat of the Antarctic Treaty 2024), the Protocol on Environmental Protection to the Antarctic Treaty, popularly known as the Madrid Protocol, entered effect. This protocol banned all development and mining of fossil fuel in Antarctica until 2048. In the same direction, in 2009, the International Maritime Organization (IMO) prohibited the use or carriage of heavy fuel oil (HFO) as fuel or cargo in the Antarctic Treaty area (Transport Canada 2024).

Human activities on the continent have occurred only over the last two centuries. They initially focused on exploration and commercial exploitation, emphasized scientific research in the second half of the twentieth century, and later incorporated tourism (Zakaria et al. 2021a, b). Access to and support for these activities require ships, aircraft, and several petroleum products (Snape et al. 2005). Fuels are predominantly aliphatic, including mineral and synthetic lubrication oil, marine gas oil (MGO), light diesel, kerosene, and petrol (Snape et al. 2005).

8.2 Hydrocarbon Pollution and Toxicity in Antarctic Environments

Crude oil is considered Earth's most complex mixture of organic chemicals (Kostka et al. 2014). Petroleum crude oil is an oily, flammable liquid composed mainly of hydrocarbons (50–98%). The remainder comprises organic compounds containing oxygen, nitrogen, or sulfur and trace amounts of organometallic compounds (Gad 2014). The hydrocarbon types in petroleum are paraffins (alkanes), naphthenes (cycloalkanes), and aromatics. The complex mixture of hydrocarbons is generally used as an energy source and raw material, and it can be refined into more than 340 different products (Salari et al. 2022). Over 50% of its components are pollutants, including mineral oils, polycyclic aromatic hydrocarbons (PAHs), and volatile aromatic compounds like benzene, toluene, ethylbenzene, and xylenes. These compounds, commonly found in fuels such as gasoline, diesel, and lubricants, pose a significant ecological threat due to their destructive impact (Kuppusamy et al. 2019). However, Antarctic fuel and diesel are predominant (Fig. 8.1).

Environmental concerns regarding petroleum hydrocarbons are many and varied. Petroleum hydrocarbons (PHs) can pose risks to human health and ecosystems upon discharge into the environment (Marigómez 2014). PHs have moderate to high acute toxicity to aquatic organisms, and their toxicity is primarily affected by metabolism and photooxidation. These compounds alter cell membranes by dissolving their lipids, which impairs photosynthesis. In chloroplasts, the organelles where photosynthesis occurs, hydrocarbons interfere with the lipid phase, preventing chlorophyll molecules from interacting correctly with one another. Furthermore, this same alteration also affects mitochondria, where cellular energy is produced, disrupting key processes like the tricarboxylic acid cycle and oxidative phosphorylation, which are essential for cell energy production (Salari et al. 2022).

Hydrocarbon pollution from the accidental release and spilling of petroleum hydrocarbons into marine and terrestrial environments is an important concern in Antarctica (Secretariat of the Antactic Treaty 2024). Table 8.1 and Fig. 8.2 represent the main spills recorded in the maritime and terrestrial compartments. The first identified report mentioning hydrocarbon presence in Antarctica was published in the first issue of the Marine Pollution Bulletin in 1970 (1970). In 1989, major maritime and terrestrial spills occurred. The largest and most damaging spill occurred in 1989 when the Argentinian ship *Bahia Paraiso* crashed in the Antarctic Archipelago,

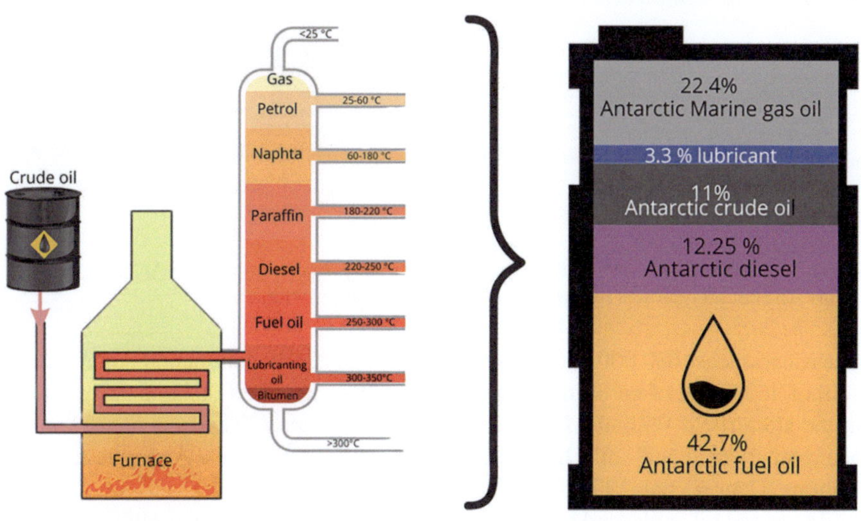

Fig. 8.1 Percentage (%) of different fuel types used in Antarctica in 2020. (Adapted from Zakaria et al. (2021a, b))

Table 8.1 Reported marine and terrestrial hydrocarbon spills in Antarctic environments

Year	Type of environment	Identifier	Volume	References
1970	Maritime and terrestrial	First report	ND	Cleaning up Oil Spills (Clark 1970)
1989	Maritime	*Bahia Paraíso*	680.000 L diesel fuel	Karl (1992)
1989	Terrestrial	Williams Field	260.000 L kerosene	Zakaria (2021a, b)
1990	Terrestrial	Casey Station	ND	Deprez et al. (1999)
1992	Terrestrial	United Kingdom Faraday Research Station	1.000 L diesel fuel	CEP-IV-IP62 (2021)
2001	Maritime	*Patriarche*	1.500 L diesel fuel	Stark et al. (2017)
2007	Maritime	MS Explorer	190.000 L of marine gas oil, 24.000 L of lubricant oil, and 1.200 L of petroleum oil	Stewart et al. (2007)

ND No data to show

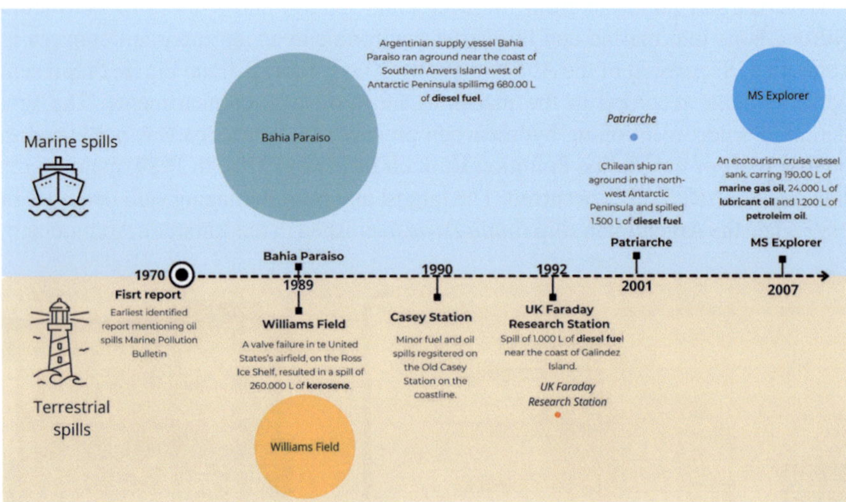

Fig. 8.2 Timeline of the reported marine and terrestrial hydrocarbon spills in Antarctic environments. The circles represent the comparative percentage of the volumes

releasing at least 680.000 L of diesel fuel (Karl 1992). In continental Antarctica, a spill of 260.000 L of kerosene occurred on the Ross Ice Shelf due to a valve failure at the US airfield Williams Field (Zakaria 2021a, b *apud* Zakaria et al. 2021a, b). Many small-scale fuel spills have occurred near the coastline near Antarctic research stations (Deprez et al. 1999).

Terrestrial oil spills tend to be found at sites of significant human activity (Cripps and Shears 1997). As Table 8.1 shows, large oil spills in terrestrial environments are rare but not infrequent (Hughes and Stallwood 2006), occasionally happening in populated and frequented areas, ranging from vehicle refueling to potentially significant incidents (e.g., as in Williams Field in 1989). Studies also mention the possibility of long-range atmospheric transport of PAHs and other pollutants (Szopińska et al. 2019). Approximately 22–47% of the oil discharges into the marine environment worldwide are caused by ship spills (Chilvers et al. 2021). In Antarctica, this percentage increases to 77% of the recorded and reported spills (Table 8.1). In this scenario, climate change considerably affects the risk of accidental fuel spills, increasing ship traffic in these environments (Góngora et al. 2022).

Marine spills have increased pollution potential over terrestrial spills. The movement of seawater can spread the oil out over vast areas, affecting a much larger area than the initial spread, and spills are more likely to intersect with animal and plant habitats (Gad 2014). Terrestrial fuel spills also threaten the marine environment due to the likelihood of fuel runoff in the nearby marine ecosystems (Zakaria et al. 2021a, b). In terrestrial environments, the heavier fraction stains in the rock and the soil or sinks below the ground surface and accumulates in depressions above permafrost. Annual soil flushing by meltwater removes some of these residues, and the effect is variable oil contamination across an impacted site in distribution and time (Pettersson and Nobes 2003). When crude oil coats surfaces, including sediment surfaces, rocky shores, and vegetation, the primary effect is to create sediment anoxia, preventing water exchange between the sediment surface and the water column (Marigómez 2014).

Hydrocarbons can accumulate in ice, releasing toxins gradually as the ice melts, exacerbating pollution impacts over time. One of the outcomes of marine oil spills is the formation of marine oil snow (MOS), which significantly affects marine ecosystems. This impact can occur directly or indirectly through mechanisms such as ingestion by marine organisms, microbial activity, smothering, creation of low-oxygen (suboxic) or oxygen-free (anoxic) zones, transfer through the marine food web, immunotoxicity through gills, absorption through the skin, and bioaccumulation in the tissues of marine species (Kuppusamy et al. 2019).

In the marine environment, following an oil spill, a major fraction of the hydrocarbons is volatilized and/or dissolved (Tarr et al. 2016). The low-molecular-weight volatile fractions quickly evaporate, leaving high-molecular-weight aliphatic, aromatic, and poly-aromatic hydrocarbons (PAHs) (Green et al. 1992). In the same way, modeling crude oil evaporation rates estimates that up to 75% of the volume evaporated within the first days after an oil spill (Fingas 1995). These are supported by the fact that after a year of the important diesel fuel spill in *Bahia Paraiso*, most of the fuel was not detectable and heavily degraded (Góngora et al. 2022). Even though more complex hydrocarbons such as PAHs are chemicals of emerging concern, they are ubiquitous and persistent in Antarctic environments (Bhardwaj et al. 2021; Góngora et al. 2022; Iriarte et al. 2023).

Low temperatures, prolonged sun irradiation, intrinsic nutrient limitation, and poor soil water holding, coupled at the same time with expensive logistic support and high energy costs to assess remediation technologies (Snape et al. 2005; Tin et al. 2009; Roslee et al. 2021), significantly slow down the natural degradation of hydrocarbons, allowing pollutants to persist for extended periods. This contamination threatens marine biodiversity, disrupts food chains, and poses long-term risks to keystone species, such as phytoplankton and krill, which serve as the foundation of polar food webs. These impacts extend beyond the primary and secondary producers of the food chain (Shahidul Islam and Tanaka 2004). Affectation has been reported in a cascade pattern throughout broad trophic levels, from phytoplankton, zooplankton, fish, aquatic birds, mammals, and benthic organisms (Kuppusamy et al. 2019).

Many natural processes come into play to eliminate oil and PHs from the environment: evaporation, weathering, oxidation, and biodegradation (Gad 2014). Among them, biodegradation is a major process that removes hydrocarbons released into both soil and aquatic environments (Das and Chandran 2011). Despite the promise of microbial remediation, significant challenges remain, including low temperatures, limited nutrients, the growing knowledge derived from genomic tools of native microorganisms, and the climate change scenario.

8.3 Microbial Communities and Their Bioremediation Potential in Polar Environments

Polar environments pose significant challenges for microbial communities (Atlas 1991). Attention has been paid to the changes caused by climate change in polar regions, such as the record of warming, water freshening, greenhouse gas emissions, and green algae blooms in snow (Morison et al. 2012; Purkey and Johnson 2013; Elder et al. 2018; Watts 2020). Over the last decades, East Antarctica has experienced minimal climatic changes (Turner et al. 2009), unlike the Antarctic Peninsula, where major temperature increases were recorded (Turner et al. 2007). On a continental scale, previous studies pointed out that the rise in climate severity from the Antarctic Peninsula to the Dry Valley regions shows a decrease in microbiome community complexity (Lawley et al. 2004; Barrett et al. 2006). In contrast to the terrestrial realm, the seas around Antarctica are highly productive and diverse because of the abundance of nutrients in the water (Pearce 2012).

There are important differences in the physical setting of the two polar regions—the Arctic, an ocean surrounded by land, and the Antarctic, a continent surrounded by an ocean (Intergovernmental Panel on Climate Change (IPCC) 2022). The different physical settings have led to the evolution of a unique polar microbiome (Cao et al. 2020a). Adaptations of polar microbes enable them to thrive under severe stressors, including psychrophilic—organisms with a requirement for low growth temperatures—and psychrotolerant, organisms that are tolerant of low growth temperatures from −15 °C to +10 °C—characteristics that enhance their metabolic functions at low temperatures (Pearce 2012). Psychrophilic bacteria thrive in low-temperature conditions and demonstrate specific protein adaptations and enzyme secretion that allow them to survive and function in these harsh climates (Pearce

2012; Cao et al. 2020a). Studies have shown that the stability and adaptability of these microorganisms are linked to factors such as bonding types, amino acid composition, genomic G + C content, and protein folding patterns (Durán et al. 2019; Pandolfo et al. 2023; Ramasamy et al. 2023). Most cellular adaptations to low temperatures and the underlying molecular mechanisms are not fully understood (Pearce 2012; Vigneron et al. 2023). Research is progressing in understanding their adaptation strategies, including freezing tolerance, regulation of protein synthesis, and membrane fluidity (Ramasamy et al. 2023).

In polar regions, the difficulty of achieving sufficient oil removal by physical washing and collection makes bioremediation a prime candidate (Aislabie et al. 2006). Hydrocarbon biodegradation can be performed by bacteria, archaea, algae, and fungi (Leahy and Colwell 1990; Das and Chandran 2011; Dell' Anno et al. 2021). Bioremediation significantly diminishes contamination in the polluted site using specific biochemical pathways to convert hydrocarbons into simply more straightforward harmful products, such as carbon dioxide, water, and bioenergy (de Melo Carlos et al. 2024). In cold regions, it is a relatively slow process. The degree of success depends on several factors, such as the properties of the oil spilled and the significant microbial and environmental limitations, e.g., the freeze-thaw seasonal cycles in cold environments between winter and summer limits the bioavailability of the spilled oil compounds, reducing the microbial biodegradation efficiency (Yang et al. 2009).

When crude oil is spilled into the sea, a thin, oily layer is formed on the surface (Fig. 8.3); currents, winds, and waves transport this oil over large distances. At the same time, in the early stages of the oil spill, natural attenuation processes, biodegradation, photooxidation, and oil weathering processes (OWP) such as wave spreading, evaporation, dispersion, emulsification, and dissolution, change contaminants'

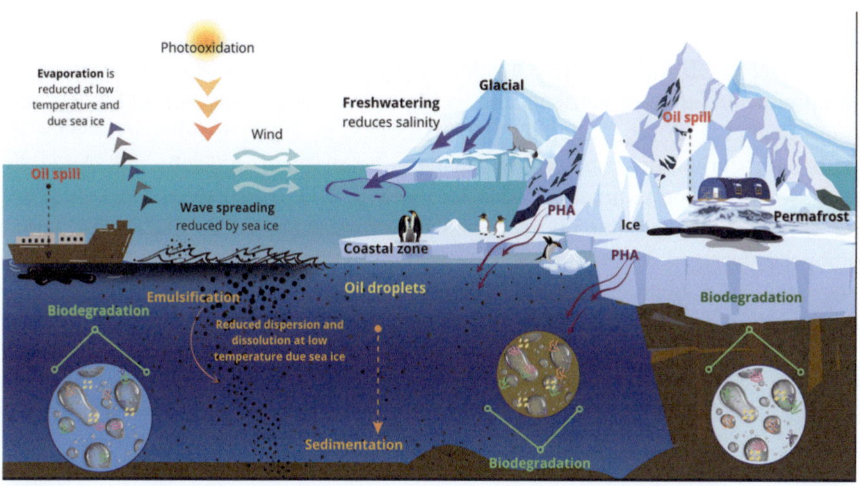

Fig. 8.3 Main natural oil spill attenuation processes in Antarctic marine and terrestrial compartments. (Image adapted from Keramea (2021) and Iriarte et al. (2023))

physical-chemical characteristics and concentration (Fingas 1995). Sea ice is practically impermeable at low temperatures, a characteristic enforced by high oil viscosity, evaporation, slow physical OWPs, and reduced wind spreading (Keramea et al. 2021; Lofthus et al. 2021). On the other hand, in terrestrial environments, following an oil spill, the volatile fractions quickly evaporate, leaving the heavier fraction consistent in high-molecular-weight aliphatic, aromatic, and poly-aromatic hydrocarbons (PAHs) (Green et al. 1992; Aislabie et al. 2006). This residue stains the rock and soil or sinks below the ground surface and accumulates in permafrost. In summer, meltwater flushes the soil, removing the oil residues (Pettersson and Nobes 2003).

Hydrocarbon-degrading microorganisms are ubiquitous in the world's oceans (Yakimov et al. 2007; Teske 2015), representing approximately 0.1% of the total microbial community, especially after a spill, when their number, activity, and dominance increase and, in some cases, can represent 100% of the community (Aislabie et al. 2006; Garneau et al. 2016; Mahjoubi et al. 2017). For example, one of the most valuable tools of alkanes, one of the main constituents of fuels, practically insoluble in water and resistant to UV radiation (Brakstad and Bonaunet 2006; Marigómez 2014), suffers mainly biodegradation through biological surface-active compounds (surfactants) produced by bacteria, which improves the bioavailability of hydrocarbons compounds for degrading metabolic pathways (Pacwa-Płociniczak et al. 2014).

Biodegradation does not proceed uniformly in all polar environmental compartments. For most of the year, ice-free regions are buried under snow and ice, and the ground is frozen. The surface soils thaw in the summer, and the underlying soil is permanently frozen (permafrost) (Deprez et al. 1999). Microbial conditions also depend on thaw conditions that allow the availability of liquid water (Aislabie et al. 2006). Some well-aerated portions of soil with access to water and nutrients will degrade more extensively than fuel trapped in micropores or isolated domains of poor connectivity (Snape et al. 2005). Additionally, freshwatering is caused by glacial freshwater discharge from several locations around Antarctica over recent decades (JACOBS 2004); these glacial and larger snow maltings produce changes in salinity and inputs of nutrients and PAHs that may negatively impact biodegradation (Tang et al. 2006; Cao et al. 2020b; Iriarte et al. 2023). Cold-adapted bacteria with hydrocarbon-degrading activities have been found in Arctic and Antarctic soils, seawater, and sediments (Table 8.2 and Fig. 8.3). On the other hand, there are a few reports of negligible oil degradation for ice compartments like sea ice, snow, and permafrost. (Atlas et al. 1978; Lofthus et al. 2021). Some authors found higher degradation of hydrocarbons by bacterial cultures inoculated from permafrost (Børresen et al. 2003) and sub-ice seawater (Garneau et al. 2016).

Table 8.2 Bacteria involved in hydrocarbon bioremediation in Arctic and Antarctic habitats and their reported hydrocarbon bioremediation efficiency

Organisms	Genus	Biodegradation	Source	Temperature	Type	References
Bacteria	*Oleispira* + *Thalassolituus*	17–43%	AR seawater	4 °C	DF and CO	
Bacteria	*Oleispira* + *Cycloclasticus*	17.80%	AR seawater	0–20 °C	NO	Ribicic et al. (2018)
	Arthrobacter	41–47.5%	AN soil	10–15 °C	DF	Abdulrasheed et al. (2020a, b)
	Flavobacterium	50–60%	AR soil	10–15 °C	DF	Chaudhary et al. (2019)
	Pseudomonas + *Rhodococcus*	61.00%	AR sediment	4 °C	ULSFO	Ellis et al. (2022)
	Marinomonas s	54–87%	AN Seawater	4–22 °C	DF	Zannotti (2023)
	Arthrobacter s	56.45%	AN Soil	15–20 °C	DF	Abdulrasheed et al. (2020a, b)
	Rhodococcus	90.39%	AN Soil	23–25 °C	DF	Roslee et al. (2021)
	Sphingomonas	95%	AN Soil	28 °C	PHE	Gran-Scheuch (2017)
	Pseudomonas	86%	AN Soil	28 °C	PHE	Gran-Scheuch (2017)
Bacterial community (microcosms experiments)		3–39%	AR soil	5 °C	DF	Børresen et al. (2003)
		39.35%	AN Seawater	15 °C	DF	Zakaria et al. (2021a, b)
		20–47%	AN sediment	1 °C	SAD	Powell et al. (2005)
		30%	AR sediment	4 °C	CO	Chen et al. (2024)
		52.23%	AN pristine seawater	10 °C	DF	Zakaria et al. (2020)
		61–100%	AN soil	5 °C	DF	Brakstad and Bonaunet (2006)
		75%	AR sediment	4 °C	DF	Rizzo et al. (2019)
		100%	AN sediment	4 °C	DF	Rizzo et al. (2019)

Temp temperature, *Type* type of hydrocarbon, *AR* Arctic, *AN* Antarctic, *DF* diesel fuel, *CO* crude oil, *NO* naphthenic oil, *ULSFO* ultralow ultra-sulfur fuel oil, *SAD* special Antarctic diesel blend, *PHE* phenanthrene

8.4 Hydrocarbon-Degrading Bacteria and Fungi

The degradation of hydrocarbons in the environment relies much more on hydrocarbon-degrading bacteria (Teske 2015). Hydrocarbon-degrading bacteria are known as hydrocarbon clastic bacteria (HCB), which include obligate HCB, bacteria that only grow using hydrocarbons as carbon and energy sources, and facultative HCB, those able to grow with alternative carbon sources (Nikolova and Gutierrez 2022).

The ability to degrade hydrocarbons is a common characteristic of the polar microbiome, possibly influenced by the natural presence of these compounds in the environment, which may originate from thermogenic, biogenic, or abiotic sources (Silva et al. 2018). Antarctic microorganisms have an excellent catabolic capacity for efficiently removing these pollutants, with few or no side effects (Wong et al. 2021. The complete degradation of hydrocarbon components requires a large consortium of indigenous bacteria with broad enzymatic capabilities and adequate metabolic potential (Leahy and Colwell 1990; Mori and Kanaly 2020). Several studies have identified and characterized various naturally occurring hydrocarbonoclastic bacteria genera in polar environments. The ability of these bacteria to degrade hydrocarbons has been widely documented in the literature, with notable genera such as *Alcanivorax*, *Anoxybacillus*, *Brevibacillus*, *Cycloclasticus*, *Geobacillus*, *Marinobacter*, *Oleiphilus*, *Oleispira*, *Planococcus*, and *Thalassolituus* (e.g., (Kasai et al. 2002; Dell' Anno et al. 2021; Peeb et al. 2022; Schultz et al. 2022).

The use of autochthones microorganisms is expected to be more effective, since these are better adapted to the specific local environmental conditions. In Antarctic and Arctic environments, HCB bacteria have been previously isolated from contaminated and pristine soil habitats among them, *Oleispira*, *Rhodococcus, Nocardia*, *Sphingomonas* spp., *Pseudomonas* spp. *Planktomarina*, *Pseudoalteromonas*, and *Sulbobacter* (Aislabie et al. 2000; Ribicic et al. 2018; Ellis et al. 2022; Peeb et al. 2022). The reported efficiencies of biodegradation are described in Table 8.2, briefly, for bacteria ranging from 17% to 82% and for bacterial communities 3 to 100%. Bacterial communities with mixed populations generally reached higher degradation efficiencies at lower temperatures than single bacteria.

Besides bacteria, fungi have been described for their ability to produce enzymes to degrade hydrocarbons, e.g., *Phialophora* spp. and *Hormoconis resinae* (Aislabie et al. 2000). Several microorganisms were isolated and characterized based on microbial isolation and cultivation of hydrocarbon-degrading microbes from cold habitats. However, this method can detect a small fraction of those communities and typically results in the isolation of bacteria (Aislabie et al. 2006; Zhang et al. 2019).

Fungi plays a significant role in the biodegradation of hydrocarbons, particularly in environments where these compounds pose environmental hazards, such as oil spills. The metabolic processes involved in fungal hydrocarbon degradation are complex and involve various enzymatic pathways that enable fungi to utilize hydrocarbons as carbon and energy sources. This section will delve into the specific enzymes involved, the mechanisms of action, and the overall degradation process.

Enzymatic mechanisms of hydrocarbon degradation employ a range of enzymes to facilitate the breakdown of hydrocarbons. Key enzymes include: (*i*) cytochrome P450 monooxygenases: these enzymes are crucial for the oxidation of hydrocarbons, allowing for the introduction of oxygen into hydrocarbon molecules. The cytochrome P450 system in fungi is versatile, enabling various substrates' regio- and stereospecific oxidation, including aliphatic and aromatic hydrocarbons. This enzymatic activity is vital for the initial steps of hydrocarbon degradation, as it transforms hydrophobic compounds into more polar metabolites that can be further processed. (*ii*) Ligninolytic enzymes: fungi such as *Phanerochaete chrysosporium* and *Pleurotus ostreatus* produce lignin peroxidases, manganese peroxidases, and laccases, which are instrumental in the degradation of complex aromatic hydrocarbons, including polycyclic aromatic hydrocarbons (PAHs) (Kadri et al. 2017). These enzymes facilitate the breakdown of aromatic rings, a critical step in detoxifying and mineralizing these persistent pollutants. However, many bacterial strains capable of degrading lignin-like compounds have recently been discovered. (*iii*) Hydrolases and dehydrogenases: these enzymes assist in the hydrolysis of hydrocarbon chains and the oxidation of alcohols, respectively. They contribute to the further degradation of the products generated by the action of cytochrome P450 enzymes and ligninolytic enzymes (Zhang et al. 2019).

The Degradation Process

The degradation of hydrocarbons by fungi typically occurs in several stages: (*a*) uptake of hydrocarbons: hydrocarbons must first penetrate the fungal cell wall. This process is facilitated by the production of biosurfactants, which increase the bioavailability of hydrophobic compounds (Husaini et al. 2008). Once inside the cell, hydrocarbons can be metabolized by intracellular enzymes. (*b*) Initial oxidation: the first step in the degradation process often involves the action of cytochrome P450 enzymes, which introduce hydroxyl groups into hydrocarbon molecules, converting them into alcohols or phenols. This transformation is crucial for making the compounds more amenable to further degradation. (*c*) Ring cleavage and further degradation: following initial oxidation, ligninolytic enzymes act on the aromatic structures, breaking down the complex rings into simpler compounds. This step is essential for the complete mineralization of hydrocarbons, as it allows for the conversion of aromatic compounds into aliphatic acids and eventually into carbon dioxide and water (Kadri et al. 2017); (*d*) Assimilation and energy production: the final products of hydrocarbon degradation can be assimilated by the fungi for growth and energy production. Fungi can utilize these metabolites in their metabolic pathways, contributing to their biomass and facilitating the cycling of nutrients in the ecosystem (Husaini et al. 2008).

Fungal diversity plays a significant role in the effectiveness of hydrocarbon degradation. Different fungal species exhibit varying capabilities for degrading specific types of hydrocarbons. For instance, species such as *Aspergillus* sp. and *Penicillium* sp. have been frequently reported as effective degraders of crude oil and its derivatives. *Penicillium* species, a group of filamentous fungi, have garnered significant attention for their potential in degrading a wide range of polycyclic aromatic hydrocarbons (PAHs). Their fast growth rate, adaptability to various environmental conditions, and ability to utilize diverse substrates such as carbon and nitrogen sources have made them ideal candidates for biodegradation studies (Steffen et al. 2002). Some fungal species isolated from Antarctica have shown remarkable potential for hydrocarbon degradation. One example is *Penicillium* sp. CHY-2, identified in Antarctic soils, which can break down both aliphatic and aromatic hydrocarbons, making it a promising candidate for the remediation of hydrocarbon-contaminated sites (Govarthanan et al. 2017). Similarly, the strain *Aspergillus glaucum* AL1, obtained from soil samples on Livingston Island, has demonstrated high efficiency in removing naphthalene and anthracene, utilizing these compounds as its sole carbon source through its intracellular enzymatic system (Stoyanova et al. 2022). The ability of these fungi to thrive in hydrocarbon-contaminated environments is often linked to their production of specialized enzymes and their morphological adaptations, which enhance their biodegradation efficiency (Husaini et al. 2008). In Antarctic ecosystems, where conditions are extreme, the adaptation of fungi to low temperatures and nutrient-poor environments is critical. Cold-adapted fungi have evolved mechanisms to maintain enzymatic activity at low temperatures, making them effective agents for bioremediation in polar regions. The isolation and characterization of these fungi are essential for developing targeted bioremediation strategies that leverage their unique metabolic capabilities. The biodegradation of hydrocarbons by Antarctic fungi is not well described compared to the information available on Antarctic bacteria (Wong et al. 2021). A few Antarctic fungal species have been characterized to have tolerance or ability to degrade hydrocarbons as shown in Table 8.3.

8.5 Mechanisms of Microbial Biodegradation

In Antarctic environments, several predominant microorganisms have been identified as being involved in the degradation of hydrocarbons. Studies have characterized a range of bacteria capable of hydrocarbon degradation in these extreme conditions (Varjani 2017). In this context, oleophilic microorganisms, such as bacteria and fungi, emerge as promising agents in bioremediation efforts. Through this innovative process, microorganisms convert toxic organic compounds into harmless substances, like CO_2, CH_4, and H_2O (Varjani 2017).

Due to their structure, complex hydrocarbons are less susceptible to microbial degradation. To achieve degradation, they must undergo cascades of complex catabolism processes mediated by multifunctional enzymes. Furthermore, exploiting enzymes and bioactive compounds produced by these microorganisms, such as laccases and biosurfactants, has proved a valuable strategy for biotechnology, with applications in bioremediation processes and industry. Enrichment methods to isolate microbial strains that tolerate and degrade hydrocarbons are key to expanding knowledge about microbial diversity and their functional capabilities in polar environments (Álvarez et al. 2015).

Table 8.3 Antarctic filamentous fungi and their tolerance or ability to degrade hydrocarbons

Fungal species	Isolation place	Incubation temperature (°C)	Pollutant	Hydrocarbon tolerance and degradation	REF
Pichia caribbica	Carlini, Potter Cove, King George Island	15	Undecane (nC11), dodecane (nC12), tridecane (nC13), (nC14), and Antarctic diesel fuel	Capable of growth using various hydrocarbons. Production of lipase and esterase observed, indicating the ability to degrade aliphatic hydrocarbons	Martorell et al. (2019), Joshi-Navare et al. (2014)
Mortierella sp.	Rothera POINT, Adelaide Island	17	Dodecane	Biomass increased when cultured on dodecane	Hughes et al. (2007)
Penicillium sp. CHY-2	Unspecified Antarctic soil	20	Aliphatic and aromatic hydrocarbons, including octane, decane, dodecane, ethylbenzene, butylbenzene, naphthalene, acenaphthene, and benzo[a] pyrene.	Degradation percentages: Decane 49%, Butylbenzene 42%, Dodecane 33%, Naphthalene 15%, Acenaphthene 10%, Octane 8%, Ethylbenzene 4.0%, and benzo[a] pyrene 2%.	Govarthanan et al. (2017)
Exophiala macquariensis sp. Nov.	Macquarie Island	10	Toluene	50% degradation of toluene.	Zhang et al. (2019)
Cladosporium sp. FAR18.	King George Island	15	Pyrene	Showed greater tolerance to pyrene (0.2152 g mL^{-1}/d)	de Melo Carlos et al. (2024)

Bacteria play a crucial role in the biodegradation of hydrocarbons. They are widely recognized as the primary and most active agents in breaking down these pollutants (Atlas 1991; Varjani 2017). These microorganisms catabolize petroleum hydrocarbons to generate energy or incorporate them into cellular biomass, making the process energetically favorable (Abbasian et al. 2015). In polar environments, understanding the metabolic potential of a multi-domain microbial community is essential, as microbial diversity significantly enhances the biodegradation process's effectiveness (Widdel and Rabus 2001; Varjani 2017).

Several enzymes, often encoded on plasmids, are vital for pollutant degradation in bacterial hydrocarbon catabolism. For example, species like *Acinetobacter* have plasmids that contribute to hydrocarbon degradation. Specific plasmids, such as Q15, OCT, TOL, and NAH7, encode genes involved in various biodegradation pathways, including alkA, alkM, and alkB. These genes facilitate the initial attack on hydrocarbons through bacterial cell attachment to substrates or the production of biosurfactants, which increase hydrocarbon availability for degradation (Abioye et al. 2009; Okoye et al. 2024). In addition to the aforementioned enzymes, other key enzymes involved in hydrocarbon degradation include alkane monooxygenases, which initiate the oxidation of alkanes, and phenol hydroxylases, which are crucial for the degradation of phenolic compounds. The *Rhodococcus* genus, known for its versatile catabolic capabilities, employs a variety of enzymes such as terminal and subterminal alkane hydroxylases that enable the breakdown of long-chain n-alkanes and crude oil (Okoye et al. 2024).

Furthermore, dioxygenases, such as those found in *Pseudomonas putida*, are essential for the degradation of aromatic hydrocarbons, facilitating the cleavage of aromatic rings and subsequent metabolic processing (Jiménez et al. 2002; Kim et al. 2004). Moreover, enzymes like laccases and peroxidases produced by certain bacterial strains enhance the degradation of complex aromatic compounds by catalyzing oxidative reactions (Roslee et al. 2021). The presence of biosurfactant-producing enzymes also significantly increases the bioavailability of hydrophobic pollutants, thereby enhancing the overall degradation process (Onur et al. 2015). In summary, the diversity of enzymes encoded by plasmids in hydrocarbon-degrading bacteria is critical for effective bioremediation. These enzymes not only facilitate the breakdown of various hydrocarbons but also enhance microbial interactions and community dynamics, ultimately leading to more efficient pollutant degradation.

Hydrocarbons can be broken down in aerobic biodegradation through several mechanisms, including terminal oxidation, subterminal oxidation, ω-oxidation, and β-oxidation. The process often begins with the oxidation of alkanes, where the terminal methyl group is oxidized to form an alcohol. This alcohol is then dehydrogenated into an aldehyde and converted into a corresponding carboxylic acid, which can be metabolized through the β-oxidation pathway of fatty acids (Das and Chandran 2011; Abbasian et al. 2015). Alkanes are commonly degraded via monoterminal oxidation, where molecular oxygen is introduced at the terminal position of the hydrocarbon to produce a primary alcohol. This alcohol is oxidized to form an aldehyde and a fatty acid. The fatty acid enters the β-oxidation pathway, which shortens it by two-carbon units sequentially, releasing acetyl coenzyme A (acetyl-CoA) (Ji et al. 2013).

Other oxidation pathways include di-terminal and subterminal oxidation. In di-terminal oxidation, both ends of the alkane molecule are oxidized to produce dicarboxylic acids, which are then metabolized via β-oxidation. In subterminal oxidation, alkanes are converted into secondary alcohols, which are oxidized to ketones and esters. These esters are hydrolyzed into alcohols and fatty acids, which also undergo degradation (Rojo 2009). Studies on *Pseudomonas aeruginosa* DQ8 have demonstrated the efficiency of terminal oxidation in breaking down hydrocarbons such as diesel, crude oil, n-alkanes, and polycyclic aromatic hydrocarbons (PAHs) (Zhang et al. 2019). However, aromatic hydrocarbons are less biodegradable than saturated hydrocarbons due to their complex and chemically stable structures (Abbasian et al. 2015).

The understanding of anaerobic biodegradation processes is more recent compared to aerobic biodegradation. However, it is known that compounds such as BTEX (benzene, toluene, ethylbenzene, and xylene) can be degraded at similar rates by both metabolic pathways. These aromatic compounds are initially oxidized into phenols or organic acids in anaerobic metabolism. They are then transformed into long-chain volatile fatty acids, which are eventually metabolized into methane (CH_4) and carbon dioxide (CO_2) Abbasian et al. 2015). Several electron acceptors, such as nitrate, ferrous iron, manganese, and sulfate ions, are crucial for this process (Widdel and Rabus 2001). For instance, sulfate-reducing bacteria isolated from cold hydrocarbon seeps in marine environments can degrade alkanes like propane and butane. These bacteria form a unique phylogenetic cluster associated with the *Desulfosarcina-Desulfococcus* group, within the δ-*Proteobacteria* class (Jaekel et al. 2013). In this context, two biochemical mechanisms have been proposed for the initial activation of alkanes: fumarate addition and carboxylation (Wilkes et al. 2016). In both cases, the intermediate benzoyl-CoA plays a central role in the anaerobic catabolism of many aromatic compounds (Harwood 1998).

8.6 Microbial Consortia for Enhanced Bioremediation

Developing microbial consortia represents an effective strategy for optimizing bioremediation by leveraging the metabolic interactions among microorganisms and providing greater long-term stability. These metabolic interactions play a crucial role in the application of microbial consortia. Substances secreted by specific species can either support or inhibit the growth of other species, modify their interactions, and even influence the overall functionality of the community (Fig. 8.4).

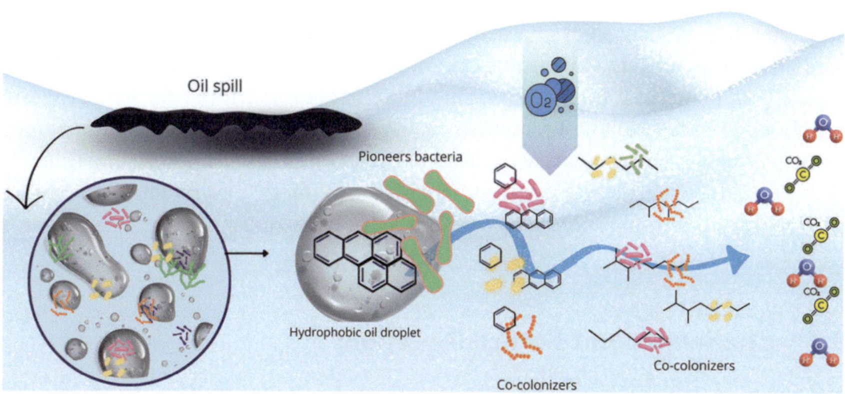

Fig. 8.4 The scheme of general multispecies bacterial oil aerobic biodegradation where first pioneer hydrocarbon-degrading bacteria biotransform hydrophobic oil compounds, and then their biotransformation products are utilized by bacterial co-colonizers. (Figure adapted from (Mori and Kanaly 2020))

Synthetic microbial consortia are artificially designed by co-culturing selected species (two or more) under a well-defined medium, particularly during the initial stages. Unlike natural microbial consortia, synthetic consortia aim to reconfigure metabolic pathways and program social interactions within the microbial community to achieve the desired functionality tailored to specific environmental conditions. This approach optimizes metabolic processes that can enhance the degradation of pollutants, such as hydrocarbons, in contaminated environments (Jia et al. 2016). Research has demonstrated that bacterial community responses to oil spill events are crucial in predicting the fate of hydrocarbon pollution in polar aquatic environments. A study assessing the effects of simulated acute oil spillage on bacterial communities from Arctic and Antarctic marine sediments highlighted the influence of hydrocarbon mixtures, incubation time, and the inherent bacterial communities in determining the selection of oil-degrading phylotypes.

The findings indicate that the effectiveness of contamination removal processes can be significantly limited by cold temperatures and low irradiance levels typical of polar systems, making understanding microbial responses vital for effective bioremediation strategies (Cai et al. 2020). The design of synthetic microbial consortia can also benefit from computational tools that model metabolic interactions among community members. For instance, tools like NetMet and OptCom allow researchers to predict the metabolic capacities of microbial species and their interactions, facilitating the selection of optimal combinations for specific bioremediation tasks (Tal et al. 2021; Zomorrodi and Maranas 2012). These models can simulate various environmental conditions and help identify the most effective microbial combinations for degrading specific pollutants. Moreover, integrating metagenomic approaches can provide insights into the functional potential of microbial communities involved in bioremediation. By analyzing the genetic material of microbial consortia, researchers can identify key metabolic pathways and genes responsible for pollutant degradation, which can inform the design of more effective synthetic consortia (Molina-Montenegro et al. 2018). This approach not only enhances our understanding of microbial interactions but also aids in developing targeted bioremediation strategies. In conclusion, microbial consortia, particularly synthetic ones, offer a promising avenue for enhancing bioremediation efforts. By harnessing the metabolic interactions among microorganisms and employing computational modeling and metagenomic analyses, researchers can optimize the design and functionality of these consortia to effectively tackle environmental pollutants, especially in challenging conditions such as those found in polar regions.

8.7 Challenges and Future Directions

Hydrocarbon contamination in polar environments presents a formidable challenge, yet the potential for bioremediation strategies utilizing local microorganisms offers a promising and sustainable solution. While microbial bioremediation has shown promise, several challenges remain, including the complexities of modeling bioremediation processes in polar conditions, the need for advancements in biotechnological applications such as enzyme engineering and multi-omics analysis, a deeper understanding of climate impacts on microbial efficacy, and inconsistencies in regulatory frameworks governing remediation practices in these sensitive regions. Continued research into the biology and ecology of these microorganisms is essential for developing effective environmental recovery methods, which will contribute to preserving polar ecosystems amidst increasing anthropogenic pressures.

Future research should focus on developing robust microbial consortia and understanding the unique metabolic capabilities of Antarctic microorganisms. For instance, studies have identified cold-adapted bacteria capable of degrading various pollutants, including phenanthrene and polychlorinated biphenyls (PCBs), which are critical for bioremediation in these environments (Gran-Scheuch et al. 2017; Papale et al. 2022). The isolation and characterization of such microorganisms, like those from diesel-contaminated Antarctic soils, have demonstrated their potential for effective hydrocarbon degradation (Gran-Scheuch et al. 2017). Furthermore, leveraging advanced techniques such as genome-resolved metagenomics can provide insights into microbial communities' functional potential, allowing for identifying key enzymes involved in hydrocarbon degradation (Abdullah et al. 2023).

In addition to exploring microbial diversity, applying synthetic biology approaches to engineer microbial consortia with tailored metabolic pathways could significantly enhance bioremediation outcomes. For example, optimizing oxygenase enzymes through genetic modifications can improve their activity and substrate specificity, thereby enhancing the degradation of hydrocarbons (Gong et al. 2016), Rylott and Bruce 2020). Moreover, integrating biostimulation techniques, such as the addition of organic amendments to enhance microbial activity, has been shown to improve the efficiency of bioremediation efforts in Antarctic soils (Primitz et al. 2021; Álvarez et al. 2015). The effectiveness of different nutrient sources as biostimulation agents has been highlighted in recent studies, emphasizing the need for tailored strategies that consider the unique environmental conditions of polar regions (Primitz et al. 2021).

Additionally, the role of extremophiles in bioremediation cannot be overlooked. Research has indicated that extremophiles, including psychrophilic bacteria isolated from Antarctic environments, possess unique biochemical pathways that can be harnessed for pollutant degradation (Orellana et al. 2018; de Melo Carlos et al. 2024). Exploring these organisms could lead to the discovery of novel bioremediation strategies specifically adapted to the harsh conditions of polar ecosystems.

In summary, addressing the challenges of hydrocarbon contamination in polar environments requires a multifaceted approach that combines ecological understanding, technological innovation, and regulatory alignment. By focusing on developing effective microbial consortia, leveraging advancements in biotechnology, and exploring the unique capabilities of extremophiles, we can enhance the efficacy of bioremediation strategies and ensure the protection of fragile polar ecosystems against the impacts of pollution.

References

Abbasian F, Lockington R, Mallavarapu M, Naidu R (2015) A comprehensive review of aliphatichydrocarbon biodegradation by bacteria. Appl Biochem Biotechnol, 176(3):670–699.

Abdullah K, Wilkins D, Ferrari B (2023) Utilization of -omic technologies in cold climate hydrocarbon bioremediation: a text-mining approach. Front Microbiol 14. https://doi.org/10.3389/fmicb.2023.1113102

Abdulrasheed M, Zakaria NN, Roslee AFA, Shukor MY, Zulkharnain A, Napis S, Convey P, Alias SA, Gonzalez-Rocha G, Ahmad SA (2020a) Biodegradation of diesel oil by cold-adapted bacterial strains of Arthrobacter spp. from Antarctica. Antarct Sci 32(5):341–353. https://doi.org/10.1017/S0954102020000206

Abdulrasheed M et al (2020b) Response surface methodology optimization and kinetics of diesel degradation by a cold-adapted Antarctic bacterium, Arthrobacter sp. strain AQ5-05. Sustainability 12(17):6966. https://doi.org/10.3390/su12176966

Abioye O, Raman A, Agamuthu P (2009) Enhanced biodegradation of used engine oil in soil amended with organic wastes. Water Air Soil Pollut 209(1–4):173–179. https://doi.org/10.1007/s11270-009-0189-3

Aislabie J, Foght J, Saul D (2000) Aromatic hydrocarbon-degrading bacteria from soil near Scott Base, Antarctica. Polar Biol 23(3):183–188. https://doi.org/10.1007/s003000050025

Aislabie J, Saul DJ, Foght JM (2006) Bioremediation of hydrocarbon-contaminated polar soils. Extremophiles 10(3):171–179. https://doi.org/10.1007/s00792-005-0498-4

Álvarez L, Balbo A, Cormack W, Ruberto L (2015) Bioremediation of a petroleum hydrocarbon-contaminated Antarctic soil: Optimization of a biostimulation strategy using response-surface methodology (RSM). Cold Reg Sci Technol 119:61–67. https://doi.org/10.1016/j.coldregions.2015.07.005

Atlas RM (1991) Microbial hydrocarbon degradation – bioremediation of oil spills. J Chem Technol Biotechnol 52(2):149–156. https://doi.org/10.1002/jctb.280520202

Atlas RM, Horowitz A, Busdosh M (1978) Prudhoe crude oil in Arctic marine ice, water, and sediment ecosystems: degradation and interactions with microbial and benthic communities. J Fish Res Board Can 35(5):585–590. https://doi.org/10.1139/f78-104

Barrett JE, Virginia RA, Parsons AN, Wall DH (2006) Soil carbon turnover in the McMurdo Dry Valleys, Antarctica. Soil Biol Biochem 38(10):3065–3082. https://doi.org/10.1016/j.soilbio.2006.03.025

Bhardwaj LK, Sharma S, Jindal T (2021) Occurrence of polycyclic aromatic hydrocarbons (PAHs) in the lake water at Grovnes Peninsula over East Antarctica. Chem Africa 4(4):965–980. https://doi.org/10.1007/s42250-021-00278-0

Børresen M, Breedveld GD, Rike AG (2003) Assessment of the biodegradation potential of hydrocarbons in contaminated soil from a permafrost site. Cold Reg Sci Technol 37(2):137–149. https://doi.org/10.1016/S0165-232X(02)00092-7

Brakstad OG, Bonaunet K (2006) Biodegradation of petroleum hydrocarbons in seawater at low temperatures (0–5 °C) and bacterial communities associated with degradation. Biodegradation 17(1):71–82. https://doi.org/10.1007/s10532-005-3342-8

Cai P et al (2020) Diagnosing bioremediation of crude oil-contaminated soil and related geochemical processes at the field scale through microbial community and functional genes. Ann Microbiol 70(1):1–12. https://doi.org/10.1186/s13213-020-01580-x

Cao S et al (2020a) Structure and function of the Arctic and Antarctic marine microbiota as revealed by metagenomics. Microbiome 8(1):47. https://doi.org/10.1186/s40168-020-00826-9

Cao Y et al (2020b) Microbial eco-physiological strategies for salinity-mediated crude oil biodegradation. Sci Total Environ 727:138723. https://doi.org/10.1016/j.scitotenv.2020.138723

CEP-IV-IP62 (2021) Item 1: opening of the meeting

Charman DJ, Amesbury MJ, Roland TP, Royles J, Hodgson DA, Convey P, Griffiths H (2018) Spatially coherent late Holocene Antarctic Peninsula surface air temperature variability. Geology 46(12):1071–1074. https://doi.org/10.1130/G45347.1

Chaudhary DK, Kim D-U, Kim D, Kim J (2019) Flavobacterium petrolei sp. nov., a novel psychrophilic, diesel-degrading bacterium isolated from oil-contaminated Arctic soil. Sci Rep 9(1):4134. https://doi.org/10.1038/s41598-019-40667-7

Chen Y-J et al (2024) Arctic's hidden hydrocarbon degradation microbes: Investigating the effects of hydrocarbon contamination, biostimulation, and a surface washing agent on microbial communities and hydrocarbon biodegradation pathways in high-Arctic beaches. Environ Microb 19(1):81. https://doi.org/10.1186/s40793-024-00626-w

Chilvers BL, Morgan KJ, White BJ (2021) Sources and reporting of oil spills and impacts on wildlife 1970–2018. Environ Sci Pollut Res 28(6):754–762. https://doi.org/10.1007/s11356-020-10538-0

Correa T, Abreu F (2020) Chapter 20: Antarctic microorganisms as sources of biotechnological products. In: Salwan R, Sharma V (eds) Physiological and biotechnological aspects of extremophiles. Academic Press, pp 269–284

Cripps GC, Shears J (1997) The fate in the marine environment of a minor diesel fuel spill from an Antarctic research station. Environ Monit Assess 46(3):221–232. https://doi.org/10.1023/A:1005766302869

Das N, Chandran P (2011) Microbial degradation of petroleum hydrocarbon contaminants: An overview. Biotechnol Res Int 2011:941810. https://doi.org/10.4061/2011/941810

de Melo Carlos L, Camacho KF, Duarte AW, de Oliveira VM, Boroski M, Rosa LH, Passarini MR (2024) Bioprospecting the potential of the microbial community associated to Antarctic marine sediments forhydrocarbon bioremediation. Brazilian J Microbiology, 55(1):471–485

Dell'Anno F, Rastelli E, Sansone C, Brunet C, Ianora A, Dell'Anno A (2021) Bacteria, fungi and microalgae for the bioremediation of marine sediments contaminated by petroleum hydrocarbons in the omics era. Microorganisms 9(8):1695. https://doi.org/10.3390/microorganisms9081695

Deprez PP, Arens M, Locher H (1999) Identification and assessment of contaminated sites at Casey Station, Wilkes Land, Antarctica. Polar Rec 35(193):299–316. https://doi.org/10.1017/S0032247400015655

Durán RE et al (2019) Genomic and physiological traits of the marine bacterium Alcaligenes aquatilis QD168 isolated from Quintero Bay, Central Chile, reveal a robust adaptive response to environmental stressors. Front Microbiol 10:528. https://doi.org/10.3389/fmicb.2019.00528

Elder CD et al (2018) Greenhouse gas emissions from diverse Arctic Alaskan lakes are dominated by young carbon. Nat Clim Chang 8(2):166–171. https://doi.org/10.1038/s41558-017-0066-9

Ellis M et al (2022) Hydrocarbon biodegradation potential of microbial communities from high Arctic beaches in Canada's Northwest Passage. Mar Pollut Bull 174:113288. https://doi.org/10.1016/j.marpolbul.2021.113288

Fingas MF (1995) A literature review of the physics and predictive modelling of oil spill evaporation. J Hazard Mater 42(2):157–175. https://doi.org/10.1016/0304-3894(95)00013-K

Gad SC (2014) Petroleum hydrocarbons. In: Wexler P (ed) Encyclopedia of toxicology, 3rd edn. Academic Press, pp 838–840

Garneau M-È et al (2016) Hydrocarbon biodegradation by Arctic sea-ice and sub-ice microbial communities during microcosm experiments, Northwest Passage (Nunavut, Canada). FEMS Microbiol Ecol 92(8):fiw130. https://doi.org/10.1093/femsec/fiw130

Gong T et al (2016) Metabolic engineering of Pseudomonas putida KT2440 for complete mineralization of methyl parathion and γ-hexachlorocyclohexane. ACS Synth Biol 5(5):434–442. https://doi.org/10.1021/acssynbio.6b00025

Góngora E et al (2022) Hydrocarbon bioremediation on Arctic shorelines: historic perspective and roadway to the future. Environ Pollut 305:119247. https://doi.org/10.1016/j.envpol.2022.119247

Govarthanan M, Fuzisawa S, Hosogai T, Chang Y-C (2017) Biodegradation of aliphatic and aromatic hydrocarbons using the filamentous fungus Penicillium sp. CHY-2 and characterization of its manganese peroxidase activity. RSC Adv 7(33):20716–20723. https://doi.org/10.1039/C6RA28687A

Gran-Scheuch A et al (2017) Isolation and characterization of phenanthrene degrading bacteria from diesel fuel-contaminated Antarctic soils. Front Microbiol 8:1634. https://doi.org/10.3389/fmicb.2017.01634

Green G, Skerratt JH, Leeming R, Nichols PD (1992) Hydrocarbon and coprostanol levels in seawater, sea-ice algae and sediments near Davis station in eastern Antarctica: a regional survey and preliminary results for a field fuel spill experiment. Mar Pollut Bull 25(5–8):293–302. https://doi.org/10.1016/0025-326X(92)90685-Y

Harwood CS, Burchhardt G, Herrmann H, Fuchs G (1998) Anaerobic metabolism of aromatic-compounds via the benzoyl-CoA pathway. FEMS Microbiol Rev, 22(5):439–458

Hughes K, Stallwood B (2006) Oil pollution in the Antarctic terrestrial environment. Polarforschung 75(2–3):141–144

Hughes KS, Bridge PD, Peck LS (2007) Tolerance of Antarctic soil fungi to hydrocarbons. Sci Total Environ 372(2–3):539–548. https://doi.org/10.1016/j.scitotenv.2006.09.016

Intergovernmental Panel on Climate Change (IPCC) (2022) Polar regions. In: The ocean and cryosphere in a changing climate: special report of the intergovernmental panel on climate change. Cambridge University Press, Cambridge, pp 203–320

Iriarte J et al (2023) Snow-dependent biogeochemical cycling of polycyclic aromatic hydrocarbons at coastal Antarctica. Environ Sci Technol 57(4):1625–1636. https://doi.org/10.1021/acs.est.2c05583

Jacobs SS (2004) Bottom water production and its links with the thermohaline circulation. Antarct Sci 16(4):427–437. https://doi.org/10.1017/S095410200400224X

Jaekel U, Musat N, Adam B, Kuypers M, Grundmann O, Musat F (2013) Anaerobic degradation ofpropane and butane by sulfate-reducing bacteria enriched from marine hydrocarbon cold seeps. ISME j,7(5):885–895

Ji Y, Mao G, Wang Y, Bartlam M (2013) Structural insights into diversity and n-alkane biodegradationmechanisms of alkane hydroxylases. Front Microbiol, 4:58

Jia X et al (2016) Design, analysis and application of synthetic microbial consortia. Synth Syst Biotechnol 1(2):109–117. https://doi.org/10.1016/j.synbio.2016.02.001

Jiménez J et al (2002) Genomic analysis of the aromatic catabolic pathways from Pseudomonas putida KT2440. Environ Microbiol 4(12):824–841. https://doi.org/10.1046/j.1462-2920.2002.00370.x

Joshi-Navare K, Singh PK, Prabhune AA (2014) New yeast isolate Pichia caribbica synthesizes xylolipid biosurfactant with enhanced functionality. Eur J Lipid Sci Technol 116(8):1070–1079. https://doi.org/10.1002/ejlt.201300363

Karl DM (1992) The grounding of the Bahia Paraiso: microbial ecology of the 1989 Antarctic oil spill. Microb Ecol 24(1):77–89. https://doi.org/10.1007/BF00171972

Kasai Y, Kishira H, Harayama S (2002) Bacteria belonging to the genus Cycloclasticus play a primary role in the degradation of aromatic hydrocarbons released in a marine environment. Appl Environ Microbiol 68(11):5625–5633. https://doi.org/10.1128/AEM.68.11.5625-5633.2002

Keramea P et al (2021) Oil spill modeling: a critical review on current trends, perspectives, and challenges. J Mar Sci Eng 9(2):181. https://doi.org/10.3390/jmse9020181

Kim S et al (2004) Proteome analysis of Pseudomonas sp. K82 biodegradation pathways. Proteomics 4(11):3610–3621. https://doi.org/10.1002/pmic.200400977

Kostka JE, Teske AP, Joye SB, Head IM (2014) The metabolic pathways and environmental controls of hydrocarbon biodegradation in marine ecosystems. Front Microbiol 5:1–16. https://doi.org/10.3389/fmicb.2014.00471

Kuppusamy S et al (2019) Ecological impacts of total petroleum hydrocarbons. In: Total petroleum hydrocarbons. Springer, pp 95–138. https://doi.org/10.1007/978-3-030-24035-6_5

Lawley B, Ripley S, Bridge P, Convey P (2004) Molecular analysis of geographic patterns of eukaryotic diversity in Antarctic soils. Appl Environ Microbiol 70(10):5963–5972. https://doi.org/10.1128/AEM.70.10.5963-5972.2004

Leahy JG, Colwell RR (1990) Microbial degradation of hydrocarbons in the environment. Microbiol Rev 54(3):305–315. https://doi.org/10.1128/mr.54.3.305-315.1990

Lofthus S, Bakke I, Greer CW, Brakstad OG (2021) Biodegradation of weathered crude oil by microbial communities in solid and melted sea ice. Mar Pollut Bull 172:112823. https://doi.org/10.1016/j.marpolbul.2021.112823

Mahjoubi M et al (2017) Microbial bioremediation of petroleum hydrocarbon–contaminated marine environments. In: Recent insights in petroleum science and engineering. IntechOpen

Marigómez I (2014) Oil, crude. In: Wexler P (ed) Encyclopedia of toxicology, 3rd edn. Academic Press, pp 663–669

Martorell MM, Ruberto LA, de Figueroa CL, Mac Cormack W (2019) Bioremediation abilities of Antarctic fungi. In: Tiquia-Arashiro SM, Grube M (eds) Fungi in extreme environments. Springer, pp 517–534. https://doi.org/10.1007/978-3-030-19030-9_26

Molina-Montenegro M et al (2018) Metagenomic exploration of soils microbial communities associated to Antarctic vascular plants. PeerJ Preprints 6:e26508. https://doi.org/10.7287/peerj.preprints.26508

Mori JF, Kanaly RA (2020) Multispecies diesel fuel biodegradation and niche formation are ignited by pioneer hydrocarbon-utilizing Proteobacteria in a soil bacterial consortium. Appl Environ Microbiol 87(5):e02268–e02220. https://doi.org/10.1128/AEM.02268-20

Morison J, Kwok R, Peralta-Ferriz C, Alkire M, Rigor I, Andersen R, Steele M (2012) Changing Arctic Ocean freshwater pathways. Nature 481(7379):66–70. https://doi.org/10.1038/nature10705

Nikolova C, Gutierrez T (2022) Marine hydrocarbon-degrading bacteria: their role and application in oil-spill response and enhanced oil recovery. In: Marine hydrocarbon biodegradation, pp 591–600

Okoye A et al (2024) Characterization and identification of long-chain hydrocarbon-degrading bacterial communities in long-term chronically polluted soil in Ogoniland. Environ Sci Pollut Res 31(21):30867–30885. https://doi.org/10.1007/s11356-024-33326-6

Onur G, Yilmaz F, İçgen B (2015) Diesel oil degradation potential of a bacterium inhabiting petroleum hydrocarbon contaminated surface waters. J Surfactant Deterg 18(4):707–717. https://doi.org/10.1007/s11743-015-1697-3

Orellana R et al (2018) Living at the frontiers of life: extremophiles in Chile and their potential for bioremediation. Front Microbiol 9. https://doi.org/10.3389/fmicb.2018.02309

Pacwa-Płociniczak M, Płaza GA, Poliwoda A, Piotrowska-Seget Z (2014) Characterization of hydrocarbon-degrading and biosurfactant-producing Pseudomonas sp. P-1 strain as a potential tool for bioremediation of petroleum-contaminated soil. Environ Sci Pollut Res 21(16):9385–9395. https://doi.org/10.1007/s11356-014-2872-1

Pandolfo E, Barra Caracciolo A, Rolando L (2023) Recent advances in bacterial degradation of hydrocarbons. Water 15(2):375. https://doi.org/10.3390/w15020375

Papale M et al (2022) A first glimpse on cold-adapted PCB-oxidizing bacteria in Edmonson Point Lakes (Northern Victoria Land, Antarctica). Water 14(1):109. https://doi.org/10.3390/w14010109

Pearce DA (2012) Extremophiles in Antarctica: life at low temperatures. In: Stan-Lotter H, Fendrihan S (eds) Adaption of microbial life to environmental extremes: novel research results and application. Springer, pp 87–118

Peeb A, Dang NP, Truu M, Nõlvak H, Petrich C, Truu J (2022) Assessment of hydrocarbon degradation potential in microbial communities in Arctic sea ice. Microorganisms 10(2):328. https://doi.org/10.3390/microorganisms10020328

Pettersson JK, Nobes DC (2003) Environmental geophysics at Scott Base: Ground penetrating radar and electromagnetic induction as tools for mapping contaminated ground at Antarctic research bases. Cold Reg Sci Technol 37(2):187–195. https://doi.org/10.1016/S0165-232X(03)00037-5

Powell SM, Snape I, Bowman JP, Thompson BAW, Stark JS, McCammon SA, Riddle MJ (2005) A comparison of the short term effects of diesel fuel and lubricant oils on Antarctic benthic microbial communities. J Exp Mar Biol Ecol 322(1):53–65. https://doi.org/10.1016/j.jembe.2005.02.005

Primitz J et al (2021) Bioremediation of hydrocarbon-contaminated soil from Carlini Station, Antarctica: effectiveness of different nutrient sources as biostimulation agents. Polar Biol 44(2):289–303. https://doi.org/10.1007/s00300-020-02787-z

Purkey S, Johnson G (2013) Antarctic bottom water warming and freshening: contributions to sea level rise, ocean freshwater budgets, and global heat gain. J Clim 26(17):6105–6122. https://doi.org/10.1175/JCLI-D-12-00834.1

Clark R (1970) Cleaning up oil spills. Mar Pollut Bull 1:163–164. https://doi.org/10.1016/0025-326X(70)90287-0

Ramasamy KP, Mahawar L, Rajasabapathy R, Rajeshwari K, Miceli C, Pucciarelli S (2023) Comprehensive insights on environmental adaptation strategies in Antarctic bacteria and biotechnological applications of cold adapted molecules. Front Microbiol 14. https://doi.org/10.3389/fmicb.2023.1197797

Ribicic D, Netzer R, Winkler A, Brakstad OG (2018) Microbial communities in seawater from an Arctic and a temperate Norwegian fjord and their potentials for biodegradation of chemically dispersed oil at low seawater temperatures. Mar Pollut Bull 129(1):308–317. https://doi.org/10.1016/j.marpolbul.2018.02.024

Rizzo C, Malavenda R, Gerçe B, Papale M, Syldatk C, Hausmann R, Bruni V, Michaud L, Lo Giudice A, Amalfitano S (2019) Effects of a simulated acute oil spillage on bacterial communities from Arctic and Antarctic marine sediments. Microorganisms 7(12):632. https://doi.org/10.3390/microorganisms7120632

Rojo F (2009) Degradation of alkanes by bacteria. Environ Microbiol, 11(10):2477–2490

Roslee AFA, Ahmad SA, Gomez-Fuentes C, Shaharuddin NA, Khalil KA, Zulkharnain A (2021) Scientometric analysis of diesel pollutions in Antarctic territories: a review of causes and potential bioremediation approaches. Sustainability 13(13):7064. https://doi.org/10.3390/su13137064

Rylott E, Bruce N (2020) How synthetic biology can help bioremediation. Curr Opin Chem Biol 58:86–95. https://doi.org/10.1016/j.cbpa.2020.07.004

Salari M, Rahmanian V, Hashemi SA, Chiang W-H, Lai CW, Mousavi SM, Gholami A (2022) Bioremediation treatment of polyaromatic hydrocarbons for environmental sustainability. Water 14(23):3980. https://doi.org/10.3390/w14233980

Schultz J et al (2022) Polyphasic analysis reveals potential petroleum hydrocarbon degradation and biosurfactant production by rare biosphere thermophilic bacteria from Deception Island, an active Antarctic volcano. Front Microbiol 13:885557. https://doi.org/10.3389/fmicb.2022.885557

Secretariat of the Antarctic Treaty (2024) Environmental protocol | Antarctic treaty. https://www.ats.aq/e/protocol.html

Shahidul Islam M, Tanaka M (2004) Impacts of pollution on coastal and marine ecosystems including coastal and marine fisheries and approach for management: a review and synthesis. Mar Pollut Bull 48(7–8):624–649. https://doi.org/10.1016/j.marpolbul.2003.12.004

Silva TR, Duarte AWF, Passarini MRZ, Ruiz ALTG, Franco CH, Moraes CB, de Melo IS, Rodrigues RA, Fantinatti-Garboggini F, Oliveira VM (2018) Bacteria from Antarctic environments: diversity and detection of antimicrobial, antiproliferative, and antiparasitic activities. Polar Biol 41(8):1521. https://doi.org/10.1007/s00300-018-2322-5

Snape I, Harvey PMA, Ferguson SH, Rayner JL, Revill AT (2005) Investigation of evaporation and biodegradation of fuel spills in Antarctica I. A chemical approach using GC–FID. Chemosphere 61(9):1485–1494. https://doi.org/10.1016/j.chemosphere.2005.04.108

Stark JS, Mohammad M, McMinn A, Ingels J (2017) The effects of hydrocarbons on meiofauna in marine sediments in Antarctica. J Exp Mar Biol Ecol 496:56–73. https://doi.org/10.1016/j.jembe.2017.07.009

Steffen KT, Hatakka A, Hofrichter M (2002) Removal and mineralization of polycyclic aromatic hydrocarbons by litter-decomposing basidiomycetous fungi. Appl Microbiol Biotechnol 60(1–2):212–217. https://doi.org/10.1007/s00253-002-1105-6

Stewart EJ, Howell SEL, Draper D, Yackel J, Tivy A (2007) Sea ice in Canada's Arctic: implications for cruise tourism. Arctic 60(4):370–380

Stoyanova K, Gerginova M, Dincheva I et al (2022) Biodegradation of naphthalene and anthracene by Aspergillus glaucus strain isolated from Antarctic soil. PRO 10(5):873. https://doi.org/10.3390/pr10050873

Szopińska M, Szumińska D, Bialik RJ, Dymerski T, Rosenberg E, Polkowska Ż (2019) Determination of polycyclic aromatic hydrocarbons (PAHs) and other organic pollutants in freshwaters on the western shore of Admiralty Bay (King George Island, Maritime Antarctica). Environ Sci Pollut Res 26(18):18143–18161. https://doi.org/10.1007/s11356-019-05045-w

Tal O, Bartuv R, Vetcos M, Medina S, Jiang J, Freilich S (2021) NetCom: a network-based tool forpredicting metabolic activities of microbial communities based on interpretation of metagenomics data.Microorg, 9(9):1838

Tang YJ, Carpenter SD, Deming JW, Krieger-Brockett B (2006) Depth-related influences on biodegradation rates of phenanthrene in polluted marine sediments of Puget Sound, WA. Mar Pollut Bull 52(11):1431–1440. https://doi.org/10.1016/j.marpolbul.2006.04.009

Tarr M, Zito P, Overton E, Olson G, Adkikari P, Reddy C (2016) Weathering of oil spilled in the marine environment. Oceanography 29(3):126–135. https://doi.org/10.5670/oceanog.2016.77

Teske AP (2015) The metabolic pathways and environmental controls of hydrocarbon biodegradation in marine ecosystems. Front Microbiol 6. https://doi.org/10.3389/fmicb.2015.00471

Tin T, Fleming ZL, Hughes KA, Ainley DG, Convey P, Moreno CA, Pfeiffer S, Scott J, Snape I (2009) Impacts of local human activities on the Antarctic environment. Antarct Sci 21(1):3–33. https://doi.org/10.1017/S0954102009001722

Transport Canada (2024) Implementation of the International Maritime Organization's (IMO) prohibition on the use and carriage for use as fuel of heavy fuel oil (HFO) by ships in Canadian Arctic waters – SSB No.: 17/2024

Turner J, Overland J, Walsh J (2007) An Arctic and Antarctic perspective on recent climate change. Int J Climatol 27(3):277–293. https://doi.org/10.1002/joc.1406

Turner J, Comiso JC, Marshall GJ, Lachlan-Cope TA, Bracegirdle T, Maksym T, Meredith MP, Wang Z, Orr A (2009) Non-annular atmospheric circulation change induced by stratospheric ozone depletion and its role in the recent increase of Antarctic sea ice extent. Geophys Res Lett 36(8):L08502. https://doi.org/10.1029/2009GL037524

Varjani SJ (2017) Microbial degradation of petroleum hydrocarbons. Bioresource technology, 223:277–286

Vigneron A, Cruaud P, Lovejoy C, Vincent WF (2023) Genomic insights into cryptic cycles of microbial hydrocarbon production and degradation in contiguous freshwater and marine microbiomes. Microbiome 11(1):104. https://doi.org/10.1186/s40168-023-01537-7

Watts J (2020) Climate change is turning parts of Antarctica green, say scientists. The Guardian

Wilkes H, Buckel W, Golding BT, Rabus R (2016) Metabolism of hydrocarbons in n-alkane-utilizinganaerobic bacteria. J Mol Microbiol Biotechnol, 26(1-3):138–151.

Widdel F, Rabus R (2001) Anaerobic biodegradation of saturated and aromatic hydrocarbons. Curr Opin Biotechnol, 12(3):259–276

Wong RR, Lim ZS, Shaharuddin NA, Zulkharnain A, Gomez-Fuentes C, Ahmad SA (2021) Diesel in Antarctica and a bibliometric study on its indigenous microorganisms as remediation agent. Int J Environ Res Public Health 18(4):1512. https://doi.org/10.3390/ijerph18041512

Yakimov MM, Timmis KN, Golyshin PN (2007) Obligate oil-degrading marine bacteria. Curr Opin Biotechnol 18(3):257–266. https://doi.org/10.1016/j.copbio.2007.04.006

Yang S-Z, Jin H-J, Wei Z, He R-X, Ji Y-J, Li X-M, Yu S-P (2009) Bioremediation of oil spills in cold environments: a review. Pedosphere 19(3):371–381. https://doi.org/10.1016/S1002-0160(09)60128-4

Zakaria NN, Roslee A, Gomez-Fuentes C, Zulkharnain A, Abdulrasheed M, Sabri S, Ramírez-Moreno N, Calisto-Ulloa N, Ahmad S (2020) Kinetic studies of marine psychrotolerant microorganisms capable of degrading diesel in the presence of heavy metals. Revista Mexicana de Ingeniería Química 19(4):1375–1388

Zakaria NN, Convey P, Gomez-Fuentes C, Zulkharnain A, Sabri S, Shaharuddin NA, Ahmad SA (2021a) Oil bioremediation in the marine environment of Antarctica: a review and bibliometric keyword cluster analysis. Microorganisms 9(2):419. https://doi.org/10.3390/microorganisms9020419

Zakaria NN, Gomez-Fuentes C, Abdul Khalil K, Convey P, Roslee AFA, Zulkharnain A, Sabri S, Shaharuddin NA, Cárdenas L, Ahmad SA (2021b) Statistical optimisation of diesel biodegradation at low temperatures by an Antarctic marine bacterial consortium isolated from non-contaminated seawater. Microorganisms 9(6):1213. https://doi.org/10.3390/microorganisms9061213

Zannotti M, Ramasamy KP, Loggi V, Vassallo A, Pucciarelli S, Giovannetti R (2023) Hydrocarbon degradation strategy and pyoverdine production using the salt tolerant Antarctic bacterium Marinomonas sp. ef1. RSC Adv 13(28):1–12. https://doi.org/10.1039/d3ra01704a

Zhang C, Sirijovski N, Adler L, Ferrari BC (2019) Exophiala macquariensis sp. nov., a cold adapted black yeast species recovered from a hydrocarbon contaminated sub-Antarctic soil. Fungal Biol 123(2):151–158. https://doi.org/10.1016/j.funbio.2018.11.011

Zomorrodi AR, Maranas CD (2012) OptCom: a multi-level optimization framework for the metabolicmodeling and analysis of microbial communities. PLoS computational biology, 8(2):e1002363

Chapter 9
Enzymes with Pharmaceutical Applications by Microorganisms from Polar Environments

Averlane Vieira da Silva, Lívia Soares Bezerra, Luiz Henrique da Silva, Monelly da Silva Bernardo, Taciana Freitas Alves Bandeira, Adeildo Júnior de Oliveira, Aline Cavalcanti de Queiroz, Magna Suzana Alexandre Moreira, Michel Rodrigo Zambrano Passarini, and Alysson Wagner Fernandes Duarte

Abstract Microorganisms from polar environments can play crucial roles in the biotechnological production of enzymes of pharmacological interest. In this chapter, we compile the main studies focused on the biotechnological aspects of enzymes derived from microorganisms from Antarctica and the Arctic with human pharmacological applications. Enzymes from microorganisms in these environments may exhibit catalytic activity at low and moderate temperatures. With the advancement of modern biotechnology, cold-adapted enzymes have been widely used in food production, animal feed, the textile industry, and especially in the development of new drugs. Consequently, these enzymes have garnered much attention in recent years, driven by the advent of bioinformatics and heterologous expression in genetically modified organisms (GMO). Enzymes such as L-asparaginase, β- and α-galactosidases, lipase, and superoxide dismutase stand out in the production and potential formulation of new biopharmaceuticals. Genetically manipulated strains stimulated to produce specific pharmacologically relevant enzymes, combined with bioinformatics, may become crucial in a variety of biotechnological applications. Thus, enzymes produced by polar microorganisms with pharmaceutical interest may lead to a larger annual market than thermostable enzymes in the near future.

A. V. da Silva · L. S. Bezerra · L. H. da Silva · M. da Silva Bernardo · T. F. A. Bandeira · A. J. de Oliveira · A. C. de Queiroz · A. W. F. Duarte (✉)
Laboratório de Microbiologia, Imunologia e Parasitologia, Complexo de Ciências Médicas e Enfermagem, Universidade Federal de Alagoas, Campus Arapiraca, Universidade Federal de Alagoas, Arapiraca, AL, Brazil
e-mail: alysson.duarte@arapiraca.ufal.br

M. S. A. Moreira
Laboratório de Farmacologia e Imunologia, Instituto de Ciências Biológicas e da Saúde, Universidade Federal de Alagoas, Maceió, AL, Brazil

M. R. Z. Passarini
Universidade Federal da Integração Latino Americana, Foz do Iguaçu, PR, Brazil

A. W. Fernandes Duarte et al. (eds.), *Polar Microbes and Climate Change*,
https://doi.org/10.1007/978-3-032-07096-8_9

Keywords Antarctica · Arctic · Enzymology · L-Asparaginase · α-Galactosidase · β-Galactosidase

9.1 Introduction

Polar ecosystems (Arctic and Antarctic) house most of the Earth's ice and are therefore considered extremely cold, dry environments with a distinct biological diversity compared to other terrestrial ecosystems (Rosa et al. 2019; Lauritano et al. 2020). However, what thrives in these environments and is well adapted are microscopic organisms (e.g., filamentous fungi, yeasts, bacteria, and protozoa) that withstand such environmental pressures (Menezes et al. 2019; Silva et al. 2022a, b).

To survive various pressures, microorganisms have developed adaptive strategies to different adverse conditions, such as increased molecular and structural flexibility of cell membranes, the production of antifreeze proteins, the presence of pigments with cryoprotective properties, and the production of extracellular enzymes (Garcia-Lopez et al. 2021). In this context, enzymes produced by such extremophilic microorganisms have been applied in a wide range of economic sectors, particularly in the pharmaceutical industry, aiming to find stable enzymes for human applications (Anisha 2023). Some of these enzymes have been used for medicinal purposes for several decades, such as L-asparaginase. One of the biotechnological advantages of polar microorganism enzymes, especially in the food industry, is the prevention of bacterial contamination that can occur at higher cultivation temperatures (Qian et al. 2023).

To further enhance enzyme studies, research has leveraged new technologies to discover novel enzymes produced by polar microorganisms, and with the aid of bioinformatics, researchers can identify potential enzyme targets for possible drug formulations by tracking biological samples at gene, transcription, and protein levels (Majumder et al. 2024).

Furthermore, genetically modifying microorganisms appears to be an alternative for producing new enzymatic compounds for biopharmaceuticals (Yan et al. 2023). One of the major benefits of genetic engineering applied to bacteria, such as *Escherichia coli*, is the increased expression of specific characteristics and the reduction of undesired side effects from their use (Medawar et al. 2020; Yu et al. 2020).

In this regard, this chapter presents the main aspects related to the use of enzymes produced by polar microorganisms with therapeutic potential, as well as highlighting what has been applied in biotechnological terms to improve this perspective.

9.2 Polar Microorganisms: Adaptation, Habitat, and Diversity

It is estimated that the polar cryosphere represents 13–14% of the Earth's surface (Malard and Pearce 2018). The Arctic hosts a portion of microorganisms that survive under stressful conditions, requiring them to be highly adapted or extremely resistant to extreme cold, variable temperatures, melting cycles, and other abiotic factors (Lauritano et al. 2020). Due to these adaptations, this environment is believed to host a rich microbial diversity with numerous species inhabiting different habitats (Chen et al. 2023).

With recent advances in scientific research, it is known today that microorganisms thrive in specific locations in the Arctic due to various adaptive strategies. In Gram-positive bacteria, for example, the cell wall may contain a higher content of peptidoglycan, providing additional protection against ice formation and freeze-thaw cycles (Lauritano et al. 2020; Rizzo and Lo Giudice 2022). Bacteria can tolerate most environmental conditions present in the Arctic due to modifications in the plasma membrane and intracellular components, which increase plasticity and influence cellular functions (Holmberg and Jørgensen 2023).

Similar to bacteria, fungi employ comparable strategies for survival at low temperatures. Several molecules (e.g., antifreeze compounds, pigments, and enzymes) contribute to the metabolic arsenal of these extremophilic fungi (Garcia-Lopez et al. 2021). A universal fungal response to hyperosmotic stress is the intracellular accumulation of small organic molecules, such as glycerol under high salinity conditions, while melanization of the cell wall is believed to reduce glycerol leakage (Bodnár et al. 2024).

Due to these adaptations, it is plausible that microorganisms thrive in the Arctic region and are abundant in different substrates of polar environments (e.g., snow, soils, glaciers, air, marine, and freshwater ecosystems), as these habitats and substrates host complex bacterial communities (Malard and Pearce 2018). In a study led by Girard et al. (2023), the authors reported that cyanobacteria were the most abundant on the surface of a glacier, while throughout the water column, there was a heterogeneous set of Chloroflexi, Actinobacteriota, and *Planctomycetota*. Among Arctic fungi, those found in freshwater samples along a permafrost melting gradient (Kluge et al. 2021), soils (Varsadiya et al. 2021), and vegetation (Rasmussen et al. 2022) stand out.

The Arctic has revealed microbial diversity in permafrost. Dziurzynski et al. (2023), through a combination of metabarcoding techniques (culture independent) and classical microbiology (culture dependent), found that the amplicon sequence variants (ASVs) were dominated by the phyla *Proteobacteria* ($\bar{x}$= 22.6%, s = 2.3%) and *Actinobacteria* ($\bar{x}$= 20.6%, s = 4.4%), being the most abundant. The culture-independent approach resulted in 831 genera, with *Candidatus* being the dominant genus in 5 out of 6 samples, followed by *Chryseobacterium*, which constituted 6.69% of all obtained sequences. Regarding fungi, members of the class *Dothideomycetes* dominated fungal communities across all samples.

In an integrative study, a total of 2.884 ASVs were identified, attributed to 11 phyla, and distributed among 278 known genera. The results demonstrated a strong relationship between fungal genera and habitat types. The genera *Pyxidiophora*, *Trichosporon*, *Libertasomyces*, *Infundichalara*, *Lasiodiplodia*, and *Spissiomyces* were present in sediments (Chen et al. 2023).

Antarctica has gained prominence in recent years in different research fields. This continent, located at the geographic South Pole, can reach temperatures as low as −50.0 °C (Jordan et al. 2020). In the East Antarctic Plateau, the lowest surface temperature observed on Earth was recorded at −89.2 °C (Turner et al. 2021). Given its unique characteristics—such as high radiation levels, soil type, and the absence of sunlight during winter—life in this environment requires various mechanisms to survive extreme conditions (Doytchinov and Dimov 2022). Antarctic microorganisms have developed unique strategies to cope with environmental challenges, such as extremely low temperatures, intense ultraviolet radiation, high salinity, and freeze-thaw cycles. Throughout evolution, these organisms have developed reactions that provide protection against Antarctica's peculiarities, as the absence of these strategies would lead to severe structural damage and cell death (Silva et al. 2021).

Psychrophiles are microorganisms adapted to low temperatures, and those found in Antarctica represent intriguing examples of adaptation and survival under extreme conditions (Varrella et al. 2021). Among the most significant adaptations is the synthesis of cold-environment microbial enzymes with high catalytic efficiency at low temperatures (Yusof et al. 2021). These enzymes have a more flexible molecular structure due to the prevalence of α-helix secondary structures, which reduce the activation energy required for biochemical reactions, making them useful for biotechnological processes with lower energy consumption. Amylases, cellulases, and lipases are enzymes produced by Antarctic fungi, such as *Pseudogymnoascus* and *Mortierella* species (Duarte et al. 2018). Additionally, structural flexibility is also present in cell membranes, due to unsaturated fatty acids that maintain fluidity even at extremely low temperatures, along with the production of antifreeze proteins that prevent ice crystal formation within cells, protecting cellular tissues, as is the case of yeasts (e.g., *Mrakia frigida*) (Rosa et al. 2019).

Another adaptation strategy is the use of atmospheric gases, such as molecular hydrogen (H_2) and carbon monoxide (CO), as energy and carbon sources. These tactics are effective in locations with low organic extract concentrations. This occurs because hydrogenases allow H_2, while carbon monoxide dehydrogenases break down atmospheric CO, releasing energy to maintain aerobic respiration even in nutrient-scarce environments. This ability is typical in Antarctic bacteria belonging to the phyla *Actinobacteria* and *Proteobacteria* (Ortiz et al. 2020).

Other changes involve the synthesis of heat shock proteins that stabilize vital cellular processes such as transcription and translation, as well as lipids that improve membrane fluidity at low temperatures (Wani et al. 2022). Biofilm formation is also a protective strategy against temperature fluctuations and ultraviolet radiation. These reactions are influenced by genomic and molecular changes that enable survival in adverse environments (Wani et al. 2022).

The geographic uniqueness of these territories favors highly specialized and often endemic communities (Mashamaite et al. 2023). A characterization of the Antarctic island region provided valuable information on microbial biogeography, suggesting that strong local homogenization forces may lead to unique components in soil microbial communities (Lebre et al. 2023). However, despite having lower microbial biomass than tropical regions, polar soils contain complex communities (Malcheva et al. 2020). These communities rely on external organic matter inputs, such as wind-transported material or marine biota introduction. Additionally, microorganisms play a crucial role in the constitution and preservation of biogeochemical cycles in these unique ecosystems, serving as an example for analyzing adaptation and functionality in extreme conditions (Rosa et al. 2019).

For instance, the predominance of fungi on the continent is notable, particularly the genera *Cryomyces* and *Antarctomyces*, which exhibit remarkable adaptations such as slow growth and resistance mechanisms to temperature and radiation. These fungi can be found in various habitats, ranging from ultra-oligotrophic soils to endolithic environments like McMurdo Dry Valleys, considered the harshest on the planet due to extreme aridity and high salinity (Rosa et al. 2019).

In the Antarctic Peninsula, seasonal snow serves as a microhabitat for fungi from different genera, such as *Penicillium*, *Rhodotorula*, and *Phenoliferia*. These microorganisms exhibited high resistance to antifungal agents and were classified as cosmopolitan, being transported by air masses from other areas (Menezes et al. 2019). In Lake Fryxell, a significant alteration in fungal diversity was observed as depth increased. Species such as *Glaciozyma watsonii* represented 69% of total abundance, while *Toxicocladosporium*, *Acremonium*, and *Geomyces* accounted for only 1% each. On the other hand, species like *Thelebolus globosus* were more frequent below this salt-rich layer. This pattern reflects the different chemical and physical aspects of the lake (Connell et al. 2018).

On the Antarctic continent, where the amount of organic matter is scarce, fungi associated with lichens and mosses predominate (Rosa et al. 2019). Among the endemic species, *Cryomyces antarcticus*, *Thelebolus ellipsoideus*, and *Antarctomyces psychrotrophicus* stand out, as they exhibit biochemical and metabolic adaptations to survive in environments with low nutrient availability, low temperatures, and high UV radiation. On the other hand, cosmopolitan fungi, such as *Penicillium chrysogenum* and *Aspergillus fumigatus*, exhibit psychrotolerant behavior due to their genetic flexibility. These organisms play a fundamental role in biogeochemical cycles, recycling organic matter and contributing to the ecological dynamics of Antarctic niches (Duarte et al. 2019; Rosa et al. 2019; Nikitin et al. 2022).

9.3 Importance and Biotechnological Aspects of Cold-Active Enzymes

Cold environments, which cover a large portion of the Earth's surface, are inhabited by microorganisms such as bacteria, yeasts, and fungi, whose cellular temperature reflects that of the environment (Shu and Huang 2022). Despite the challenges imposed by the cold on biochemical reactions, these organisms thrive due to structural adaptations in membranes, proteins, and enzymes. Enzymes, essential for catalyzing biochemical reactions, are fundamental to cold adaptation, and recent studies have explored the molecular mechanisms that ensure their functionality at low temperatures, highlighting their great biotechnological potential. When synthesized by polar microorganisms, they exhibit unique and valuable properties for various applications (Yusof et al. 2021).

Compared to mesophilic and thermophilic enzymes, psychrophilic enzymes demonstrate superior biocatalytic activity at lower temperatures, making them ideal for processes that preserve heat-sensitive compounds (Hamid et al. 2022). The use of psychrophilic enzymes in low-temperature processes not only improves efficiency but also contributes to sustainability by reducing heating requirements and minimizing energy consumption.

Polar microorganisms, such as bacteria, fungi, and yeasts, that produce cold-active enzymes have been widely used in various sectors, including food, beverages, and biopharmaceuticals. Among these enzymes, cellulases, L-asparaginase, lipases, and superoxide dismutase (SODs) stand out as essential in several pharmaceutical processes, as per Fig. 9.1. Additionally, β-galactosidases active at low temperatures are used in the food, cosmetic, and pharmaceutical industries and are particularly important in the production of lactose-free dairy products (Hamid et al. 2022). These enzymes are characterized by their ability to remain stable and function efficiently at low temperatures; however, they are sensitive to heat and quickly lose activity when exposed to moderate temperatures.

Dirt and fabric stains, mainly composed of lipids, proteins, and alkaline substances such as ash, are difficult to remove from textiles, and processes such as heating and friction accelerate material wear (Hindborg 2023). While enzymatic detergents are efficient in removing these stains, in cold climate regions, detergents containing mesophilic enzymes do not perform as effectively. Replacing these enzymes with chemical agents to improve cleaning results in greater pollution and environmental damage. On the other hand, cold-active enzymes, which offer greater stability and efficiency in cold environments, are becoming more popular as they help maintain fabric integrity and save energy. Moreover, these enzymes are biodegradable and safe for aquatic life, providing environmental benefits (Kuddus et al. 2024).

Lipases adapted to low temperatures are particularly effective in removing oil from wastewater in cold climate regions. Due to their efficiency in low-temperature conditions, these enzymes are frequently investigated and used for cleaning environments affected by oil spills, especially in locations where lower temperatures

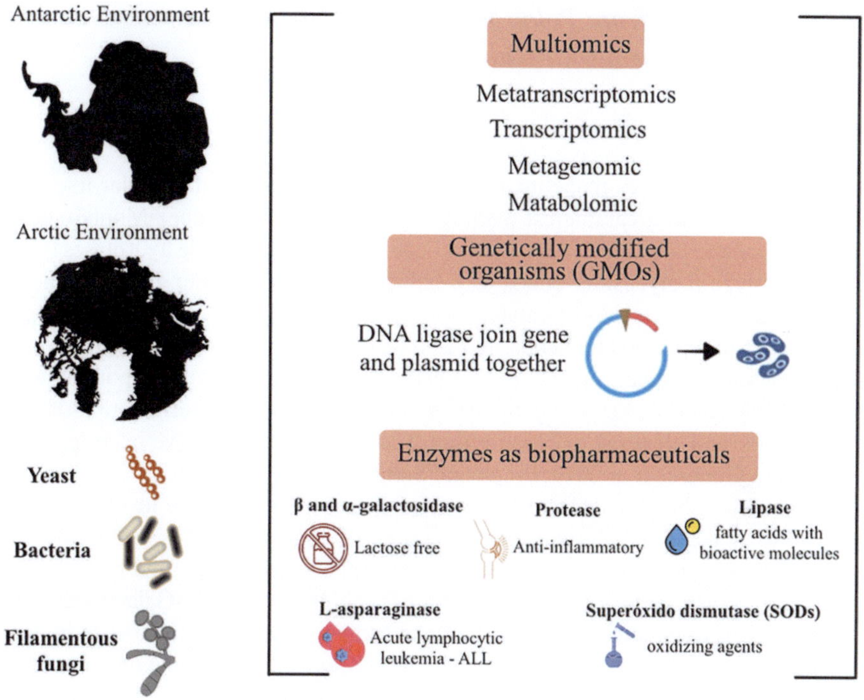

Fig. 9.1 Overview of enzymes produced by microorganisms from cold environments and the role of multiomics and GMOs in obtaining therapeutic enzymes

pose additional challenges. Additionally, other cold-adapted enzymes, such as amylases extracted from various microorganisms, are being studied for the bioremediation of waste generated from agricultural and domestic activities (Kumar et al. 2021).

Cold-adapted enzymes have shown advantages such as improved catalytic activity and substrate specificity at low temperatures, but they also present some important disadvantages. The primary limitation of these enzymes is their low thermostability (Kumar et al. 2021), which leads to activity loss when exposed to moderately higher temperatures. This makes them less effective in industrial processes that require thermal variation. The structural flexibility of these enzymes, necessary for their activity at low temperatures, can also contribute to this thermal instability. Another challenge related to the use of cold-adapted enzymes, particularly in large-scale industrial applications, is the difficulty of cultivating polar microorganisms in controlled environments. These microorganisms, although abundant in the Earth's biosphere, have slow metabolic rates, making the isolation and cultivation process more complex (Kumar et al. 2021). Overcoming these barriers with innovations in genetic engineering and cultivation techniques is crucial to further enabling the use of these enzymes in various industrial sectors.

9.3.1 Enzymes from Polar Microorganisms Applied as Biopharmaceuticals

9.3.1.1 β-Galactosidase and α-Galactosidase

β- and *α*-galactosidases make up a group of enzymes produced by polar microorganisms that are of great interest worldwide (Lee et al. 2012; Núñez-Montero et al. 2021). β-galactosidase (EC 3.2.1.23) has been defined as glycoside hydrolases (GH) that hydrolyze β-glycosidic bonds of *β*-galactosidase to generate galactose molecules. Some *β*-galactosidases catalyze the transfer of sugar fractions from a sugar donor to an acceptor (Mangiagalli and Lotti 2021). These enzymes have been produced by various species of microorganisms, and new bioactive compounds with medical applications, including biosensors for more specific determination of lactose in milk and cleavage of blood group A and B glycotopes, have been reported (Asraf and Gunasekaran 2010; Singh and Sambyal 2023).

On the other hand, *α*-galactosidase (EC 3.2.1.22) is an enzyme involved in the hydrolysis of the terminal *α*-galactosyl fractions of oligosaccharides such as melibiose, raffinose, stachyose, and verbascose, galactopolysaccharides, and glycoconjugates (e.g., glycolipids and glycoproteins) (Zhang et al. 2015). *α*-Galactosidase of microbial origin in some cases can be linked to applications of interest in human and industrial medicine (Bangoria et al. 2023) and some nonmicrobial ones involved in the treatment and prevention of intestinal gas and drug treatment of Fabry disease (Germain et al. 2016; Fabris et al. 2021).

Arthrobacter sp. 32cB, isolated from Antarctica, had its enzyme cloned and expressed in *Escherichia coli* was able to hydrolyze lactose and o-nitrophenyl-*β*-d-galactopyranoside, with Km values of 1.52 and 16.56 mM, with kcat of 30.55 and 31.84 s-1, respectively. It also showed a productivity of 42% of maximum activity when subjected to a temperature of 10.0 °C (Pawlak-Szukalska et al. 2014). In this study, the authors report that this bacterium has potential for use in the industrial removal of lactose from cow's milk, making it interesting for people who are intolerant of this carbohydrate.

The excellent therapeutic applications of *α*-galactosidases produced by microorganisms have intensified in order to identify new α-galactosidases with improved catalytic efficiency, but there is still a lack of studies reporting microorganisms from cold environments (mainly the Arctic) as potential producers of therapeutic *α*-galactosidases. In the study by Lee et al. (2012) with *Bacillus* sp. from Antarctica, good stability was observed against intestinal proteolytic enzymes such as trypsin and pancreatin, which could be a first step in the search for a stable enzyme with therapeutic potential in the food and feed industry. In addition, this enzyme showed optimum *α*-galactosidase activity at pH 7.0 and a temperature of 40.0 °C, exhibiting relatively high enzymatic activity (22% of maximum activity), with values between 1.55 ± 0.037 U/mL, activity relative to pH 7 and 1.57 ± 0.012 U/mL, equivalent to a temperature of 40.0 °C.

9.3.2 L-Asparaginase

L-Asparaginase (EC 3.5.1.1) is an amidohydrolase widely recognized for its ability to catalyze the conversion of L-asparagine into L-aspartate and ammonia. One of its main applications is in the treatment of neoplasms such as acute lymphocytic leukemia (ALL) (Table 9.1). This enzyme reduces the availability of circulating L-asparagine in the human body, an amino acid that the neoplastic cell lacks, thus compromising neoplastic cell growth. Thus, L-asparaginase is one of the main chemotherapeutic agents in the treatment of (ALL) and lymphosarcoma, standing out for its importance in interrupting the metabolism of tumor cells (Chand et al. 2020).

Advances in research into L-asparaginase have focused on the discovery of new therapeutic agents from both prokaryotic and eukaryotic sources, as well as the development of mutations that improve its functionality (Lailaja et al. 2024). Recent studies highlight the need to identify new enzyme sources, improve thermal stability, and reduce the immunogenic response, and consequently the side effects that affect cancer patients, such as hypersensitivity and bruising, and, in more severe cases, encephalopathy and hepatotoxicity. For efficient clinical use, it is essential to evaluate the affinity of the enzyme for substrates such as glutamine, with a low Michaelis constant (Km), allowing for more effective reactions. Future research should prioritize the detailed characterization of new enzymes in relation to their performance with L-asparagine and L-glutamine (Beckett and Gervais 2019).

Several efforts have been made to find a more stable enzyme with fewer adverse effects for cancer patients. Ashok et al. (2019), working with fungi from Antarctica, reported that of 55 isolates evaluated, 30 produced L-asparaginase free of glutaminase and urease in solid culture medium. *Trichosporon asahii*, IBBLA1, was highlighted as the best isolate in the research and showed maximum enzymatic activity of 20.57 U. mL^{-1} at 30.0 °C, pH 7.0, and concentrations of L-asparagine and glucose of 10 $g.L^{-1}$ and 2 $g.L^{-1}$, respectively. According to the authors of the research, the enzymatic activity obtained is comparable to that of bacterial sources (2–5 U mL^{-1}) that are currently being used on the market (Sajitha et al. 2015), which opens up prospects for its use in particular in the treatment of ALL and in the food industry.

Recently, researchers isolated fungi from Antarctic mosses, and of the 31 fungi tested, 13 morphotypes (41%) produced glutaminase- and urease-free L-ASNase in solid culture medium, with *Collariella* sp. MSA10, *Peroneutypa* sp. MSA13, and *Epicoccum* sp. MPA01 having the highest enzyme activities, with 1.29, 1.17, and 1.13 U g^{-1}, respectively. In contrast to the activity for L-ASNase, the quantifications of glutaminase and urease were relatively low. In the process of optimizing enzyme production, *Epicoccum* sp. MPA01 exhibited the highest post-optimization production values, producing 2.461 U g^{-1} (Andrade et al. 2023).

With regard to studies in the Arctic environment, researchers isolated an actinobacterium identified as *Streptomyces koyangensis* SK4 capable of producing 23.12 IU/mL on the seventh day of incubation at pH 7 and 20.0 °C. The activity of the enzyme was analyzed every 24 h for up to 9 days of incubation. Optimization of the enzymatic process by Plackett-Burman to evaluate eight variables increased

Table 9.1 Different enzymes and their pharmacological applications from Antarctic microorganisms

Microorganisms	Sample origin	OGM	Application/potential	References
		α and β-galactosidases		
Arthrobacter sp. 32cB	Soil samples	*Escherichia coli*	Use in industrial removal of lactose from milk (biocatalyst for pharmaceutical products)	Pawlak-Szukalska et al. (2014)
Bacillus sp.	Soil samples	*x*	Good stability against intestinal proteolytic enzymes such as trypsin and pancreatin	Lee et al. (2012)
Rahnella inusitata Se8.10.12	Soil samples	*x*	Possible alternatives to meet the current needs of lactose-intolerant consumers	Núñez-Montero et al. (2021)
Pseudoalteromonas sp. KMM 701	*x*	*Escherichia coli*	*Pseudoalteromonas* α-d-galactosidase of GH family 36, important in biomedical technology, demonstrates weak glycosynthase properties in vitro	Bakunina et al. (2018)
Arthrobacter sp. C2–2	fellfield soil	*Escherichia coli*	Production of lactose-reduced dairy products	Karasová-Lipovová et al. (2003)
		L-asparaginase		
Leucosporidium scottii L115	Terrestrial and marine samples	*x*	Characterization of a new ASNase free of GLNase activity produced, its application as a potential biosimilar or biobetter ASNase	Sánchez-Moguel et al. (2023)
Trichosporon asahii IBBLA1	Moss, water, soil, and snow	*x*	Extremophiles may be particularly useful in the treatment of ALL and in the food industry	Ashok et al. (2019)
Leucosporidium muscorum CRM 1648	Marine-sediments	*x*	New source of glutaminase-free antileukemic enzyme for commercial purposes	Freire et al. (2021)
Epicoccum nigrum, *Peroneutypa scoparia* and *Collariella virescens*	Antarctic mosses	*x*	Fungi that thrive in extreme environments may have applications in several domains, including the potential treatment of acute lymphoblastic leukemia	Andrade et al. (2023)

(continued)

Table 9.1 (continued)

Microorganisms	Sample origin	OGM	Application/potential	References
Leucosporidium scottii L115	Marine sediments	*x*	Potential to result in an Analysis with better properties than those currently available on the market, produced by bacteria	Sánchez-Moguel et al. (2022)
Sclerotinia borealis, Acidilobus saccharovorans, Melioribacter roseus	*x*	*Escherichia coli* BL21(DE3)	L-asparaginase from *M. roseus* may be a promising object for further study of enzymatic properties and biotechnological applications	Dumina et al. (2021)
		Lipase		
Candida antarctica	*x*	*x*	Preparation of calcimimetic Cinacalcet hydrochloride (pharmaceutical intermediate)	Wang et al. (2018)
		Superoxide dismutase (SODs)		
Aspergillus glaucus 363	Solo samples	*x*	*x*	Abrashev et al. (2016)

enzyme production to 68 IU/mL, followed by Central Composite Design, with a maximum production of 136 IU/mL at 20.0 °C. Among the nutritional variables that influence the production of L-asparaginase, the concentration of L-asparagine stands out, in addition to variables such as temperature, pH, incubation time, and agitation speed (Saleena et al. 2023).

9.3.3 *Lipase*

One of the enzymes applied to biopharmaceutical techniques is lipase (E.C. 3.1.1.3), which is commonly associated with lipid metabolism and the stabilization of therapeutic preparations, Table 9.1 (Vachher et al. 2021). Lipases belong to a group of hydrolytic enzymes, i.e., they carry out enzymatic hydrolysis, and are increasingly gaining prominence, especially in the pharmaceutical industry. In addition, their ability to process reactions in both aqueous and nonaqueous media makes them viable for use in hydrolysis and synthesis processes, and they are especially useful in modifying active pharmaceutical ingredients to increase specificity and therapeutic efficacy (Santos et al. 2022).

Lipases from *Candida antarctica* have also been of particular biotechnological interest. Enzymes derived from microorganisms have notable benefits compared to those from other sources, such as greater stability under harsh pH and temperature conditions, and are also simpler to produce in large quantities through fermentation methods (Godoy et al. 2022). Furthermore, these lipases can have their catalytic activity improved or have specific characteristics improved, such as stability and selectivity (Godoy et al. 2022; Santos et al. 2022). Methods such as cross-linked enzyme aggregation (CLEAs) have shown optimistic results, reducing costs and high catalytic efficiency in industrial uses (Santos et al. 2022).

In the pharmaceutical field, lipases play a significant role in preserving enantiomerically pure intermediates, which are essential for the manufacture of efficient drugs. For example, the enzyme called *Candida antarctica* lipase B (CALB) has been used in the kinetic resolution of racemic compounds, such as in the process of producing the intermediates of the drug pregabalin. This is because of its high enantiomeric selectivity, ensuring products with the optical purity required for therapeutic uses (Rios et al. 2018; Wang et al. 2018).

A lipase B from *Candida antarctica* was investigated in the resolution of (R,S)-1-(1-naphthyl)ethylamine ((R,S)-NEA) in ionic liquids (ILs), an important pharmaceutical intermediate for the preparation of calcimimetic Cinacalcet hydrochloride. The activity of the enzyme as a function of the cation decreased and CD spectroscopy revealed that the secondary structure of CALB at [C 6 mim][Tf 2 N] was slightly different from that of the native lipase, opening up prospects for its use as a chiral intermediate (Wang et al. 2018).

9.3.4 Superoxide Dismutase (SODs)

Superoxide dismutase (SOD) (E.C. 1.15.1.1) is characterized as a group of antioxidant metalloenzymes produced by many organisms, including humans, involved in the processes of protection against oxidizing agents—such as reactive oxygen species (ROS) (Yang et al. 2022; Hamre et al. 2023). As a result of its protective role, its deficiency is associated with a variety of diseases and comorbidities—such as inflammatory processes, asthma, cancer, cardiovascular diseases, etc.—as well as the process of senility (Yang et al. 2022; Dong et al. 2023). Its biochemical catalytic function is through the conversion of the free radical superoxide (O_2^-) into hydrogen peroxide (H_2O_2) and molecular oxygen (O_2), with H_2O_2 subsequently being reduced by other enzymes, such as catalase (Rosa et al. 2021; Ding et al. 2024).

There are three types of SODs, differentiated by their cellular location and molecular characteristics: SOD1, found in the cell cytoplasm and forms bonds with copper (Cu^{2+}) or zinc (Zn^{2+}) ions, also called ZnSOD or CuSOD; SOD2, located in the mitochondrial matrix and forming bonds with manganese ion (Mn^{3+}); and SOD3, the only one located in the extracellular matrix, forming bonds with copper and zinc ions, and can also be called EC-SOD (Zheng et al. 2023; Ding et al. 2024).

Currently, in addition to nutritional and cosmetic applications, SODs have great potential for biomedical and pharmacological applications, becoming the target of more and more studies due to their potential use in chronic diseases (e.g., diabetes), cardiovascular diseases (e.g., stroke), gastrointestinal diseases and cancers (Abdel-Monsef et al. 2023; Zheng et al. 2023). The use of recombinant human SOD species (rhSOD) as a treatment and prevention of acute rectal lesions in patients undergoing radiotherapy for cervical cancer reduced the incidence, severity, and duration of acute rectal lesions induced by radiotherapy therapy, with a reduction of around 27.8% in the incidence of diarrhea in these patients (Zhu et al. 2024). Similarly, the use of SOD2-PL (SOD2 plasmid liposomes, a mimetic of endogenous SOD2 activity) proved useful in protecting the oral mucosa of rats during radiotherapy treatment of oral tumors, with a reduction in the incidence of oral mucositis (Ding et al. 2024). In addition, the oral use of hsSOD (highly stable SOD enzyme) has shown great tissue efficacy in terms of its anti-aging capacity (Dong et al. 2023).

One of the major obstacles to the use of superoxide dismutase is its thermal instability, early renal clearance, low cell penetration capacity, as well as unfavorable pharmacokinetic characteristics and immunological consequences (Rosa et al. 2021; Dong et al. 2023; Kim et al. 2023). Therefore, due to the need and technological advances, the use of genetic engineering and chemical combinations and the search for new sources of SOD have been increasingly strengthened in order to make biomedical applications possible.

The fungus *Aspergillus glaucus* 363, isolated from Antarctica, showed high productivity of SOD, and after purification the enzyme was obtained with a specific activity of 4000 U mg^{-1}, comparable to or even higher than the activity of currently commercial mesophilic SOD (Table 9.1). These findings make the *A. glaucus* isolate interesting in research to optimize enzyme production (Abrashev et al. 2016). Thus, although it is not yet a substance used in clinical practice to treat diseases and prevent comorbidities, there is a high potential for the use of SOD in the biomedical and pharmaceutical environment.

9.4 The Role of "Omics" in the Discovery of New Therapeutic Enzymes

The term "omics" refers to a new area of biological sciences dedicated to the characterization and quantification of sets of biological molecules. It seeks to understand how these molecules influence the structure, function, and dynamics of organisms, as well as investigating how cells react to damage and alterations in cellular functional systems (Miao et al. 2023). The suffix "omics" was created to classify different fields of this science, such as genomics, which analyzes genes and their functions; transcriptomics, which studies messenger RNA; metabolomics, which investigates molecules related to cell metabolism; and proteomics, which focuses on proteins generated by the transcription of messenger RNA (Miao et al. 2023; Sanches et al. 2024).

One of the major bottlenecks in omics technologies is the production of a large volume of raw biological data, which requires specific computer tools and specialized researchers to organize and analyze it. This need is met by bioinformatics, which plays a key role in interpreting the data generated by these technologies. Through bioinformatics, researchers can identify potential target enzymes for possible drug formulations by screening biological samples at the gene, transcript, protein, and metabolite levels. In the current scenario, enzymes of microbial origin present knowledge gaps in studies involving omics, and therefore deserve special attention (Mullish et al. 2018; Majumder et al. 2024).

Genomic studies have opened up important avenues for the identification of therapeutic targets, omics data, including metagenomics, metabonomics, transcriptomics, and metatranscriptomics, and may at some point predict enzyme functions of microbial origin using different bioinformatics approaches (Aguiar-Pulido et al. 2016). Nevertheless, the development of effective drugs often requires the prioritization of small molecules that can serve as lead compounds, thus reducing the time needed to create new treatments, especially against cancer. These processes are often expensive and labor-intensive, which makes the efficient use of omics approaches even more crucial (Aguiar-Pulido et al. 2016).

Recently, one of the applications of omics has been the search for enzymes of therapeutic interest in microorganisms, as they are known to play a crucial role in the health of the host (Fig. 9.2) (Shankar and Sharma 2022). Furthermore, even with all these tools currently available, it is particularly difficult to elucidate the function of unknown enzymes in the human gut, since these enzymes catalyze diverse enzymatic reactions within this complex and distinct community existing in the gut (Jia et al. 2020, 2021).

Regarding enzymes from polar microorganisms with therapeutic potential, researchers have employed metagenomics-based approaches in an effort to isolate novel compounds from different Antarctic environments. Polar microorganisms, including bacteria from the *Actinobacteria* and *Proteobacteria* groups, are capable of synthesizing functional antimicrobial biomolecules against human pathogens (Zhang et al. 2015).

For example, researchers identified a bacterium *Rahnella inusitata* from Antarctica and studied its complete genome by characterizing the β-galactosidase enzyme based on genomic data. The results showed that three β-galactosidase genes were found in the genome of *R. inusitata*, which belong to the glycoside hydrolase families GH2 (LacZ and EbgA) and GH42 (BglY), in addition to the extract of the Antarctic strain showing greater enzymatic efficiency in the event of cold, maintaining up to 62% of its activity at 15.0 °C. Thus, such findings pave the way for a possible new alternative for the dairy industry, especially in response to the current needs of lactose-intolerant consumers (Núñez-Montero et al. 2021).

In this sense, omics techniques such as transcriptomics, epigenomics, microbiomics, metabolomics, and proteomics support existing endotyping paradigms and suggest new ways to understand diseases. In addition, they have been applied to

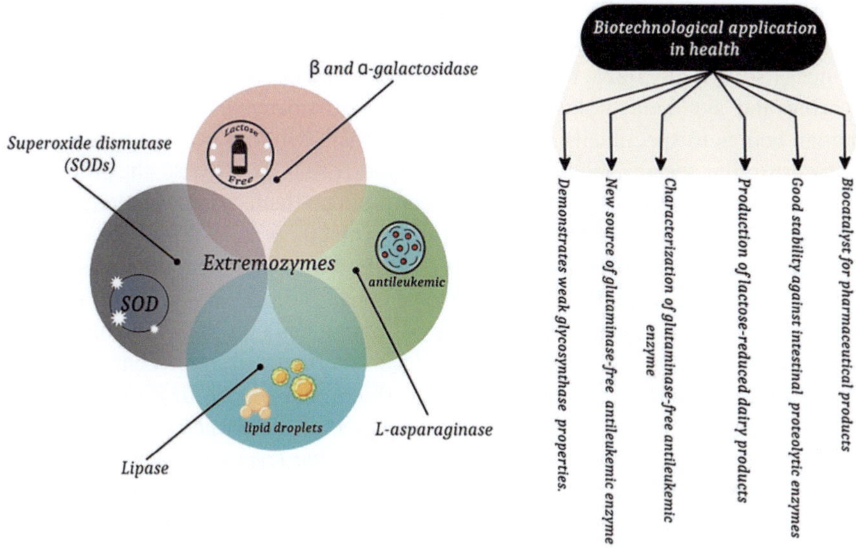

Fig. 9.2 Main pharmaceutical applications of enzymes from polar microorganisms

determine disease-specific markers and new diagnostic targets, discover functional alterations in the pathophysiology of various diseases, and uncover the relationship between the gut microbiota and host metabolisms, for example (Zhao et al. 2024). Although each of these areas provides valuable *insights*, they represent only parts of the complex pathology of disease that need to be investigated further.

9.5 Genetically Modified Organisms

Genetically modified organisms (GMOs) are characterized as any living organism that has undergone some process of artificial genetic remodeling through genetic engineering methodologies—such as insertion, deletion, mutation, activation, deactivation, amplification, etc.—resulting in one or more recombinant DNA (rDNA) molecules (Ferreira and Terra 2017; Ultchak 2018; Silva et al. 2022a, b). Although often considered synonymous, transgenic organisms (TO) are characterized as a subgroup of GMOs, in which the organisms have necessarily undergone some genetic alteration that has introduced genetic characteristics of another species of organism into their DNA, while the definition of GMOs encompasses genetic alterations with or without the participation of other species (Ferreira and Terra 2017; Silva et al. 2022a, b).

In order to obtain an organism with certain favorable properties, human beings have, for around ten millennia, used artificial selection of plants and animals to acquire species that would grow and develop under the best conditions and with the greatest productivity, such as larger animals and plants with greater resistance to pests, almost always empirically and with a long wait to observe the results of their choices (Silva et al. 2022a, b). With the discovery and development of genetic engineering and synthetic biology, this panorama has changed drastically, allowing human beings to specifically alter certain organisms without necessarily using the same or similar species, leading to the modulation of unique aspects, the production of substances by beings that initially did not produce them, and even alterations to the metabolic pathways of a living being (Silva et al. 2022a, b; Rodrigues et al. 2023).

Due to ethical and biosafety issues, GMOs and their production have been and are the subject of many discussions in the political, biological, and health spheres, which have led to national and international agreements on their modulation. An example of this is the Cartagena Protocol, an international agreement signed by Brazil in 2003, referring to the transfer, handling, and use of GMOs, in order to maintain their use within ethical and safe standards for society and the environment. Furthermore, at the national level, in order to provide legal provisions on GMOs, Law 11.105/2005 was enacted in Brazil, providing the new Brazilian Legal Framework on GMOs, which aims precisely to establish rules and conducts to promote safety and oversight on the most diverse aspects of GMOs (Brasil 2010).

Although they are best known for their applications in agriculture to produce more resistant and productive plants, GMOs have applications in a wide range of areas, with a great increase in the use of microorganisms, such as fungi and bacteria, which can be applied to the development of substances of interest, treatment of pathologies, plant protection, increased soil fertility, remediation of environments contaminated by heavy metals, etc. (Rebello et al. 2021; Rafeeq et al. 2023; Yan et al. 2023).

Currently, research into genetically altered microorganisms (mOGM), such as fungi and bacteria, has become increasingly popular, especially with regard to their potential biomedical applicability in the diagnosis and treatment of human diseases (Yan et al. 2023). An example of this is the spectrum of variations and genetic studies on *Escherichia coli*, a bacterium with a broad spectrum of action, which is already used in the treatment of diarrhea, ulcerative colitis, and acute lymphocytic leukemia, among other pathologies (Medawar et al. 2020; Yan et al. 2023). Research includes the role of *E. coli* in the potential treatment of phenylketonuria, colorectal cancer, liver diseases, tumors, etc., as well as in the diagnosis of diabetes, inflammatory bowel disease, liver metastases, etc. (Yu et al. 2020; Yan et al. 2023). One of the great benefits of genetic engineering applied to bacteria, such as *E. coli*, is the increased expression of certain characteristics, as well as the reduction of unwanted side effects produced by their use, as is the case with the application of L-asparaginase in the treatment of acute lymphocytic leukemia, as well as the use of the *E. coli* Nissle 1917 variant in the targeted treatment of tumors (Medawar et al. 2020; Yu et al. 2020).

In addition to direct health applications, mOGMs are also potentially applied in agriculture and soil management to remove pesticides, xenobiotics, and heavy metals that can contaminate food and cause harm to both the environment and consumers (Rebello et al. 2021). Among the organisms studied, there are microbial assemblages that include fungi and bacteria, as well as isolated microorganisms, such as *Caulobacter crescentus*, *Apostichopus japonicus*, and *Streptomyces coelicolor*, bacteria applied in this context (Rebello et al. 2021).

The potential application of GMOs is more than clear for the most diverse areas of knowledge, but it must be kept in mind that there are serious possible consequences of their use and study, and these possible outcomes must be understood and controlled. Among these various aspects are the ecological impact and alterations to wild populations, unintended deleterious effects, allergenicity and toxicity, the development of resistant species and uncontrollable pests, ethical, health and environmental issues, as well as the loss of genetic diversity due to extreme artificial control (Rodrigues et al. 2023). Therefore, the debate on GMOs must always be pleaded for in the most diverse spheres of discussion, in order to provide a balance between safety, legality, technological development, and human and environmental health.

9.6 Conclusion

Enzymes derived from microorganisms in cold environments could be the key to finding new therapeutic compounds. Developments in microbial genetics and physiology have had a profound impact on enzyme production technologies, and this is an important area in the engineering of biotechnological processes. However, it is of great value to have ideal tools and techniques to interpret the numerous data currently available. In this sense, new studies should be carried out in the search for enzymes from polar microorganisms, as these can offer versatile tools for the development of new biopharmaceuticals. It is imperative that studies continue to be carried out in the field of enzymology and in the search for microorganisms that can offer competent alternatives to pharmaceutical therapy. Finally, it is thought that the enzymes mentioned here, from cold-adapted microorganisms, could be an emerging class of therapeutics in the future.

References

Abdel-Monsef MM, Darwish DA, Zidan HA, Hamed AA, Ibrahim MA (2023) Characterization, antimicrobial and antitumor activity of superoxide dismutase extracted from *Egyptian honeybee* venom (*Apis mellifera lamarckii*). J Genet Eng Biotechnol 21:21. https://doi.org/10.1186/s43141-023-00470-4

Abrashev R, Feller G, Kostadinova N, Krumova E, Alexieva Z, Gerginova M et al (2016) Production, purification, and characterization of a novel cold-active superoxide dismutase from the Antarctic strain *Aspergillus glaucus* 363. Fungal Biol 120:679–689. https://doi.org/10.1016/j.funbio.2016.03.002

Aguiar-Pulido V, Huang W, Suarez-Ulloa V, Cickovski T, Mathee K, Narasimhan G (2016) Metagenomics, metatranscriptomics, and metabolomics approaches for microbiome analysis:

supplementary issue: bioinformatics methods and applications for big metagenomics data. Evol Bioinforma:12s1. https://doi.org/10.4137/EBO.S36436

Andrade GAK, Vargas MVM, Goulart SNB, Bernandes BM, Bezerra JDP, Lemos RPM, Victoria FC, Albuquerque MP (2023) Screening of endophytic fungi from Antarctic mosses: Potential production for L-asparaginase free of glutaminase and urease activity. J Biotechnol 377:1–12. https://doi.org/10.1016/j.jbiotec.2023.10.001

Anisha GS (2023) Biopharmaceutical applications of α-galactosidases. Biotechnol Appl Biochem 70:257–267. https://doi.org/10.1002/bab.2349

Ashok A, Doriya K, Rao JV, Qureshi A, Tiwari AK, Kumar DS (2019) Microbes producing l-asparaginase free of glutaminase and urease isolated from extreme locations of Antarctic soil and moss. Sci Rep 9:1423. https://doi.org/10.1038/s41598-018-38094-1

Asraf SS, Gunasekaran P (2010) Current trends of β-galactosidase research and application. In: Current research, technology and education topics in applied microbiology and microbial biotechnology, Microbiology book series. Formatex Research Center, pp 880–890

Bakunina I, Slepchenko L, Anastyuk S, Isakov V, Likhatskaya G, Kim N, Tekutyeva L, Son O, Balabanova L (2018) Characterization of properties and Transglycosylation abilities of recombinant α-galactosidase from cold-adapted marine bacterium *Pseudoalteromonas* KMM 701 and Its C494N and D451A mutants. Mar Drugs 16:349. https://doi.org/10.3390/md16100349

Bangoria P, Patel A, Shah AR (2023) Characterization of a fungal α-galactosidase and its synergistic effect with β-mannanase for hydrolysis of galactomannan. Carbohydr Res 531:108893. https://doi.org/10.1016/j.carres.2023.108893

Beckett A, Gervais D (2019) What makes a good new therapeutic l-asparaginase? World J Microbiol Biotechnol 35:152. https://doi.org/10.1007/s11274-019-2731-9

Bodnár V, Antal K, de Vries RP, Pócsi I, Emri T (2024) *Aspergillus nidulans gfdB*, encoding the hyperosmotic stress protein glycerol-3-phosphate dehydrogenase, disrupts Osmoadaptation in *Aspergillus wentii*. J Fungi 10(4):291. https://doi.org/10.3390/jof10040291

Brasil (2010) Ministério da Saúde. Organização Pan-Americana da Saúde. Marco Legal Brasileiro Sobre Organismos Geneticamente Modificados / Ministério da Saúde, Organização Pan-Americana da Saúde, – Brasília : Ministério da Saúde, 218 p.

Chand S, Mahajan RV, Prasad JP, Sahoo DK, Mihooliya KN, Dhar MS et al (2020) A comprehensive review on microbial l-asparaginase: bioprocessing, characterization, and industrial applications. Biotechnol Appl Biochem 67:619–647. https://doi.org/10.1002/bab.1888

Chen X, Yan D, Yu L, Zhang T (2023) An integrative study of mycobiome in different habitats from a high Arctic region: diversity, distribution, and functional role. J Fungi 9:437. https://doi.org/10.3390/jof9040437

Connell L, Segee B, Redman R, Rodriguez RJ, Staudigel H (2018) Biodiversity and abundance of cultured microfungi from the permanently ice-covered Lake Fryxell, Antarctica. Life 8:37. https://doi.org/10.3390/life8030037

Ding D, Li N, Ge Y, Wu H, Yu J, Qiu W et al (2024) Current status of superoxide dismutase 2 on oral disease progression by supervision of ROS. Biomed Pharmacother 175:116605. https://doi.org/10.1016/j.biopha.2024.116605

Dong L, Chen Y, Gu L, Gan M, Carrier A, Oakes K et al (2023) Oral delivery of a highly stable superoxide dismutase as a skin aging inhibitor. Biomed Pharmacother 164:114878. https://doi.org/10.1016/j.biopha.2023.114878

Doytchinov VV, Dimov SG (2022) Microbial Community Composition of the Antarctic Ecosystems: Review of the Bacteria, Fungi, and Archaea Identified through an NGS-Based Metagenomics Approach. Life 12:916. https://doi.org/10.3390/life12060916

Duarte AWF, Santos JA, Vianna MV, Vieira JMF, Mallagutti VH, Inforsato FJ, Wentzel LCP, Lario LD, Rodrigues A, Pagnocca FC, Junior AP, Sette LD (2018) Cold-adapted enzymes produced by fungi from terrestrial and marine Antarctic environments. Crit Rev Biotechnol 38(4):600–619. https://doi.org/10.1080/07388551.2017.1379468

Duarte AWF, Centurion VB, Oliveira VM (2019) Uncultivated fungi from Antarctica. In: Rosa LH (ed) Fungi of Antarctica. Springer, Cham, pp 9–41. https://doi.org/10.1007/978-3-030-18367-72
Dumina M, Zhgun A, Pokrovskaya M, Aleksandrova S, Zhdanov D, Sokolov N, El'darov M (2021) Highly active Thermophilic L-Asparaginase from *Melioribacter roseus* represents a novel large group of type II bacterial L-Asparaginases from Chlorobi-Ignavibacteriae-Bacteroidetes clade. Int J Mol Sci 22(24):13632. https://doi.org/10.3390/ijms222413632
Dziurzynski M, Gorecki A, Pawłowska J, Istel L, Decewicz P, Golec P, Styczynski M, Poszytek K, Rokowska A, Gorniak D, Dziewit L (2023) Revealing the diversity of bacteria and fungi in the active layer of permafrost at Spitsbergen island (Arctic) – Combining classical microbiology and metabarcoding for ecological and bioprospecting exploration. Sci Total Environ 856:159072. https://doi.org/10.1016/j.scitotenv.2022.159072
Fabris E, Bulfoni M, Nencioni A, Nencioni E (2021) Intra-laboratory validation of alpha-galactosidase activity measurement in dietary supplements. Molecules 26(6):1566. https://doi.org/10.3390/molecules26061566
Ferreira MPR, Terra RBMRB (2017) Organismos geneticamente modificados: por uma obediência aos ditames constitucionais e à legislação consumerista. Rev Dir Da Cid 10.12957/rdc.2017.29188:9
Freire RKB, Mendonça CMN, Ferraro RB, Moguel IS, Tonso A, Lourenço FR, Santos JHPM, Sette LD, Junior AP (2021) Glutaminase-free L-asparaginase production by *Leucosporidium muscorum* isolated from Antarctic marine-sediment. Prep Biochem Biotechnol 51(3):277–288. https://doi.org/10.1080/10826068.2020.1815053
Garcia-Lopez EG, Alcazar P, Cid C (2021) Identification of biomolecules involved in the adaptation to the environment of cold-loving microorganisms and metabolic pathways for their production. Biomolecules 11:1155. https://doi.org/10.3390/biom11081155
Germain DP, Hughes DA, Nicholls K, Bichet DG, Giugliani R, Wilcox WR et al (2016) Treatment of Fabry's disease with the pharmacologic chaperone migalastat. N Engl J Med 375:545–555. https://doi.org/10.1056/NEJMoa1510198
Girard C, Vincent WF, Culley AI (2023) Arctic bacterial diversity and connectivity in the coastal margin of the Last Ice Area. ISME Commun 3:105. https://doi.org/10.1038/s43705-023-00313-w
Godoy CA, Pardo-Tamayo JS, Barbosa O (2022) Microbial lipases and their potential in the production of pharmaceutical building blocks. Int J Mol Sci 23:9933. https://doi.org/10.3390/ijms23179933
Hamid B, Bashir Z, Yatoo AM, Mohiddin F, Majeed N, Bansal M et al (2022) Cold-active enzymes and their potential industrial applications: a review. Molecules 27:5885. https://doi.org/10.3390/molecules27185885
Hamre AG, Al-Sadawi R, Johannesen KM, Bisarro B, Kjendseth ÅR, Leiros H-KS et al (2023) Initial characterization of an iron superoxide dismutase from *Thermobifida fusca*. J Biol Inorg Chem 28:689–698. https://doi.org/10.1007/s00775-023-02019-9
Hindborg H (2023) A pilot study on stain reduction effects of stains, cleaning agents, and cleaning methods on polyester textiles. University of Gothenburg Department of Conservation, p 23
Holmberg SM, Jørgensen NOG (2023) Insights into abundance, adaptation and activity of prokaryotes in arctic and Antarctic environments. Polar Biol 46:381–396. https://doi.org/10.1007/s00300-023-03137-5
Jia B, Park D, Hahn Y, Jeon CO (2020) Metagenomic analysis of the human microbiome reveals the association between the abundance of gut bile salt hydrolases and host health. Gut Microbes 11:1300–1313. https://doi.org/10.1080/19490976.2020.1748261
Jia B, Park D, Chun BH, Hahn Y, Jeon CO (2021) Diet-related alterations of gut bile salt hydrolases determined using a metagenomic analysis of the human microbiome. Int J Mol Sci 22:3652. https://doi.org/10.3390/ijms22073652
Jordan TA, Riley TR, Siddoway CS (2020) The geological history and evolution of West Antarctica. Nat Rev Earth Environ 1:117–133. https://doi.org/10.1038/s43017-019-0013-6
Karasová-Lipovová P, Strnad H, Spiwok V, Malá Š, Králová B, Russell NJ (2003) The cloning, purification and characterisation of a cold-active β-galactosidase from the psychrotolerant

Antarctic bacterium *Arthrobacter* sp. C2-2. Enzym Microb Technol 33:836–844. https://doi.org/10.1016/S0141-0229(03)00211-4

Kim NH, Kim HY, Lee JH, Chang I, Heo S-H, Kim J et al (2023) Superoxide dismutase secreting *Bacillus amyloliquefaciens* spores attenuate pulmonary fibrosis. Biomed Pharmacother 168:115647. https://doi.org/10.1016/j.biopha.2023.115647

Kluge M, Wauthy M, Clemmensen KE, Wurzbacher C, Hawkes JA, Einarsdottir K, Rautio M, Stenlid J, Peura S (2021) Declining fungal diversity in Arctic freshwaters along a permafrost thaw gradient. Glob Change Biol 27:5889–5906. https://doi.org/10.1111/gcb.15852

Kuddus M, Roohi BN, Sheik GB, Joseph B, Hamid B et al (2024) Cold-active microbial enzymes and their biotechnological applications. Microb Biotechnol 17:e14467. https://doi.org/10.1111/1751-7915.14467

Kumar A, Mukhia S, Kumar R (2021) Industrial applications of cold-adapted enzymes: challenges, innovations and future perspective. 3 Biotech 11:426. https://doi.org/10.1007/s13205-021-02929-y

Lailaja VP, Hari V, Sumithra TG, Anusree VN, Suresh G, Sanil NK, Sharma SR, Gopalakrishnan A (2024) In vitro and *in silico* analysis unravelled clinically desirable attributes of *Bacillus altitudinis* L-asparaginase. J Appl Microbiol 135:3. https://doi.org/10.1093/jambio/lxae062

Lauritano C, Rizzo C, Lo Giudice A, Saggiomo M (2020) Physiological and molecular responses to main environmental stressors of microalgae and bacteria in polar marine environments. Microorganisms 8:1957. https://doi.org/10.3390/microorganisms8121957

Lebre PH, Bosch J, Coclet C, Hallas R, Hogg ID, Johnson J et al (2023) Expanding Antarctic biogeography: microbial ecology of Antarctic island soils. Ecography 2023:e06568. https://doi.org/10.1111/ecog.06568

Lee J, Park I, Cho J (2012) Production and partial characterization of α-galactosidase activity from an Antarctic bacterial isolate, *Bacillus* sp. LX-1. Afr J Biotechnol 11:12396–12405. https://doi.org/10.5897/AJB12.1219

Majumder D, Dey A, Ray S, Bhattacharya D, Nag M, Lahiri D (2024) Use of genomics & proteomics in studying lipase producing microorganisms & its application. Food Chem Mol Sci 9:100218. https://doi.org/10.1016/j.fochms.2024.100218

Malard LA, Pearce DA (2018) Microbial diversity and biogeography in Arctic soils. Environ Microbiol Rep 10:611–625. https://doi.org/10.1111/1758-2229.12680

Malcheva B, Nustorova M, Zhiyanski M, Sokolovska M, Yaneva R, Abakumov E (2020) Diversity and activity of microorganisms in Antarctic polar soils. One Ecosyst 5:e51816. https://doi.org/10.3897/oneeco.5.e51816

Mangiagalli M, Lotti M (2021) Cold-active β-galactosidases: insight into cold adaptation mechanisms and biotechnological exploitation. Mar Drugs 19:43. https://doi.org/10.3390/md19010043

Mashamaite L, Lebre PH, Varliero G, Maphosa S, Ortiz M, Hogg ID et al (2023) Microbial diversity in Antarctic Dry Valley soils across an altitudinal gradient. Front Microbiol 14:1203216. https://doi.org/10.3389/fmicb.2023.1203216

Medawar CV, Mosegui GBG, Vianna CMDM, Costa TMAD (2020) PEG-asparaginase and native *Escherichia coli* L-asparaginase in acute lymphoblastic leukemia in children and adolescents: a systematic review. Hematol Transfus Cell Ther 42:54–61. https://doi.org/10.1016/j.htct.2019.01.013

Menezes GCA, Amorim SS, Gonçalves VN, Godinho VM, Simões JC, Rosa CA et al (2019) Diversity, distribution, and ecology of fungi in the seasonal snow of Antarctica. Microorganisms 7:445. https://doi.org/10.3390/microorganisms7100445

Miao BB, Dong W, Gu YX, Han ZF, Luo X, Ke CH, You WW (2023) OmicsSuite: a customized and pipelined suite for analysis and visualization of multi-omics big data. Hortic Res:10. https://doi.org/10.1093/hr/uhad195

Mullish BH, Osborne LS, Marchesi JR, McDonald JA (2018) The implementation of omics technologies in cancer microbiome research. Ecancermedicalscience:12. https://doi.org/10.3332/ecancer.2018.864

Nikitin DA, Sadykova VS, Kuvarina AE, Dakh AG, Biryukov MV (2022) Enzymatic and antimicrobial activities in polar strains of microscopic soil fungi. Dokl Biol Sci 507:380–393. https://doi.org/10.1134/S0012496622060151

Núñez-Montero K, Salazar R, Santos A, Gómez-Espinoza O, Farah S, Troncoso C, Hoffmann C, Melivilu D, Scott F, Barrientos Díaz L (2021) Antarctic *Rahnella inusitata*: a producer of cold-stable β-galactosidase enzymes. Int J Mol Sci 22(8):4144. https://doi.org/10.3390/ijms22084144

Ortiz M, Leung PM, Shelley G, Van Goethem MW, Bay SK, Jordaan K et al (2020) A genome compendium reveals diverse metabolic adaptations of Antarctic soil microorganisms. https://doi.org/10.1101/2020.08.06.239558

Pawlak-Szukalska A, Wanarska M, Popinigis AT, Kur J (2014) A novel cold-active β-d-galactosidase with transglycosylation activity from the Antarctic *Arthrobacter* sp. 32cB – gene cloning, purification and characterization. Process Biochem 49:2122–2133. https://doi.org/10.1016/j.procbio.2014.09.018

Qian YF, Yu JY, Xie J, Yang SP (2023) A mini-review on cold-adapted enzymes from psychrotrophic microorganisms in foods: benefits and challenges. Curr Res Biotechnol 6:100162. https://doi.org/10.1016/j.crbiot.2023.100162

Rafeeq H, Afsheen N, Rafique S, Arshad A, Intisar M, Hussain A et al (2023) Genetically engineered microorganisms for environmental remediation. Chemosphere 310:136751. https://doi.org/10.1016/j.chemosphere.2022.136751

Rasmussen PU, Abrego N, Roslin T, Öpik M, Sepp S, Blanchet FG et al (2022) Elevation and plant species identity jointly shape a diverse arbuscular mycorrhizal fungal community in the High Arctic. New Phytol 236:671–683. https://doi.org/10.1111/nph.18342

Rebello S, Nathan VK, Sindhu R, Binod P, Awasthi MK, Pandey A (2021) Bioengineered microbes for soil health restoration: present status and future. Bioengineered 12:12839–12853. https://doi.org/10.1080/21655979.2021.2004645

Rios NS, Pinheiro BB, Pinheiro MP, Bezerra RM, Santos JCS, Gonçalves LRB (2018) Biotechnological potential of lipases from *Pseudomonas*: sources, properties and applications. Process Biochem 75:99–120. https://doi.org/10.1016/j.procbio.2018.09.003

Rizzo C, Lo Giudice A (2022) Life from a Snowflake: diversity and adaptation of cold-loving bacteria among ice crystals. Crystals 12:312. https://doi.org/10.3390/cryst12030312

Rodrigues RC, Pereira HS, Senra RL, Ribon ADOB, Mendes TADO (2023) Understanding the emerging potential of synthetic biology for food science: Achievements, applications and safety considerations. Food Chem Adv 3:100476. https://doi.org/10.1016/j.focha.2023.100476

Rosa LH, Zani CL, Cantrell CL, Duke SO, Van Dijck P, Desideri A et al (2019) Fungi in Antarctica: diversity, ecology, effects of climate change, and bioprospection for bioactive compounds. In: Rosa LH (ed) Fungi of Antarctica. Springer, Cham, pp 1–17. https://doi.org/10.1007/978-3-030-18367-7_1

Rosa AC, Corsi D, Cavi N, Bruni N, Dosio F (2021) Superoxide dismutase administration: a review of proposed human uses. Molecules 26:1844. https://doi.org/10.3390/molecules26071844

Sajitha S, Vidya J, Varsha K, Binod P, Pandey A (2015) Cloning and expression of l-asparaginase from *E. coli* in eukaryotic expression system. Biochem Eng J 102:2–5. https://doi.org/10.1016/j.bej.2015.02.027

Saleena SK, Johnson JI, Joseph JK, Padinchati KK, Abdulla MHA (2023) Production and optimization of l-asparaginase by *Streptomyces koyangensis* SK4 isolated from Arctic sediment. J Basic Microbiol 63:417–426. https://doi.org/10.1002/jobm.202200116

Sanches PHG, de Melo NC, Porcari AM, de Carvalho LM (2024) Integrating molecular perspectives: strategies for comprehensive multi-omics integrative data analysis and machine learning applications in transcriptomics, proteomics, and metabolomics. Biology 13:848. https://doi.org/10.3390/biology13110848

Sánchez-Moguel I, Yamakawa CK, Brumano LP, Pessoa AJ, Mussatto SI (2022) Selection and optimization of medium components for the efficient production of L-Asparaginase by

Leucosporidium scottii L115—a Psychrotolerant Yeast. Fermentation 8(8):398. https://doi.org/10.3390/fermentation8080398

Sánchez-Moguel I, Costa-Silva TA, Pillaca-Pullo OS, Flores-Santos JC, Freire RKB, Carretero G et al (2023) Antarctic yeasts as a source of L-asparaginase: Characterization of a glutaminase-activity free L-asparaginase from psychrotolerant yeast *Leucosporidium scottii* L115. Process Biochem 129:121–132. https://doi.org/10.1016/j.procbio.2023.03.003

Santos MR, Hirata DB, Angelotti JAF (2022) Lipases: sources of acquisition, ways of production, and recent applications. Catal Res 2:1. https://doi.org/10.21926/cr.2202013

Shankar A, Sharma KK (2022) Fungal secondary metabolites in food and pharmaceuticals in the era of multi- omics. Appl Microbiol Biotechnol 106:3465–3488. https://doi.org/10.1007/s00253-022-11945-8

Shu WS, Huang LN (2022) Microbial diversity in extreme environments. Nat Rev Microbiol 20:219–235. https://doi.org/10.1038/s41579-021-00648-y

Silva TRE, Silva LCF, Queiroz AC, Moreira MSA, Fraga CAC, Menezes GCA et al (2021) Pigments from Antarctic bacteria and their biotechnological applications. Crit Rev Biotechnol 41:809–826. https://doi.org/10.1080/07388551.2021.1888068

Silva DNLD, Nunes GP, Santiago VDS, Freitas FMNDO, Ferrreira JCDS (2022a) Alimentos transgênicos: impactos na saúde humana e ambiental. Res Soc Dev 11:e494111436511. https://doi.org/10.33448/rsd-v11i14.36511

Silva AV, Silva MK, de Oliveira AJ et al (2022b) Phosphate solubilization by Antarctic yeasts isolated from lichens. Arch Microbiol:204. https://doi.org/10.1007/s00203-022-03310-0

Singh RV, Sambyal K (2023) β-galactosidase as an industrial enzyme: production and potential. Chem Pap 77:11–31. https://doi.org/10.1007/s11696-022-02507-3

Turner JH, Lu J, King GJ, Marshall T, Phillips D, Bannister CS (2021) Extreme temperatures in the Antarctic. J Clim 34:2653–2668. https://doi.org/10.1175/JCLI-D-20-0538.1

Ultchak AADMS (2018) Organismos geneticamente modificados: a legalização no Brasil e o desenvolvimento sustentável. R Internac Interdisciplinar Interthesis 15:125–142. https://doi.org/10.5007/1807-1384.2018v15n2p125

Vachher M, Sen A, Kapila R, Nigam A (2021) Microbial therapeutic enzymes: a promising area of biopharmaceuticals. Curr Res Biotechnol 3:195–208. https://doi.org/10.1016/j.crbiot.2021.05.006

Varrella S, Barone G, Tangherlini M, Rastelli E, Dell'Anno A, Corinaldesi C (2021) Diversity, ecological role and biotechnological potential of Antarctic marine fungi. J Fungi 7:391. https://doi.org/10.3390/jof7050391

Varsadiya M, Urich T, Hugelius G, Bárta J (2021) Fungi in permafrost-affected soils of the canadian Arctic: horizon and site-specific keystone taxa revealed by co-occurrence network. Microorganisms 9:1943. https://doi.org/10.3390/microorganisms9091943

Wang B, Zhang C, He Q, Qin H, Liang G, Liu W (2018) Efficient resolution of (R,S)-1-(1-naphthyl) ethylamine by *Candida Antarctica* lipase B in ionic liquids. Mol Catal 448:116–121. https://doi.org/10.1016/j.mcat.2018.01.026

Wani AK, Akhtar N, Sher F, Navarrete AA, Américo-Pinheiro JHP (2022) Microbial adaptation to different environmental conditions: molecular perspective of evolved genetic and cellular systems. Arch Microbiol 204:144. https://doi.org/10.1007/s00203-022-02757-5

Yan X, Liu X, Zhao C, Chen G-Q (2023) Applications of synthetic biology in medical and pharmaceutical fields. Signal Transduct Target Ther 8:199. https://doi.org/10.1038/s41392-023-01440-5

Yang Y, Wang W, Liu K, Zhao J (2022) Immobilization of superoxide dismutase in mesoporous silica and its applications in strengthening the lifespan and healthspan of *Caenorhabditis elegans*. Front Bioeng Biotechnol 10:795620. https://doi.org/10.3389/fbioe.2022.795620

Yu X, Lin C, Yu J, Qi Q, Wang Q (2020) Bioengineered *Escherichia coli* Nissle 1917 for tumour-targeting therapy. Microb Biotechnol 13:629–636. https://doi.org/10.1111/1751-7915.13523

Yusof NA, Hashim NHF, Bharudin I (2021) Cold Adaptation Strategies and the Potential of Psychrophilic Enzymes from the Antarctic Yeast, *Glaciozyma antarctica* PI12. J Fungi (Basel):30. https://doi.org/10.3390/jof7070528

Zhang W, Du F, Wang L, Zhao L, Wang H, Ng T (2015) Hydrolysis of oligosaccharides by a thermostable α-galactosidase from *Termitomyces eurrhizus*. Int J Mol Sci 16:29226–29235. https://doi.org/10.3390/ijms161226159

Zhao Z, Qin J, Qian Y, Huang C, Liu X, Wang N, Li L, Chao Y, Tan B, Zhang N, Qian M, Li D, Liu M, Du B (2024) FFAR2 expressing myeloid-derived suppressor cells drive cancer immunoevasion. J Hematol Oncol 17(1):9. https://doi.org/10.1186/s13045-024-01529-6

Zheng M, Liu Y, Zhang G, Yang Z, Xu W, Chen Q (2023) The applications and mechanisms of superoxide dismutase in medicine, food, and cosmetics. Antioxidants 12:1675. https://doi.org/10.3390/antiox12091675

Zhu J, Li X, Huang M, Zhu H, Tan Y, He X et al (2024) Application of recombinant human superoxide dismutase in radical concurrent chemoradiotherapy for cervical cancer to prevent and treat radiation-induced acute rectal injury: a multicenter, randomized, open-label, prospective trial. Int J Radiat Oncol Biol Phys 120:720–729. https://doi.org/10.1016/j.ijrobp.2024.04.070

Chapter 10
Antarctic Continent as a Natural Reservoir for Prospecting New Antifungal Agents Produced by Filamentous Fungi

Cledna Kaline dos Santos Duarte, Maria Nicolle Pereira da Silva, Mariana dos Santos, Vanessa Daiany Vieira Medeiros, Driele Crispiniano Santos, Adeildo Júnior de Oliveira, Magna Suzana Alexandre Moreira, Aline Cavalcanti de Queiroz, Luiz Henrique Rosa, Michel Rodrigo Zambrano Passarini, and Alysson Wagner Fernandes Duarte

Abstract Fungal infections are a public health problem on the global scene, which is worsening with the increasing resistance to antifungal drugs and the contemporary climate change scenario. These infections continue to be underestimated, with insufficient economic efforts and few classes of effective antifungal drugs, with increasing reports of resistance by fungi such as the genera *Aspergillus*, *Candida*, *Cryptococcus*, *Fusarium*, *Histoplasma*, and *Rhodotorula*, among others. In addition, many mycoses are treated with drugs for an extended period. The need for new antifungal therapies requires the investigation of alternative routes, using natural sources of antifungals and which are economically viable and sustainable. In this context, despite its geographic isolation and peculiar climatic conditions, Antarctica is home to a fungal community that produces antifungal drugs, including *Aspergillus sydowii* (versicone A), *Purpureocillium lilacinum*, *Penicillium*, *Pseudogymnoascus*, *Aspergillus flavus*, *Oidiodendron truncatum* (quetracin), *Chrysosporium* sp., *Cladosporium* sp., *Acrostalagmus luteoalbus* (luteoalbusin A and T988 C), *Geomyces* sp. (Geomycin B), *Penicillium nalgiovense* Laxa (amphotericin B), *Pseudogymnoascus*, *Guehomyces pullulans*, *Metschnikowia australis*, *Cladosporium*

C. K. dos Santos Duarte · M. N. P. da Silva · M. dos Santos · V. D. V. Medeiros · D. C. Santos · A. J. de Oliveira · A. C. de Queiroz · A. W. F. Duarte (✉)
Universidade Federal de Alagoas, Campus Arapiraca, Arapiraca, AL, Brazil
e-mail: alysson.duarte@arapiraca.ufal.br

M. S. A. Moreira
Universidade Federal de Alagoas, Campus AC Simões, Maceió, AL, Brazil

L. H. Rosa
Universidade Federal da Integração Latino Americana, Foz do Iguaçu, PR, Brazil

M. R. Z. Passarini
Universidade Federal de Minas Gerais, Belo Horizonte, MG, Brazil

A. W. Fernandes Duarte et al. (eds.), *Polar Microbes and Climate Change*,
https://doi.org/10.1007/978-3-032-07096-8_10

sp., *Trichoderma polysporum*, and *Penicillium chrysogenum*. In general, fungal extracts have shown activity against pathogenic fungi of global importance, such as *Candida albicans*, *Fusarium oxysporum*, *Paracoccidioides brasiliensis*, *Aspergillus fumigatus*, *Candida krusei*, *Cladosporium sphaerospermum*, and *Cryptococcus neoformans*. More detailed studies on the chemical characterization of the extracts, toxicity, and production optimization are relevant to elevate Antarctic fungi as an alternative antifungal therapy.

Keywords *Candida* · Cold-active proteins · Extremophiles · Psychrophiles · Resistance · Secondary metabolites

10.1 Introduction

Fungi are associated with 6.5 million invasive fungal infections annually and cause more than 3.8 million deaths per year (Denning 2024). These data place fungal infections with a mortality rate similar to that of widely studied diseases such as tuberculosis and malaria (Bongomin et al. 2017). However, fungi continue to be underestimated in the context of human infections, especially when compared to bacterial and viral pathogens, which usually receive more attention and financial resources (Janbon et al. 2019). According to the World Health Organization (2022), species belonging to the genera *Candida*, *Cryptococcus*, *Aspergillus*, *Pneumocystis*, *Fusarium*, and *Paracoccidioides* stand out among emerging pathogens, as they are included in the list of priority fungal pathogens. These organisms can cause anything from superficial infections (skin, hair, nails, eyes) to invasive infections that affect multiple organs and organ systems (Shi et al. 2024; Wang et al. 2024).

Added to that, the increase in antifungal resistance, the limitation of available antifungals, and climate change are exacerbating the problem of Invasive Fungal Infections (IFI). These conditions make treatment difficult and favor the spread of these infections (Minarini et al. 2020). Currently, several pathogenic fungi have shown silent resistance to different classes of antifungal drugs, one example is *Candida glabrata*, which causes candidemia and invasive candidiasis, especially in immunocompromised patients; this species shows worrying resistance to the echinocandins class (Posteraro et al. 2020; Aldejohann et al. 2021). In addition, new emerging species, such as *Candida auris*, have attracted attention due to their multidrug resistance (Carolus et al. 2021; Hong et al. 2023).

Additionally, climate change contributes to the worsening of fungal infections, influencing the modes of transmission of some existing pathogens (Mora et al. 2022). A notable example is *Coccidioides immitis* and *Coccidioides posadasii*, which cause valley fever, a respiratory disease. Rising temperatures have contributed to sandstorms and periods of drought, which in turn intensify cases of the disease, because the winds disperse the endospores present in the dust, facilitating their transmission (Howard et al. 2024). Despite the negative impact that fungi have on human health, treatment options are still limited, partly due to the fact that there are

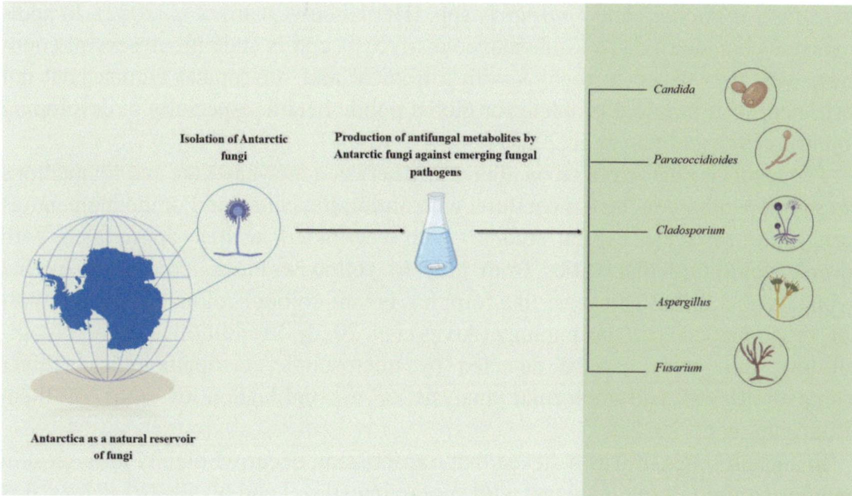

Fig. 10.1 Antarctica as a source of fungi with potential biotechnological activity against emerging pathogenic fungi

only four classes of drugs available on the market, including azoles, echinocandins, polyenes, and the pyrimidine analog 5-flucytosine, which limits the pharmacological options for treating serious infections (Houšť et al. 2020).

Clearly, the search for new antifungals is necessary in order to mitigate the current and growing fungal threats. Thus, filamentous fungi represent an invaluable source for the discovery of new drugs, a legacy that dates back to the twentieth century, with the discovery of *penicillin* (Correia et al. 2023). This importance is due to their vast repertoire of secondary metabolites, characterized by complex structures and a wide diversity of bioactive properties (Lombardi et al. 2023; Iacovelli et al. 2024). In this context, Antarctica functions as a veritable natural biofactory, housing fungi capable of producing unique secondary metabolites in response to adverse environmental conditions (Camacho et al. 2014).

Therefore, this chapter aims to report on Antarctic fungi that produce metabolites with antifungal potential, highlighting known antimicrobial agents, emerging fungi in human health (Fig. 10.1) and the challenges and opportunities related to the bioprospecting of these organisms in the development of new therapeutic treatments.

10.2 Emerging Fungi in Human Health

10.2.1 Sporothrix spp.

Sporotrichosis is an infection caused by members of the *Sporothrix* genus, the species of clinical interest being *Sporothrix schenckii* sensu stricto, *Sporothrix brasiliensis*, *Sporothrix globosa*, *Sporothrix schenckii*, *Sporothrix schenckii* complex,

Sporothrix mexicana, and *Sporothrix* spp. (Hernández-Castro et al. 2022). In addition, it is characterized as a subcutaneous mycosis and is endemic in several countries, with prevalence in regions with a tropical and subtropical climate, but not restricted, resulting in a problem for global public health, especially in developing countries (Schechtman et al. 2022).

The genus is characterized by alternating between the yeast-like and filamentous phases, depending on the temperature, with dimorphic behavior. For the filamentous form, *Sporothrix* can exhibit colonies with a wrinkled, whitish appearance, with reverse coloration that varies from gray to yellowish-black, containing hyphal hyphae and conidia. The yeast-like form has cream-colored colonies, with the typical "cigar-shaped" cell morphology (Alves et al. 2020). In addition to the mycological test, the genus can be detected by microscopic, cytopathological, direct histopathological and molecular analysis or a combination of some of them (Rodrigues et al. 2022).

In ancient times, it was believed that transmission occurred mainly through skin trauma acquired through contact with contaminated soil and plants. Nowadays, it is known that its transmission is more widespread, as it can also occur via zoonotic means (Schechtman et al. 2022). Corroborating its cosmopolitan nature, several studies have reported cases of sporotrichosis in Europe (Monno et al. 2020), Asia (Lv et al. 2022), Africa (Vismer and Hull 1997), North America (Hernández-Castro et al. 2022), and South America (Rabello et al. 2022).

As a representative of South America, Brazil has described numerous cases of human sporotrichosis, with Rio de Janeiro as the main endemic area, but with occurrence in all regions of the country, mainly affecting women (55.9%) and patients aged between 30 and 40 years, where 79.8% of cases reported contact with infected animals, such as domestic cats (79.3%), according to a systematic review (Rabello et al. 2022). In agreement, Morgado et al. (2022) showed a greater number of sporotrichosis in cats and dogs related to South America, especially Brazil and Argentina, with a predominance of *S. brasiliensis* and *S. schenckii*. In addition, cases of animal sporotrichosis have been reported in Asia, Europe, North America, Africa, Central America, and Oceania (Morgado et al. 2022).

In humans, the manifestation of sporotrichosis can occur through the formation of erythematous and papular lesions with close or individualized distribution on the skin. In addition, there are lymphocutaneous and disseminated forms, with the latter occurring in an ulcerative, granulomatous, or crusted form with multiple lesions (Lv et al. 2022). It is important to note that sporotrichosis can also manifest as extracutaneous, related to bone, eye, and lung lesions, through inhalation of conidia, especially in immunocompromised patients (Fichman et al. 2022; Ramírez-Soto et al. 2021; Ramos et al. 2021). In cats, sporotrichosis can manifest as ulcerated and nodular lesions in a disseminated or localized form, with the possibility of fatality if left untreated or if treatment is abandoned (Nakasu et al. 2021).

Successful treatment of sporotrichosis requires medical follow-up and the use of antifungal drugs for between 3 and 6 months, such as itraconazole, fluconazole, ketoconazole, terbinafine, and amphotericin B, in different concentrations. In addition to the long treatment period, these antifungal drugs can result in side effects, hindering the adoption of the correct treatment, and high cost (Sharma et al. 2022). It is also important to note that there are recent studies describing *Sporothrix brasiliensis* isolated from domestic cats that are resistant to itraconazole, with a high minimum inhibitory concentration (MIC) ($\geq$ 4 μg/mL), which highlights the need to search for new efficient therapies (Nakasu et al. 2021; Wall and Lopez-Ribot 2020).

10.2.2 Dermatophytes

Dermatophytes are a group of fungi characterized by their ability to produce keratinases, enzymes responsible for degrading keratin in the skin, hair, nails, and feathers. Their traditional division includes three genera: *Trichophyton*, *Microsporum*, and *Epidermophyton*. They are among the most prevalent causes of animal and human mycoses, and their epidemiology is influenced by various factors, such as geographical and climatic conditions, lifestyle changes, and even changes in the pattern of human interaction with animals (Segal and Elad 2021).

The *Microsporum canis* complex consists of one zoophilic species, *M. canis*, and two anthropophilic species – *Microsporum audouinii* and *Microsporum ferrugineum*. These species are the most widespread zoonotic pathogens causing dermatophytosis in cats and humans. *M. canis* is prevalent in dogs and cats and, when transmitted to humans, causes tinea capitis (which can present an often severe inflammatory response to the host, resulting in pruritic and scaly areas of alopecia), tinea faciei, and tinea corporis. In contrast, tinea capitis caused by the anthropophilic species *M. audouinii* and *M. ferrugineum*, usually transmitted from human to human, shows limited inflammation and minimal symptoms (Zhou et al. 2023). *M. canis*, macroscopically, presents velvety white colonies, with a reverse yellowish pigmentation when cultivated in vitro, on most fungus-specific culture media. Microscopically, it has fusiform, numerous, fast-growing macroconidia with thick, septate cell walls (Da Cunha et al. 2019).

Trichophyton mentagrophytes is the second most common causative agent of dermatophytosis in animals and humans, after *Trichophyton rubrum.* This fungus is characterized morphologically by the development of micro- and macroconidia with smooth walls. The macroconidia originate laterally on the hyphae or on short pedicles with thin or thick walls, fusiform or club-shaped. Microconidia are often abundant, spherical, pyriform, or irregularly shaped. The most consistent characteristic of *T. mentagrophytes* is the production of globose microaluriospores arranged in 45 groups (like a bunch of grapes). The taxonomy of *T. mentagrophytes* is complex due to the changes it has undergone in recent years. Until 2017, the *T. mentagrophytes* series included seven species: *Trichophyton tonsurans*, *T. mentagrophytes*, *Trichophyton interdigitale*, *Trichophyton equinum*, *Trichophyton quinckeanum*, *Trichophyton schoenleinii*, and *Trichophyton simii*. Conventionally, *T. mentagrophytes* is identified on the basis of its macro- and microscopic characteristics and sometimes by its physiological characteristics (capillary perforation and urease activity); however, the results are often uncertain due to phenotypic variations between isolates. In clinical practice, it is common for velvety anthropophilic isolates obtained from the human foot to be referred to as *T. interdigitale*, while powdery isolates, regardless of their origin, are categorized as *T. mentagrophytes*. However, this strategy is insufficient and inadequate to differentiate *T. mentagrophytes* from the other species that can cause infection in humans; moreover, like *T. interdigitale*, it also produces velvety colonies. Therefore, to ensure accurate identification, the use of molecular methods in combination with morphological analysis is recommended (Frías-De-León et al. 2020).

Indian dermatologists have already warned about the alarming increase in extensive, chronic, and difficult-to-treat anthropophilic dermatophytosis due to the *Trichophyton mentagrophytes* species complex – probably due to the widespread misuse of topical antifungals and corticosteroids. Among this species complex, *T. interdigitale* is considered anthropophilic and responsible for tinea pedis, while *T. mentagrophytes* is considered zoophilic and associated with cases of tinea corporis. The dermatophyte responsible for the Indian outbreak was classified as a particular genotype (i.e., genotype VIII) within the *T. mentagrophytes* species complex, and its anthropophilic characteristics (human-to-human transmission without direct contact with infected animals) raised a taxonomic debate, which led to the recent naming of a new species, *Trichophyton indotineae*. Although separating *T. indotineae* from *T. mentagrophytes* as a new species is taxonomically imprecise, the new name is clinically relevant and can help raise awareness. The worrying feature of *T. indotineae* infections is that they are difficult to treat and standard first-line treatment is often not enough to completely eradicate the infection. In some cases, treatment failure can be explained by in vitro resistance of the fungus to terbinafine and/or azoles. There is currently no consensus on the most effective treatment approach for infections caused by *T. indotineae*, including antifungal agent, dose, duration, and infection control measures needed to reduce future transmission (Jabet et al. 2023).

10.2.3 Aspergillus spp.

The genus *Aspergillus* is home to various fungi related to the development of human diseases, also known as aspergillosis. The main species of clinical interest are *Aspergillus fumigatus*, *Aspergillus niger*, and *Aspergillus flavus*, isolated from the respiratory, cardiovascular, and nervous systems (Valério et al. 2022; Memória et al. 2020; Ragozzino et al. 2022). In addition, the genus in question has been isolated from various locations around the world, such as Europe, North America, South America, Asia, Oceania, and Africa (Morrissey et al. 2024; Buyi et al. 2023).

Damage to the respiratory system occurs when *Aspergillus conidia*, when inhaled, adhere to the lung, generating a high inflammatory response (allergic bronchopulmonary aspergillosis), especially in patients with asthma or cystic fibrosis, and chronic infection (chronic pulmonary aspergillosis) and invasive infection (invasive pulmonary aspergillosis) in patients with risk factors, such as those with human immunodeficiency virus (HIV) (Russo et al. 2020). In this context, a multicenter prospective cohort study (Ragozzino et al. 2022) described the prevalence of *Aspergillus* in samples such as sputum (59.3%), tracheobronchial aspirate (23.5%), bronchoalveolar lavage (9.5%), upper respiratory samples (throat or sinuses) (4.5%), and bronchial/lung biopsy (3.3%) obtained from 365 patients in 7 Swiss hospitals.

In addition, among the samples investigated, there was a predominance of *A. fumigatus* (88.8%), *A. niger* (5.0%) and *A. flavus* (3.0%), where chronic lung diseases (49.7%) and cystic fibrosis (28.4%) were the main adjacent conditions related to aspergillosis among hospitalized patients, where 11.7% were diagnosed with invasive pulmonary aspergillosis (Ragozzino et al. 2022).

Infections by *Aspergillus* in the cardiovascular system are considered rare; however, some species have been associated with infective endocarditis (n = 4528 patients), especially by *A. fumigatus* (47.5%) and *A. flavus* (24.6%), and mainly after surgical procedures such as valve prosthesis (34.4%) and solid organ transplantation (SOT) (19.7%), with a mortality rate of 52.5%, with greater aggravation in immunocompromised patients (Valerio et al. 2022). In another perspective, infection can occur through direct inoculation of the fungus, as described by Born et al. (2022), where a patient (39 years old) developed endocarditis probably due to intravenous drug use with contaminated material.

The patient in question underwent an aortic valve procedure 10 years ago due to problems related to *Enterococcus faecalis* endocarditis which, in turn, was replaced and evaluated, confirming the presence of *Aspergillus tubingensis* (Born et al. 2022). Subsequently, studies described cerebral aspergillosis related to *A. fumigatus* (Memória et al. 2020) and *Aspergillus terreus* (Rayón-López et al. 2023). This infection can result in intracranial aneurysms, infarction, meningitis, hemorrhages, and other damage to the central nervous system (Xiao et al. 2012).

In this sense, a recent case report described an immunocompetent patient who presented with a hyperdense mass in the left frontal region, in the left maxillary sinus, and other lesions in the left frontal lobe, with isolation of *A. fumigatus*. The patient studied had no history of previous craniofacial trauma or frequent medication use, although he did have a history of smoking and drug use (Memória et al. 2020). In another case, Rayón-López et al. (2023) isolated *A. terreus* from a pediatric patient with cerebral aspergillosis capable of forming biofilms in vitro, starting in the first 24 h (Early maturation stage), with maximum formation in 72 h (Late maturation stage) and reduced production of the extracellular matrix in 96 h (Dispersion stage), with the greatest release of conidia to enable the cycle to begin.

It is also worth noting the existence of intestinal aspergillosis, related to infection by *Aspergillus* spp. which, according to data from a systematic review by Yelika et al. (2021), affected patients with an average age of 44.9 ± 20.5 years, men and women (29:27), with restriction to the small intestine in 61%, with a mortality rate close to 39% in 56 cases. Similar to other pathogenic microorganisms, there is evidence of resistance related to the genus, especially to antifungals of the Azol class and amphotericin B. Alarmingly, resistance to amphotericin B (AmBR) was predominant in *A. terreus* (36.8%), *A. flavus* (14.9%), *A. niger* (5.2%), and *A.*

fumigatus (2.01%), described by a systematic review of the literature carried out between 2010 and 2020, warning of the importance of more studies on AmBR trends in this genus (Fakhim et al. 2022).

10.2.4 *Candida* spp.

Candidiasis is an infection caused by fungi belonging to the *Candida* genus, especially *Candida albicans*, *Candida auris*, *Candida glabrata*, and *Candida krusei*, which are clinically relevant and have been isolated from various samples found in Central America (Araúz et al. 2018), South America (Riera et al. 2022), North America (Lyman et al. 2023), Asia (Chowdhary et al. 2020), and Africa (Abrantes et al. 2014). This infection can cause damage to the surface of the skin and mucous membranes and has the capacity to spread through the bloodstream (Riad et al. 2021; Taverne-Ghadwal et al. 2022; Bilal et al. 2022). It is important to note that *C. albicans* can be characterized by its dimorphism in relation to the yeast and mycelial version and its ability to form biofilms, just like *C. auris* (Bing et al. 2023), which can function as virulence factors (Sachivkina et al. 2021). For cutaneous candidiasis, the *C. albicans* species is frequently isolated in clinical samples (95.68%), followed by *C. glabrata* (2.59%), *C. krusei* (1.15%), *C. tropicalis* (0.43%), and *C. parapsilosis* (0.14%), as evidenced by a retrospective study carried out by Bilal et al. (2022).

Similarly, *C. albicans* can also be isolated from samples related to human oral mucosa in HIV-positive patients, resulting in the development of a whitish tongue and median rhomboid glossitis, in particular. In addition, the patients investigated, in the absence of antiviral therapy, had a higher prevalence of moderate to severe symptoms and yeast load, presenting itself as an opportunistic fungus (Taverne-Ghadwal et al. 2022). In addition, Hato et al. (2022) described that *C. albicans* was associated with *C. glabrata* in 23 patients with oral candidiasis investigated, with statistically significant associations with the use of dentures (OR, 8.44; 95% CI, 1.99–28.3) and a history of immunosuppressive therapies (OR, 9.20; 95% CI, 1.19–62.0).

In addition, the genus *Candida* can also infect the vaginal mucosa, causing vulvovaginal candidiasis, related to *C. albicans* and *C. krusei*. In this sense, Wu et al. (2020a) demonstrated in a study in mice that the adherence of C. albicans to the formation of biofilms can result in pro-inflammatory responses, which can consequently cause damage to the host's tissue. In addition, the authors identified that biotic biofilms (in mice) showed high resistance to the antifungals nystatin and clotrimazole and also promoted a greater number of persistent cells compared to the formation of abiotic and planktonic biofilms (in microplates), indicating that this is an important factor in the recurrence of the infection (Wu et al. 2020a). Similarly, *C. krusei*, isolated from patients with vulvovaginal candidiasis, was resistant to fluconazole (MIC $\geq$16 μg/mL) (Wu et al. 2020b).

Another important species for the genus is *C. auris*, where, in the United States alone, there was a 95% increase in cases in 2021 compared to 2019, which showed

44% (Lyman et al. 2023). In this sense, some risk factors have been associated with invasive candidiasis, such as broad-spectrum antibiotics (OR, 5.6; 95% CI, 3.6–8.8), blood transfusion (OR, 4.9; 95% CI, 1.5–16.3), *Candida* colonization (OR, 4.7; 95% CI, 1.6–14.3), central venous catheter (OR, 4.7; 95% CI, 2.7–8.1), total parenteral nutrition (OR, 4.6; 95% CI, 3.3–6.3), mechanical ventilation (OR, 4.4; 95% CI, 2.6–7.4), and HIV (OR, 3.9; 95% CI, 1.4–10.3) mainly (Thomas-Rüddel et al. 2022).

In addition, with the Sars-CoV-2 pandemic, it was possible to isolate *C. auris* as a co-infection in patients with severe manifestations by a Brazilian study, which, in turn, warned about the challenges faced by the health system, especially related to the prolonged use of personal protective equipment (PPE), providing favorable conditions for the colonization of medical devices by *C. auris* (Almeida Jr et al. 2021).

In this perspective, isolates of *C. auris* associated with COVID-19 obtained from blood samples of ten patients showed resistance to fluconazole (MIC >32 mg/L), where 40, 30, and 60% of the isolates showed resistance to voriconazole (MIC >2 mg/L), amphotericin B (MIC >2 mg/L), and 5-flucytosine (MIC >32 mg/L), respectively. In addition, 70% showed resistance to multidrugs such as the combination of azoles, amphotericin B, 5-flucytosine, azoles and 5-flucytosine, and azoles and amphotericin B, showing an emergence in relation to resistance in this species (Chowdhary et al. 2020).

10.2.5 Cryptococcus neoformans

Cryptococcus neoformans is a dimorphic fungus associated with lethal meningoencephalitis, especially in immunocompromised individuals such as HIV patients, transplant patients and those affected by hematological diseases. It has unique virulence factors that facilitate its survival and intracellular dissemination, such as the presence of a capsule made up essentially of mucopolysaccharides, 90% of which are glucuronoxylomannans (GXM) (Casadevall et al. 2018). The initial infection, latency, and dissemination of the pathogen are determined by its specific morphological characteristics, such as capsule size, melanin production, and biofilm development, among others (Rathore et al. 2022).

Cryptococcus spp. are widely distributed in nature in a saprophytic form, living in soil, trees, and dried fruit and isolated in bird droppings, especially pigeons. They grow as budding yeast cells (~5 μm in diameter) under nutrient-rich conditions. A characteristic feature of *Cryptococcus* is the polysaccharide capsule, the production of which is stimulated by conditions that mimic human physiology, including elevated CO_2, low iron content, and neutral/alkaline pH. The ability to use various phenolic substrates, including the neurotransmitter dopamine, to produce melanin increases its survival in the host and distinguishes it from other pathogenic yeasts (Bahn et al. 2020).

Cell surface features that support pathogenesis include the deposition of melanin on the cell wall and the attachment of polysaccharides forming a capsule layer. Melanin gives the wall protection against oxidative stress, interfering with the efficacy of antifungal drugs. The polysaccharide capsule, in turn, is regulated by a series of conditions related to the host, including iron deprivation, which impairs its immune response. The still insufficient arsenal of antifungal drugs, coupled with antifungal resistance, limits the effectiveness of therapies for cryptococcosis (Xue et al. 2023).

Bisso et al. (2023) investigated biofilm formation and phospholipase and proteinase production as virulence factors in *C. neoformans* strains, and most of the isolates studied exhibited strong phospholipase and proteinase production when subjected to stress; in addition, all isolates produced biofilms with optical density values ranging from 0.16 to 0.89.

10.2.6 Rhodotorula spp.

While *Cryptococcus* spp. infections have, over the years, attracted a substantial amount of research, infections caused by *Rhodotorula* spp. have not traditionally been highlighted as potential disease-causing agents. However, recent research has revealed an increasing prevalence of this yeast in fungal infections with symptoms and characteristics common to those caused by *Cryptococcus* spp. which can lead to misdiagnosis and inadequate treatment, underlining the need for more sensitive tests capable of distinguishing between closely related yeast species (Dornelles et al. 2023).

The genus *Rhodotorula* is an ancient polyphyletic genus with basidiomycete yeasts that colonize plants, humans, and other mammals. It is a common environmental yeast that can act as an opportunistic pathogen and infect susceptible patients. The genus *Rhodotorula* contains several clinically important species including *Rhodotorula mucilaginosa* and *Rhodotorula glutinis* and *Rhodotorula minuta* (current nomenclature *Cystobasidium minutum*). *Rhodotorula mucilaginosa* is associated with cutaneous fungal infections, especially among immunocompromised individuals, and is an emerging pathogen that causes life-threatening infections. Other *Rhodotorula* species, such as *R. glutinis* and *R. minuta* are responsible for diseases in patients taking corticosteroids or cytotoxic drugs and are associated with meningeal, cutaneous, ocular, and peritoneal diseases (Naik et al. 2024).

In a systematic review, Cespedes et al. (2022) point to the resistance of *Rhodotorula* to fluconazole and caspofungin. The absence or significant reduction of ergosterol in the fungal membrane, as well as its replacement by other sterols, is a potential mechanism of resistance to azoles; some strains have also been shown to be resistant to echinocandins, which reduces the therapeutic options for controlling infections caused by yeast. Hardaha et al. (2023) conducted an observational study in the dermatology department of Bundelkhand Medical College, Sagar, India. During the period between June 2021 and June 2022, they observed a total of 97 cases of superficial mycoses and observed that, of the 61% of cases confirmed for fungemia, non-dermatophyte fungi were isolated in 54% of cases, led by *Aspergillus* sp. and *Candida* sp., and 15% of them are caused by *Rhodotorula*—a yeast considered non-dermatophyte, showing a tendency for dominance of *Rhodotorula* sp. in superficial mycoses.

Caetano et al. (2023), in a systematic review, corroborate the indications of an increase in human fungal infections by *Rhodotorula* spp. They also point out that the species *R. mucilaginosa* is the one most frequently associated with human infection, followed by *R. glutinis* and *R. minuta*.

10.3 Antarctic Environment

Antarctica covers approximately 14 million km^2, with only 0.44% remaining ice (de Menezes et al. 2019). Most of the ice-free areas emerge from the retreat of ice in marine Antarctica and glacial erosion (Lee et al. 2017), and other parts are free because they harbor desert valleys and geothermal environments including marine fumaroles (which generate a contrast of extremely cold and hot environments) (Burton et al. 2016). The continent is almost entirely within the Antarctic polar circle and has no continental shelf connection with another continent, having a coastline surrounded by the Antarctic or Southern Ocean of about 45,317 km, and of this, only 5468 km are ice-free during the summer and less significant portions during the winter (Peck 2018). The extent and unique territorial location of the Antarctic reflects a variety of climates (Singh et al. 2018).

It is considered one of the most extreme environments on the planet. Its southern location enables it to have sunnier days in the summer months, while darkness prevails in the winter and higher altitude (Yusof et al. 2021), with scarce rainfall (Gonzalez-Herrero et al. 2023), and a high rate of strong winds, with 327 km/h already recorded in the coastal area, with different levels of surface temperature per region, but with stable environments at −60 °C. The lowest temperature ever recorded was −93.2 °C in the regions between Dome Argus (4093 m) and Dome Fuji (3810 m), a region more than 1000 km away from the coastal region (https://svs.gsfc.nasa.gov/4126/), with a previous record of −89.6 °C at the Vostok station (Lopes et al. 2020). Although just as extreme in terms of freezing temperatures, in February 2020, the Antarctic Peninsula recorded 18.3 °C at Esperanza Station (Gonzalez-Herrero et al. 2023).

With persistent and frequent superior cold, wind, and dry conditions, they have created unique ecosystems, which have enabled their biological diversity when compared to other ecosystems around the world relatively limited in species and with simple food webs (Fraser et al. 2018), which are almost exclusively formed by microorganisms (Rosa et al. 2019), in different habitats, as a result of natural selection and evolutionary changes reflected in genetic characteristics that shape metabolism and physiology (Cowan et al. 2014), and are considered extremophile microorganisms, which vary according to the different environments they inhabit on the continent. There are terrestrial communities (in soil, rocks, mats, sediments, cryoconite holes, glacial ice, and snow communities), in aquatic environments (inland waters, glacial, thaw lakes, and subglacial), and marine environment, and there are microorganisms that are airborne, which contributes to microbial diversity in different environments (Doytchinov and Dimov 2022).

Among microorganisms, fungi have a diverse biota on the continent with an important role in the recycling and decomposition of nutrients and are well adapted to survive those conditions, even with a direct impact on the typical molecular structure of a fungus and its genome (Ibrar et al. 2020). The synthesis and production of psychrophilic enzymes (enzymes developed by microorganisms that colonize cold environments) is one of the most investigated strategies (Zucconi et al. 2020). In order to survive in cold environments, they adapt by reducing the activation energy through the production of psychrophilic enzymes that maintain a more flexible structure, which makes them more active in carrying out catalysis, allowing chemical reactions to occur more efficiently (Zucconi et al. 2020). The potential of cold-active enzymes and their fungal producers has been evaluated in recent years, revealing a wide range of present and future industrial and biotechnological applications (Kuddus et al. 2024; Hamid et al. 2022; Duarte et al. 2018).

10.4 Antarctic Fungi as Antimicrobial Agents

Several studies have shown that the Antarctic continent is a promising source for the discovery of compounds with antifungal properties, harboring a diversity of fungi capable of metabolizing secondary compounds with potent antifungal properties against clinically relevant pathogens (Table 10.1, Figs. 10.2 and 10.3).

In this context, a number of fungi isolated from soil samples have proved to be valuable sources of bioactive compounds. Cong et al. (2020) evaluated the extract produced by *Aspergillus sydowii* MS-19, which was effective in inhibiting the growth of the pathogenic strains *Fusarium oxysporum* and *Candida albicans*. In addition, the authors tested the compounds obtained individually, and among the four molecules isolated, versicone A stood out for its strong activity against *C. albicans*, with an MIC of 3.91 μg/mL. It is worth noting that this is the first time that versicone A has been extracted from the isolate *Aspergillus sydowii* MS-19.

Similarly, extracts from the genera *Penicillium*, *Pseudogymnoascus*, and *Purpureocillium* were able to inhibit *Paracoccidioides brasiliensis*, with *Purpureocillium lilacinum* extract UFMGCB 1510 having the highest activity, with an MIC of 31.2 μg/mL (Gonçalves et al. 2015). *P. brasiliensis* is of great interest in Latin America, especially in Brazil, as it is a dimorphic fungus responsible for causing systemic mycosis. This pathogen has a high incidence in the country, where it is considered endemic, and the search for drug development remains neglected (Martinez 2017), with pharmacological treatment posing significant challenges, such as the scarcity of therapeutic options, serious side effects, and long periods of pharmacological therapy (Lima et al. 2022). In this scenario, the extract of the fungus UFMGCB 1510 has emerged as a therapeutic alternative to overcome the limitations of conventional treatments. However, further studies are needed to purify the

Table 10.1 Antifungal activity of compounds produced by filamentous fungi isolated from Antarctica against emerging human pathogens

Fungi	Compound	Target pathogen	Evaluation method	CIM	Substrate	References
Chrysosporium sp. HSXSD-11-1	–	*C. albicans*	CIM	75.0 ± 3.9/71.4 ± 0.7	Soil/Fields Peninsula	Shi et al. (2022)
Cladosporium sp. HSXSD-12	–	*C. albicans*	CIM	76.0 ± 3.1	Soil/Fields Peninsula	Shi et al. (2022)
Acrostalagmus luteoalbus CH-6	Luteoalbusin A, T988 C	*C. albicans*	CIM	78.7 ± 1.2/72.3 ± 0.8	Soil/Fields Peninsula	Shi et al. (2022)
Aspergillus sydowii (MS-19)	Versicone A	*Fusarium oxysporum; Candida albicans*	CIM	62.5 μg/mL	Soil/Fields Peninsula	Cong et al. (2020)
Cladosporium sp. 1 UFMGB 13968	–	Cladosporium sphaerospermum CCT 1740	CIM	–	Lakes/Antarctic Peninsula	Ogaki et al. (2020)
Trichoderma polysporum UFMGB 13968	–	Cladosporium sphaerospermum CCT 1740	CIM	–	Lakes/Antarctic Peninsula	Ogaki et al. (2020)
Pseudogymnoascus appendiculatus UFMGB 13023	–	Cladosporium sphaerospermum CCT 1740	CIM	–	Lakes/Antarctic Peninsula	Ogaki et al. (2020)
Pseudogymnoascus destructans UFMGB 13966	–	Cladosporium sphaerospermum CCT 1740	CIM	–	Lakes/Antarctic Peninsula	Ogaki et al. (2020)
Pseudogymnoascus verrucosus UFMGB 13979	–	Cladosporium sphaerospermum CCT 1740	CIM	–	Lakes/Antarctic Peninsula	Ogaki et al. (2020)
Mortierella antarctica UFMGB 13969	–	Cladosporium sphaerospermum CCT 1740	CIM	–	Lakes/Antarctic Peninsula	Ogaki et al. (2020)

(continued)

Table 10.1 (continued)

Fungi	Compound	Target pathogen	Evaluation method	CIM	Substrate	References
Mortierella sp. 1 UFMGB 113012	–	Cladosporium sphaerospermum CCT 1740	CIM	–	Lakes/Antarctic Peninsula	Ogaki et al. (2020)
Mortierella sp. 2 UFMGB 13008	–	Cladosporium sphaerospermum CCT 1740	CIM	–	Lakes/Antarctic Peninsula	Ogaki et al. (2020)
Thelebolus globosus UFMGB 13001	–	Cladosporium sphaerospermum CCT 1740	CIM	–	Lakes/Antarctic Peninsula	Ogaki et al. (2020)
Aspergillus flavus GW31–9	–	*C. albicans*	Disk diffusion	16.5/17.1 mm	Soil/Fildes Peninsula	Ding et al. (2016)
Oidiodendron truncatum GW25–13	Quetracins	*C. albicans*	Disk diffusion	15.9/10.3 mm	Soil/Fildes Peninsula	Ding et al. (2016)
Penicillium nalgiovense Laxa	Amphotericin B	*C. albicans*	CIM	0.125 μg/mL	Soil/Fildes Peninsula	Svahn et al. (2015)
Pseudogymnoascus sp. 2 UFMGCB 5936	–	*C. albicans; C. krusei; Cladosporium sphaerospermum*	CIM	58 ± 20; 29 ± 2; 95 ± 11	Macroalgae/Antarctic Peninsula	Furbino et al. (2014)
Pseudogymnoascus sp. 1 UFMGCB 5941	–	*C. albicans; C. krusei*	CIM	29 ± 13; 61 ± 1	Macroalgae/Antarctic Peninsula	Furbino et al. (2014)
Guehomyces pullulans UFMGCB MH33.1	–	*C. albicans; C. krusei; Cladosporium sphaerospermum*	CIM	41.5 ± 6; 64 ± 0; 29 ± 4	Macroalgae/Antarctic Peninsula	Furbino et al. (2014)
Metschnikowia australis UFMGCB MH47.1	–	*C. albicans; C. krusei*	CIM	34.5 ± 1; 61 ± 4	Macroalgae/Antarctic Peninsula	Furbino et al. (2014)
Dipodascus australiensis (6031)	–	*C. albicans; C. krusei*	CIM	71.0; 41.5	Macroalgae/ Antarctic Peninsula	Furbino et al. (2014)

Aspergillus clavatus IWW 447	–	*C. albicans; C. neoformans*	CIM	128.0 μg/mL; 2048.0 μg/mL	Larsemann lakes/hills, Vestfold hills and McMurdo dry valleys	Brunati et al. (2009)
Penicillium sp. IWW 1054	–	*C. albicans; C. neoformans*	CIM	8.0 μg/mL; 16.0 μg/mL	Larsemann lakes/hills, Vestfold hills and McMurdo dry valleys	Brunati et al. (2009)
Penicillium chrysogenum IWW 1055	–	*A. fumigatus*	CIM	16.0 μg/mL	Larsemann lakes/hills, Vestfold hills and McMurdo dry valleys	Brunati et al. (2009)
Penicillium sp. IWW 1059	–	*C. albicans; C. neoformans*	CIM	16.0 μg/mL; 16.0 μg/mL	Larsemann lakes/hills, Vestfold hills and McMurdo dry valleys	Brunati et al. (2009)
Beauveria sp. IWW 1017	–	*C. neoformans*	CIM	8.0 μg/mL	Larsemann lakes/hills, Vestfold hills and McMurdo dry valleys	Brunati et al. (2009)
Geomyces sp.	Geomycin B	*Aspergillus fumigatus*	CIM	0.86/29.5 μM	Soil/ King George Island's Fildes Peninsula	Li et al. (2008)
Purpureocillium lilacinum UFMGCB 1510	–	*Paracoccidioides brasiliensis*	CIM	31.2 μg/mL	Soil/King George Island	Gonçalves et al. (2015)

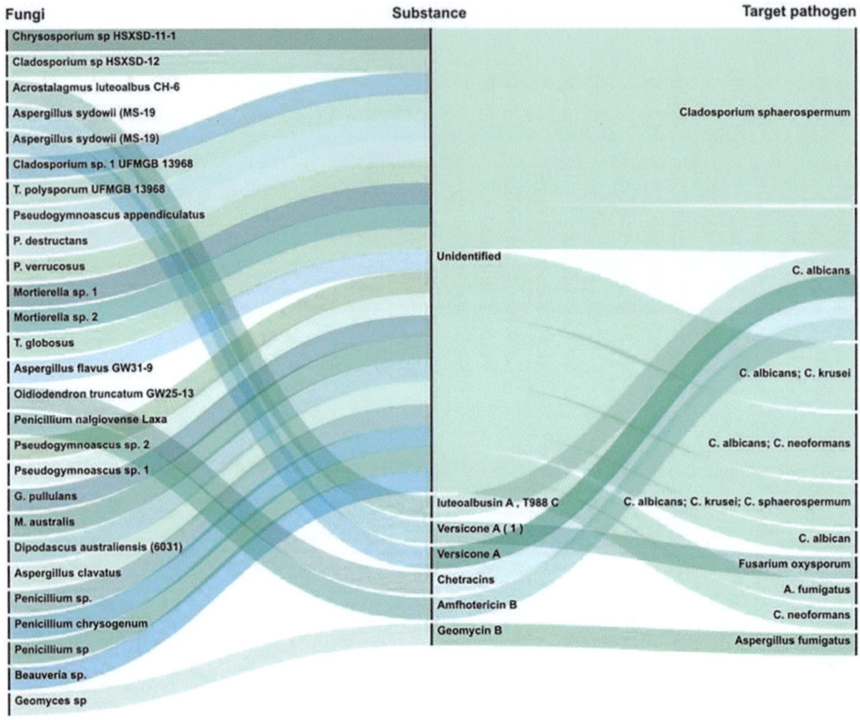

Fig. 10.2 Illustration of the connection between fungi isolated from Antarctica, the bioactive compounds they produce and the pathogenic fungi sensitive to these compounds
***Column 1: Represents the fungi isolated from Antarctica**
***Column 2: Indicates the compounds produced by Antarctic fungi**
***Column 3. Shows emerging pathogenic fungi sensitive to these compounds**

active substance and confirm its efficacy and safety in the management of these infections.

Extracts produced by *Aspergillus flavus* GW31–9 and *Oidiodendron truncatum* GW25-13 revealed an antifungal profile against *C. albicans* using the disk diffusion method and the latter, because it exhibits other different biological properties, was subjected to HPLC-DAD-UV, with the main component being quetracin, which explains its antimicrobial potential (Ding et al. 2016). Additionally, three Antarctic fungi *Chrysosporium* sp. HSXSD-11-1, *Cladosporium* sp. HSXSD-12, and *Acrostalagmus luteoalbus* CH-6 exhibited an antifungal profile against *C. albicans*, with the last isolate standing out, with a more intense activity, surpassing the extracts of the other isolates, inhibiting around 72.3/84.8% of the growth of the aforementioned pathogen. For this reason, the authors carried out a chemical characterization of the CH-6 extract, which resulted in the isolation of several compounds: Acrostalapirone A (1), Acrostalapirone B (2), multiforisin G (3), luteoalbusin A (4), gliocladin C (5), and T988 C (6). They then evaluated the antifungal activity of the

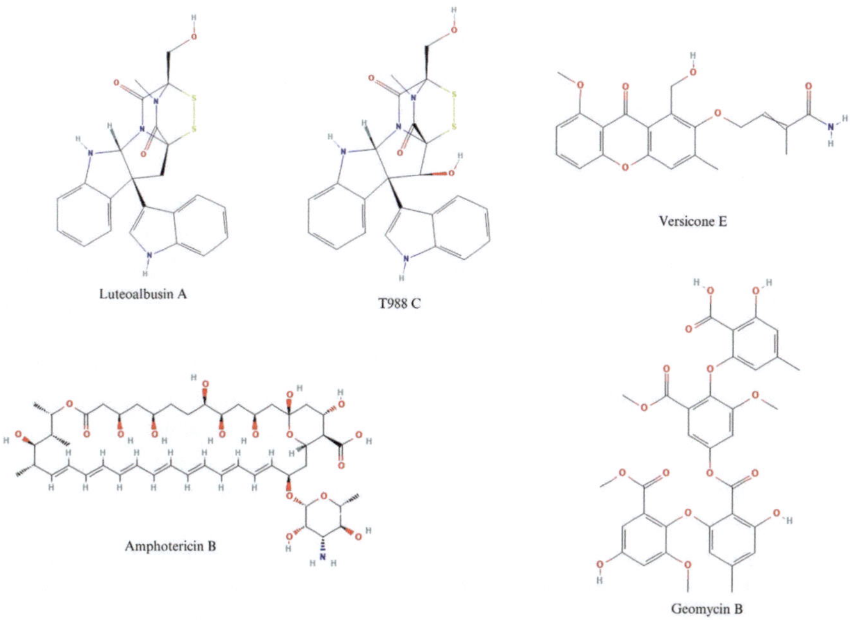

Fig. 10.3 Main bioactive compounds identified in Antarctic fungi with potential against emerging fungal pathogens, with prospects for new antifungal therapies

compounds in isolation, and, among them, compounds 4 and 6 stood out, with MIC values of 12.5 and 6.25, respectively, as shown in Fig. 10.3.

It was found that the interesting *Geomyces* sp., grown in solid fermentation culture, showed the ability to synthesize compounds against the growth of *Aspergillus fumigatus*, with the compound identified as Geomycin B having the most pronounced activity, with MIC values ranging from 0.86/29.5 μM, compared to the positive control; in this specific case, the standard antifungal fluconazole was used as a comparison, showing IC50/MIC values of 7.35/163.4 μM (Li et al. 2008). The results suggest that Geomycin B, in particular, has greater potency, surpassing fluconazole, a drug widely used in fungal infections (Xie et al. 2023).

In the study by Svahn et al. (2015), a new source of the antifungal amphotericin B was reported from the fungus *Penicillium nalgiovense* Laxa. Amphotericin B extracted from this isolate was compared with the reference amphotericin and the antifungal from the Antarctic isolate showed a minimum inhibitory concentration of 0.125 μg/mL, while the reference antifungal exhibited a MIC of 0.250, both tested against *C. albicans*, indicating that the test antifungal was more effective compared to the reference. It is significant to report that this was the first study to isolate amphotericin B from *P. nalgiovense* Laxa, which may be linked to the specific approaches that were used in the research, including cultivation, extraction, and isolation.

Bioactive substances have also been recovered from fungi isolated from Antarctic macroalgae. Extracts from *Pseudogymnoascus* sp. (UFMGCB 5941; UFMGCB 5936), *Guehomyces pullulans* UFMGCB MH33.1 and *Metschnikowia australis* UFMGCB MH47.1 revealed a selective antifungal profile against *C. albicans* and *C. krusei*. At the same time, *Dipodascus australiensis* UFMGCB 6031 selectively suppressed the growth of *C. albicans*, while *G. pullulans*, *M. australis*, and *Pseudogymnoascus* sp. 1 (5941) were selective against *C. krusei*. Extracts of *Pseudogymnoascus* sp. 2 (5936) showed an antifungal effect of 95% against *Cladosporium sphaerospermum*, achieving a performance equivalent to the control drug benomyl, with 94.5% inhibition (Furbino et al. 2014).

C. sphaerospermum is a saprophytic fungus that plays a significant ecological role' however, it can also act as a disease-causing agent. It causes a variety of infections, including respiratory problems such as upper airway infections and pneumonia, as well as subcutaneous infections (Kong et al. 2021; Villanueva et al. 2022). The pathogen in question can also cause brain abscess in immunocompromised patients, further highlighting its ability to seriously affect human health (Batra et al. 2019), which reinforces the need to seek new efficient therapeutic strategies.

Similarly, Antarctic lakes have also proved to be a rich substrate for fungi with compounds of pharmacological interest. The study by Ogaki et al. (2020) on the diversity and bioprospecting of cultivable fungal assemblages in lake sediments on the Antarctic Peninsula reported that 14 extracts reduced the growth of the infectious agent *C. sphaerospermum*, with emphasis on the isolates *Cladosporium* sp. 1 UFMGB 13968 and *T. polysporum* UFMGB 13968, which had the highest values, followed by the extracts of *Pseudogymnoascus appendiculatus* UFMGB 13023, *P. destructans* UFMGB 13966, *P. verrucosus* UFMGB 13979, *M. antarctica* UFMGB 13969, *Mortierella* sp. 1 UFMGB 113012, *Mortierella* sp. 2 UFMGB 13008, and *Thelebolus globosus* UFMGB 13001.

In the same way Brunati et al. (2009) evaluated isolates from lake samples for their antifungal activity against *C. albicans*, *C. neoformans*, and *A. fumigatus*. The genera *Aspergillus* and *Penicillium* stood out as showing the best results. *Aspergillus* extracts were effective against *C. albicans* and *C. neoformans*, but not against *A. fumigatus*. In the case of the isolate *Penicillium chrysogenum* IWW 1055, it showed strong inhibitory action against *C. albicans* and *A. fumigatus,* while *Penicillium chrysogenum* IWW 1053 showed activity against *C. albicans*. Additionally, *Beauveria* sp. IWW 1017 showed activity against *C. neoformans*.

10.5 Known Antimicrobial Agents: A Current Overview

Currently, four classes of antifungal drugs are used in clinical and veterinary medicine, each of which targets different parts of the fungal cell. The polyene class of drugs interacts with ergosterol – the main constituent of the fungal cell membrane. Only three polyenes remain in therapeutic use to this day: amphotericin B, used to treat systemic mycoses; nystatin, applied to mucosal infections such as oral and vulvovaginal candidiasis; and natamycin, indicated for the treatment of ophthalmic infections. In general, polyenes have an unusual mode of action compared to other classes of antifungal drugs, since they do not target a specific enzyme but ergosterol. Natamycin, for example, inhibits several ergosterol-dependent cell membrane proteins, disrupting essential cellular processes such as glucose and amino acid transport (Carolus et al. 2021).

Azoles are widely used as first-line antifungals for the treatment of systemic and topical infections, due to their relatively low cost and efficacy for most fungal infections. They act on the enzyme 14-α lanosterol demethylase, which catalyzes the biosynthesis of ergosterol, resulting in the accumulation of intermediates that are toxic to membrane integrity. Most azoles are fungistatic and have a broad spectrum of activity against yeasts and filamentous fungi, with fluconazole being the most commonly used azole in systemic fungal infections, particularly those caused by yeasts of the genus *Candida* and *Cryptococcus*. Some fungi, including the genus *Aspergillus*, as well as other emerging fungi, show intrinsic resistance to fluconazole; in response, other azol derivatives have been developed, for example, itraconazole and voriconazole. The fungistatic nature of azoles makes the development of resistance during treatment using this class of antifungal drugs a relatively common occurrence, mainly through mutations in the ERG11 target gene, but also because of the overexpression of efflux pumps that transport azole molecules out of the cell. Despite growing resistance, azoles are still some of the most commonly used antifungal drugs for the treatment of a variety of fungal infections (Wall and Lopez-Ribot 2020).

Polyenes, azoles, and flucytosine were the main classes of drugs used to treat fungal infections until the end of the twentieth century. Since then, there has been concern about the side effects related to the hepatotoxicity and nephrotoxicity of these drugs. Nevertheless, many fungal strains have developed resistance to these antifungals, reducing their effectiveness. It is worth noting that in some cases, although rare, cross-resistance to polyenes and azoles can occur, which raises concerns about the future of antifungal drugs that target membrane ergosterol (polyenes) or sterol synthesis (azoles), which has led, and continues to lead, to the search for alternative agents to combat fungal infections. Echinocandins are a class of antifungal drugs that are fungicidal against a range of fungi, including yeasts of the *Candida* genus, but fungistatic for the *Aspergillus* genus, acting by blocking the biosynthesis of β-(1, 3)-D-glucan—an important component in the maintenance of the fungal cell wall, but absent in animal cells—which leads to structural abnormalities in the fungal cell wall, causing inhibition of its growth, or death due to an imbalance in osmotic pressure (Szymański et al. 2022).

Among the four classes of antifungal drugs currently available, echinocandins are a clinically important class of non-ribosomal antifungal lipolipids produced from secondary metabolites of filamentous fungi. Their structure is characterized by numerous hydroxylated non-proteinogenic amino acids, and their chemical synthesis led to the development of the active ingredients of caspofungin, micafungin, and anidulafungin—drugs used as first-line treatment against invasive fungal infections, especially those caused by *Candida* and in hemodynamically unstable patients after previous treatment with triazoles. The molecular target of echinocandins is the enzyme β-(1,3)-D-glucan synthase, which is responsible for synthesizing β-(1,3-D)-glucan, an important component for conferring the integrity of the cell wall of many fungi. Among the four classes of antifungal drugs currently available, echinocandins are a clinically important class of non-ribosomal antifungal lipolipids produced from secondary metabolites of filamentous fungi. Their structure is characterized by numerous hydroxylated non-proteinogenic amino acids, and their chemical synthesis led to the development of the active ingredients of caspofungin, micafungin, and anidulafungin—drugs used as first-line treatment against invasive fungal infections, especially those caused by *Candida* and in hemodynamically unstable patients after previous treatment with triazoles. The molecular target of echinocandins is the enzyme β-(1,3)-D-glucan synthase, which is responsible for synthesizing β-(1,3-D)-glucan, an important component for conferring the integrity of the cell wall of many fungi (Mroczyńska and Brillowska-Dąbrowska 2020).

Flucytosine, also known as 5-fluorocytosine or 5FC, is a family of drugs used to treat systemic fungal infections, especially those caused by yeasts, and is typically effective against *Candida* and *Cryptococcus*. It is a synthetic fluorinated analog of cytosine. Each fungal cell imports 5FC through an energy-dependent process, in which the enzyme cytosine permease plays a key role, from there, 5-flurocytosine is converted into its active form, 5-fluorouracil (5FU) which is then metabolized into 5-fluorouridine triphosphate, which in turn is incorporated into fungal RNA and replaces uridylic acid in its RNA, thus inhibiting protein synthesis. In addition, the conversion of 5 FU into fluorodeoxyuridine monophosphate blocks the action of thymidylate synthase, resulting in the inhibition of DNA synthesis. Both mechanisms are responsible for its antifungal activity. Unlike prokaryotic and fungal cells, human cells lack the cytosine deaminase enzyme that converts 5FC into the active form, and therefore 5FC is safe in humans at pharmacological doses. Despite this, adverse effects can be observed when exposure exceeds 100 μg/ml in plasma or accumulates with renal dysfunction. This toxicity is explained by the discovery that the microflora of the human intestine converts 5FC into 5-FU (Sigera and Denning 2023).

10.6 Challenges and Opportunities

The extreme conditions of Antarctica make the Antarctic environment an extremely rich source for scientific research, especially in the field of microbiology (Duarte et al. 2018). It is a rich place with different ecological niches that can be used as substrates/samples for scientific studies, for example, ice samples, atmospheric air, marine and continental sediments, soil, rocks, minerals, and living organisms (Ayele et al. 2021; Centurion et al. 2022; Lee et al. 2019; Menezes et al. 2019). Since this environment provides a high selectivity of organisms, it may be strategically relevant to study genetic resources in the search for biotechnological demands (Coutinho et al. 2021; Costa et al. 2023), such as antifungals, for example.

However, when taking into account the extreme environmental conditions of this region, it is possible to list some of the difficulties that researchers encounter in this environment. The first important point to consider is the geographical location of Antarctica, as it is a geographically isolated environment, i.e., difficult to access, which requires long journeys (Giesbrecht and Jeffrey 2022; Kuwabara et al. 2021). These require high costs and complex logistics. This also hinders and limits the infrastructure for scientific bases and the resources allocated to them (drinking water and food supplements). These are places that are difficult to access, which requires spending on fuel and transportation suitable for ice/snow. These transports can also malfunction due to unstable terrain or mechanical failures, as the low temperatures can freeze moving parts and damage batteries, electronic instruments, and cameras. This can also hinder communication between groups of researchers in the field, as satellite communication can be unstable due to weather conditions (Petenko et al. 2019). Therefore, access logistics must be strategically planned in order to minimize transport and communication difficulties in Antarctic research.

Among these extreme conditions that make it difficult for researchers to collect, we can also mention the climate, which is characterized by extremely low temperatures and katabatic winds. Antarctica's katabatic winds originate from the high-density cold air that accumulates on the ice, generating great gravitational energy that results in strong winds with high speeds (Grazioli et al. 2017). In addition, storms can arise quickly, causing a sudden change in the climate which can significantly hinder collections. In this way, the climate of Antarctica and its abrupt changes can generate risks to human health, such as dehydration, frostbite, exhaustion, and damage beyond the physical, such as psychological damage (Tortello et al. 2018). In addition, due to the isolated location and the difficulty of ground transportation, medical access is limited. In the event of possible emergencies in this location, teams of scientists have limited access to medical care due to difficulties in getting around, and in the event of an emergency, evacuations can take days.

Another important point is the polar winter, which is characterized by the scarcity of sunlight, with months of total darkness, which hampers navigation and the work of researchers in the field (Petenko et al. 2019). This requires groups of researchers to plan their research schedule so as to avoid this season or to prepare thoroughly in order to face the difficulties and possible adversities that exist in the intense cold of the polar winter, which makes it practically unfeasible to carry out research at this time of year. However, the fundamental pillar of all these difficulties reported here is the lack of funding for research in Antarctica, especially when it comes to Brazilian research in this region, since Brazilian funding for research in Antarctica is outdated and has worsened considerably with the COVID-19 pandemic and the territorial conflicts generated by Russia's war with Ukraine (Kellner 2022).

In addition, Antarctica is protected by the Antarctic Treaty and strict environmental protocols, which is extremely necessary to avoid contamination and environmental impact in this place; however, it should also be noted that it can complicate the collection and storage of samples, and it can also be added, the conflict between geopolitical and scientific interests which further aggravates the situation of scientific funding (Câmara et al. 2021).

10.7 Conclusion

Despite the extreme conditions of the Antarctic environment presenting challenges for the development of life, a diversity of filamentous fungi are not only able to metabolize compounds that aid survival in this habitat but also exhibit potent antifungal activity against different emerging pathogens, including the genera *Candida*, *Aspergillus*, *Paracoccidioides*, *Fusarium*, *Cladosporium*, and *Cryptococcus*. This capacity suggests the presence of a powerful arsenal of bioactive compounds that could offer a hopeful alternative in the development of new pharmacological therapies against threatening pathogenic fungi. However, it is important to note that there are still limitations in the chemical characterization of the compounds obtained from Antarctic fungi, as well as in the description of the mechanism of action, making further studies necessary to elucidate these aspects.

Acknowledgments The authors would like to acknowledge the National Council for Scientific and Technological Development (CNPq—Process #314775/2023-4, #433388/2018-8, and #442258/2018-6) and the Foundation for Research Support of the State Alagoas (FAPEAL - Process #60030.0000000161/2022).

References

Abrantes PMS, Mcarthur CP, Africa CWJ (2014) Multi-drug resistant oral Candida species isolated from HIV-positive patients in South Africa and Cameroon. Diagn Microbiol Infect Dis 79:222–227. https://doi.org/10.1016/j.diagmicrobio.2013.09.016

Aldejohann AM et al (2021) Emergence of resistant Candida Glabrata in Germany. JAC-Antimicrob Resist 3:dlab122. https://doi.org/10.1093/jacamr/dlab122

Almeida JN Jr et al (2021) Emergence of Candida auris in Brazil in a COVID-19 intensive care unit. J Fungi (Basel) 7:220. https://doi.org/10.3390/jof7030220

Alves MM et al (2020) Fatal pulmonary sporotrichosis caused by Sporothrix brasiliensis in Northeast Brazil. PLoS Negl Trop Dis 14:e0008141. https://doi.org/10.1371/journal.pntd.0008141

Araúz AB et al (2018) Isolation of Candida auris from 9 patients in Central America: importance of accurate diagnosis and susceptibility testing. Mycoses 61:44–47. https://doi.org/10.1111/myc.12709

Ayele A et al (2021) Comparative utilization of dead and live fungal biomass for the removal of heavy metal: a concise review. Sci World J 2021:558811. https://doi.org/10.1155/2021/5588111

Bahn YS et al (2020) Microbe profile: Cryptococcus neoformans species complex. Microbiology 166:797–799. https://doi.org/10.1099/mic.0.000973

Batra N et al (2019) Cladosporium sphaerospermum causing brain abscess, a saprophyte turning pathogen: case and review of published reports. J Mycol Med 29:180–184. https://doi.org/10.1016/j.mycmed.2019.04.005

Bilal H et al (2022) Antifungal susceptibility pattern of Candida isolated from cutaneous candidiasis patients in eastern Guangdong region: a retrospective study of the past 10 years. Front Microbiol 13:981181. https://doi.org/10.3389/fmicb.2022.981181

Bing J et al (2023) Clinical isolates of Candida auris with enhanced adherence and biofilm formation due to genomic amplification of ALS4. PLoS Pathog 19:e1011239. https://doi.org/10.1371/journal.ppat.1011239

Bisso BN et al (2023) Biofilm formation and phospholipase and proteinase production in cryptococcus neoformans clinical isolates and susceptibility towards some bioactive natural products. Sci World J 2023:6080489. https://doi.org/10.1155/2023/6080489
Bongomin F et al (2017) Global and multi-national prevalence of fungal diseases – estimate precision. J Fungi 3:57. https://doi.org/10.3390/jof3040057
Born T et al (2022) Aspergillus tubingensis endocarditis: a case report and review of the literature. Mycopathologia 187:249–258. https://doi.org/10.1007/s11046-022-00621-0
Brunati M et al (2009) Diversity and pharmaceutical screening of fungi from benthic mats of Antarctic lakes. Mar Genomics 2:43–50. https://doi.org/10.1016/j.margen.2009.04.002
Burton-Johnson A et al (2016) An automated methodology for differentiating rocks from snow, clouds and sea in Antarctica from Landsat 8 imagery: a new rock outcrop map and area estimate for the entire Antarctic continent. Cryosphere 10:1665–1677. https://doi.org/10.5194/tc-10-1665-2016
Buyi J, Mwambi B, Taremwa IM (2023) Prevalence of pulmonary aspergillosis and drug susceptibility testing amongst adult HIV patients at Kisugu Health Centre III, Makindye Division, in Uganda. SJHR Africa 4:9. https://doi.org/10.51168/sjhrafrica.v4i3.319
Caetano CP et al (2023) The role of yeasts in human health: a review. Life 13:924–924. https://doi.org/10.3390/life13040924
Camacho KF et al (2014) Antarctic marine sediment as a source of filamentous fungi-derived antimicrobial and antitumor compounds of pharmaceutical interest. Extremophiles 28:21. https://doi.org/10.1007/s00792-024-01339-1
Câmara PEAS et al (2021) Brazil in Antarctica: 40 years of science. Antarct Sci 33:30–38. https://doi.org/10.1017/S0954102020000449
Carolus H et al (2021) Genome-wide analysis of experimentally evolved Candida auris reveals multiple novel mechanisms of multidrug resistance. MBio 12:10–1128. https://doi.org/10.1128/mbio.03333-20
Casadevall A et al (2018) The capsule of Cryptococcus neoformans. Virulence 10:822–831. https://doi.org/10.1080/21505594.2018.1431087
Centurion VB et al (2022) Comparing resistome profiles from anthropogenically impacted and non-impacted areas of two South Shetland Islands–Maritime Antarctica. Environ Pollut 304:119219. https://doi.org/10.1016/j.envpol.2022.119219
Cespedes LD et al (2022) Rhodotorula an emergent pathogen. J Exp Clin Microbiol 6
Chowdhary A et al (2020) Multidrug-resistant Candida auris infections in critically ill coronavirus disease patients, India, April–July 2020. Emerg Infect Dis 26:2694. https://doi.org/10.3201/eid2611.203504
Cong B et al (2020) Diversity of cultivable microbes from soil of the Fildes Peninsula, Antarctica, and their potential application. Front Microbiol 11:570836. https://doi.org/10.3389/fmicb.2020.570836
Correia J et al (2023) Beyond penicillin: the potential of filamentous fungi for drug discovery in the age of antibiotic resistance. Antibiotics 12:1250. https://doi.org/10.3390/antibiotics12081250
Costa V et al (2023) Diferenciação de ambientes de sedimentação glaciais e vulcânicos na ilha Deception (Arquipélago Shetland do Sul, Antártica). Rev Bras Geomorfol 24:3. https://doi.org/10.20502/rbgeomorfologia.v24i3.2274
Coutinho JOPA et al (2021) In vitro and in vivo evaluation of the probiotic potential of Antarctic yeasts. Probiotics Antimicrob Proteins 13:1338–1354. https://doi.org/10.1007/s12602-021-09758-8
Cowan DA et al (2014) Microbial ecology and biogeochemistry of continental Antarctic soils. Front Microbiol 5:154. https://doi.org/10.3389/fmicb.2014.00154
Da Cunha MM et al (2019) Epidemiological investigation and molecular typing of dermatophytosis caused by Microsporum canis in dogs and cats. Prev Vet Med 167:39–45. https://doi.org/10.1016/j.prevetmed.2019.03.019
De Menezes GC et al (2019) Fungi in snow and glacial ice of Antarctica. In: Rosa L (ed) Fungi of Antarctica. Springer, Cham, pp 127–146

Denning DW (2024) Global incidence and mortality of severe fungal disease. Lancet Infect Dis 24:e269. https://doi.org/10.1016/S1473-3099(24)00103-8
Ding Z et al (2016) Richness and bioactivity of culturable soil fungi from the Fildes Peninsula, Antarctica. Extremophiles 20:425–435. https://doi.org/10.1007/s00792-016-0833-y
Dornelles G et al (2023) Comparative analysis of capsular and secreted polysaccharides produced by Rhodotorula mucilaginosa and Cryptococcus neoformans. J Fungi 9:1124–1124. https://doi.org/10.3390/jof9111124
Doytchinov VV, Dimov SG (2022) Microbial community composition of the Antarctic ecosystems: review of the bacteria, fungi, and archaea identified through an NGS-based metagenomics approach. Life 12:916. https://doi.org/10.3390/life12060916
Duarte AWF et al (2018) Cold-adapted enzymes produced by fungi from terrestrial and marine Antarctic environments. Crit Rev Biotechnol 38:600–619. https://doi.org/10.1080/07388551.2017.1379468
Fakhim H et al (2022) Trends in the prevalence of amphotericin B-resistance (AmBR) among clinical isolates of Aspergillus species. Med Mycol J 32:101310. https://doi.org/10.1016/j.mycmed.2022.101310
Fichman V et al (2022) Pulmonary sporotrichosis caused by Sporothrix brasiliensis: a 22-year, single-center, retrospective cohort study. J. Fungus 8:536. https://doi.org/10.3390/jof8050536
Fraser CI et al (2018) Antarctica's ecological isolation will be broken by storm-driven dispersal and warming. Nat Clim Chang 8:704–708. https://doi.org/10.1038/s41558-018-0209-7
Frías-De-León MG et al (2020) Molecular identification of isolates of the Trichophyton mentagrophytes complex. Int J Med Sci 17:45–52. https://doi.org/10.7150/ijms.35173
Furbino LE et al (2014) Diversity patterns, ecology and biological activities of fungal communities associated with the endemic macroalgae across the Antarctic Peninsula. Microb Ecol 67:775–787. https://doi.org/10.1007/s00248-014-0374-9
Giesbrecht GG, Jeffrey RB (2022) Death after crevasse rescue in Antarctica. Wilderness Environ Med 33:239–244. https://doi.org/10.1016/j.wem.2022.02.0
Gonçalves N et al (2015) Antibacterial, antifungal and antiprotozoal activities of fungal communities present in different substrates from Antarctica. Polar Biol 38:1143–1152. https://doi.org/10.1007/s00300-015-1672-5
Gonzalez-Herrero S et al (2023) Extreme precipitation records in Antarctica. Int J Climatol 43:3125–3138. https://doi.org/10.1002/joc.8020
Grazioli J et al (2017) Katabatic winds diminish precipitation contribution to the Antarctic ice mass balance. Proc Natl Acad Sci 114:10858–10863. https://doi.org/10.1073/pnas.1707633114
Hamid B et al (2022) Cold-active enzymes and their potential industrial applications – a review. Molecules 27:5885. https://doi.org/10.3390/molecules27185885
Hardaha A et al (2023) Rhodotorula: changing trend of superficial mycosis in Central India. Int J Life Sci Biotechnol Pharm Res 12
Hato H et al (2022) Factor associated with oral candidiasis caused by co-infection of Candida albicans and Candida glabrata: a retrospective study. J Dent Sci 17:1458–1461. https://doi.org/10.1016/j.jds.2021.10.020
Hernández-Castro R et al (2022) Epidemiology of clinical sporotrichosis in the Americas in the last ten years. J Fungus 8:588. https://doi.org/10.3390/jof8060588
Hong H et al (2023) Candida auris infection; diagnosis, and resistance mechanism using high-throughput sequencing technology: a case report and literature review. Front Cell Infect Microbiol 13:1211626. https://doi.org/10.3389/fcimb.2023.1211626
Houšť J, Spížek J, Havlíček V (2020) Antifungal drugs. Meta 10:106. https://doi.org/10.3390/metabo10030106
Howard M et al (2024) Valley fever under a changing climate in the United States. Environ Int 193:109066. https://doi.org/10.1016/j.envint.2024.109066
Iacovelli R et al (2024) Genome sequencing and molecular networking analysis of the wild fungus Anthostomella pinea reveal its ability to produce a diverse range of secondary metabolites. Fungal Biol Biotechnol 11:1. https://doi.org/10.1186/s40694-023-00170-1

Ibrar M et al (2020) Fungi from the extremes of life: an untapped treasure for bioactive compounds. Appl Microbiol Biotechnol 104:2777–2801. https://doi.org/10.1007/s00253-020-10399-0
Jabet A et al (2023) Trichophyton indotineae, from epidemiology to therapeutic. J Med Mycol 33:101383. https://doi.org/10.1016/j.mycmed.2023.101383
Janbon G et al (2019) Studying fungal pathogens of humans and fungal infections: fungal diversity and diversity of approaches. Microbes Infect 21:237–245. https://doi.org/10.1016/j.micinf.2019.06.011
Kellner AWA (2022) Research in Antarctica–challenging but necessary. An Acad Bras Cienc 94:e202294S1. https://doi.org/10.1590/0001-37652022202294S1
Kong QT et al (2021) Subcutaneous infection caused by Cladosporium sphaerospermum: a case report. Mycopathologia 186:135–136. https://doi.org/10.1007/s11046-020-00503-3
Kuddus M et al (2024) Cold-active microbial enzymes and their biotechnological applications. Microb Biotechnol 17:e14467. https://doi.org/10.1590/0001-37652022202294S1
Kuwabara T et al (2021) Human change and adaptation in Antarctica: Psychological research on Antarctic wintering-over at Syowa station. Int J Circumpolar Health 80:18867042. https://doi.org/10.1080/22423982.2021.1886704
Lee JR et al (2017) Climate change drives expansion of Antarctic ice-free habitat. Nature 547:49–54. https://doi.org/10.1038/nature22996
Lee CK et al (2019) Biotic interactions are an unexpected yet critical control on the complexity of an abiotically driven polar ecosystem. Commun Biol 2:62. https://doi.org/10.1038/s42003-018-0274-5
Li Y et al (2008) Bioactive asterric acid derivatives from the Antarctic ascomycete fungus Geomyces sp. J Nat Prod 71:1643–1646. https://doi.org/10.1021/np8003003
Lima RM et al (2022) A structure-based approach for the discovery of inhibitors against methylcitrate synthase of Paracoccidioides lutzii. J Biomol Struct Dyn 40:9361–9373. https://doi.org/10.1080/07391102.2021.1930584
Lombardi A et al (2023) Phenolic-rich extracts from circular economy: chemical profile and activity against filamentous Fungi and dermatophytes. Molecules 28:4374. https://doi.org/10.3390/molecules28114374
Lopes D et al (2020) Estudos geográficos sobre a Antártica: geopolítica, solos, relevo e clima. Revista Ponto de Vista 9:2–17. https://doi.org/10.47328/rpv.v9i1.10237
Lv S et al (2022) Clinical epidemiology of sporotrichosis in Jilin province, China (1990–2019): a series of 4969 cases. Infect Drug Resist 15:1753–1765. https://doi.org/10.2147/IDR.S354380
Lyman M et al (2023) Worsening spread of Candida auris in the United States, 2019 to 2021. Ann Intern Med 176:489–495. https://doi.org/10.7326/M22-3469
Martinez R (2017) New trends in paracoccidioidomycosis epidemiology. J Fungi 3:1. https://doi.org/10.3390/jof3010001
Memória JRJR et al (2020) Brain aspergilloma in an immunocompetent individual: a case report. Surg Neurol Int 11:211. 10.25259/SNI_321_2020
Minarini LADR et al (2020) Antimicrobial resistance as a global public health problem: how can we address it? Front Public Health 8:61284. https://doi.org/10.3389/fpubh.2020.612844
Monno R et al (2020) Infection caused by Sporothrix schenckii: an autochthonous case in Bari, Southern Italy. Eur J Clin Microbiol 39:2457–2460. https://doi.org/10.1007/s10096-020-03939-z
Mora C et al (2022) Over half of known human pathogenic diseases can be aggravated by climate change. Nat Clim Chang 12:869–875. https://doi.org/10.1038/s41558-022-01426-1
Morgado DS et al (2022) Global distribution of animal sporotrichosis: a systematic review of Sporothrix sp. identified using molecular tools. Curr Res Microb Sci 3:100140. https://doi.org/10.1016/j.crmicr.2022.100140
Morrissey CO et al (2024) Aspergillus fumigatus – a systematic review to inform the World Health Organization priority list of fungal pathogens. Med Mycol 62:myad129. https://doi.org/10.1093/mmy/myad129

Mroczyńska M, Brillowska-Dąbrowska A (2020) Review on Current Status of Echinocandins Use. Antibiotics (Basel) 9:227. https://doi.org/10.3390/antibiotics9050227

Naik B et al (2024) Fungal coexistence in the skin mycobiome: a study involving Malassezia, Candida, and Rhodotorula. AMB Express 14:20. https://doi.org/10.1186/s13568-024-01674-8

Nakasu CCT et al (2021) Feline sporotrichosis: a case series of itraconazole-resistant Sporothrix brasiliensis infection. Braz J Microbiol 52:163–171. https://doi.org/10.1007/s42770-020-00290-5

Ogaki MB et al (2020) Diversity and bioprospecting of cultivable fungal assemblages in sediments of lakes in the Antarctic Peninsula. Fungal Biol 124:601–611. https://doi.org/10.1016/j.funbio.2020.02.015

Peck LS (2018) Antarctic marine biodiversity: adaptations, environments and responses to change. Oceanogr Mar Biol Annu Rev 56:105–236. https://doi.org/10.1201/9780429454455

Petenko I et al (2019) Stable surface-based turbulent layer during the polar winter at Dome C, Antarctica: Sodar and in situ observations. Boundary-Layer Meteorol 171:101–128. https://doi.org/10.1007/s10546-018-0419-6

Posteraro B et al (2020) Pan-echinocandin-resistant Candida glabrata bloodstream infection complicating COVID-19: a fatal case report. J Fungi 6:163. https://doi.org/10.3390/jof6030163

Rabello VBS et al (2022) The historical burden of sporotrichosis in Brazil: a systematic review of cases reported from 1907 to 2020. Braz J Microbiol 53:231–244. https://doi.org/10.1007/s42770-021-00658-1

Ragozzino S et al (2022) Distribution of Aspergillus species and prevalence of azole resistance in respiratory samples from Swiss tertiary care hospitals. Open Forum Infect Dis 9:ofab638. https://doi.org/10.1093/ofid/ofab638

Ramírez-Soto MC, Tirado-Sánchez A, Bonifaz A (2021) Ocular sporotrichosis. J Fungus 7:951. https://doi.org/10.3390/jof7110951

Ramos V et al (2021) Bone sporotrichosis: 41 cases from a reference hospital in Rio de Janeiro, Brazil. PLoS Negl Trop Dis 15:e0009250. https://doi.org/10.1371/journal.pntd.0009250

Rathore SS et al (2022) A holistic review on Cryptococcus neoformans. Microb Pathog 166:105521. https://doi.org/10.1016/j.micpath.2022.105521

Rayón-López G et al (2023) Lipid-like biofilm from a clinical brain isolate of Aspergillus terreus: quantification, structural characterization and stages of the formation cycle. Mycopathologia 188:35–49. https://doi.org/10.1007/s11046-022-00692-z

Riad A et al (2021) Oral candidiasis of COVID-19 patients: case report and review of evidence. J Cosmet Dermatol 20:1580. https://doi.org/10.1111/jocd.14066

Riera FO et al (2022) Invasive candidiasis: update and current challenges in the management of this mycosis in South America. Antibiotics 11:877. https://doi.org/10.3390/antibiotics11070877

Rodrigues AM et al (2022) Current progress on epidemiology, diagnosis, and treatment of sporotrichosis and their future trends. J Fungus 8:776. https://doi.org/10.3390/jof8080776

Rosa LH et al (2019) Fungi in Antarctica: diversity, ecology, effects of climate change, and bioprospection for bioactive compounds. In: Rosa L (ed) Fungi of Antarctica. Springer, Cham

Russo A et al (2020) Pulmonary aspergillosis: an evolving challenge for diagnosis and treatment. Infect Dis Ther 9:511–524. https://doi.org/10.1007/s40121-020-00315-4

Sachivkina N, Podoprigora I, Bokov D (2021) Morphological characteristics of Candida albicans, Candida krusei, Candida guilliermondii, and Candida glabrata biofilms, and response to farnesol. Vet World 14:1608. https://doi.org/10.14202/vetworld.2021.1608-1614

Schechtman RC et al (2022) Sporotrichosis: hyperendemic by zoonotic transmission, with atypical presentations, hypersensitivity reactions and greater severity. An Bras Dermatol 97:1–13. https://doi.org/10.1016/j.abd.2021.07.003

Segal E, Elad D (2021) Human and zoonotic dermatophytoses: epidemiological aspects. Front Microbiol 12:7135326. https://doi.org/10.3389/fmicb.2021.713532

Sharma B, Sharma AK, Sharma U (2022) Sporotrichosis: a comprehensive review on recent drug-based therapeutics and management. Curr Dermatol Rep 11:110–119. https://doi.org/10.1007/s13671-022-00358-5

Shi T et al (2022) Bioactivity-guided screening of antimicrobial secondary metabolites from Antarctic cultivable fungus Acrostalagmus luteoalbus CH-6 combined with molecular networking. Mar Drugs 20:334. https://doi.org/10.3390/md20050334

Shi L, Feng Y, Shi D (2024) Two cases of superficial fungal infection caused by Non-Albicans Candida species manifest Greenish-black discoloration. Infect Drug Resist 17:665–672. https://doi.org/10.2147/IDR.S438079

Sigera LSM, Denning DW (2023) Flucytosine and its clinical usage. Ther Adv Infect Dis 10:2049936123116138. https://doi.org/10.1177/20499361231161387

Singh J, Singh RP, Khare R (2018) Influence of climate change on Antarctic flora. Pol Sci 18:94–101. https://doi.org/10.1016/j.polar.2018.05.006

Svahn KS et al (2015) Penicillium nalgiovense Laxa isolated from Antarctica is a new source of the antifungal metabolite amphotericin B. Fungal Biol Biotechnol 2:1–8

Szymański M et al (2022) Echinocandins–structure, mechanism of action and use in antifungal therapy. J Enzyme Inhib Med Chem 37:876–894. https://doi.org/10.1080/14756366.2022.2050224

Taverne-Ghadwal L et al (2022) Epidemiology and prevalence of oral candidiasis in HIV patients from chad in the post-HAART era. Front Microbiol 13:844069. https://doi.org/10.3389/fmicb.2022.844069

Thomas-Rüddel DO et al (2022) Risk factors for invasive Candida infection in critically ill patients: a systematic review and meta-analysis. Chest 161:345–335. https://doi.org/10.1016/j.chest.2021.08.081

Tortello C et al (2018) Psychological adaptation to extreme environments: Antarctica as a space analogue. https://doi.org/10.19080/PBSIJ.2018.09.555768

Valerio M et al (2022) Aspergillus endocarditis in the recent years, report of cases of a multicentric national cohort and literature review. Mycoses 65:362–373. https://doi.org/10.1111/myc.13415

Villanueva DM, Venkatesan B, Figueroa N (2022) Cladosporium sphaerospermum as a rare cause of pneumonia. Cureus 14:e26256. https://doi.org/10.7759/cureus.26256

Vismer HF, Hull PR (1997) Prevalence, epidemiology and geographical distribution of Sporothrix schenckii infections in Gauteng, South Africa. Mycopathologia 137:137–143. https://doi.org/10.1023/A:1006830131173

Wall G, Lopez-Ribot JL (2020) Current antimycotics, new prospects, and future approaches to antifungal therapy. Antibiotics 9:445. https://doi.org/10.3390/antibiotics9080445

Wang S et al (2024) The first child with mixed invasive pulmonary Mucor and Aspergillus infection: a case report and literature review. Front Med 11:1387278. https://doi.org/10.3389/fmed.2024.1387278

WHO fungal priority pathogens list to guide research, development and public health action (2022) World Health Organization. Geneva. Licence: CC BY-NC-SA 3.0 IGO. https://www.who.int/publications/i/item/9789240060241. Accessed 21 Nov 2024

Wu X et al (2020a) Biofilm formation of Candida albicans facilitates fungal infiltration and persister cell formation in vaginal candidiasis. Front Microbiol 11:1117. https://doi.org/10.3389/fmicb.2020.01117

Wu Y et al (2020b) Antifungal activity and mode of action of miltefosine against clinical isolates of Candida krusei. Front Microbiol 11:854. https://doi.org/10.3389/fmicb.2020.00854

Xiao A et al (2012) Invasive intracranial aspergillosis spread by the pterygopalatine fossa in an immunocompetent patient. BJID 16:192–195. https://doi.org/10.1016/S1413-8670(12)70305-7

Xie J, Zeng J, Zheng S (2023) The efficacy and safety of fluconazole in preventing invasive fungal infection in very low birth weight infants: a systematic review and meta-analysis. Ital J Pediatr 49:51. https://doi.org/10.1186/s13052-023-01460-5

Xue P et al (2023) Metals and the cell surface of Cryptococcus neoformans. Curr Opin Microbiol 74:102331. https://doi.org/10.1016/j.mib.2023.102331

Yelika SB, Tumati A, Denoya P (2021) Intestinal aspergillosis: systematic review on patterns of clinical presentation and management. Surg Infect 22:326–333. https://doi.org/10.1089/sur.2020.157

Yusof NA, Hashim NHF, Bharudin I (2021) Cold adaptation strategies and the potential of psychrophilic enzymes from the Antarctic Yeast, Glaciozyma Antarctica PI12. J Fungi 7:528. https://doi.org/10.3390/jof70705282021
Zhou X et al (2023) Human adaptation and diversification in the Microsporum canis complex. IMA Fungus 14:14. https://doi.org/10.1186/s43008-023-00120-x
Zucconi L et al (2020) Extracellular enzymes and bioactive compounds from Antarctic terrestrial fungi for bioprospecting. Int J Environ Res Public Health 17:6459. https://doi.org/10.3390/ijerph17186459

Chapter 11
Soil, Climate, and Food Security: Strategies and Sustainable Solutions for Global Nourishment

Susmita Shukla, Shreedhanya Dattatray Patil, G. Pravin, Sparsh Phutela, and Shiv Kant Shukla

Abstract Soil health is necessary for productivity of food systems at global level, ecological balance, and climate harmony. Intensive agricultural practices for many decades have caused degradation of soil leading to food security and sustainable development challenges. This chapter evaluates relationship between soil health, sustainability, and food resilience, focusing on soil's function of carbon sink, nutrient and water cycle, and management. As the global population is expected to reach 10 billion by the year 2050, there is a need for formulated solutions to curb productivity and ecological imbalance. This chapter analyses traditional approaches such as conservation agriculture, crop rotation, agroforestry, and use of biofertilizers for their ability to recover soil organic matter, improve biodiversity, and reduce chemical dependency. At the same time, new technologies that serve as the foundation for innovative ways to maximize resource efficiency, mitigate climate change, and increase crop resistance are beginning to attract attention. These include CRISPR-engineered soil microbes, biochar, precision agriculture, and AI soil management systems. As identified in socio-economic studies, the double dividend of sustainable soil management includes economic sustainability for farmers due to saved costs and premium market access and equitable rural development through the creation of jobs and policy interventions such as India's Soil Health Card scheme. This chapter integrates indigenous knowledge and technology and advocates for a soil stewardship paradigm that requires collaboration among policymakers, farmers, and researchers. Finally, protecting soil health transcends agricultural obligation, representing a basic responsibility towards achieving the Sustainable Development Goals and providing an equitable, nourished, and resilient world for future generations.

S. Shukla (✉) · S. D. Patil · G. Pravin · S. Phutela
Applied Plant Biotechnology Research Lab, Centre for Plant and Environment Biotechnology, Amity Institute of Biotechnology, Amity University, Noida, Uttar Pradesh, India
e-mail: sshukla3@amity.edu; shreedhanya.patil@s.amity.edu; g.pravin@s.amity.edu

S. K. Shukla
Kaushalaya- The Skill University, Ahmedabad, Gujarat, India
e-mail: shivkant.shukla@kaushalyaskilluniversity.ac.in

A. W. Fernandes Duarte et al. (eds.), *Polar Microbes and Climate Change*,
https://doi.org/10.1007/978-3-032-07096-8_11

Keywords Soil health · Sustainability · Conservation agriculture · Biodiversity · Sustainable development goals

11.1 Introduction

Soil serves as an organ between the atmosphere and lithosphere. It is crucial to the Earth as it contains organic life and functions as a carbon-driven world where solid, liquid, and gaseous things critical fishes of food interaction. These interactions occur at several scales and foster the variety of ecosystem services and ecological goods required. Environmentally, soil is used in the complex global food system. It can be considered the basis of all life forms on Earth. Its primary constituents comprise an organic substance, minerals, water, and a bio-domain composed of trillions of microbes. Thus, soil serves as a natural storehouse of water and minerals for the growth of plants. It also assists in augmenting 95% of food crops planted in the Earth, filters 60% of the freshwater on Earth preventing the supply from getting contaminated, and captures thrice the carbon in the environment helping to alleviate climate change. Soil aids in harbouring countless life forms ranging from microscopic fungi and bacteria to insects and worms. Nonetheless, a vast majority of soil combined contributes greatly in assisting in nutrient circulation, carbon capture, and disease mitigation (Lal 2016). Figure 11.1 highlights key soil functions, including carbon sequestration, nutrient cycling, water filtration, biodiversity support, waste decomposition, food production, disease regulation, and structural support, all essential for sustainability and agriculture.

The United Nations estimates that between 2015 and 2030, the global population will grow at an annual rate of 0.96%, and for the 2030 to 2050 period, it will be 0.63% per year. It estimates that the global population is likely to grow from 8

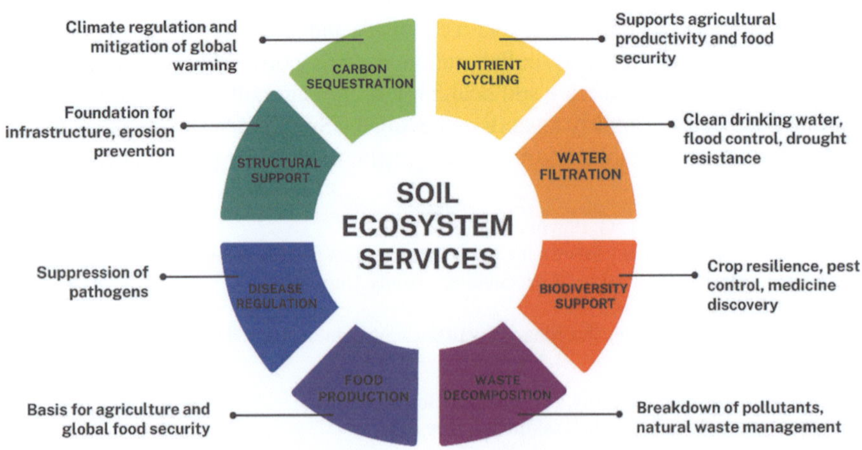

Fig. 11.1 Soil ecosystem services

billion to 10 billion by 2050, which means that food consumption per person is also projected to increase. This increase is expected to take place mostly in low income and lesser developed countries, which have historically struggled more than developed countries with food security issues. The Food and Agriculture Organization (FAO) defines food security as a condition in which every individual has easy access at all times to safe food that is enough to meet their nutritional needs and food choices. Food security is concerned with the imbalance of food availability which is exacerbated by the ever-growing global population. With a predicted growth of 1.7 billion people by 2050, humanity is increasing the strain on already scarce resources available for food production. A model of endlessly growing population and limited resources is not sustainable in the long term. This calls for "resource efficiency" fused with a "circular economy" (Carthy et al. 2022).

While it is true that technology has advanced in the field of agriculture and in society as a whole, soil remains a key component of our food systems that is often overlooked. There is a direct relationship between soil and food. Good soil yields crops, which are necessary for human's nutrition and food security. It is most efficient and effective towards addressing the issue of food security and malnutrition to try to improve the quality of food offered by improving health of the soil and promoting diversification of diet than relying on external amendments. Without the requisite nutrients in the soil, crops grown in those regions will have low nutrient content (Das et al. 2022). According to an estimate, around 2 billion hectares of land are degraded, and an additional 12 million hectares are lost each year. This degradation results in the annual loss of 24 billion tons of fertile soil. Soil degradation adversely affects the well-being of more than 3 billion people, costing over 10% of the annual global GDP due to the loss of ecosystem services and reducing the productivity of 23% of the global terrestrial area (Wang et al. 2023).

Intensive agricultural methodologies have negatively influenced the health of soils on a global level. Monocultural cropping, excessive tilling, and over-fertilization and pesticide applications have all resulted in the reduction of soil structure, organic matter, and the balance of soil ecosystems. The consequences of this degradation is alarming and needs urgent attention—loss of soil fertility translates to lower crop yields and, ultimately, increased area under cultivation, which further accelerates environmental degradation. Soil erosion is detrimental because it depletes the precious topsoil, contaminating water bodies and rendering the nutritious water unable to support vegetation. As a result, soil associated biodiversity is lost further diminishing the soils ability to stand against pests and diseases, meaning greater dependence on chemicals and emission of carbon gases from derelict soils worsens climate change. These contribute to a cycle of food insecurity and environmental deterioration.

Therefore, sustainable management of soil ecosystems is essential for global food, water, and energy security, particularly because of increasingly unpredictable weather patterns caused by climate change. These methods help in outlining the critical synergy between soil health, the environment, and food security. Societies can achieve long-term food security by improving the soil's ability to sequester carbon, retain water, and support crop productivity. This shift towards regenerative practices displays the direct link between soil health and basic sustainability goals such as the ability of ecosystems and communities to exist in a hostile climate and demographic environment. The land-water-food-energy nexus is central to sustainable development, and soil serves as a crucial link between these vital domains (Wang et al. 2023). Additionally, it eliminates opportunities for agricultural production and disrupts the essential services provided by soil ecosystems.

This chapter will look into the newest research, practices, and policy gaps on how to effectively promote integrated efforts for the enhancement of soil health and resilience. We will analyse the 'so what' and the 'now what' of scaling up sustainable soil management practices and use these as a foundation to emphasize the need for collaboration and active participation from farmers, researchers, policy makers, and even consumers.

11.2 Inter-Relationship Between Soil, Sustainability, and Food Resilience

The Sustainable Development Goals (SDGs) are a total of 17 objectives set in 2015 by the United Nations in relation to combating global challenges such as poverty, inequality, climate change, and environmental degradation with the target year of 2030. The SDGs are intended to form an essential part of the new international development paradigm for all countries which will significantly affect national development planning efforts in the middle of the twenty-first century (Allen et al. 2016). Soil sustainability requires ensuring soil resources are used in such a way that they remain healthy, fertile, and productive for years to come. As a result, negative environmental consequences are minimized. It means ensuring that soils are still able to facilitate agricultural endeavours, ecological systems, and human requirements in the long term.

Soil science might not be referred to definitively in the SDGs, but it underpins a lot of them. Soil sustainability is the ability of soils to be kept healthy, fertile, and productive for use by future generations while causing the least harm to the environment. (Keesstra et al. 2016; Tóth et al. 2018). Table 11.1 illustrates important links between sustainable soil management practices and achievement of key United Nations SDGs.

11.2.1 Examining the Relationship of Food Resilience with Soil Health

Resilience and sustainability are two concepts that are closely related. Sustainability is keeping a system at equilibrium which is also a characteristic of a resilient system. Resilience is the ability to sustain critical functions overtime irrespective of

Table 11.1 Comprehensive overview of how soil plays a role in achieving each SDG

S. no.	SDG	Role of sustainable soil practices in achieving SDG	References
SDG 1	No Poverty	Sustainable soil management boosts agricultural productivity, essential for reducing poverty. Healthy soils lead to higher yields and greater resilience, ensuring reliable and consistent income for farmers in developing regions where agriculture is a primary livelihood	Lal et al. (2021)
SDG 2	Zero Hunger	Soil health is essential for food security, with nearly 95% of food production relying on it. Improving soil fertility through sustainable practices enhances food availability and nutritional quality, playing a critical role in eradicating hunger	Bouma and Montanarella (2016), Lal et al. (2021)
SDG 3	Good Health and Well-Being	Good health and well-being are intertwined with soils because healthy soils create healthy crops that in turn sustain humans and animals, allowing for their health and productivity	Lal et al. (2021)
SDG 4	Quality Education	Educating people about soils, especially at basic and non-formal levels, is a powerful way to improve education quality and outreach. It enables a comprehensive approach to tackling environmental challenges	Lal et al. (2021)
SDG 5	Gender Equality	Women play a crucial role in agriculture, often managing family food production. Access to resources and information on sustainable soil practices empowers them and boosts food security	Lal et al. (2021)
SDG 6	Clean Water and Sanitation	Healthy soils filter water, minimizing runoff and preventing contamination from agricultural activities. Sustainable land management strategies assist sustain water quality by limiting erosion and nutrient leaching into waterways	Lal et al. (2021)
SDG 7	Affordable and Clean Energy	Sustainable soil management supports bioenergy production by cultivating energy crops on healthy soils, providing clean energy solutions while preserving soil integrity	Lal et al. (2021), Smith et al. (2021)
SDG 8	Decent work and economic growth	Improving soil health leads to increased agricultural productivity, creating jobs in rural areas and contributing to economic growth through enhanced farm profitability	Lal et al. (2021)
SDG 9	Industry, Innovation and Infrastructure	Innovations in soil science, like precision agriculture and soil carbon sequestration, optimize resource use, enhance soil health, and promote sustainable agricultural practices	Lal et al. (2021)
SDG 10	Reduced Inequalities	Soil degradation harms marginalized communities that rely on agriculture. Improving soil health can help reduce inequalities and provide better access to productive land	Lal et al. (2021)

(continued)

Table 11.1 (continued)

S. no.	SDG	Role of sustainable soil practices in achieving SDG	References
SDG 11	Sustainable Cities and Communities	Urban soils play a crucial role in green infrastructure, supporting community gardens and urban agriculture initiatives that enhance food security while promoting biodiversity	Lal et al. (2021)
SDG 12	Responsible Consumption and Production	Sustainable agricultural practices that focus on improving soil health can reduce the reliance on chemical fertilizers and pesticides, promoting responsible consumption patterns while maintaining ecosystem balance	Lal et al. (2021)
SDG 13	Climate Action	Healthy soils are key carbon sinks that help combat climate change through carbon sequestration. Implementing no-till agriculture and proper grazing management boosts their carbon storage capacity. Furthermore, sustainable land management reduces greenhouse gas emissions from degraded soils	Lal et al. (2021), Lal (2009), Griscom et al. (2017)
SDG 14	Life Below Water	Soil erosion contributes to sedimentation in aquatic ecosystems, harming marine life. By promoting healthy soils, we can reduce runoff and protect water quality in oceans and rivers	Lal et al. (2021)
SDG 15	Life on Land	Soil health is fundamental for maintaining terrestrial ecosystems. Sustainable land management practices help combat desertification, restore degraded lands, and promote biodiversity	Lal et al. (2021), Keesstra et al. (2016)
SDG 16	Peace, Justice, and Strong Institutions	Land disputes often arise from degraded soils that diminish agricultural productivity. Promoting equitable access to healthy soils can foster stability and peace within communities	Lal et al. (2021)
SDG 17	Partnerships for the Goals	Collaboration among governments, NGOs, and private sectors is essential for advancing soil health initiatives globally. Multi-stakeholder partnerships can enhance knowledge sharing and resource mobilization for sustainable land management practices	Bouma and Montanarella (2016) , Biermann et al. (2017)

obstacles. It is true that while sustainability is an indicator of a system's functioning, resilience is how that functioning is achieved during disturbance (Tendall et al. 2015). Soil system resilience is the ability of soil to endure periods of environmental variability. This includes the immune responses of the soil towards biotic invasions and its resistances/tolerances to abiotic stresses, including pollution and the meddling by microorganisms. The relationships among a pathogen, pollutant, and moist soil microbes can be depicted using the disease triangle that soil microbiome is subjected to. Food resilience concerns the ability of food systems to continue being productive, as well as to maintain the availability and nutrition of food during environmental, economic, or social disruptions. To achieve food resilience, we need to

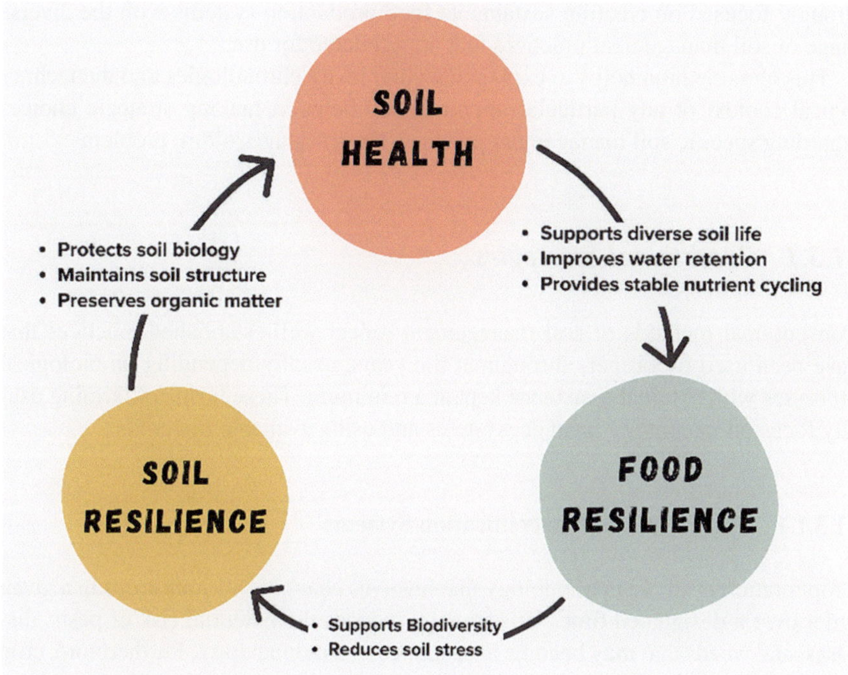

Fig. 11.2 The soil-health-food-resilience-soil resilience cycle

begin with healthy soils. Figure 11.2 represents relationships between soil health, food resilience, and soil resilience. Each component contributes to and reinforces the others, creating a sustainable system for agricultural and ecological balance.

In supporting a varied soil biota, improving water holding capacity and nutrient cycling, all of which are critical and stable, soil health is imperative. This reinforces soil food web resilience, which in turn helps to maintain biodiversity and alleviate stress to soil. Consequently, resilience of soil is improved by safeguarding biologicals of soil, organic matter, and its structural integrity. This cyclic process, which is interrelated in a broader prospective, is beneficial for sustainability of agriculture and stability of the ecosystem.

11.3 Integrated Strategies for Soil Management

The most effective approach to soil management may be to combine elements of all soil management approaches: traditional methods, modern methods, and new technologies. There is a worrying tendency in agriculture towards treating more and more land, as a response to a changing climate, depletion of natural resources, and increasing demand for food. Consequently, soil management systems must be more strongly focused on creating sustainable food production systems with the diverse range of soil management practices that are available for use.

This classification helps to explain the complexity chronologies and the technological context of any particular approach and helps in making strategic choices regarding specific soil management practices for given agriculture problems.

11.3.1 Traditional Strategies

Conventional methods of soil management reflect well-established practices that have been used by farmers throughout the years, usually depending on biological processes with external assistance kept at a minimum. These forms of farming usually focus on integrating local ecosystems and using available materials.

11.3.1.1 Crop Rotation-Diversification Systems

Crop rotation is an old methodology that involves changing various crops in a given order over a designated time. This strategy reduces the potential risk of pests, diseases, and weeds that may become accustomed to a monoculture. Furthermore, crop rotation has positive impacts on soil fertility by using cover crops and legumes, thus reducing the reliance on chemical fertilizers. According to a meta-analysis of 54 studies across 6 continents, implementing more diverse crop rotations achieved a 49% reduction in weed density but did not noticeably affect total weed biomass (Tamburini et al. 2020). Increased crop planting dates and crop management as a plant diversification strategy help to explain the increased variability in weed density. Under a zero-tillage strategy, this type of diversification performed the best, where weed density was lowered by 65% rather than the 41% that was reported for tilled conditions. The research suggests that changing the timing and intensity of management practices such as tillage or herbicide application during crop rotation helps limit the spread of weeds (Tamburini et al. 2020). Figure 11.3 demonstrates how crop rotation improves soil health through having increased nutrient supply, lowered fertilizer applications, and increased microbial activity. Most legumes enrich the soil by fixing nitrogen, while organic matter decomposition amplifies fertility. It assists in pest control biologically and improves the soil structure due to the activity of earthworms.

The integration of different plant types with varying root forms and nutritional values increase the rate at which nutrients are cycled. Different crops have distinct rhizodeposits which influence the structure of the soil microbial community and nutrient cycling in the rhizosphere. The role of soil biota in determining the specifics of crop rotation is, especially, pointed out by Dias et al., who recommend enhancing the relations between crops and soil microorganisms for better yields and soil health. Many of these more complex rotations shift some of the burden of pests and pathogens while maintaining a balance of microbial communities which

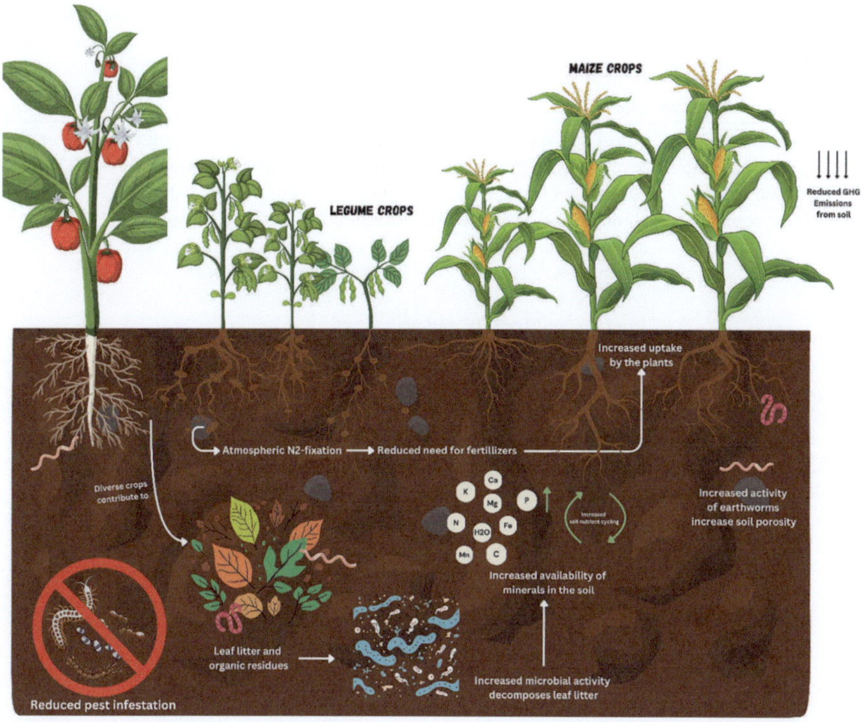

Fig. 11.3 Crop rotation and soil health

promote nutrient cycling as well as soil structure. The greatest crop diversity produced the greatest metabolically diverse and active soil microbial communities which were associated with increased crop biomass and higher soil pH and suggested a positive feedback loop from crop diversity to soil microbial activity to soil fertility. Table 11.2 summarises effect of diverse crop rotations on soil health and plant growth.

11.3.1.2 Intercropping

Similarly, other crops can be cultivated on the same piece of land which results in the higher yield and reduced pest resistance. This system is more resource efficient and productive because intercropping improves competition between legumes and other crops (Tamburini et al. 2020). Including legumes with other crops significantly enhances nutrient availability and water utilization, mitigates pest influx, escalates organic matter content of soil, and increases crop output. In addition, intercropping reduces the overall weed growth because of the competitive advantage gained through capturing resources (Garland et al. 2021). Weeds are less difficult to manage because intercropped plants greatly exceed them in harvest index

Table 11.2 Comprehensive overview of impact of diverse crop rotations on soil health and plant growth

Combination of crops	Benefits to the soil and plant	Reason	Refrences
Crop rotations that incorporate legume crops, such as soybeans, peas	Boost soil nitrogen levels, offering nutrients for future crops Reduces the need for synthetic fertilizers and greenhouse gas emissions Improve soil fertility and microbial activity for subsequent crops	Legumes have the ability to fix atmospheric nitrogen through symbiotic relationships with nitrogen-fixing rhizobia bacteria Add organic matter through leaf litter and root biomass	Zou et al. (2024)
Crop rotations that incorporate deep rooted crops such as alfalfa	Improved nitrogen mobility Better nutrient supply and efficiency for plants Increases carbon transfer to soil and enhances carbon sequestration, improving soil structure Higher density of vertical bio-pores promotes root growth and nutrient uptake	Deep roots access deep soil layers, improving the efficiency of nitrate (NO3-) uptake Also boost the activity of anecic earthworms, which thrive in the presence of deep roots and organic matter	Zou et al. (2024)
Crop rotations that incorporate certain cover crops	Provide ground cover, prevent nutrient leaching, and improve soil structure and functions Mitigate soil temperature fluctuations Greater resistance against erosion	Cover crops add organic matter to the soil, thus enhancing the soil's organic carbon content, through biomass incorporation, further improving soil fertility and structure	Zou et al. (2024)
Crop rotation that incorporates perennial plants, like fruit orchards	Reduction of emissions generated by field crops due to soil disturbance Reduced release of greenhouse gases	Perennial plants effectively capture atmospheric carbon dioxide and methane. They enhance carbon storage in both above-ground and below-ground biomass	Zou et al. (2024)
Crop rotations with broadleaf crops, rye, sorghum, rice, sunflower, rape seed, wheat, and legumes	Inhibit weed growth Enhance soil microbial activity	These crops can release allelochemicals into the environment, which can either inhibit or stimulate nearby organisms. These chemicals can affect various processes, including cell division, photosynthesis, and hormone regulation	Zou et al. (2024)

and also exhibit natural weed suppression mechanism such as allelopathy (Cheng 2015). Practices aimed at maintaining or increasing soil fertility and productivity through crop rotation and diversification (CRD) use also reduce the need for the use of herbicides and pesticides, improving overall soil health. Use of fewer herbicides leads to less environmental contamination resulting from excess chemicals and soil pollution from the beneficial organisms. Moreover, crops suffer less stress resulting from competitive weed interference which improves their productivity and the nutrient resources available to plants. Over time, these conditions lead to improved soil health and make farming practices more sustainable (Zou et al. 2024).

It is also important to diversify agricultural landscapes in order to enhance the sustainability and resilience of farming systems. This type of diversification might shield farmers from market volatility and price variations, while improving dietary diversity and addressing nutritional deficits within local communities. Still, it must be pointed out that the benefits of crop rotation are not uniform and vary with soil type and environmental conditions (Zou et al. 2024). Findings from Norway over time suggested that different crop rotations and ways of applying fertilizers had very little influence on yield and soil fertility, accompanied by a gradual decline in soil organic carbon (SOC) but an increase in soil nitrogen over the years. On the other hand, some research conducted on alluvial soils pointed out that crop rotations, especially including maize-mung bean and mung bean-chilli, enhanced soil organic carbon, total nitrogen, and available phosphorus (Zou et al. 2024).

11.3.1.3 Agroforestry

Agroforestry is the incorporation of trees and shrubs to farmlands transforming them into a multifaceted productive agricultural system that is both sustainable and efficient. This method of farming permits the simultaneous growth of crops, trees, and livestock, which is why it is becoming more widely accepted for its climate change adaptation and mitigation benefits. Apart from economic returns, this practice assists in carbon capturing, soil productivity restoration, and biodiversity protection, thus propelling farmers towards achieving the UN Sustainable Development Goals in 2030 (Castle et al. 2021).

Increases in nutrient cycling due to litterfall and fine root turnover AFS, coupled with nutrient retrieval from deep-rooted trees during nutrient leaching, assist in the prevention of nutrient loss. Water body pollution and nutrient runoff are simultaneously diminished. Agroforestry outcompetes traditional mono-cropped fields for carbon capture and storage in fossil fuels, sequestering annually between 0.29 Mg and 15.21 Mg of carbon per hectare in biomass and up to 300 megagrams per hectare in soil (Nair 2011). Figure 11.4 explains a sustainable farming ecosystem with an agroforestry-based framework aimed at improving soil health, enhancing food security, and fostering environmental sustainability. It demonstrates how trees, livestock, and crops work together to achieve a balanced ecosystem and greater productivity in agriculture.

Fig. 11.4 Agroforestry-based sustainable farming ecosystem

According to research, the trees in agroforestry systems enhance the soil microbial population. For example, research demonstrated that poplar trees cultivated in alley cropping systems increased microbial biomass carbon (MBC) and microbial biomass nitrogen (MBN) significantly relative to associated crop rows. MBC and MBN had also been greater in tree rows where MBC increased from 365 mg kg^{-1} in crop rows to 572 mg kg^{-1} close to the trees. The surrounding crop regions can also experience this increase (Beule et al. 2022).

The experiences related to the local context within agroforestry systems have long been appreciated. With funding based on conjectures and anticipations, now it is being targeted through a holistic scientific framework. To realize these benefits, detailed and wider studies at the local level need to be conducted which have been bestowed upon farmers, in order to nurture the vital research funding that has been greatly neglected, but which hides significant resources and opportunities.

11.3.1.4 Biochar and Carbon Sequestration

Biochar is an organic matter carbon-rich material created from agricultural waste using a pyrolysis process, which is the thermal breakdown of biomass into various carbon components without oxygen (Lehmann and Joseph 2015). Meyer et al. (2021) call it an effective long-term carbon sequestration measure that can remain in soils for centuries while outperforming organic materials due to its ability to reduce greenhouse emissions. Organic materials, on the other hand, decompose and release carbon dioxide (CO_2).

Biochar has a multiplicity of advantages for agriculture, such as enhancing soil health, microbial populations, and moisture content. It also improves leaching: the process whereby biosoluble compounds can be removed from the soil by the action of water. This is made possible by the porous biochar structure that improves cation exchange capacity while retaining fundamental nutrients such as nitrogen, phosphorus, and potassium (Biederman and Harpole 2013). In low-fertility regions, Jeffery et al. (2011) claim that it can also expand crop production: it does so by neutralizing acidic soils.

Another benefit is that biochar's hydrophilic nature enables it to retain water, which in combination with its ability to provide a habitat for beneficial microbes, which enhances the breakdown of organic matter, available nutrients, makes the biochar ideal for arid regions.

Studies have highlighted Masto et al. (2015), who noted increased crop yields and microorganism activity from water hyacinth biochar kudzu. What is striking is that Kammann et al. (2017) showed that biochar has the potential to improve biomass production while simultaneously decreasing greenhouse gases. Other trials conducted in sub-Saharan Africa and Southeast Asia showed improved soil health while reducing the need for chemical fertilizers (Mashamaite et al. 2024). All these examples underscore in clear terms biochar's place in sustainable agriculture.

Biochar has been identified as one of the most novel technologies that aid in environmental conservation because of the wide array of benefits it has to offer. These include:

1. Carbon Sequestration: Biochar assists in reducing CO_2 emissions by locking carbon in the soil for an exceedingly long time.
2. Limiting Greenhouse Gases: Biochar helps in reducing CH_4 and N_2O emissions from soil.
3. Soil Improvement: Biochar increases soil fertility and helps in nutrient cycling, erosion control, and drought tolerance.
4. Trash Disposal: The production of biochar assists in reducing biomass agricultural waste by incorporating forestry and agricultural residues.

Even with all of these advantages, there are some restrictions when it comes to the more expansive use of biochar. These stem from shifts in the kind of feedstock employed, the pyrolysis processes used, and particular soil conditions (Lehmann and Joseph 2015). Others, like farmers' lack of information regarding biochar and the high production costs further reduce its adoption.

11.3.2 Modern Strategies

Modern soil management strategies incorporate scientific understanding and technological advancements while balancing productivity with sustainability concerns. These approaches typically represent adaptations or refinements of traditional methods, enhanced with contemporary knowledge and equipment.

11.3.2.1 Biofertilizers

Biofertilizers are living microorganisms or dormant cells that can live in symbiosis with the roots of crop plants. In the opinion of Abdo et al. (2022), biofertilizers promote growth by making nutrients available. Their biological activities are responsible for the decomposition of organic matter to yield growth-promoting hormones and also help in soil fertility enhancement through nitrogen and phosphorus fixation from the atmosphere. Biofertilizers can also be incorporated into composite formulations for formulation and application as soil inoculants. These powdered or dry forms of biofertilizers can be packed with suitable carriers like clay minerals, rice bran, soil peat, brown coal, and wheat. The use of these materials enhances ease of handling and storage of microbial bioinoculants due to their long shelf life. It can be soil applied to seeds, roots, or soil or even mixed with water for root dipping and foliar spraying. The action of these microorganisms on soil makes it more fertile and improves the health and yield of crops. As bioinoculants, biofertilizers proficiently mitigate the salinity negative impacts on soil, thus enhancing crop production. In addition, biofertilizers also serve as biocontrol agents which are less harmful than chemicals when used against certain bacteria, fungal, and viral pathogens as well as aphids and nematodes. Biofertilizers as endophytes are able to colonize in plant tissues, so they do not cause any disease to the plants, rather help the plant. Biofertilizers also have parasitic, symbiotic, or mutual beneficial relationships with the host. The advantageous aspects learned from the use of biofertilizers were low financial expense and increased nutrient availability within the soil, as well as fertility improvement, soil pathogens protection, sustainable agricultural production, and plants' tolerance to abiotic and biotic stress, which leads to phytohormone production, reducing pollution and improving soil health, all of which drastically increased soil fertility over time (Chaudhary et al. 2022). Moreover, biofertilizers are recommended to be used as inputs for sustainable agricultural production due to being more environmentally friendly and cheaper (Abdo et al. 2022). Figure 11.5 categorizes key biofertilizers involved in soil health and plant growth based on their functions.

11.3.2.2 Conservation Tillage

Farmers have tracked intensive cultivation since the eighteenth century and its subsequent effects on erosion and land productivity. As a remedial measure, practices such as cropping rotations with winter crops were introduced. In the 1930s, extension researchers pioneered the first strategy of conservation tillage. Use of no-till farming multiplied in the 1960s with herbicide atrazine and paraquat and again in the 1990s with glyphosate-resistant crops (Bilen et al. 2010). Conservation tillage refers to a method of soil management that is novel and one that is able to minimize the challenges created by over-tillage efficiently. Throughout varying regions of the world taking care of farmland is done differently, but science has shown that conservation tillage systems can help nurture soil.

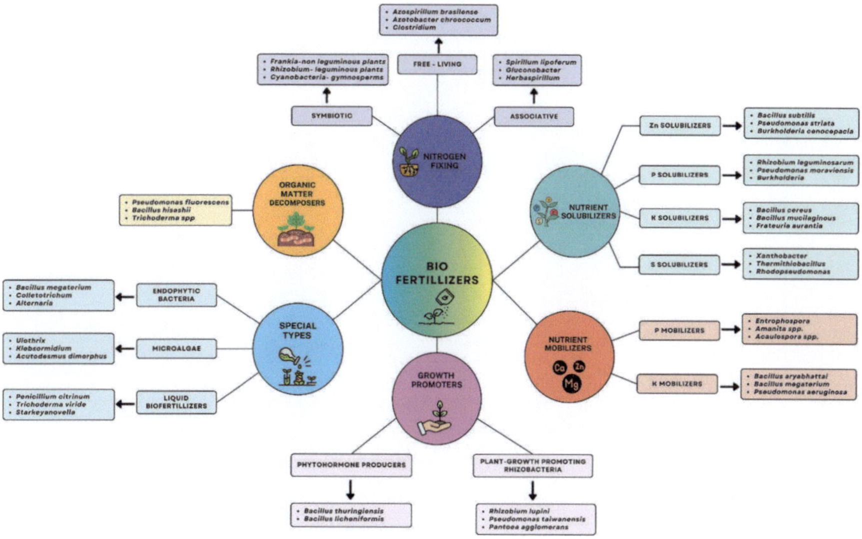

Fig. 11.5 Classification and functions of biofertilizers

In general, conservation tillage refers to agricultural methods that involve no plowing, saves tilling to the minimum necessary, and places shelter for crops on the land to preserve it and restrain water evaporation. Its purpose is to retain soil moisture and fertility, boost carbon retention, increase yield, and lower GHG emissions (Bilen et al. 2010). Even though the various regions of the world have developed their own particular methods for land preparation, there exist some principles and practices developed through years of research on conservation tillage methods around the world. The technique refers to several tillage methods that aim to integrate harvest of crops and care for soil in an eco-friendly manner. In this modern era, everything from saving water to increasing biomass gasses from the soil is captured where farming is done (Bilen et al. 2010).

These approaches can further be subdivided into straw incorporation into the field, including no tillage, mulch tillage, straw-retention with no-tillage systems, minimum tillage, ridge tillage, and some more (Bilen et al. 2010) as shown in Fig. 11.6.

A recent research conducted in the semiarid regions of Dingxi belt of the Gansu province, China, reveals the advantages of conservation tillage practices in spring wheat agroecosystems. This study notes that conservation tillage, particularly no-tillage systems with stubble return, outperforms conventional tillage methods by increasing wheat yields alongside enhanced soil while decreasing greenhouse gas emissions. In particular, these practices have the potential to reduce greenhouse gas intensity by as much as 29.96%, and the reduction in global warming potential is by up to 23.44%.

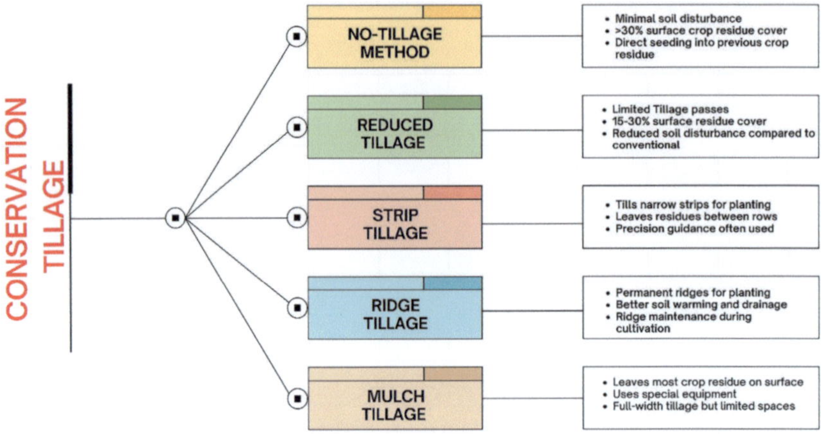

Fig. 11.6 Types of conservation tillage methods

The study explored four different tillage systems and analysed the results concerning a number of agricultural and environmental factors (Bilen et al. 2010).

Table 11.3 summarizes these practices and the measured benefits as compared to the conventional practice of tillage.

11.3.2.3 Biofortification

Introducing micronutrients into crop genome has alleviated "hidden hunger" through biofortification (Sen et al. 2024; Paul et al. 2024). It has the potential to address the struggles of challenging dietary intake value of food crops concerning the malnutrition problem and food security (Raut et al. 2024). Certain routinely cultivated crops are deficient in crucial nutrients essential for human health like iron, vitamin A, and zinc. Likely, biofortification will help alleviate this problem (Galili et al. 2016; Fitzpatrick and Chapman 2020). This is significant for developing countries where nearly two billion people are malnourished by improving health and enhancing food security (Reddy et al. 2024). These methods also involve the use of genetics, traditional breeding, and agronomy to specially tailor the miniature food components of a plant (Huey et al. 2024; Raut et al. 2024).

These methods strive to solve the problem of important nutrients which are situated within the soil. In addition, biofortification also improves soil health (Soorya et al. 2024; Ahmed et al. 2024). There was notable increase in the engagement with healthy diets, especially in the households within the sub-Saharan Africa region (Philipo et al. 2021), alongside the plant cultivars which are comparatively more tolerant to environmental stress (M et al. 2020; Raut et al. 2024). Key issues surrounding biofortification are the crop genetic modifications involving the zestful

Table 11.3 Comparison of benefits of the tree conservation tillage practices

Benefit	NTS method	NT method	Straw incorporation +CTS	References
Yield parameters	Greatest improvements in the height of plants, spike, seeds per square meter, root yield, above ground biomass, thousand grain weight, and grains yields to total	Significant improvements in yield components and overall productivity, though generally less than NTS	Moderate improvements in yield parameters compared to conventional tillage but less effective than no-tillage approaches	Sadiq et al. (2024)
Soil physical properties	Maximum increase in soil water content, water-filled pore space, water holding capacity, soil pores, and reduction in soil bulk density	Substantial improvements in soil physical characteristics, particularly water-related properties and soil structure	Moderate improvements in soil physical properties through organic matter addition but limited by continued soil disturbance	Sadiq et al. (2024)
Soil chemical properties	Maximum increases in soil organic carbon, light fraction organic carbon, carbon pools, total nitrogen, N pools, various nutrient parameters	Significant improvements in soil chemical properties, particularly carbon sequestration and nutrient availability	Some enhancement of soil chemical properties through residue incorporation, but less effective than no-tillage systems	Sadiq et al. (2024)
Soil biological properties	Optimal conditions for biological activity of soil results in maximum increase in microbial biomass carbon, microbial biomass nitrogen, microbial population, and enzyme activities	Substantial enhancement of soil biological properties through reduced disturbance and improved habitat conditions	Limited improvement in biological properties due to continued soil disturbance offsetting benefits of residue addition	Sadiq et al. (2024)
Greenhouse gas reduction	Maximum reduction of CO_2, CH_4, and N2O emissions	Significant reductions in greenhouse gas emissions, though generally less than NTS	Modest reductions in emissions compared to conventional tillage, but less effective than no-tillage approaches	Sadiq et al. (2024)
Climate impact metrics	Decreased global warming potential by 23.44%. Decreased greenhouse gas intensity by 23.20%	Reduced global warming potential by 19.57%; reduced greenhouse gas intensity by 29.96%	Reduced global warming potential by 16.54%; reduced greenhouse gas intensity by 18.72%	Sadiq et al. (2024)

genes and changeable processes of genetic engineering. Iron and folate concentration within the crops' DNA is now possible with the use of clustered regularly interspaced short palindromic repeats (CRISPR)-Cas9 techniques (Chaudhary et al. 2022).

The 2020 Nobel Prize in Chemistry was awarded for genome editing, which is now known to be simple and efficient. There are a multitude of remarkable cases of biofortified crops. Table 11.4 summarizes the most important biofortified crops.

Table 11.4 Key biofortified crops: traits, mechanisms, and nutritional impact

Crop name	Biofortified trait	Key genes/ components	Mechanism/ pathway	Nutritional impact	References
Golden rice	Vitamin A (β-carotene)	Daffodil's *psy* gene *Pantoea ananatis crtI* gene	*psy* converts GGPP to phytoene *crtI* converts phytoene to lycopene Lycopene → β-carotene via endogenous enzymes	β-carotene metabolized to retinol (vitamin A) in the intestine Targets vitamin A deficiency (blindness, immune issues)	Paine et al. (2005)
Iron-rich rice	Iron	Soybean ferritin gene Phytase gene (co-expressed)	Ferritin stores iron in grains Phytase breaks phytic acid to enhance iron bioavailability	Iron content: 2 μg/g → 15 μg/g in polished grains - Reduces iron deficiency anemia	Vasconcelos et al. (2003) (Plant Sci.)
Multi-nutrient potato	Protein, essential amino acids	*AmA1* gene (*Amaranthus hypochondriacus*) Tuber-specific promoter	*AmA1* produces seed storage proteins rich in lysine, methionine, cysteine Protein stability ensured by proper folding	Protein content increased by 35–60% Enhances essential amino acids in edible tubers	Das et al. (2022)
Orange-fleshed sweet potatoes	Vitamin A (beta-carotene)	High beta-carotene (natural) Varieties like Beauregard	Beta-carotene → vitamin A in the body Orange flesh acts as a visual indicator of nutrient content	Improves vitamin A status in children and women Reduces deficiency-related health risks	Sheoran et al. (2022), Elemike et al. (2019), Rajput et al. (2021)
Iron-rich pearl millet	Iron	Enhanced iron content via selective breeding	Bioavailable iron absorbed efficiently in the body	Effectively reduces iron deficiency anaemia	Kumar et al. (2022), Yadava and Hossain (2018)

11.3.2.4 Precision Agriculture

By employing remote sensing, GPS, IoT devices, and machine learning, precision agriculture (PA) increases farming output and sustainability while aiming to provide farmers with real time data for informed decision-making. Furthermore, PA aims to minimize resource consumption and environmental harm.

Remote sensing paired with unmanned drones has increased the efficiency of soil and crop monitoring. For example, drones equipped with multispectral cameras can identify nitrogen deficiencies with an accuracy of 92%, which makes nitrogen fertilizer application 20–30% more efficient (Chen et al. 2023). Additionally, IoT-based soil sensors can also determine pH, temperature, and humidity, thereby enabling the sensor to reduce water utilization by 15–25% without affecting the crop yield (Wolfert et al. 2017). Moreover, machine learning algorithms in PA facilitate yield predictions and identify diseases at early stages, resulting in 34% improvement in predictions accuracy (Basso and Antle 2020).

From Multidisciplinary Perspectives of AI in Agriculture with Machine Learning's Edge:

1. Random Forest and Support Vector Machines: These models create management zone objectives by mapping terrain fertility and classifying land forms with geospatial data (MZOs) (Hengl et al. 2017).
2. Deep Learning: The use of drones to obtain multispectral images enables early detection of high salinity in soils at an impressively high accuracy of 89% when conditions are arid (Kamilaris and Prenafeta-Boldú 2018).
3. Hybrid Models: The application of reinforcement learning to the usage of fertilizer improves the efficiency of nitrogen fertilizer (Liakos et al. 2018).

Variable rate technology (VRT) forms a vital component in precision agriculture (PA) because it allows the application of seeds, fertilizers, and pesticides to be done in specific areas. This differs from these methods which tend to be more resource-intensive. The adoption of VRT led to an 18% reduction of herbicides applied while growing maize without undermining the level of control over weeds (Schimmelpfennig 2019). Using precision agriculture also helps to mitigate the emission of greenhouse gases which results from overusing nitrogen fertilizers as it increases the efficiency of inputs (Liakos et al. 2018). PA has increased net benefits to smallholder farmers by 10–15% owing to increase in productivity and decrease in expenditures.

Figure 11.7 depicts the application of modern technologies in agriculture for the purpose of achieving soil sustainability and food resilience. This shows the evolution of a smart farming ecosystem wherein agriculture is practiced with the use of information technology and automation. It also highlights the role of technology in achieving food security on a global scale without causing detrimental harm or straining the Earth's resources.

Yet, for economically poorer regions, the lack of resources to purchase newer technologies, insufficient knowledge of these technologies, and lack of Internet access pose a serious challenge. Furthermore, other technologies have their own

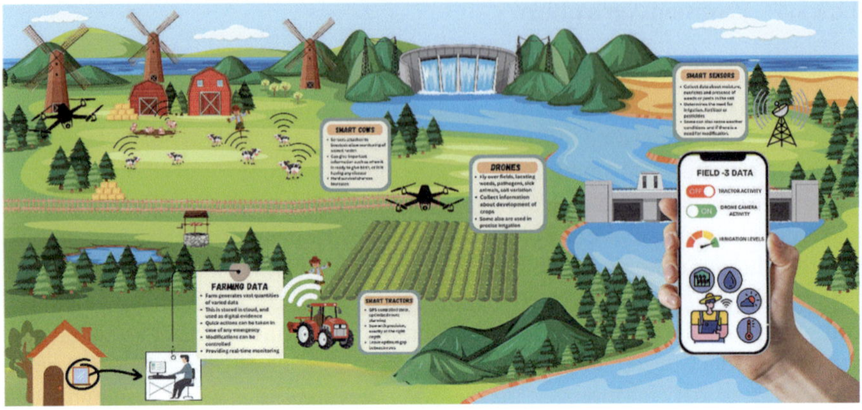

Fig. 11.7 Precision agriculture: technology-driven soil sustainability and food resilience

hurdles such as issues of data privacy regarding its adoption. Therefore, inclusion barriers are always present (Kamilaris and Prenafeta-Boldú 2018). But there is room for improvement through increased utilization of blockchain for improving transparency in the supply chain and 5G networks to solve the issue of slow dissemination of information.

11.3.3 Innovative Strategies

This approach combines new technologies with innovative strategies for soil management. These techniques often push the limits of conventional farming by applying modern scientific methods to tackle complex challenges.

11.3.3.1 CRISPR-Engineered Soil Health

CRISPR (clustered regularly interspaced short palindromic repeats) technology attempts to improve the fertility of soil, as well as the growth of plants and their environmental stress tolerance, through the microbial engineering of soils and plant genomes (Prabhukarthikeyan et al. 2020). These methods are directed towards building more amicable interactions between plants and microbes, directing ecosystem nutrient processes, and crop health issue management. This enables agriculture to be less harmful to the environment than the conventional method of farming (Araujo 2022). Raut et al. (2024) remarked that the process of nitrogen fixation and phosphate solubilization by some beneficial microbes is likely to be regulated by the CRISPR-Cas systems owing to the antagonistic interaction of the microbiomes with the synthetic fertilizers which would help to develop sustainable agriculture. Some specific populations of microorganisms can also be created using some plant growth-promoting bacteria that have biodegrading abilities (Araujo 2022). Using CRISPR technology, beneficial microbes associated with plants may be modified in order to increase the efficiency of their interaction with plants. An example is CRISPRi, a gene silencing technology that silences genes in nitrogen-fixing bacteria such as *Azotobacter vinelandii* (Russell et al. 2024). Increasing synergisms by mutating OsS5H genes associated with flavone biosynthesis in rice enhances disease resistance (Bilen et al. 2010; Liu et al. 2023). Some other plants are already known to have the ability to carry out phytoremediation (Rao and Sudhakar 2014). Examples are *Typha* and *Saccharum spontaneum.* According to C. M. Rao's citations from 2014, the tailings ponds from bauxite aluminium extracters were shown to be vegetatively remediated by *latifolia*, *Pteris vittata*, and *Cyperus compressus* which are known to bioaccumulate heavy metals and radionuclides. *Lavandula angustifolia* has been shown to accumulate and translocate heavy metals (Vidican et al. 2023). *Hibiscus cannabinus-Sedum plumbizincicola* rotation is another case study for trace-metal-polluted farmland (Gao et al. 2023). Table 11.5 highlights innovative applications of CRISPR technology in agricultural biotechnology.

CRISPR is revolutionizing soil management because it employs an entirely different approach compared to the traditional civilizations' agricultural techniques. The prospects of gene editing CRISPR give promise for a controllable approach with minimal off-target effects. Moreover, CRISPR provides relatively faster advancements to soil health in comparison to other methods. The self-sustaining processes in agriculture are enhanced by novel methods of CRISPR, which decreases the dependency on chemicals and foster an agricultural practice that is more environmentally sustainable. In addition, CRISPR permits the crafting of comprehensive approaches to promote health for specific soil challenges which enables better control of soil health in farming.

Ethics, legality, and environmental implications pose a risk to the promise CRISPR holds for soil fertility and food security. There are issues of safety regarding CRISPR's health-engineered organisms for off-target effects and unintended consequences. "Social equity" and "environmental justice" are two of the ethical disputes dealt with agricultural gene editing focus on (Shozi 2020). Such organisms require proper CRISPR regulation scrutiny to mitigate risks (Bu 2019). After

Table 11.5 CRISPR applications in nitrogen fixation and plant engineering

Organism/plant	Gene/modification	Effect/application	References
Azotobacter vinelandii	CRISPRi system	Silencing nitrogen fixation genes	Russell et al. (2024)
Rice (flavone biosynthesis)	Modified flavone biosynthesis	Amplified biofilm development and diethanolamine nitrogen fixation in soil bacteria	–
Burkholderia contaminans	CRISPR-Cas9 editing of PKS gene clusters	Loss of antibacterial activity against plant pathogens	Bilen et al. (2010)
Rice	Mutated *OsS5H* (salicylic acid 5-hydroxylase)	Broad-spectrum disease resistance	X. Liu et al. (2023)

organism CRISPR technology application, the impact on soil ecosystems and biodiversity will need to be looked into. This technology, like any other, needs to be monitored and managed responsibly, to avoid risks and accumulate knowledge on its effects over time (Piergentili et al. 2021).

11.3.3.2 AI and Machine Learning in Soil Management

The use of AI and ML has transformed soil management with the help of data-based precision agriculture. AI models improve the allocation of nutrients, irrigation, and crop choice by using large datasets from remote sensing and IoT sensors (Pantazi et al. 2016; Basso and Antle 2020). CNNs and random forests, for example, achieve 85–92% accuracy in predicting soil organic carbon and nitrogen concentrations without the need for laboratory analyses.

A few benefits of these new implementations are highlighted below:

- The Digital Soil Health Card (DSHC) initiative in India analyses 12 million soil samples every year which has resulted in lower urea overuse by 15% and improved wheat production by 8–12% (Patil and Kumar 2021).
- The initiative "eSoilCare" in Africa identified areas vulnerable to drought, which decreased crop loss by 25% (Mwangi et al. 2022).
- The precision fertilization system in Brazil employed federated learning to optimize the application of phosphorus and achieved an 18% boost in soybean yields (Dias et al. 2020).

The latest developments such as Edge AI make it possible to perform soil diagnostics on the IoT devices in real time which is more insightful for farmers. The demand for trust too is contributing to the development of Explainable AI (XAI), which offers interpretable predictions (Sagan et al. 2021; Roscher et al. 2020). In addition, incorporating machine learning to climate models may enhance the effectiveness of climate adaptation and help mitigate soil erosion, which fosters sustainable agricultural practices within the context of climate change (Tzachor et al. 2022). Figure 11.8 demonstrates the contribution of artificial intelligence (AI) towards the different stages of soil management by depicting its applications within the cycle of agricultural activities. In the middle, the bulbed light which is augmented with AI technology represents invention and innovation that comes with the use of AI. AI components show how it helps to achieve efficient soil management and enhanced agricultural productivity.

While these improvements are laudable, they are still very much lacking. One of the biggest challenges remains in the data itself with the non-existence or lack of labelled datasets to train more advanced models. In tandem, advanced sensors have extremely high infrastructure costs that could be of great strain to less resourced farmers (Shepherd et al. 2020; Bronson and Knezevic 2021).

There are other ethical gaps especially in the AI model bias towards local indigenous crops which are less economically profitable and harmful to biodiversity and local agriculture cornucopia. If these issues are not properly handled, they will act

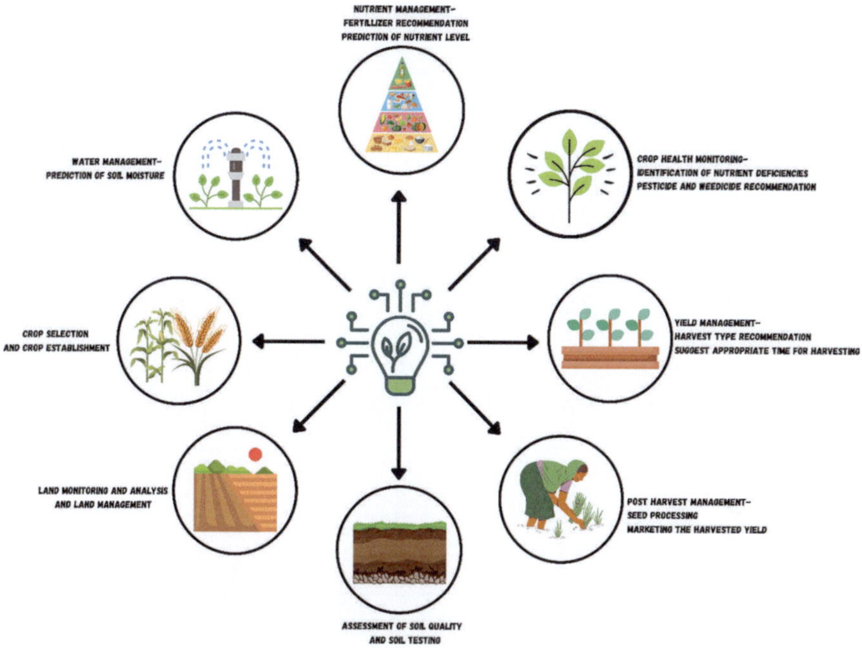

Fig. 11.8 AI-driven innovations in soil management

as barriers against successfully deploying AI technology into agriculture. Table 11.6 outlines innovative applications of AI in enhancing soil health and advancing sustainable practices.

11.3.3.3 Bioplastics

Plants, microorganisms, and even agricultural waste are sources for renewables used in bioplastics. Even though some bioplastics are not compostable, their aim is to reduce environmental damages by pollution (Haider et al. 2019). In addition, bioplastics are meant to tackle the global challenge of plastic pollution. Bioplastics as an advancement in the agricultural sector enhances soil health, encourages zero waste, supports circular economies, and reduces pollution all at once.

There are three categories of bioplastics:

1. Bio-based plastics such as bio-polyethylene are derived from renewable sources but do not guarantee compost ability.
2. Biodegradable plastics are those that are decomposable without the help of controlled conditions.
3. Polylactic acid (PLA) and polyhydroxyalkanoates (PHAs) pose as examples for bio-based and biodegradable plastics since they can both renew from and decompose into simpler substances (Narancic et al. 2020).

Table 11.6 AI-driven strategies for soil health assessment and sustainable agriculture

Category	Strategy	Key application	Benefits	Limitations	References
Soil nutrient analysis	CNNs, random forest	Quantify soil organic carbon (SOC) and nitrogen via hyperspectral imagery	85–92% accuracy; reduces lab dependency	Requires high-resolution sensors	Pantazi et al. (2016), Basso and Antle (2020)
Soil moisture prediction	LSTM networks	Predict soil moisture using weather and evapotranspiration data	Reduces water use by 20–30%	Dependent on real time weather data	Chlingaryan et al. (2018)
Crop recommendation	AI-driven platforms (e.g., DSHC)	Real-time nutrient advisories based on soil health data	Reduced urea overuse (15%); increased yields (8–12%)	Requires farmer training for adoption	Patil and Kumar (2021)
Soil salinity detection	Deep learning (CNNs)	Analyse drone-captured multispectral images to identify saline soils	89% precision in arid regions	Limited effectiveness in cloudy conditions	Kamilaris and Prenafeta-Boldú (2018)
Precision fertilization	Federated learning	Optimize phosphorus application using aggregated farm data	18% increase in soybean yields; preserves data privacy	Requires robust IoT infrastructure	Dias et al. (2020)
Drought prediction	Satellite ML models	Identify drought-prone areas for crop resilience planning	25% reduction in crop failure (Kenya pilot)	Limited accuracy in regions with sparse satellite coverage	Mwangi et al. (2022)
Edge AI	Lightweight ML on IoT devices	Real-time soil diagnostics without cloud dependency	Low-latency decision-making	High upfront sensor costs	Sagan et al. (2021)
Explainable AI (XAI)	Interpretable ML frameworks	Transparent predictions for farmer trust (e.g., soil health insights)	Enhances adoption of AI tools	Complexity in balancing accuracy and interpretability	Roscher et al. (2020)
Climate resilience	ML-climate model integration	Predict soil degradation under climate scenarios	Proactive mitigation of erosion	Computational intensity	Tzachor et al. (2022)
Data scarcity	Transfer learning	Adapt pretrained models to low-data regions	Improves generalizability	Limited performance in highly variable soils	Shepherd et al. (2020)

The production of bioplastics seeks to address challenges posed by traditional plastics, especially those that persist within a given environment like farmlands where the presence of microplastics severely threatens the quality of soil and food production (Bandopadhyay et al. 2018).

However, bioplastics have certain limitations, one of them being that their degradation efficiency depends on specific factors, such as temperature, diversity of microbes, and humidity levels in the region. Some areas may not have the required conditions for bioplastics to perform effectively (Haider et al. 2019).

To begin with, we note the primary uses of bioplastics, which are mulch films, coating of seeds, packaging materials, and controlled release fertilizers. For some of the applications, however, away from the polyethylene films, it is noticeable that these biodegradable films have gained preference. This is because polyethylene is frequently reduced to microplastics which is detrimental to the ecosystem and food chain (Hayes et al. 2019). Research has demonstrated that the use of bioplastic mulches improves soil health, which can be altered to non-toxic, moisture retaining, and allows for removal which by itself is quite painstaking (Sintim et al. 2020). In particular, it is the incorporation of polylactic acid (PLA) and polyhydroxyalkanoate (PHA) films that is of interest because these materials get metabolized by microorganisms. They therefore serve to increase soil fertility and benefit the microbiota (Bandopadhyay et al. 2018).

In addition to mulching, bioplastics are increasingly used in seed coating applications because they enhance the prospects of germination and safeguard seedlings from pathogens. Starch-based coatings that contain adequate nutrients, or even biocontrol agents, offer an excellent opportunity to promote crop robustness while simultaneously decreasing the reliance on pesticides or fertilizers (Accinelli et al. 2018). Moreover, bioplastic packaging has an important role in reducing postharvest waste and fossil fuel consumption in alignment with the circular economy.

Furthermore, bioplastics support the concept of plasticulture which involves integrating plastic materials into agriculture to boost output while saving resources. Bioplastics that are manufactured to biodegrade are different from sorghum and cane husk bioplastics that are made to create waste plastic bioproducts. Such materials need revision in the sense that they claim to replace traditional plastic in performing diverse agricultural activities without compromising agroecosystems health and productivity (Riggi et al. 2012). In addition, the implementation of bioplastic technologies using agricultural byproducts such as corn husks or sugarcane bagasse emerges as another important avenue. These residues can serve as raw materials for synthesizing valuable biopolymers by microbial fermentation, which transforms organic substrates into PHAs. This new technology will significantly help in waste reduction and promote a circular economy in agriculture (Mandal et al. 2024). Figure 11.9 highlights the step of a biorefinery where biopolymers are produced from organic feedstock and processed into bioplastic products, which eventually are biodegraded.

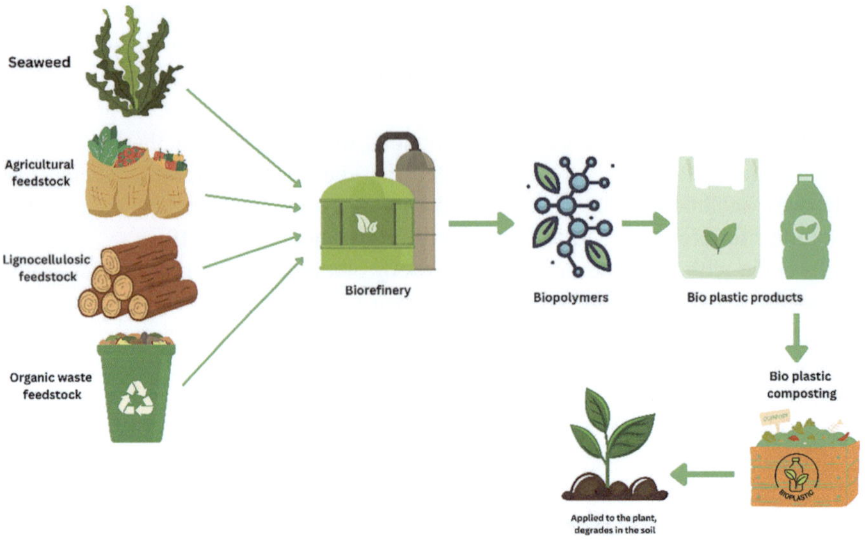

Fig. 11.9 Bioplastic production and decomposition

11.4 Socio-economic Implications and Policy Frameworks

11.4.1 Employment Opportunities in Sustainable Agriculture

Sustainable agriculture is vital for generating employment opportunities in both rural and urban settings (Sandeep and Mittal 2018; Abhiram et al. 2018). It serves as a pillar of food and nutrition security, economic development, and environmental conservation (Dawn et al. 2023). Besides providing food for a nation, it contributes to the economy by creating employment and supporting the export of agricultural products. The advent of modern technologies such as drone flying, automated farming, and precision farming is opening up employment opportunities that can benefit millions of people around the world (Dutta and Goswami 2020).

The introduction of soil conservation, agroforestry, and modern regenerative agricultural practices influences the level of employment, particularly in farming, agribusiness, and food industry. Sustainable agricultural practices are productive over the long term and require skilled labour for soil health management, conservation planning, and sustainable crop cultivation. Many developing countries are experiencing a rapid growth in the horticultural industry and, therefore, are able to offer significant employment to the smallholder farmers and agricultural workers (Wahome 2016). Also, some new models of agriculture, like the care farms in the European Union, combine social welfare aspects with farming and, thus, create employment opportunities for people with disabilities and other disadvantaged groups (Seong Goo Kim and Bae 2024).

A wide range of global endeavours showcase the possibility of creating jobs through sustainably managing and conserving soil. For example, the Mahatma Gandhi National Rural Employment Guarantee Act (MGNREGA) in India has specific projects aimed at soil conservation which not only employ rural workers but also increase the fertility and productivity of farming (Dawn et al. 2023). Other agroforestry employment models in Kenya also show that growing trees together with crops not only improves soil nutrition but also increases income-generating activities, thus providing employment in both forestry and farming (Dawn et al. 2023).

11.4.2 Rural Development Through Soil Health and Sustainable Farming

The health of the soil as well as sustainable farming practices has a tremendous impact on rural development. The investment made in soil conservation projects pays off in form of enhanced agricultural productivity, further development of rural infrastructure, and an improved quality of life among farming communities (Nikolov et al. 2024). In addition, sustainable farming is likely to raise incomes, access to education, and healthcare for the rural population (Regmi 2023; Mamatha et al. 2024). Sustainable agricultural practices are necessary in order to improve rural infrastructure, including irrigation, storage, and transportation facilities (Dawn et al. 2023; Wahome 2016). The use of digital twin technology, which creates virtual replicas of real farms through the utilization of sensors and weather stations for real-time data collection, has deeply changed the way resources are optimized and yields are enhanced (Dawn et al. 2023). Digital farming solutions improve irrigation and soil management, which contributes greatly to the rural infrastructural development.

Effective management of soil contributes significantly towards the quality of life of rural populations by improving agricultural productivity and meeting the food requirements of the locals (Susma Regmi 2023; Mamatha et al. 2024). Greater farm productivity leads to greater income levels, which allows farmers to spend more money on education and health-related issues. Soil-saving methods, sustainable farming, and resource conservation education are being promoted through funding campaigns and policy changes aimed at farmers by governments and agro-based organizations all over the globe (Regmi 2023).

The European Union Common Agricultural Policy (CAP) and India's Soil Health Card (SHC) scheme are, certainly, two different models that aim to achieve a common goal of sustainable agriculture and rural development. As examples, The European Union's Common Agricultural Policy (CAP) and India's Soil Health Card (SHC) scheme both seek to promote sustainable agriculture and rural development. Through endless reforms, the EU CAP tries to mitigate farmers' food security problems and the drastic socio-economic deterioration of the rural population while striving to be green (García et al. 2024). In India, the SHC scheme that has been running since 2015 is aimed at reducing costs for farmers, improving soil health and increasing crop yields through regionally tailored recommendations for soil nutrients (Reddy 2019).

Addressing agricultural sustainability and the rural development resource management gap is the goal of almost all initiatives (Singh et al. 2018). It is true that both policies, the EU CAP and Indian SHC scheme, have significantly improved the life quality of farmers and the rural economy. Following Oliinyk (2022), adoption of practices farmers took to improve soil health and reduce degradation is considered to increase with the eco-sustainability oriented EU CAP reforms. Singh et al. (2018) point out the socio-economic advancement of subsidized farmers in India's SHC programme with regard to rationalized fertilizer application and cost-effective production. Despite some constraints at the implementation and adoption stage, both programmes have been successful in educating farmers and providing them the necessary tools to promote sustainable agricultural practices and rural development.

11.4.3 Economic Viability and Farmer Profitability

Sustainable economic development in agriculture depends on optimal soil management practices. Sustainability in agriculture entails the use of organic farming methodology, which is more cost-effective, productivity maximizing, and high-yielding (Mamatha et al. 2024; Omer et al. 2024). Sustainable farming promoted by policies from various agricultural institutions and governments around the world aid economic development and has greater environmental benefits (Nikolov et al. 2024). Sustainable agriculture achieves long-term sustainability through the retention of building blocks, including improved soil structure, beneficial microbial populations, and nutrient cycling. Thus, a change in the resource use efficiency such as water, fertilizers, and pesticide through precision agriculture increases farm efficiencies and the resilience of food production systems (Mamatha et al. 2024). Adoption of systematic crop rotation, application of organic aids, conservation tillage, and other sustainable agricultural practices result in considerably higher profit margins to farmers (Mamatha et al. 2024; Omer et al. 2024). Degradation of soil fertility in conjunction with reduced chemical fertilizer usage increases soil productivity and crop yield all while decreasing operational costs. Carbon dioxide fixation and water retention agriculture also enhance sustainability and profitability measures (Neupane et al. 2024).

Adopting new farming practices might require some capital outlay initially, but over time this rationale will be economically sound. The economic return of farmers will increase due to higher productivity from improved soil health (Omer et al. 2024; Mamatha et al. 2024). By implementing cover crops, crop rotation, and conservation tillage, soil erosion is reduced, while soil retention is increased, thereby improving the sustainability and economic viability of agriculture on Omer et al. (2024)'s logic.

11.5 Conclusion

Healthy soil, sustainability, and food resilience need immediate collective action. Unsustainable practices lead to soil degradation which threatens water, climate, and food security across the globe. This chapter showcases how lasting solutions can be achieved through traditional integration-driven innovation. Conservation agriculture, agroforestry, and biofertilizers contribute to soil organic matter accumulation, foster microbial diversity, and reduce erosion, thus regenerating soil ecosystems. Modern innovations such as biochar, CRISPR-edited microbes, and precision agriculture practice all further these benefits through precise nutrient management, carbon capture, and drought resistance.

Alleviating input cost, enhancing yield, and opening access to premium markets make sustainable practices economically attractive, especially under initiatives like the EU's Common Agricultural Policy or the Indian Soil Health Card Scheme which give knowledge and subsidy support to farmers. These and other socio-economic transformations like rural jobs developed livelihoods, boosted standards of the life, and created opportunities highlight the power of soil-centric approaches. High cost of technology adoption, ethical issues of genetic engineering, and dominance of global policies are some of the challenges that still persist, highlighting the need for collaboration.

In the future, it is essential to shift from exploitative systems to regenerative ones. Multi-stakeholder collaboration, investment in farmer training, and effective governance will be crucial. Focusing on soil health will allow humans to sever the degradation cycle so that ecosystems and communities continue to prosper in the face of climate instability. The cycle of global nutrition and sustainability starts with the care of the soil that nourishes life. This chapter should serve as a challenge to all of us, to grow not just food but also sustainability for those that come after us.

References

Abdo AI, El-Sobky ESEA, Zhang J (2022) Optimizing maize yields using growth stimulants under the strategy of replacing chemicals with biological fertilizers. Front Plant Sci 13(November):1–12. https://doi.org/10.3389/fpls.2022.1069624

Abhiram P, Sandhya N, Chaitanya KV, Reddy SS (2018) Comparison of uterus in Heifer and she buffalo by corrosion cast technique. J Pharmacogn Phytochem 7:2897–2898

Accinelli C, Abbas HK, Shier WT (2018). A bioplastic-based seed coating improves seedling growth and reduces production of coated seed dust. J Crop Improv 32(3):318–330. https://doi.org/10.1080/15427528.2018.1425792

Ahmed N, Deng L, Narejo MUN, Baloch I, Deng L, Chachar S, Li Y, Li J, Bozdar B, Chachar Z, Hayat F, Chachar M, Gong L, Tu P (2024) Bridging agro-science and human nutrition: zinc nanoparticles and biochar as catalysts for enhanced crop productivity and biofortification. Front Plant Sci 15(August):1–19. https://doi.org/10.3389/fpls.2024.1435086

Allen C, Metternicht G, Wiedmann T (2016) National pathways to the Sustainable Development Goals (SDGs): a comparative review of scenario modelling tools. Environ Sci Policy 66:199–207. https://doi.org/10.1016/j.envsci.2016.09.008

Araujo R (2022) Advances in soil engineering: sustainable strategies for rhizosphere and bulk soil microbiome enrichment. Front Biosci Landmark 27(6):10.31083/j.fbl2706195

Bandopadhyay S, Martin-Closas L, Pelacho AM, DeBruyn JM (2018) Biodegradable plastic mulch films: impacts on soil microbial communities and ecosystem functions. Front Microbiol 9(April):1–7. https://doi.org/10.3389/fmicb.2018.00819

Basso B, Antle J (2020) Digital agriculture to design sustainable agricultural systems. Nat Sustain 3(4):254–256. https://doi.org/10.1038/s41893-020-0510-0

Beule L, Vaupel A, Moran-Rodas VE (2022) Abundance, diversity, and function of soil microorganisms in temperate alley-cropping agroforestry systems: a review. Microorganisms 10(3):1–14. https://doi.org/10.3390/microorganisms10030616

Biederman LA, Harpole WS (2013) Biochar and its effects on soil fertility and crop yields: a meta-analysis. https://doi.org/10.1111/gcbb.12037

Biermann F, Kanie N, Kim RE (2017) Global governance by goal-setting: the novel approach of the UN Sustainable Development Goals. Curr Opin Environ Sustain 26–27:26–31. https://doi.org/10.1016/j.cosust.2017.01.010

Bilen S, Çelik A, Altikat S (2010) Conservation tillage systems in the Southeast. Afr J Biotechnol 9(38):6312–6319. pp 167–170. www.sare.org

Bouma J, Montanarella L (2016) Facing policy challenges with inter- and transdisciplinary soil research focused on the un Sustainable Development Goals. Soil 2(2):135–145. https://doi.org/10.5194/soil-2-135-2016

Bronson K, Knezevic I (2021) The digital divide: how COVID-19 widened the gap between small and large farms. Agric Hum Values 38(3):803–814

Bu Q (2019) Reassess the law and ethics of heritable genome editing interventions: lessons for China and the world. Issues Law Med. https://www.semanticscholar.org/paper/9045c3aed2e10cdda6a5f83bb5ac1ca3adfc0ec4

Carthy UM, Uysal I, Badia-Melis R, Mercier S, O'Donnell C, Ktenioudaki A (2022) Corrigendum to 'global food security – issues, challenges and technological solutions' [Trends in Food Science & Technology 77 (2018) 11–20](S0924224417305125)(10.1016/j.tifs.2018.05.002). Trends Food Sci Technol 123(August 2017):404. https://doi.org/10.1016/j.tifs.2022.03.029

Castle SE, Miller DC, Ordonez PJ, Baylis K, Hughes K (2021) The impacts of agroforestry interventions on agricultural productivity, ecosystem services, and human well-being in low- and middle-income countries: a systematic review. Campbell Syst Rev 17(2). https://doi.org/10.1002/cl2.1167

Chaudhary P, Singh S, Chaudhary A, Sharma A, Kumar G (2022) Overview of biofertilizers in crop production and stress management for sustainable agriculture. Front Plant Sci 13. https://doi.org/10.3389/fpls.2022.930340

Cheng Z (2015) Research progress on the use of plant Allelopathy in agriculture and the physiological and ecological mechanisms of Allelopathy. 6(November):1–16. https://doi.org/10.3389/fpls.2015.01020

Chen F. Zhang H, Li H, Lian L, Wei Y, Lin Y, Wang L, He W, Cai Q, Xie H, Zhang H, Zhang, J (2023). IPA1 improves drought tolerance by activating SNAC1 in rice. BMC Plant Bol 23(1). https://doi.org/10.1186/s12870-023-04062-9

Chlingaryan A, Sukkarieh S, Whelan B (2018) Machine learning approaches for crop yield prediction and nitrogen status estimation in precision agriculture: a review. Comput Electron Agric 151:61–69. https://doi.org/10.1016/j.compag.2018.05.012

Das BS, Wani SP, Benbi DK, Muddu S, Bhattacharyya T, Mandal B, Santra P, Chakraborty D, Bhattacharyya R, Basak N, Reddy NN (2022) Soil health and its relationship with food security and human health to meet the sustainable development goals in India. Soil Secur 8(June):100071. https://doi.org/10.1016/j.soisec.2022.100071

Dawn N, Ghosh S, Ghosh T, Guha S, Sarkar S, Saha A, Mukherjee P, Sanyal T (2023) A review on digital twins technology: a new frontier in agriculture. Artif Intell Appl. https://www.semanticscholar.org/paper/4d3ac61a5032825f2d4498f16012c4943b3678c0

Dias PA et al (2020) Federated learning for sustainable agriculture: a case study on soybean yield prediction. IEEE Access 8:180700–180710

Dutta G, Goswami P (2020) Application of drone in agriculture: a review. Int J Chem Stud. https://www.semanticscholar.org/paper/b0e937f933bf892b775cdae06be2f29b9ad345b2

Elemike EE, Uzoh IM, Onwudiwe DC, Babalola OO (2019) The role of nanotechnology in the fortification of plant nutrients and improvement of crop production. Appl Sci. https://www.mdpi.com/2076-3417/9/3/499

Fitzpatrick TB, Chapman LM (2020) The importance of thiamine (vitamin B1) in plant health: from crop yield to biofortification. J Biol Chem 295(34):12002–12013. https://doi.org/10.1074/jbc.REV120.010918

Galili G, Amir R, Fernie AR (2016) The regulation of essential amino acid synthesis and accumulation in plants. Annu Rev Plant Biol 67:153–178. https://doi.org/10.1146/annurev-arplant-043015-112213

Gao S, Guo Y, Cao X, Qiu H, Zhao X (2023) Enhanced phytoremediation for trace-metal-polluted farmland with hibiscus Cannabinus–sedum plumbizincicola rotation: a case study in Hunan, China. Agronomy. https://www.mdpi.com/2073-4395/13/5/1231

García FMM, Alonso GC, Masot AN (2024) Impact of the Common Agricultural Policy (CAP) on the development of rural territories: principal component analysis in SW Spain, extremadura (2007–2020). Agriculture. https://www.semanticscholar.org/paper/667f404ae3cd47365239e1cd08dcb095537d2456

Garland G, Edlinger A, Banerjee S, Degrune F, García-palacios P, Pescador DS, Herzog C, Romdhane S, Saghai A, Spor A, Wagg C, Hallin S, Maestre FT, Philippot L, Rillig MC, Van Der Heijden MGA (2021) Yields in European cropping systems. Nat Food 2(January). https://doi.org/10.1038/s43016-020-00210-8

Griscom BW, Adams J, Ellis PW, Houghton RA, Lomax G, Miteva DA, Schlesinger WH, Shoch D, Siikamäki JV, Smith P, Woodbury P, Zganjar C, Blackman A, Campari J, Conant RT, Delgado C, Elias P, Gopalakrishna T, Hamsik MR et al (2017) Natural climate solutions. Proc Natl Acad Sci USA 114(44):11645–11650. https://doi.org/10.1073/pnas.1710465114

Haider TP, Völker C, Kramm J, Landfester K, Wurm FR (2019) Plastics of the future? The impact of biodegradable polymers on the environment and on society. Angew Chem Int Ed 58(1):50–62. https://doi.org/10.1002/anie.201805766

Hengl T, De Jesus JM, Heuvelink GBM, Gonzalez MR, Kilibarda M, Blagotić A, Shangguan W, Wright MN, Geng X, Bauer-Marschallinger B, Guevara MA, Vargas R, MacMillan RA, Batjes NH, Leenaars JGB, Ribeiro E, Wheeler I, Mantel S, Kempen B (2017). SoilGrids250m: Global gridded soil information based on machine learning. PLoS ONE 12(2):e0169748. https://doi.org/10.1371/journal.pone.0169748

Huey SL, Mehta NH, Konieczynski EM, Bhargava A, Friesen VM, Krisher JT, Mbuya MNN, Monterrosa E, Nyangaresi AM, Boy E, Mehta S (2024) Bioaccessibility and bioavailability of biofortified food and food products: current evidence. Crit Rev Food Sci Nutr 64(14):4500–4522. https://doi.org/10.1080/10408398.2022.2142762

Hayes DG, Anunciado MB, DeBruyn JM, Bandopadhyay S, Schaeffer S, English M, Ghimire S, Miles C, Flury M, Sintim HY (2019). Biodegradable plastic mulch films for sustainable specialty crop production. In Springer eBooks (pp. 183–213). https://doi.org/10.1007/978-3-030-19416-1_11

Jeffery S et al (2011) A quantitative review of the effects of biochar on soil properties and crop yield. https://doi.org/10.1016/j.agee.2011.08.015

Kamilaris A, Prenafeta-Boldú FX (2018) Deep learning in agriculture: a survey. Comput Electron Agric 147:70–90. https://doi.org/10.1016/j.compag.2018.02.016

Kammann C et al (2017) Influence of biochar on soil microbial communities and carbon cycling. https://doi.org/10.2134/jeq2011.0132

Keesstra SD, Bouma J, Wallinga J, Tittonell P, Smith P, Cerdà A, Montanarella L, Quinton JN, Pachepsky Y, Van Der Putten WH, Bardgett RD, Moolenaar S, Mol G, Jansen B, Fresco LO

(2016) The significance of soils and soil science towards realization of the United Nations sustainable development goals. Soil 2(2):111–128. https://doi.org/10.5194/soil-2-111-2016

Kumar A, Sengar RS, Sudheer S (2022) Sustainable management and production of biofortified crops through green technology. https://www.igi-global.com/chapter/sustainable-management-and-production-of-biofortified-crops-through-green-technology/303691

Lal R (2009) Soil degradation as a reason for inadequate human nutrition. Food Secur 1(1):45–57. https://doi.org/10.1007/s12571-009-0009-z

Lal R (2016) Soil health and carbon management. Food Energy Secur 5(4):212–222. https://doi.org/10.1002/fes3.96

Lal R, Bouma J, Brevik E, Dawson L, Field DJ, Glaser B, Hatano R, Hartemink AE, Kosaki T, Lascelles B, Monger C, Muggler C, Ndzana GM, Norra S, Pan X, Paradelo R, Reyes-Sánchez LB, Sandén T, Singh BR et al (2021) Soils and sustainable development goals of the United Nations: an international union of soil sciences perspective. Geoderma Reg 25. https://doi.org/10.1016/j.geodrs.2021.e00398

Lehmann J, Joseph S (2015) Biochar for environmental management: science, technology and implementation. https://doi.org/10.4324/9780203762264

Liakos K, Busato P, Moshou D, Pearson S, Bochtis D (2018). Machine Learning in Agriculture: A review. Sensors 18(8):2674. https://doi.org/10.3390/s18082674

Liu X, Yu Y, Yao W, Yin Z, Wang Y, Huang Z, Zhou JH, Liu J, Lu X, Wang F, Zhang G, Chen G, Xiao Y, Deng H, Tang W (2023) CRISPR/Cas9-mediated simultaneous mutation of three salicylic acid 5-hydroxylase (OsS5H) genes confers broad-spectrum disease resistance in rice. Plant Biotechnol J. https://onlinelibrary.wiley.com/doi/10.1111/pbi.14099

M, R. K. B. Government B, & Grade F (2020) Biofortification strategies for enhancing nutrient content in staple crops: progress, challenges, and future directions. 5(1)

Mamatha B, Mudigiri C, Ramesh G, Saidulu P, Meenakshi N, Prasanna CL (2024) Enhancing soil health and fertility management for sustainable agriculture: a review. Asian J Soil Sci Plant Nutr. https://www.semanticscholar.org/paper/d9b59450e542ed1a08bb5f4f62f211bd08be1c27

Mandal M, Roy A, Mitra D, Sarkar A (2024) Possibilities and prospects of bioplastics production from agri-waste using bacterial communities: finding a silver-lining in waste management. Curr Res Microb Sci 7(September):100274. https://doi.org/10.1016/j.crmicr.2024.100274

Mashamaite CV, Motsi H, Manyevere A, Poswa SB (2024) Assessing the potential of biochar as a viable alternative to synthetic fertilizers in Sub-Saharan Africa smallholder farming: a review. Agronomy 14(6):1215. https://doi.org/10.3390/agronomy14061215

Masto RE, Kumar S, Rout T, Sarkar P, George J, Ram L (2015). Biochar from water hyacinth (Eichornia crassipes) and its impact on soil biological activity. CATENA 111:64–71. https://doi.org/10.1016/j.catena.2013.06.025

Meyer S et al (2021) The role of biochar in climate change mitigation. https://doi.org/10.1007/s42773-024-00391-6

Mwangi M et al (2022) Satellite-driven machine learning for drought prediction in Sub-Saharan Africa. Remote Sens 14(5):1120

Nair PKR (2011) Agroforestry systems and environmental quality: introduction. J Environ Qual 40(3):784–790. https://doi.org/10.2134/jeq2011.0076

Narancic T, Cerrone F, Beagan N, O'Connor KE (2020) Recent advances in bioplastics: application and biodegradation. Polymers 12(4). https://doi.org/10.3390/POLYM12040920

Neupane B, Bhattarai B, Gurung L, Rawal JS, Joshi GR (2024) Integrating climate-smart agriculture for sustainable agriculture: opportunities, challenges and future directions. Arch Agric Environ Sci. https://www.semanticscholar.org/paper/e6af7cc23e6d33d6f00b743edb736dffa3ed6295

Nikolov D, Boevsky I, Banov M, Tzvetanova E, Kostenarov K, Todorova K (2024) Methodological approach for assessing new incentives for soil health business models in EU. Bulg J Agric Econ Manage. https://www.semanticscholar.org/paper/197d908d74d1498827ba4da7202f107b7109542d

Oliinyk L (2022) Peculiarities, and directions of implementation of common agricultural policy in EU. Ukr J Appl Econ Technol. https://www.semanticscholar.org/paper/a32c9a91903944d061405eccbd9c28f2fa2d56d6

Omer E, Szlatenyi D, Csenki S, Alrwashdeh J, Czakó I, Láng V (2024) Farming practice variability and its implications for soil health in agriculture: a review. Agriculture. https://www.semanticscholar.org/paper/8655609c2ec583edac7ab747cfbc68e8cb58d9be

Paine JA, Shipton CA, Chaggar S, Howells RM, Kennedy MJ, Vernon G, Wright SY, Hinchliffe E, Adams JL, Silverstone AL, Drake R (2005) Improving the nutritional value of golden rice through increased pro-vitamin A content. Nat Biotechnol 23(4):482–487. https://doi.org/10.1038/nbt1082

Pantazi XE et al (2016) Wheat yield prediction using machine learning and advanced sensing techniques. Comput Electron Agric 121:57–65. https://doi.org/10.1016/j.compag.2015.11.018

Patil SS, Kumar R (2021) India's digital soil health card initiative: impacts on fertilizer use and crop yields. Agric Syst 190:103102

Paul M, Thriveni V, Niharika M, Upadhyay DK, Kiro D, Verma A (2024) A review on biofortification of crops: a nutritional strategy for combating malnutrition. 16(11):63–77

Philipo M, Ndakidemi PA, Mbega ER (2021) Importance of common bean genetic zinc biofortification in alleviating human zinc deficiency in sub-Saharan Africa. Cogent Food Agric 7(1). https://doi.org/10.1080/23311932.2021.1907954

Piergentili R, del Rio A, Signore F, Umani Ronchi F, Marinelli E, Zaami S (2021) CRISPR-Cas and its wide-ranging applications: from human genome editing to environmental implications, technical limitations, hazards and bioethical issues. Cells. https://www.mdpi.com/2073-4409/10/5/969/pdf

Prabhukarthikeyan SR, Parameswaran C, Keerthana U, Teli B, Jagannadham PTK, Cayalvizhi B, Panneerselvam P, Senapati A, Nagendran K, Kumari S, Yadav MK, Aravindan S, Sanghamitra S (2020) Understanding the plant-microbe interactions in CRISPR/Cas9 era: indeed a sprinting start in marathon. Curr Genomics 21(6):429–443. https://doi.org/10.2174/1389202921999200716110853

Rajput P, Rajput V, Singh A, Sharma R, Rawat S, Tomar RS, Singh A, Minkina T (2021) Impact of emerging tools for solving 'hidden hunger' in human

Rao CM, Sudhakar G (2014) Phytoremediation studies of forest plants growing in and around uranium tailing ponds in Jaduguda, Jharkhand, India – a case study. J Appl Geochem. https://www.semanticscholar.org/paper/11e41da1a9acbfec2f245a27d6f5d230ad0cce9b

Raut DA, Afrayeem S, Singh V, Jejal AD, Sachan P, Sahoo S, Pandey SK (2024) Enhancing nutrition, crop resilience, and food security through biofortification. Int J Environ Clim Chang 14(2):241–253. https://doi.org/10.9734/ijecc/2024/v14i23942

Reddy A (2019) The soil health card scheme in India: lessons learned and challenges for replication in other developing countries. J Nat Resour Policy Res. https://www.semanticscholar.org/paper/55d2459d8b6b9401440e33d4b2349ae086260a7c

Reddy BK, Vijayreddy D, Chandra KB, Emmi SR, Hegde SS, Majumder B, Kumar DP (2024) A comprehensive review on biofortification in vegetable crops. J Adv Biol Biotechnol 27(8):1448–1458. https://doi.org/10.9734/jabb/2024/v27i81267

Regmi S (2023) Review on organic farming in Nepal: a feasible option for agricultural sustainability. Agric Ext Dev Cries. https://www.semanticscholar.org/paper/1712f65deef1961ebb538202ad9e27a59f90e1a2

Riggi E, Santagata G, Malinconico M (2012) Bio-based and biodegradable plastics for use in crop production. Recent Pat Food Nutr Agric 3(1):49–63. https://doi.org/10.2174/2212798411103010049

Roscher R et al (2020) Explainable machine learning for scientific insights and discoveries. IEEE Access 8:42200–42216

Russell SJ, Garcia AK, Kaçar B (2024) A CRISPR interference system for engineering biological nitrogen fixation. MSystems 9(3). https://doi.org/10.1128/msystems.00155-24

Sadiq M, Rahim N, Tahir MM, Alasmari A, Alqahtani MM, Albogami A, Ghanem KZ, Abdein MA, Ali M, Mehmood N, Yuan J, Shaheen A, Shehzad M, El-Sayed MH, Chen G, Li G (2024) Conservation tillage: a way to improve yield and soil properties and decrease global warming potential in spring wheat agroecosystems. Front Microbiol 15(June):1–21. https://doi.org/10.3389/fmicb.2024.1356426

Schimmelpfennig D (2019). Improvements in on-farm resource stewardship with profitable information technologies in rice production. J Environ Econ Policy 8(3): 250–267. https://doi.org/10.1080/21606544.2018.1561329

Sagan V et al (2021) Edge computing for real-time soil health monitoring. Sensors 21(4):1234

Sandeep S, Mittal A (2018) Evaluation of antimicrobial and antioxidant activity of isolated constituents from areal part of cuscuta reflexa roxb. Plant. Asian J Pharm Clin Res 11(3):187–190. https://doi.org/10.22159/ajpcr.2018.v11i3.22475

Sen H, Kumar A, Janeja HS (2024) Biofortification of major crops through conventional and modern biotechnological approaches to fight hidden hunger: an overview. J Adv Biol Biotechnol 27(7):96–113. https://doi.org/10.9734/jabb/2024/v27i7970

Seong Goo Kim, Bae J (2024) To expand employment opportunities for the mentally disabled in Korea A study on the introduction of care farms in the European Union. Korean Society of Vocational Rehabilitation https://www.semanticscholar.org/paper/8770267eb6ff6628ed088646591c9861072b2e75

Sheoran S, Kumar S, Ramtekey V, Kar P, Meena RS (2022) Current status and potential of biofortification to enhance crop nutritional quality: an overview. Sustainability. https://www.mdpi.com/2071-1050/14/6/3301

Shepherd KD et al (2020) Soil spectral diagnostics: Unleashing the potential of machine learning for global soil mapping. Geoderma 375:114540

Shozi B (2020) A critical review of the ethical and legal issues in human germline gene editing: considering human rights and a call for an African perspective. S Afr J Bioethics Law. https://www.semanticscholar.org/paper/b26ae0ee360954f0ce3d1cd5ea1a4cb9c478de0a

Singh V, Ganguly S, Dakshinamurthy V (2018) Evaluation of India's soil health card from users' perspectives. https://www.semanticscholar.org/paper/73b86cc7e60e77ac4caf076dad73135312e4a642

Sintim HY, Bary AI, Hayes DG, Wadsworth LC, Anunciado MB, English ME, Bandopadhyay S, Schaeffer SM, DeBruyn JM, Miles CA, Reganold JP, Flury M (2020) In situ degradation of biodegradable plastic mulch films in compost and agricultural soils. Sci Total Environ 727:138668. https://doi.org/10.1016/j.scitotenv.2020.138668

Smith J, Farmer J, Smith P, Nayak D (2021) The role of soils in provision of energy. Philos Trans R Soc B Biol Sci 376(1834). https://doi.org/10.1098/rstb.2020.0180

Soorya M, Suganya S, Elayarajan M, Davamani V, Chitra P (2024) e – Publishing Group. 11(3):1–17

Tamburini G, Bommarco R, Wanger TC, Kremen C, Van Der Heijden MGA, Liebman M, Hallin S (2020) Agricultural diversification promotes multiple ecosystem services without compromising yield. November

Tendall DM, Joerin J, Kopainsky B, Edwards P, Shreck A, Le QB, Kruetli P, Grant M, Six J (2015) Food system resilience: defining the concept. Glob Food Sec 6:17–23. https://doi.org/10.1016/j.gfs.2015.08.001

Tóth G, Hermann T, da Silva MR, Montanarella L (2018) Monitoring soil for sustainable development and land degradation neutrality. Environ Monit Assess 190(2). https://doi.org/10.1007/s10661-017-6415-3

Tzachor A et al (2022) Artificial intelligence in the food system: risks and opportunities. Glob Food Sec 33:100641

Vasconcelos M, Datta K, Oliva N, Khalekuzzaman M, Torrizo L, Krishnan S, Oliveira M, Goto F, Datta SK (2003) Enhanced iron and zinc accumulation in transgenic rice with the ferritin gene. Plant Sci 164(3):371–378. https://doi.org/10.1016/s0168-9452(02)00421-1

Vidican R, Mihăiescu T, Pleşa A, Malinas A, Pop B (2023) The phytoremediation potential of Lavandula Angustifolia mill. Grown in soils historically polluted with heavy metals: a case study from Baia Mare, Romania. J Appl Life Sci Environ. https://www.semanticscholar.org/paper/06da487040e09be97a94ed7920cda70021ea0bda
Wahome PK (2016) Paul K. Wahome: recent advances in hydroponics production of horticultural crops: a review. https://www.semanticscholar.org/paper/99091fd31a7c544bf814f2b687c7592c468087f0
Wang G, Liu Y, Yan Z, Chen D, Fan J, Ghezzehei TA (2023) Soil physics matters for the land–water–food–climate nexus and sustainability. Eur J Soil Sci 74(6):1–6. https://doi.org/10.1111/ejss.13444
Wolfert S, Ge L, Verdouw C, Bogaardt M (2017). Big Data in Smart Farming – A review. Agric Syst 153:69–80. https://doi.org/10.1016/j.agsy.2017.01.023
Yadava DK, Hossain F (2018) Nutritional security through crop biofortification in India: status & future prospects. https://journals.lww.com/ijmr/fulltext/2018/48050/nutritional_security_through_crop_biofortification.16.aspx
Zou Y, Liu Z, Chen Y, Wang Y, Feng S (2024) Crop rotation and diversification in China: enhancing sustainable agriculture and resilience. Agriculture 14(9):1465. https://doi.org/10.3390/agriculture14091465

Chapter 12
An Overview of Microbes Present in the Extreme Environment

Aleena Taufiq, Saurabh Singh, and Taru Singh

Abstract Microorganisms capable of withstanding the harshest conditions of Earth are described as extremophiles: they include organisms adapted to grow and thrive in high pressure, radiation, and pH, salinity, and temperatures far above or below ambient. These microbial forms include thermophiles, psychrophiles, halophiles, acidophiles, alkaliphiles, and piezophiles, which are all adapted to their environment in a specialized physiological, biochemical, and genetic part. Due to their uniqueness in having unique metabolic pathways and biomass, such as extremozymes, they possess immense biotechnological potential, as that was derived from pharmaceuticals, biofuels, food processing, or environmental bioremediation. Extremophiles are also models to understand the origins and evolution of life and elucidate early Earth conditions and the search for extraterrestrial life in astrobiology. Further expansion of the exploration and engineering of extremophiles for industrial and scientific innovations is enabled by advances in omics technologies and synthetic biology. We review the diversity of extremophiles and their adaptive mechanisms in addition to emphasizing their applications, specifically because they enable the biotechnology and space exploration frontiers to be reached. This knowledge expands our understanding of life's resilience and simultaneously opens doors to groundbreaking work in numerous related scientific areas.

Keywords Extremophiles · Adaptation Mechanisms · Extremozymes · Biotechnology · Astrobiology

A. Taufiq · T. Singh (✉)
Centre for Medical Biotechnology, Amity Institute of Biotechnology, Amity University, Noida, Uttar Pradesh, India
e-mail: tsingh17@amity.edu

S. Singh
Delhi Technological University, Delhi, India

A. W. Fernandes Duarte et al. (eds.), *Polar Microbes and Climate Change*,
https://doi.org/10.1007/978-3-032-07096-8_12

12.1 Introduction

In what has proven to be an extraordinary feat for microorganisms (and life), they have thrived under some of Earth's harshest conditions. Such extreme habitats include high-temperature hydrothermal vents, hypersaline lakes, acidic and alkaline hot springs, deep sea trenches, and the exposure to radiation. Extremophiles are microbes that dwell in such environments, and they have adapted molecular, structural, and physiological features that enable them to survive and even thrive, where life forms that are not extremophiles would die otherwise (Cavicchioli et al. 2011). They are exploited to study microbial evolution, biotechnological applications, as well as astrobiology, due to their potential for extraterrestrial life (Rothschild and Mancinelli 2001).

There is remarkable adaptability that extreme environment microorganism shows, and they usually were classified according to the conditions they grow well. Thermophiles live at high temperatures, which they find in geothermal springs and hydrothermal vents; most of the species are from the domain *Archaea*. This ability is due to heat resistant enzymes and specialized lipid membranes keeping this cellular integrity at temperatures of greater than 80 (Stetter 2006). On the other hand, psychrophiles tolerate subzero temperatures such as those found in Antarctic ice sheets and deep-sea sediments and have adapted by expressing antifreeze proteins and enzyme plasticity to keep metabolic activity (Merino et al. 2019). There is another, elaborate, better known group, halophiles, who inhabit hypersaline conditions, such as salt flats, saline lakes, and salt mines. Osmotic pressure is regulated within these microbes, and they can accumulate compatible solutes, such as glycine betaine and ectoine, and unique ion pumps to prevent cellular dehydration (Williams and Hallsworth 2009; Oren 2002). Due to acquisition of proton pumps which maintain intracellular pH balance and modification of their cell membranes, acidophiles and alkaliphiles have adapted to extreme pH conditions such as sulfuric hot springs and alkaline soda lakes (Pikuta et al. 2007).

Among them are radiation-resistant microbes such as *Deinococcus radiodurans* that can survive high doses of ionizing radiation by virtue of highly efficient DNA repair mechanisms and cells' protective cellular antioxidants (Rampelotto 2013). Also, deep-sea barophiles (barophiles; piezophiles) live deep at the bottom of sea trenches, where fluidized cell membranes and pressure-resistant proteins allow their function in this extreme level (Seckbach and Oren 2010). They are valuable in biotechnological applications due to the unique survival strategies of extremophiles. Examples are thermophilic enzymes like those from *Thermus aquaticus* used to make polymerase chain reaction (PCR) using the hot temperature amplified DNA (Stetter 2006). Furthermore, the stable biomolecules produced by halophiles are used as medications or are used in pharma, food preservation, and industrial processes using high salt concentrations (Williams and Hallsworth 2009). Extremophiles are important beyond Earth in astrobiology simply for expanding our knowledge of what constitutes planetary habitability. This discovery of microbial life in extreme terrestrial environments is compatible with the existence of similar organisms on Mars, Europa, or Enceladus (where water is subglacial, radiation is high, and low temperatures abound on Earth, too) (Merino et al. 2019). The persistence of life under various extreme ecological pressures also applies to microbial communities, like those in stratified microbial mats (Seckbach and Oren 2010).

Exploring extremophiles is not only pushing the boundary of life understanding but also has finding of great importance for biotechnology and biochemistry and space research. These microorganisms are models for early life forms on Earth and potential extraterrestrial life on which much is known about evolutionary biology and about the search for habitable environments beyond our planet. Future exploration of microbial ecology, genomics, and astrobiology will reveal increasingly remarkable capabilities of extremophiles which may be used in science and industry (Rampelotto 2010). The table below displays how factors like temperature, pressure, salinity, and radiation interact with microbes and have physiological and biochemical effects on their survival metabolisms and cellular adaptations to the environments (Table 12.1).

Table 12.1 Effect of extreme environments on microbes

Extreme environment	Effects on microbes
High temperature (hyperthermophiles)	Protein denaturation risk, enhanced enzyme thermostability, lipid monolayer formation for membrane stability (Cavicchioli et al. 2011; Stetter 2006)
Low temperature (psychrophiles)	Reduced metabolic rates, increased membrane fluidity, production of cold-active enzymes to maintain biochemical reactions (Cavicchioli et al. 2011; Rothschild and Mancinelli 2001)
High salinity (halophiles)	Water loss due to osmotic stress, accumulation of compatible solutes, specialized ion pumps for salt regulation (Williams and Hallsworth 2009; Oren 2002)
High pressure (barophiles/piezophiles)	Protein structure alteration, increased membrane rigidity, expression of pressure-adapted enzymes (Cavicchioli et al. 2011; Pikuta et al. 2007)
Extreme pH (acidophiles/ alkaliphiles)	Risk of proton influx (acidic environments) or proton loss (alkaline environments), modification of membrane lipids, proton pumps for pH homeostasis (Rothschild and Mancinelli 2001; Rampelotto 2013)
Desiccation (xerophiles)	DNA and protein damage due to water loss, accumulation of protective sugars, activation of efficient DNA repair mechanisms (Merino et al. 2019; Rampelotto 2010)
High radiation (radioresistant microbes)	DNA damage and mutation risks, activation of rapid DNA repair mechanisms, production of antioxidant molecules (Seckbach and Oren 2010)

12.2 Types of Extreme Environments and Microbial Inhabitants

12.2.1 High-Temperature Environments (Thermophiles and Hyperthermophiles)

Thermophilic and hyperthermophilic microorganisms are supported by a very varied habitat in high-temperature environments such as hot springs and hydrothermal vents. Encompassed by these extreme habitats are elevated temperatures of over 80 °C and, in some cases, as high as 121 °C requiring special adaptations for survival (Stetter 2006). But these settings are inhabited by bacteria, archaea, fungi, and even viruses, and every one of these microbial lifeforms has a long, sordid history of biochemical and physiological mechanisms that can allow them to survive and thrive in the face of extreme thermal conditions (Fig. 12.1). *Thermus aquaticus*, from among the bacteria, is one of the most studied thermophiles, which was isolated originally from hot springs in Yellowstone National Park, United States. This species received a great deal of scientific attention because of the thermostable DNA polymerase, Taq polymerase, derived from this species, was applied to molecular biology in a new capacity for the polymerase chain reaction (PCR). A further bacterial genus of note is *Geobacillus* featured with many thermophile species able to produce heat stable enzymes such as amylases and proteases with important applications in industrial biotechnology (Bertoldo and Antranikian 2002). The metabolic versatility of such bacteria in geothermal environments offers potential for biotechnological exploitation.

High-temperature ecosystems depend on hyperthermophilic archaea that can survive at temperatures greater than 80 °C. A strictly anaerobic archaeon, *Pyrococcus*

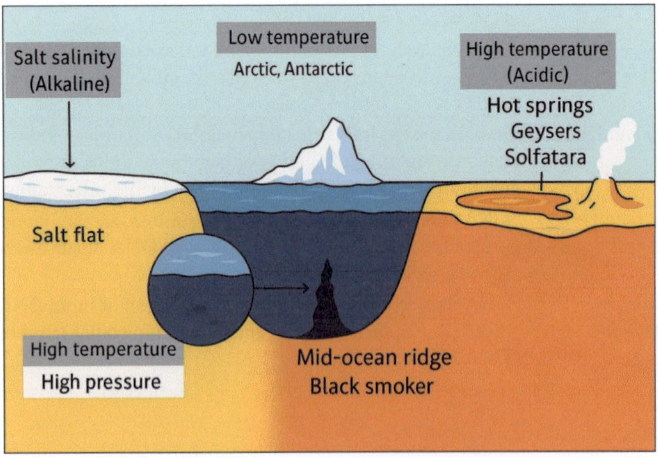

Fig. 12.1 Diverse extreme environments on Earth: exploring habitats characterized by high salinity, temperature, pressure, and acidity

furiosus is optimally grown at high temperature, around 100 °C, and synthesizes an array of thermostable enzymes of industrial use, such as for biofuel production and biocatalysis (Vieille and Zeikus 2001; Huber and Stetter 1998). *Methanopyrus kandleri* has also been found in deep-sea hydrothermal vents, exemplifying their extraordinary ability of producing methane at temperatures as close as 110 °C while it participates in carbon cycling in extreme environments (Stetter 2006). These unique membrane lipid composition and special protein structure attributed to the induced resilience of these archaea to the thermostability and enzymatic efficiency. Some thermophilic fungal species are also presented, the best-known being *Thermomyces lanuginosus*. Because of its production of thermostable enzymes xylanases and cellulases, it is known to be widely used in biofuel and paper industries (Gomes et al. 2016). Its finding in the geothermal habitats demonstrates the extensibility of eukaryotic microorganisms to extreme conditions and their ecological importance in the organic matter decomposition. In addition, thermophilic viruses infecting archaea play a role in the microbial ecology of high-temperature environments, in addition to cellular life forms. Specific structural adaptations of archaeal viruses enable them to remain stable under hostile conditions, which inform evolutionary processes within the viruses as well as host-virus interactions in high-temperature environments (Prangishvili et al. 2006). This is necessary to understand the dynamics of coevolution between extremophiles and their viral counterparts.

However, research on microbial diversity in high-temperature ecosystems is being increased by the work of researchers using advanced metagenomic methods to examine microbial communities in hot springs located in different location (Menzel et al. 2015). Finally, these studies show varying composition of microbes according to temperature, pH, and geochemical conditions, which other studies have shown adaptations of thermophiles and hyperthermophiles (Hedlund et al. 2015a). These microorganisms are actively being explored both to elucidate life under so extreme conditions and for emerging use in biotechnology, especially enzyme engineering, industrial processing, and bioenergy production (Huber and Stetter 1998).

12.2.2 Low-Temperature Environments (Psychrophiles)

A great variety of psychrophilic microorganisms live in low-temperature environments, like polar ice caps and deep-sea habitats, adapted to survive and to thrive in extreme cold. Bacteria, archaea, fungi, algae, and viruses are these microorganisms of which each different contributes to ecological balance as well as biogeochemical cycles of these environments. Some of the most well-studied microbial inhabitants of Antarctic Sea ice are psychrophilic bacteria including *Psychrobacter* and *Colwellia*. The production of cold-adapted enzymes, antifreeze proteins, and membrane modifications permit fluidity at subzero temperatures and specialized cellular mechanisms, to ensure survival of these bacteria (Bowman et al. 1997). Genomic and proteomic analyses of *Colwellia psychrerythraea* have shown that it can survive under extreme cold, as demonstrated by key genes related to cryoprotection, lipid metabolism, and stress response required for the survival in such conditions (Methé et al. 2005).

Like the other major microbial group, archaea, extreme cold environments provide a life raft giving this group extraordinary resilience. For instance, *Methanococcoides burtonii*, a methanogenic archaeon, survives in Antarctic lakes where it takes advantage of metabolic pathways to do methane production at freezing temperatures (Saunders et al. 2003). At low temperatures, its genome carries cold-shock proteins, compatible solutes, and membrane adaptations for cellular function in near freezing waters. This indicates the role of such archaea in life evolution in extreme habitats and the global methane cycle (Saunders et al. 2003). In Arctic and Antarctic aquatic environments, psychrophilic fungi have been discovered in the phylum *Chytridiomycota*, which are essential in decomposing organic matter and recycling nutrients. Because these fungi have evolved functional enzymatic systems at low temperatures, they are critical to microbial food webs in glacial and subglacial ecosystems (Wang et al. 2017). Like these cold-adapted algae, such as *Chlamydomonas nivalis*, snow algae, fresh adapted algae commonly survive in frozen habitats using photoprotective pigments, ice-binding proteins, or metabolic adjustment to withstand high UV radiation and freezing stress. Carbon cycling in the snowfields is dependent on this microalga which is responsible for the characteristic red and green coloration seen in snow fields (Remias et al. 2005). Cold-adapted bacteriophages are central to the ecology of microbial communities in polar environments. The properties of these phages allow them to infect, regulate, and influence the populations of bacteria, with consequences for genetic exchange, microbial diversity and nutrient cycling (Sullivan et al. 2010). Oceanic myoviruses have been studied to show their great distribution in the deep cold marine environments where they participated in evolution and adaptation of cold marine microbial hosts (Sullivan et al. 2010).

The ability to study psychrophiles has enormous scope for answering questions about life's resilience in the extreme and for biotechnological application. The psychrophilic bacteria and fungi have given out cold-adapted enzymes that have had great potential in industrial processes like bioremediation, pharmaceutical, and in food processing where low-temperature function is required (Margesin and Miteva 2011). Investigating these extremophiles answers to astrobiology, as icy extraterrestrial environments, like the subsurface oceans of Europa and Enceladus could harbor similar microbial life forms.

12.2.3 High-Salinity Environments (Halophiles)

Extremely osmotic stress is imposed on microbes living in high-salinity environments, for example, salt lakes and salt mines. Though living in such a harsh environment, a wide range of halophilic microorganisms have adapted and diversely evolved that are capable of surviving in such habitats. They are key to ecosystem stability and nutrient cycling along with playing an important role in biotechnological applications.

Especially halophilic bacteria, belonging to the genus of *Halomonas*, are rather tolerant to saline environment. To overcome osmotic stress these bacteria, use various tactics that include accumulation of compatible solutes such as ectoine and betaine which facilitate the maintenance of cell turgor without affecting metabolic functions (Oren 2002). Metabolically versatile *Halomonas* species, utilizing a wide variety of carbon sources and normalizing to a wide range of salinity, allow their application as bioremediating agents or industrial enzyme production (Ventosa and Nieto 1995). Members of the genera *Halobacterium* and *Haloferax* are dominant in extreme salinity environments because of their highly specialized physiological and structural adaptations, and it is these groups of archaea that are studied in the most detail. The potassium concentrating mechanism in the archaea is directed to counteract the external osmotic pressure which would otherwise cause cellular dehydration (Ma et al. 2010). Moreover, many archaebacteria are also halophilic and possess gas vesicles, which give them buoyancy and enable their positioning optimally in water columns to capture nutrients and undergo phototrophic growth. *Halobacterium salinarum*, for example, has use of bacteriorhodopsin, a light-driven proton pump that allows the species to utilize phototrophy even in the absence of nutrient (Antón et al. 2002).

Unlike bacteria and archaea, halophilic fungi such as *Wallemia ichthyophaga* exhibit unique strategies for coping with hypersaline environments. The main mode of osmotic balance in this case is the accumulation of intracellular glycerol, a powerful osmoprotectant which stabilizes cellular structures and protects these organisms from desiccation (Gunde-Cimerman et al. 2009). Especially adapted to extreme salinity, *Wallemia ichthyophaga* will grow in NaCl (gravity!) around 20% and is known as one of the most halotolerant fungi. These properties suggest these enzymes' potential applications in biotechnology (especially in the creation of salt-tolerant enzymes for industrial processes). The green unicellular alga *Dunaliella salina* is one of the most "well-adapted" eukaryotic organism to hypersaline conditions. To flourish in environments that are a Salt Lake within a salt desert, it regulates intracellular glycerol concentrations to allow the balancing of osmotic pressure and of plasmolysis (Oren 2014). Furthermore, *D. salina* is a beta carotene prolific producer or more exactly a very strong producer of the powerful antioxidant which preserves the cell against oxidative damage brought by high salinity and the light intensity. And because of it, this particular trait has been used extensively in the production of β-carotene in the commercial arena for use in the food, cosmetic, and pharmaceutical industries. High-salinity environments also depend upon viruses to shape microbial communities. Archeal viruses are infecting some species of *Haloferax* and *Halorubrum* and modulation of microbial population dynamics and genetic exchange (Atanasova et al. 2012). These viruses facilitate horizontal transfer of genes that confer increased halophilicity among halophilic archaea. By their presence, they highlight the ecological complexity that occurs in hypersaline ecosystems in which viruses help regulate the evolution of microbial populations.

Halophile studies in high-salinity environments are a valuable tool to study microbial adaptation, ecological interaction, and their possible applications in biotechnology. Halophiles are remarkable in that they possess osmoprotective mechanisms in bacteria and archaea, or the production of bioactive compounds in fungi and algae. The potentials for both such innovation and discovery are boundless in the field of extremophile exploitation given further exploration of these extremophiles.

12.2.4 Acidic and Alkaline Environments (Acidophiles and Alkaliphiles)

There are microorganisms that live in very acidic or alkaline environments such as acid mine drainage and soda lake that have very big biochemical and physiological adaptations to live in such an extreme pH environment. These include acidophiles that can thrive in an environment of very low pH and where they fulfill key ecological, especially, metal cycling, geochemical transformations, and bioleaching roles. *Acidithiobacillus ferrooxidans* is one of the best studied acidophilic bacteria that oxidize iron and sulfur compounds in acid mine drainage and play a major role in acidogenesis and metal solubilization (Johnson and Hallberg 2003). *Picrophilus torridus* is one of the most acid-tolerant organisms known and can live at a pH as low as 0.06 in the domain of archaea. It is extremely unique with its cellular membrane composition and intracellular pH regulation mechanisms because these mechanisms are enabling it to maintain steady, stable in extreme acidity (Golyshina 2011). Also, *Acidomyces* and other fungal species have been identified in acid conditions, where they degrade organic matter and cycle nutrients, playing along with other microorganisms in shaping the microbial communities (Baker and Banfield 2003). Extreme acid tolerance in protozoa, such as the acidified waters that are colonized by the Euglena mutabilis, contributes to the primary production in such extreme ecosystems (Baker and Banfield 2003).

In contrast to the natural environment in which their cousins live, alkaliphiles carry out key roles in CO(2) fixed carbon, nitrogen, and sulfur cycles in soda lakes. Bacteria with this extraordinary level of pH homeostasis, include *Bacillus alcalophilus*, for example, possess sophisticated pH homeostasis mechanisms, the concerted activity of which allows this microorganism to maintain an optimal intracellular pH under highly alkaline external conditions (Jones et al. 1998). Haloalkaliphilic microorganisms found in soda lakes live in high pH and high salinity conditions. Sulfur-oxidizing bacteria found in these lakes contribute to the lake's unique chemistry oxidation of reduced sulfur compounds by facilitating sulfur biogeochemical cycles (Sorokin and Kuenen 2005). As well, archaeal communities have been shown to be remarkably adapt to alkaline conditions: phylogenetic studies of alkaline salterns have discovered archaeal phylotypes capable of growth in extreme salinity and high pH, further expanding the knowledge of extremophilic microbial diversity (Grant et al. 1999). Extreme environments of pH at acid levels also host acidophilic archaeal viruses that have adapted to infect acidophilic hosts. The structural and genetic properties of these viruses are in fact unusual and enable these viruses to be stable in extremely acidic environment, which explains some aspects of viral evolution in very harsh conditions (Golyshina 2011). It also marks the discovery of such viruses, which highlights the complicate microbial interactions that take place in extreme habitats.

Apart from being ecological significance, acidophiles and alkaliphiles are of biotechnological importance. Acidophiles, such as the *Acidithiobacillus ferrooxidans*, are commonly applied in biomining and by bioleaching, for metal extraction from ores, being an ecofriendly replacement for classical mining processes (Johnson and Hallberg 2003). On the contrary, production of industrially relevant enzymes, i.e., alkaline proteases and cellulases, by alkaliphiles are used in the detergent, textile, and paper industries (Zavarzin et al. 1999). Due to their ability to function under extreme conditions, these microorganisms are good candidates for use in applications of environmental remediation, in industrial biotechnology, but also for space exploration (Niederberger et al. 2010).

12.2.5 High-Pressure Environments (Barophiles/Piezophiles)

Extreme microorganisms have adapted to high-pressure environments (e.g., depths of subglacial lakes and deep-sea trenches) where hydrostatic pressure can reach enormous values. Living in these environments, microbial life is diverse and consists of bacteria, archaea, and viruses, and each plays its part in biogeochemical cycles, stability of ecosystems (Fang et al. 2010).

Shewanella violacea and *Psychromonas* are uniquely barophilic bacteria and among the bacterial species that thrive in these extreme situations. These bacteria have specially prepared membrane structures and pressure adapted enzymes which allow these bacteria to perform cellular function under extreme conditions (Nogi et al. 1998). They can maintain survival in deep sea trenches where the pressure reaches 100 MPa by their ability to regulate membrane fluidity and solve stability (Kato et al. 2008). Like piezophilic archaea such as *Methanococcus jannaschii*, which adapt to high-pressure environments by developing metabolic pathway to produce energy through methanogenesis, essentially the process of carbon cycling in the deep sea (DeLong and Pace 2001). Furthermore, viruses, in particular, deep-sea bacteriophages, are also highly important for the establishment and maintenance of microbial communities in these extremely inhospitable surroundings. They participate in regulating bacterial populations via viral predation, horizontal gene transfer, and increase of genetic diversification in high-pressure environments (Sogin et al. 2006). Advantage was taken of the presence of viral particles in deep-sea sediments and hydrothermal vents to show their role in microbial ecology and the adaptation of *piezophilic* bacteria and archaea (Eloe 2011).

They are important for deep-sea ecology and for biotechnology, enzyme engineering, astrobiology, etc. Extreme conditions encountered in the deep sea provide valuable insights into the outer limits of life on Earth and the ability for microbial life to exist beyond the Earth in high-pressure environments such as subsurface oceans on icy moons (Jørgensen and Boetius 2007). Further research on piezophiles is important for the study of novel biochemical pathways, biomolecules, and possible applications of them in medicine and industry (Fang et al. 2010).

12.2.6 Dry and Radiation-Intensive Environments (Xerophiles and Radioresistant Microbes)

Due to living in dry and radiation intensive environments like deserts, outer space, and nuclear waste sites, microorganisms that live there have developed the ability to withstand extreme conditions. These are xerophiles and radioresistant microbes that have highly tuned physiological and genetic mechanisms for the very survival (Fig. 12.2). *Deinococcus radiodurans* belongs to this group of bacteria, which is widely considered one of the most radiation-resistant bacteria, the ability of which to survive doses of ionizing radiation that would be lethal for most organisms. It's highly efficient DNA repair systems with homologous recombination and nonhomologous end joining and the ability to neutralize oxidative damage by protective proteins and enzymatic antioxidant protein to explain its extraordinary ability to survive wide variations (Slade and Radman 2011). Desiccation stress also allows *Streptomyces* species that occur in arid environments to develop strategies for

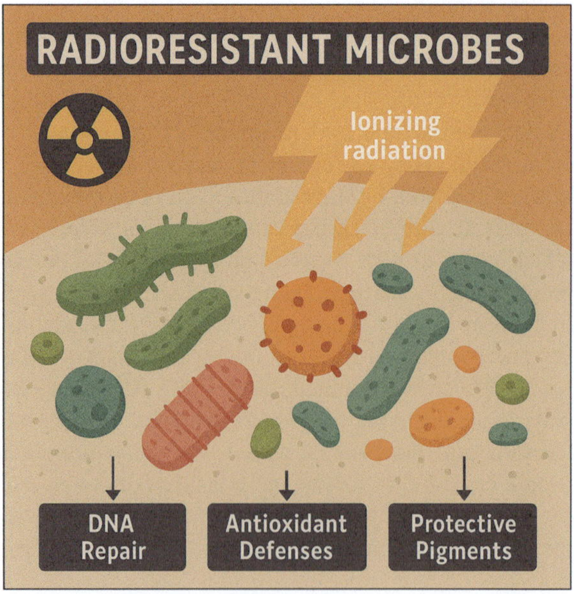

Fig. 12.2 Mechanisms of radioresistant microbes for DNA repair, antioxidant defense, and radiation protection

survival, such as producing spores with thick protective coatings and the production of secondary metabolites that promote survival in nutrient deprived environments such as that found in arid environments (Rainey et al. 2005).

Furthermore, adaptations for facilitating living in extreme habitats are found among archaea, as *Halococcus dombrowskii* is adapted to survive in hypersaline environments that would also be exposed to intense radiation. The production of compatible solutes supports its resilience by maintaining cellular integrity and enzymatic function under stress (Oren 2002). Fungi, namely, *Aspergillus niger*, have been demonstrated to survive on space conditions on spacecraft surfaces such as the International Space Station (ISS). It is presumed that melanin in the fungal cell walls is responsible for the radiation protection as it absorbs and dissipates dangerous radiation thus minimizing the DNA (Dadachova and Casadevall 2008; Horneck et al. 2010). Despite often getting overlooked in extreme environment studies, viruses, too, have proven remarkably resistant to UV radiation. Prolonged exposure to high-energy radiation proves to be particularly friendly toward certain bacteriophages which are known to tolerate strong radiation, survive in hostile environments, and shape the microbial community dynamics. Particularly importantly for the role they play in horizontal gene transfer, extremophiles use them to facilitate genetic exchange of genetic material between other extremophiles; they may facilitate adaptation and survival in extreme conditions (Nicholson et al. 2005). Thus, the study of these extremophiles not only enhances our knowledge of the ecology of life on our planet but also has implications for astrobiology, biotechnology, and research on radiation protection. Microbial survival strategies in extreme conditions can provide relevant information on the possibility of extraterrestrial life, development of radiation-resistant biomaterials, and bioremediation of contaminated environments (Rothschild and Mancinelli 2001).

12.3 Molecular and Physiological Adaptations

Specialized molecular and physiological adaptations to low or high temperatures, radiation, and osmotic stress allow psychrophiles and (hyper)thermophiles to survive in environments from extreme negative to extreme positive temperatures (Cavicchioli et al. 2002; DiRuggiero et al. 1997). Reduced hydrophobic interactions and weaker stabilizing forces allow psychrophilic enzymes to have structural flexibility and therefore improve catalytic efficiency at low temperatures. Due to these adaptations, activation energy barriers are reduced, such as in the case of cold-active chitobiase, as well as β galactosidase from *Arthrobacter*, whose operations are optimized by the use of electrostatic interactions during substrate binding (Csonka 1989; Feller and Gerday 2003). In contrast, hyperthermophiles such as *Pyrococcus furiosus* utilize increased intramolecular interactions along with the presence of thermostable solutes, namely, di-myo-inositol phosphate (DIP) and mannosyl glycerate to maintain enzyme stability (DiRuggiero et al. 1997; Santos and Da Costa 2002) (Fig. 12.3).

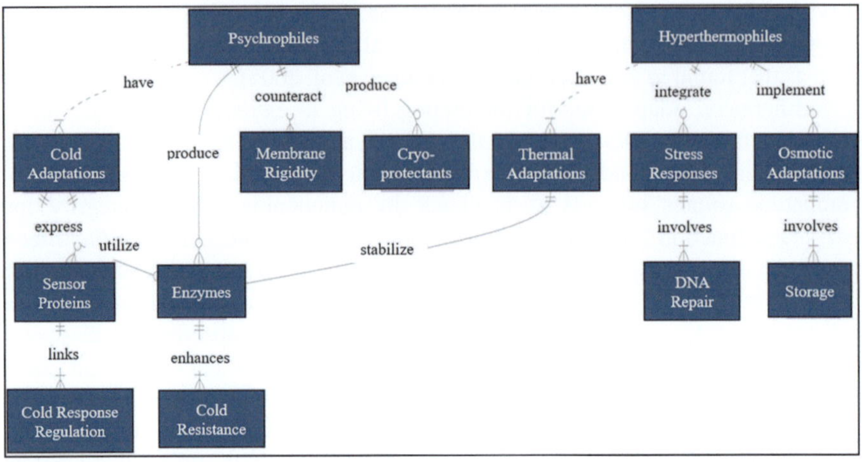

Fig. 12.3 Molecular and physiological adaptations of psychrophiles and hyperthermophiles

Psychrophiles include unsaturated and short-chain fatty acids in membranes to counteract membrane rigidity in cold environments in *Micrococcus roseus* and *Sphingobacterium antarcticus* and some bacteria convert cis fatty acids to trans-forms to further stabilize membranes. Membrane integrity also includes carotenoids (Cavicchioli et al. 2002; Feller and Gerday 2003). Transfer of ether linking provides thermal stability preventing membrane collapse at extreme temperatures, in hyperthermophiles (Santos and Da Costa 2002). Such RNA chaperones, helicases, and cold-shock proteins (CSPs) as CspA of *Escherichia coli* are produced by cold-adapted bacteria all over to assure efficient transcription and translation at low temperatures. Cold stress also sustains expression of genes through ribosomal protein modifications (Cavicchioli et al. 2002; Chattopadhyay 2006). Conversely, *P. furiosus* repairs DNA fragmentation caused by radiation and thermal damage via highly efficient recombination repair as shown for *Deinococcus radiodurans*, and genome restoration occurs within 20 h postradiation (DiRuggiero et al. 1997). Psychrophiles also produce intracellular cryoprotectants such as trehalose, glycerol, and glycine betaine to prevent intracellular ice formation. For example, *Marinomonas primoryensis* or *Moraxella* sp. AFPs inhibit ice recrystallization and help bacteria survive in freezing environments (Cavicchioli et al. 2002; Chattopadhyay 2006). Compatible solutes, for example, MG, DIP, and diglycerol phosphate (DGP) stabilize proteins and protect them from osmotic or even thermal stress and increase thermal resistance in (hyper)thermophiles. Unlike mesophiles, they have interior solute synthesis (Santos and Da Costa 2002; Feller and Gerday 2003).

Solutes can be accumulated or ions regulated to adapt to osmotic stress, depending on the bacterium. While MG and mannosylglyceramide are stored in thermophiles such as *Rhodothermus marinus*, mannosylglyceramide production in the two-step pathway is essential during heat production in *Pyrococcus* species. Thermophiles such as *Thermotoga thermarum* have evolved without the accumulation of solutes suggesting that osmotic balance is their main adaptation (Santos and Da Costa 2002). Periplasmic oligosaccharides play a role in both the osmolarity regulation in gram-negative bacteria, as well as ions' dependent uptake. The *Salmonella typhimurium* proU operon encodes proteins involved in the uptake and efflux of osmoprotectants, while the EnvZ-OmpR system tunes ingress of solutes through outer membrane porins (Feller and Gerday 2003). Membrane-associated phosphorylation is used by cold-adapted bacteria to sense temperature fluctuations seen in *Pseudomonas syringae.* Cold response is also regulated in *Synechocystis* and *Listeria monocytogenes* by histidine kinase Hik33, and it mediates the cold adaptation links with oxidative and heat stress responses through HtpG (Hsp90 homolog), respectively (Chattopadhyay 2006). Other as processes of hyperthermophiles, stress-regulated mechanism are integrated including responses to temperature fluctuations and to damage induced in DNA by radiation (DiRuggiero et al. 1997).

12.4 Evolutinary and Ecological Significance

Microorganisms that inhabit extreme environments including high-altitude stratosphere ecosystems, acid mine drainage systems (AMD), polyextreme settings, and deep-sea hydrothermal vents possess remarkable survival tactics that redefine what we know about life. Their adaptation methods reveal important conclusions about ecological resilience as well as evolutionary mechanisms together with possible space life implications. The stratosphere environment subjects microorganisms to crippling ultraviolet rays along with dehydrating conditions and freezing temperatures and reduced atmospheric pressure. Survival mechanisms that allow microorganisms to persist include the ability to sporulate (*Bacillus subtilis* as an example) as well as biofilm formation by *Deinococcus aetherius* and DNA repair effectiveness in *Halobacterium* sp. NRC-1 (DasSarma and DasSarma 2018). These survival methods affect how microorganisms spread through the atmosphere as well as possible life transport across interplanetary space.

The extremely acidic conditions (pH <2) and elevated metals and oxidative stress within AMD ecosystems host various extremophilic microbes including *Acidithiobacillus*, *Leptospirillum, Ferroplasma*, and *Thermoplasmatales* bacteria along with eukaryotic microorganisms such as fungi and algae (Méndez-García et al. 2015). The toxic environmental conditions allow microorganisms to perform vital biogeochemical cycles that include iron and sulfur metabolism to survive in such extreme conditions. Handicaps of AMD create adaptive pressure which encourages variable genetic adaptation and horizontal gene exchange and metabolic evolutionary abilities. The ability of polyextremophiles extends beyond individual harsh environments as these microorganisms demonstrate success in locations that contain multiple stressors (Harrison et al. 2013). The extreme heat and pressure do not stop *Methanopyrus kandleri* from thriving, yet additional microbes survive through antifreeze proteins together with osmoprotectants and membrane lipids that specifically adapt to harsh environmental conditions. These adjustments redefine current life limits while building our comprehension of microbial survival techniques.

One example of extreme environments inhabited by piezophilic and thermophilic archaea is deep-sea hydrothermal vents, where *Pyrolobus fumarii* survives extreme temperatures and pressures and their chemosynthetic metabolism (Stetter 2006). Without sunlight the deep sea has complex ecosystems supported on top of unique food webs dependent on the mixing and oxidation of hydrogen sulfide and methane by these microorganisms. It has important ramifications for biotechnology, biomining, environmental remediation, and the search for extraterrestrial life since extremophiles study can be carried out across these diverse environments. Unique enzymes and metabolic pathways mean they could also be useful in industrial applications, and resilience could help instills clues about the possibility of life elsewhere in the solar system such as on exoplanets or icy moons. Uncovering the techniques used to allow microbial life to survive under the most extreme conditions allows researchers a better understanding of the evolutionary innovation, planetary habitability, and the basic mechanisms by which life supports itself under the most extreme conditions.

12.5 Biotechnological and Industrial Applications

Extremozymes have been derived from extremophilic microorganisms from the extremely high temperatures, pH, salinity, and pressure and have been proven the ones with an extreme stability and extreme efficiency in the industrial and biotechnological purposes (Elleuche et al. 2014). Extremozymes are very important in biofuel production process and, for instance, in bioethanol production from biomass sources such as sugarcane, molasses, and lignocellulose. Extremozymes enhance lignocellulose breakdown and fermentation efficiency mainly through optimization of yeast fermentation using *Saccharomyces cerevisiae* controlled by temperature and oxygen levels (Sanchez and Cardona 2008). Further advancements in genetic engineering in yeast strains and enzyme performance continue to enhance degrees of utilization of pentose sugars and ethanol yields (Elleuche et al. 2014; Sanchez and Cardona 2008).

Other than biofuels, extremozymes play an important role in other industries. Both biofungicides, bioinsecticides, and pectinases that improve fruit juice and wine production are chitinases from *Sulfolobus tokodaii* (Elleuche et al. 2014). The food, detergent, leather, and textile industries, as well as fatty acid synthesis and bioethanol production, rely on microbial proteases and lipases (Elleuche et al. 2014; Coker 2016; Kumar and Nussinov 2001). In animal feed, phytases improve the utilization of phosphorus, a benefit to poultry and fish farming (Elleuche et al. 2014). Tremendous thermostability of thermophilic enzymes is particularly important for use in biofuel production, waste treatment, and molecular biology (e.g., PCR) (Kumar and Nussinov 2001). Sustainable innovations from extremophiles include biomining, the environmentally friendly extraction of metals such as gold or copper (Coker 2016). Thus, some halophilic microorganisms (*Halobacterium salinarum*) produce bacteriorhodopsin (BR), used in optical computing, holography, and artificial retinas, while halotolerant microorganisms contribute to biodegradation, waste treatment, and soil salinity improvement (Margesin and Schinner 2001). Thermophilic proteins, extremozymes, and halotolerant microorganisms lead biotechnology and industry toward engagement with robust, sustainable, and efficient developments with a very potential of further innovation.

12.6 Challenges

Research and application on polyextremophiles, microorganisms living under extreme conditions including high temperature, acidity, or salinity are difficult. Many of these organisms are unculturable using traditional methods, and culturing and identifying them from harsh environments, including geothermal vents and hypersaline areas, are difficult (Siddiqui and Cavicchioli 2006; Zhang et al. 2021). In addition, environmental factors such as high radiation, fluctuating temperatures, and high pH add to the difficulties involved in isolation of hyperthermophiles, which are not yet cultivated despite metagenomics and single cell genomics advances (Hedlund et al. 2015b). Research is also hindered by genetic and biotechnological challenges (Fig. 12.4). We lack well-annotated genome sequences or functional characterization of the genes required for their metabolic pathways (Choi et al. 2019). In addition, CRISPR/Cas tools fail in supercritical conditions, and thus the

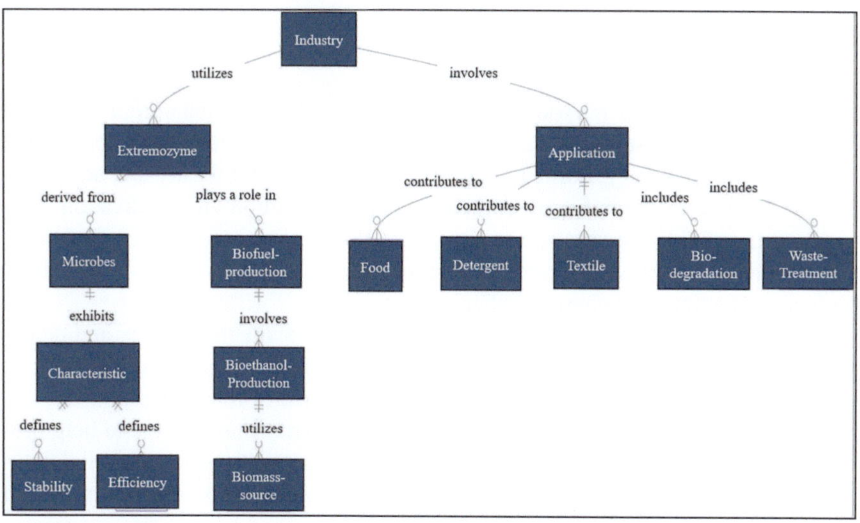

Fig. 12.4 Industrial applications of extremozymes: roles and characteristics

implementations of genetic modifications and biotechnological applications are very difficult (Choi et al. 2019).

There are additional obstacles for DNA extraction and sequencing from extreme environments. Contaminants such as humic acids and minerals can cause contamination that can affect the purity and later sequencing accuracy of the DNA (Zhang et al. 2021). High-throughput sequencing methods continue to be expensive and technically complex, and the lack of functional reference genomes makes it difficult to identify unrepresented taxa (Merino et al. 2019; Zhang et al. 2021; Davenport and Tümmler 2013). Polyextremophile communities are highly complex ecologically. In these microorganisms it is hard to find growth conditions that mimic what they would normally require to grow in the lab (Zhang et al. 2021; Hedlund et al. 2015b). While functional metagenomics has been successful in some areas, the discovery of active genes and bioactive compounds has been relatively unsuccessful (Merino et al. 2019). Further complicating the study of microbial communities in such extreme habitats as deep-sea hydrothermal vent is environmental variability such as changes in salinity and water activity (Siddiqui and Cavicchioli 2006).

However, polyextremophiles present a remarkable opportunity to play a crucial role in biotechnology, medicine, and environmental management despite these challenges. They also have the potential for industrially applicable heat-resistant enzymes and other valuable bioproducts (Siddiqui and Cavicchioli 2006). Studying extremophiles of a similar nature as Mars or Europa could also shine light on extraterrestrial life (Siddiqui and Cavicchioli 2006). The barriers to overcome the present limitations and to further expand our understanding of life in extreme environments will depend on the improvement of sequencing technologies, genetic tools, and biotechnological innovations (Merino et al. 2019; Choi et al. 2019).

12.7 Future Perspective

Polyextremophiles and microorganisms in extreme environments are a wealth of information for the study of space exploration, environmental sustainability, and industrial biotechnology. Further deciphering of microbial diversity and adaptability will depend on expanding microbial databases and improving cultivation techniques. Microfluidic devices and in situ cultivation will be used for isolating novel extremophiles with novel metabolic capabilities for applications in bioremediation, biofuel production, and pharmaceuticals.

Microbial resilience is synergistically improved by synthetic biology and bioengineering. By using CRISPR-based genome editing along with advanced bioreactor systems, engineered extremophiles that optimally perform for industrial usage can be developed (Zhang et al. 2021; Choi et al. 2019). They may be able to help provide sustainable biofuel and bioplastic production as well as environmental solutions such as carbon capture and pollutant degradation, including oil spills and plastic waste. Combining multi-omics approaches which include metagenomics together with transcriptomics and proteomics and metabolomics will establish a thorough understanding of microbial metabolic processes and adaptive mechanisms (Siddiqui and Cavicchioli 2006). Through the implementation of artificial intelligence along with machine learning techniques, the multi-omics methods will speed up valuable microbial product research and help make predictions for microbial functions while enhancing industrial and pharmaceutical metabolic pathway optimization (Merino et al. 2019).

Astrobiologists use extremophiles to study extraterrestrial life potential because these organisms demonstrate survival strategies in cold and hot hostile environments. Research on their adaptations on Earth helps create better techniques for detecting extraterrestrial life during Martian and Europa space exploration and future missions beyond these worlds (Rothschild 2016). The investigation of harsh environmental zones generates new insights about microbial richness while also creating methods to identify extraterrestrial life.

The study of extremophiles could help revolutionize biotechnology, foster better understanding of microbial resilience, and address problems of climate change, space exploration, and environmental sustainability (Cavicchioli et al. 2019). We can take full advantage of the capabilities of extremophiles for scientific and industrial application by advancing techniques in cultivation, utilizing multi-omics data and combining it with synthetic biology methods.

12.8 Conclusion

Extremophiles help scientists gain important knowledge about how life responds to environments that humans consider extremely harsh. Extremophiles maintain existence in extremely hot and cold temperatures alongside severe hydraulic pressure alongside salt-loaded aquatic conditions and radiation storms due to their unique physiological coping mechanisms and biochemical adaptations and genetic traits. The microorganisms show important industrial relevance since their extremozymes serve various functions in industrial processes while supporting biofuel manufacturing and pharmaceutical development as well as environmental cleanup operations. Their research improves our knowledge about Earth's initial chemical state and lunar existence which supports studies in astrobiology. Research developments in synthetic biology along with multi-omics methods and AI applications in microbial studies will enable scientists to overcome challenges of cultivation and sequencing as well as genetic manipulation for realizing the full potential of microorganisms. Studies about extremophiles help us understand life's limits while simultaneously developing new technological solutions for space exploration and sustainable development and industrial processes.

References

Antón J, Oren A, Benlloch S, Rodríguez-Valera F, Amann R, Rosselló-Mora R (2002) Salinibacter ruber gen. nov., sp. nov., a novel, extremely halophilic member of the Bacteria from saltern crystallizer ponds. Int J Syst Evol Microbiol 52(2):485–491

Atanasova NS, Roine E, Oren A, Bamford DH, Oksanen HM (2012) Global network of specific virus–host interactions in hypersaline environments. Environ Microbiol 14(2):426–440

Baker BJ, Banfield JF (2003) Microbial communities in acid mine drainage. FEMS Microbiol Ecol 44(2):139–152

Bertoldo C, Antranikian G (2002) Starch-hydrolyzing enzymes from thermophilic archaea and bacteria. Curr Opin Chem Biol 6(2):151–160

Bowman JP et al (1997) Diversity and association of psychrophilic bacteria in Antarctic Sea Ice. Appl Environ Microbiol 63:3–068

Cavicchioli R, Siddiqui KS, Andrews D, Sowers KR (2002) Low-temperature extremophiles and their applications. Curr Opin Biotechnol 13(3):253–261

Cavicchioli R, Amils R, Wagner D, McGenity T (2011) Life and applications of extremophiles. Environ Microbiol 13(8):1903–1907

Cavicchioli R, Ripple WJ, Timmis KN, Azam F, Bakken LR, Baylis M et al (2019) Scientists' warning to humanity: microorganisms and climate change. Nat Rev Microbiol 17(9):569–586

Chattopadhyay MK (2006) Mechanism of bacterial adaptation to low temperature. J Biosci 31:157–165

Choi KR, Jang WD, Yang D, Cho JS, Park D, Lee SY (2019) Systems metabolic engineering strategies: integrating systems and synthetic biology with metabolic engineering. Trends Biotechnol 37(8):817–837

Coker JA (2016) Extremophiles and biotechnology: current uses and prospects. F1000Res 5: F1000-Faculty

Csonka LN (1989) Physiological and genetic responses of bacteria to osmotic stress. Microbiol Rev 53(1):121–147

Dadachova E, Casadevall A (2008) Ionizing radiation: how fungi cope, adapt, and exploit with the help of melanin. Curr Opin Microbiol 11(6):525–531

DasSarma P, DasSarma S (2018) Survival of microbes in Earth's stratosphere. Curr Opin Microbiol 43:24–30

Davenport CF, Tümmler B (2013) Advances in computational analysis of metagenome sequences. Environ Microbiol 15(1):1–5

DeLong EF, Pace NR (2001) Environmental diversity of bacteria and archaea. Syst Biol 50(4):470–478

DiRuggiero J, Santangelo N, Nackerdien Z, Ravel J, Robb FT (1997) Repair of extensive ionizing-radiation DNA damage at 95 degrees C in the hyperthermophilic archaeon Pyrococcus furiosus. J Bacteriol 179(14):4643–4645

Elleuche S, Schröder C, Sahm K, Antranikian G (2014) Extremozymes – biocatalysts with unique properties from extremophilic microorganisms. Curr Opin Biotechnol 29:116–123

Eloe EA (2011) Molecular diversity and adaptations of microorganisms from the deep ocean. Doctoral dissertation, UC San Diego

Fang J, Zhang L, Bazylinski DA (2010) Deep-sea piezosphere and piezophiles: geomicrobiology and biogeochemistry. Trends Microbiol 18(9):413–422

Feller G, Gerday C (2003) Psychrophilic enzymes: hot topics in cold adaptation. Nat Rev Microbiol 1(3):200–208

Golyshina OV (2011) Environmental, biogeographic, and biochemical patterns of archaea of the family Ferroplasmaceae. Appl Environ Microbiol 77(15):5071–5078

Gomes E, de Souza AR, Orjuela GL, Da Silva R, de Oliveira TB, Rodrigues A (2016) Applications and benefits of thermophilic microorganisms and their enzymes for industrial biotechnology. In: Gene expression systems in fungi: advancements and applications. Springer, pp 459–492

Grant S, Grant WD, Jones BE, Kato C, Li L (1999) Novel archaeal phylotypes from an East African alkaline saltern. Extremophiles 3:139–145

Gunde-Cimerman N, Ramos J, Plemenitaš A (2009) Halotolerant and halophilic fungi. Mycol Res 113(11):1231–1241

Harrison JP, Gheeraert N, Tsigelnitskiy D, Cockell CS (2013) The limits for life under multiple extremes. Trends Microbiol 21(4):204–212

Hedlund BP, Reysenbach AL, Huang L, Ong JC, Liu Z, Dodsworth JA et al (2015a) Isolation of diverse members of the Aquificales from geothermal springs in Tengchong, China. Front Microbiol 6:157

Hedlund BP, Murugapiran SK, Alba TW, Levy A, Dodsworth JA, Goertz GB et al (2015b) Uncultivated thermophiles: current status and spotlight on 'Aigarchaeota'. Curr Opin Microbiol 25:136–145

Horneck G, Klaus DM, Mancinelli RL (2010) Space microbiology. Microbiol Mol Biol Rev 74(1):121–156

Huber H, Stetter KO (1998) Hyperthermophiles and their possible potential in biotechnology. J Biotechnol 64(1):39–52

Johnson DB, Hallberg KB (2003) The microbiology of acidic mine waters. Res Microbiol 154(7):466–473

Jones BE, Grant WD, Duckworth AW, Owenson GG (1998) Microbial diversity of soda lakes. Extremophiles 2:191–200

Jørgensen BB, Boetius A (2007) Feast and famine – microbial life in the deep-sea bed. Nat Rev Microbiol 5(10):770–781

Kato C, Nogi Y, Arakawa S (2008) Isolation, cultivation, and diversity of deep-sea piezophiles. High-Pressure Microbiol:203–217

Kumar S, Nussinov R (2001) How do thermophilic proteins deal with heat? Cell Mol Life Sci 58:1216–1233

Ma Y, Galinski EA, Grant WD, Oren A, Ventosa A (2010) Halophiles 2010: life in saline environments. Appl Environ Microbiol 76(21):6971–6981

Margesin R, Miteva V (2011) Diversity and ecology of psychrophilic microorganisms. Res Microbiol 162(3):346–361

Margesin R, Schinner F (2001) Potential of halotolerant and halophilic microorganisms for biotechnology. Extremophiles 5(2):73–83

Méndez-García C, Peláez AI, Mesa V, Sánchez J, Golyshina OV, Ferrer M (2015) Microbial diversity and metabolic networks in acid mine drainage habitats. Front Microbiol 6:475

Menzel P, Gudbergsdóttir SR, Rike AG, Lin L, Zhang Q, Contursi P et al (2015) Comparative metagenomics of eight geographically remote terrestrial hot springs. Microb Ecol 70:411–424

Merino N, Aronson HS, Bojanova DP, Feyhl-Buska J, Wong ML, Zhang S, Giovannelli D (2019) Living at the extremes: extremophiles and the limits of life in a planetary context. Front Microbiol 10:780

Methé BA, Nelson KE, Deming JW, Momen B, Melamud E, Zhang X et al (2005) The psychrophilic lifestyle as revealed by the genome sequence of Colwellia psychrerythraea 34H through genomic and proteomic analyses. Proc Natl Acad Sci 102(31):10913–10918

Nicholson WL, Schuerger AC, Setlow P (2005) The solar UV environment and bacterial spore UV resistance: considerations for Earth-to-Mars transport by natural processes and human spaceflight. Mutat Res 571(1–2):249–264

Niederberger TD, Perreault NN, Tille S, Lollar BS, Lacrampe-Couloume G, Andersen D et al (2010) Microbial characterization of a subzero, hypersaline methane seep in the Canadian High Arctic. ISME J 4(10):1326–1339

Nogi Y, Kato C, Horikoshi K (1998) Taxonomic studies of deep-sea barophilic Shewanella strains and description of Shewanella violacea sp. nov. Arch Microbiol 170:331–338

Oren A (2002) Diversity of halophilic microorganisms: environments, phylogeny, physiology, and applications. J Ind Microbiol Biotechnol 28(1):56–63

Oren A (2014) The ecology of Dunaliella in high-salt environments. J Biol Res-Thessalon 21:1–8

Pikuta EV, Hoover RB, Tang J (2007) Microbial extremophiles at the limits of life. Crit Rev Microbiol 33(3):183–209
Prangishvili D, Forterre P, Garrett RA (2006) Viruses of the Archaea: a unifying view. Nat Rev Microbiol 4(11):837–848
Rainey FA, Ray K, Ferreira M, Gatz BZ, Nobre MF, Bagaley D et al (2005) Extensive diversity of ionizing-radiation-resistant bacteria recovered from Sonoran Desert soil and description of nine new species of the genus Deinococcus obtained from a single soil sample. Appl Environ Microbiol 71(9):5225–5235
Rampelotto PH (2010) Resistance of microorganisms to extreme environmental conditions and its contribution to astrobiology. Sustainability 2(6):1602–1623
Rampelotto PH (2013) Extremophiles and extreme environments. Life 3(3):482–485
Remias D, Lütz-Meindl U, Lütz C (2005) Photosynthesis, pigments and ultrastructure of the alpine snow alga Chlamydomonas nivalis. Eur J Phycol 40(3):259–268
Rothschild LJ (2016) Synthetic biology meets bioprinting: enabling technologies for humans on Mars (and Earth). Biochem Soc Trans 44(4):1158–1164
Rothschild LJ, Mancinelli RL (2001) Life in extreme environments. Nature 409(6823):1092–1101
Sanchez OJ, Cardona CA (2008) Trends in biotechnological production of fuel ethanol from different feedstocks. Bioresour Technol 99(13):5270–5295
Santos H, Da Costa MS (2002) Compatible solutes of organisms that live in hot saline environments. Environ Microbiol 4(9):501–509
Saunders NF, Thomas T, Curmi PM, Mattick JS, Kuczek E, Slade R et al (2003) Mechanisms of thermal adaptation revealed from the genomes of the Antarctic Archaea Methanogenium frigidum and Methanococcoides burtonii. Genome Res 13(7):1580–1588
Seckbach J, Oren A (eds) (2010) Microbial mats: modern and ancient microorganisms in stratified systems, vol 14. Springer
Siddiqui KS, Cavicchioli R (2006) Cold-adapted enzymes. Annu Rev Biochem 75(1):403–433
Slade D, Radman M (2011) Oxidative stress resistance in Deinococcus radiodurans. Microbiol Mol Biol Rev 75(1):133–191
Sogin ML, Morrison HG, Huber JA, Welch DM, Huse SM, Neal PR et al (2006) Microbial diversity in the deep sea and the underexplored "rare biosphere". Proc Natl Acad Sci 103(32):12115–12120
Sorokin DY, Kuenen JG (2005) Haloalkaliphilic sulfur-oxidizing bacteria in soda lakes. FEMS Microbiol Rev 29(4):685–702
Stetter KO (2006) Hyperthermophiles in the history of life. Philos Trans R Soc B 361(1474):1837–1843
Sullivan MB, Huang KH, Ignacio-Espinoza JC, Berlin AM, Kelly L, Weigele PR et al (2010) Genomic analysis of oceanic cyanobacterial myoviruses compared with T4-like myoviruses from diverse hosts and environments. Environ Microbiol 12(11):3035–3056
Ventosa A, Nieto JJ (1995) Biotechnological applications and potentialities of halophilic microorganisms. World J Microbiol Biotechnol 11:85–94
Vieille C, Zeikus GJ (2001) Hyperthermophilic enzymes: sources, uses, and molecular mechanisms for thermostability. Microbiol Mol Biol Rev 65(1):1–43
Wang M, Tian J, Xiang M, Liu X (2017) Living strategy of cold-adapted fungi with the reference to several representative species. Mycology 8(3):178–188
Williams JP, Hallsworth JE (2009) Limits of life in hostile environments: no barriers to biosphere function? Environ Microbiol 11(12):3292–3308
Zavarzin GA, Zhilina TN, Kevbrin VV (1999) Alkaliphilic microbial community and its functional diversity. Микробиология 68(5):579–599
Zhang L, Chen F, Zeng Z, Xu M, Sun F, Yang L et al (2021) Advances in metagenomics and its application in environmental microorganisms. Front Microbiol 12:766364

Chapter 13
Pigment Production by Fungi from Polar Environments: From Cold to Industry

Vannêssa Rodrigues Teles Maia, Emanuelly Beatriz Tenório Sampaio, Adeildo Júnior de Oliveira, Leonardo da Silva Santos, Mayanne Karla da Silva, Aline Cavalcanti de Queiroz, Magna Suzana Alexandre Moreira, Luiz Henrique Rosa, Thaissa Lúcio Silva, Michel Rodrigo Zambrano Passarini, and Alysson Wagner Fernandes Duarte

Abstract The production of natural pigments for use in the food, cosmetics, pharmaceuticals, textiles, paints, and different material industries constitutes an environmentally sustainable alternative to chemical synthesis, with a considerable increase in demand. While synthetic pigments have the potential to cause harm to the human body and the environment, the consumption of pigments of natural origin has increased significantly in recent decades. In this chapter, we explore the potential of filamentous fungi from polar regions for pigment production. These organisms are notable for their advantageous characteristics, including metabolic diversity, rapid growth, ease of cultivation, and susceptibility to genetic tractability. With each new fungal species discovered, the potential for biotechnological applications increases. Among the filamentous fungi from polar environments, Antarctic genera include *Alternaria*, *Antarctomyces*, *Arthrobotrys*, *Aspergillus*, *Cladosporium*, *Penicillium*, *Pseudogymnoascus*, *Sporobolomyces*, *Talaromyces*, and *Thelebolus*. Among the Arctic fungi, *Alternaria*, *Cadophora*, and *Cladosporium* are the main pigment producers. This chapter explores the potential of polar filamentous fungi, presenting an innovative compilation of the main pigment-producing genera. Filamentous fungi

V. R. T. Maia · E. B. T. Sampaio · A. J. de Oliveira · L. da Silva Santos · M. K. da Silva · T. L. Silva · A. W. F. Duarte (✉)
Universidade Federal de Alagoas, Campus Arapiraca, Arapiraca, AL, Brazil
e-mail: alysson.duarte@arapiraca.ufal.br

A. C. de Queiroz · M. S. A. Moreira
Universidade Federal de Alagoas, Campus AC Simões, Arapiraca, AL, Brazil

L. H. Rosa
Universidade Federal de Minas Gerais, Instituto de Ciências Biológicas, Belo Horizonte, MG, Brazil

M. R. Z. Passarini
Universidade Federal da Integração Latino Americana, Foz do Iguaçu, PR, Brazil

A. W. Fernandes Duarte et al. (eds.), *Polar Microbes and Climate Change*,
https://doi.org/10.1007/978-3-032-07096-8_13

from polar environments are highlighted for their diverse metabolic resources, which promote development and produce bioactive essential and high performance in the biotechnology industries, especially in the textile, food, pharmaceutical, and cosmetic areas. These characteristics align with the global demand for sustainable natural products that are safe for both humans and the environment.

Keywords Fungal bioactive · Natural pigments · Bioproducts · Antarctica · Arctic

13.1 Polar Regions

The Arctic and Antarctic regions are considered polar environments, a designation that stems from their status as the geographic extremes of planet Earth. These regions are distinguished by limited water and nutrient availability, extremely low temperatures, strong winds, and extended periods of darkness during winter. They play a crucial role in regulating the global climate and preserving marine and terrestrial ecosystems (Dotta et al. 2022; Grum-Grzhimaylo et al. 2024; Rosa et al. 2019).

In the southern hemisphere, the region below the 60th parallel corresponds to the Antarctic Continent, a region characterized by elevated altitudes and substantial ice cover, giving rise to mountain ranges, valleys, and glaciers. This topography results in a significant reflection of incoming solar energy back into the atmosphere, thereby generating a high albedo value (Lopes et al. 2024). The Antarctic territory is divided into three biogeographic zones: continental, maritime, and sub-Antarctic (Convey 2011; Ding et al. 2024). In contrast, the Arctic Circle, located north of 66.7° N, is an oceanic region surrounded by continental land (Dotta et al. 2022).

Polar regions (Fig. 13.1) play a pivotal role in the interactions between atmospheric and oceanic systems, regulating the global climate (Robinson 2022; Grum-Grzhimaylo et al. 2024). Concurrently, global warming exerts a direct influence on the glacial coverage of the poles, leading to an increase in marine acidity in Antarctica and significant ecosystem impacts (Robinson 2022). According to the sixth Assessment Report of the Intergovernmental Panel on Climate Change (IPCC AR6), there is a high probability that the Arctic will experience at least one ice-free summer before the year 2050 (Chen et al. 2023).

13.2 Pigmented Fungi in Polar Regions

The polar regions, encompassing both the Arctic and Antarctica, present significant obstacles to life. Despite the presence of substantial amounts of water, it is predominantly found in the form of ice and snow. The water in liquid form, which is essential for all life forms, is seasonal and limited. Despite these harsh conditions, polar ecosystems harbor highly adapted organisms, including microorganisms, which play essential roles in ecological balance (Dziurzynski et al. 2023; Nikitin 2023).

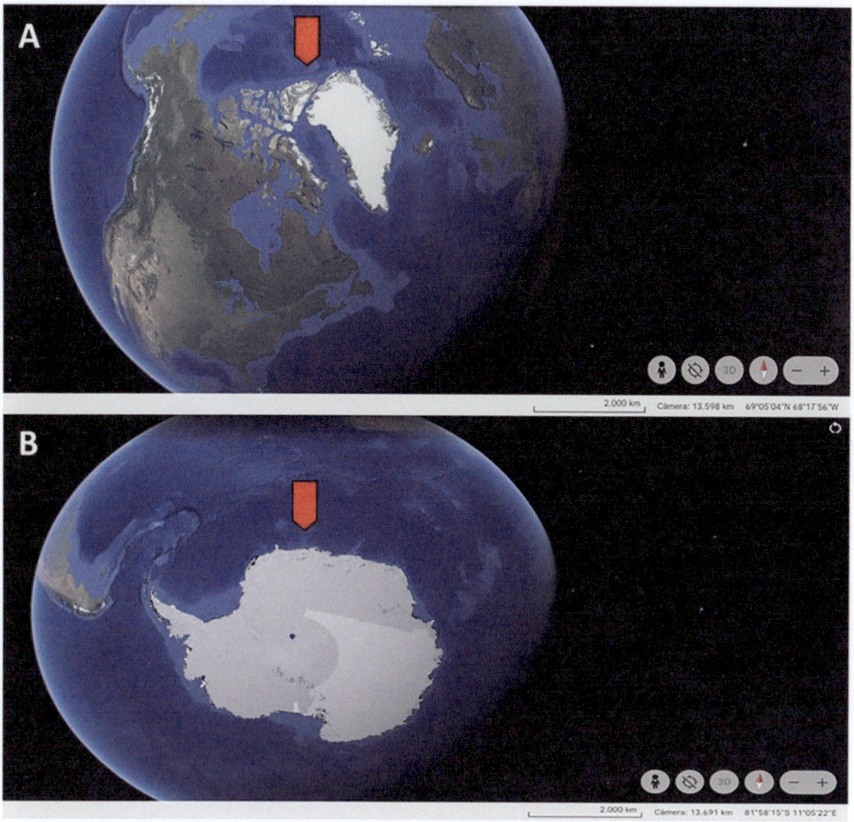

Fig. 13.1 Map of the Arctic (**a**) and Antarctica (**b**) using Google Earth

Among these resistant organisms, polar fungi are of particular significance in maintaining ecological balance. Fungi are present in soil biomass and also colonize rocks, performing important functions for the region, such as the decomposition of organic matter, the transformation of minerals, and the provision of nutrients for other organisms. These functions facilitate the proliferation of diverse ecological communities, including lichens and endophytic microorganisms, underscoring the pivotal role of fungi in polar ecological dynamics (Zalar et al. 2011; Cavalcante et al. 2023).

The fungi present in this region have evolved adaptive characteristics to extreme environmental conditions, including the production of pigments as a protective strategy against UV radiation and dehydration. Pigmentation mitigates oxidative damage by neutralizing free radicals generated under stress, a phenomenon that is exacerbated by the region's environmental conditions (Dziurzynski et al. 2023).

In Antarctica, there are several well-adapted fungi that dominate the local mycobiome and that present pigmentation, both intracellular and extracellular (Figs. 13.2 and 13.3). Among them are yeasts belonging to the phylum *Basidiomycota*, such as *Sporobolomyces* and *Rhodotorula*, and filamentous fungi belonging to the phylum

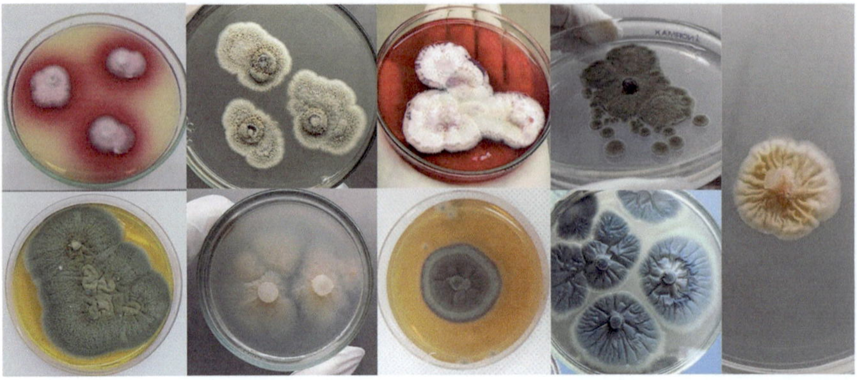

Fig. 13.2 Colonies macroscopic characteristics of pigmented filamentous fungi isolated from different ecosystems of Antarctica

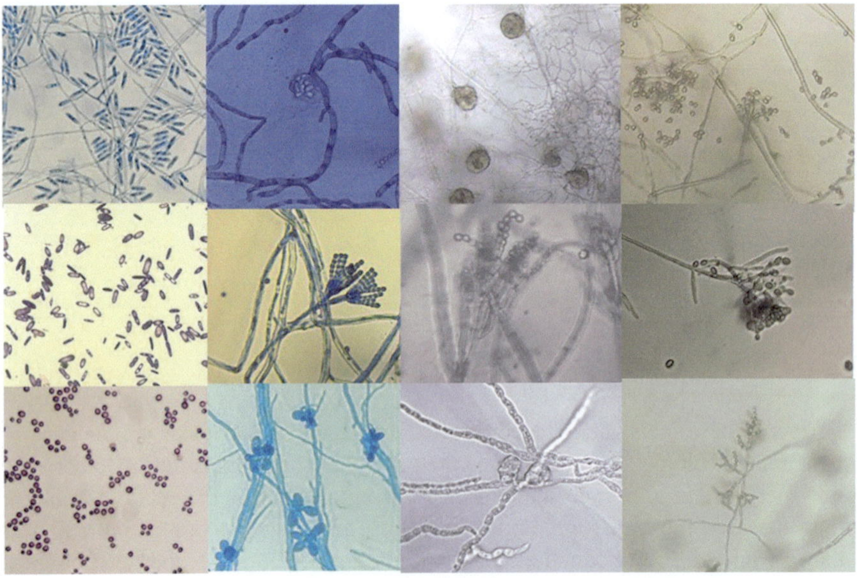

Fig. 13.3 Microscopic features of fungi isolated from the Antarctic environment

Ascomycota, such as *Antarctomyces*, *Penicillium*, *Cladosporium*, *Sporobolomyces*, *Alternaria*, *Aspergillus*, *Arthrobotrys*, *Talaromyces*, *Pseudogymnoascus*, and *Thelebolus* (de Menezes et al. 2017; Duarte et al. 2018; Nikitin 2023; Cavalcante et al. 2024). These fungal species possess hyphal structures that confer the ability to resist cold, thereby enabling survival and multiplication in these conditions. For instance, *Cladosporium herbarum* and *Cryomyces antarcticus*, which have been observed to exhibit dark pigmentation resulting from the production of melanin on exposed surfaces, are subjected to ice and sun exposure (Pacelli et al. 2020; Marchetta et al. 2023).

In contrast to the Antarctic, the fungal diversity of the Arctic is not as well studied, although the Arctic has a unique biodiversity. Despite few studies, some pigmented fungal species have already been recorded, such as the following genera: *Alternaria*, *Cladosporium*, and *Cadophora*, belonging to the phyla *Ascomycota*, *Basidiomycota*, these fungi show specific adaptations to survive the peculiarities of the environment. Some of the survival mechanisms described in the literature are the production of pigments such as melanin, which protects them from the harmful effects of UV radiation, as occurs with some species of fungi of the genera *Alternaria* and *Cladosporium*, which are known for their darkened colonies (Dziurzynski et al. 2023; Marchetta et al. 2023).

13.3 Pigment Classes in Polar Fungi: Diversity and Functions

A diverse range of pigments is produced by fungi in cryophilic environments. These pigments are secondary metabolites that act as a response to environmental stresses to which these organisms are exposed, thus allowing them to survive in diverse stressful conditions (Saijad et al. 2020). The color of the pigments varies according to their class, encompassing a wide spectrum of hues. The predominant pigment classes produced by Antarctic fungi include carotenoids and polyketides, such as melanins and anthraquinones, which display a range of colors, from dark to vibrant hues, and fulfill distinct functions (Afroz Toma et al. 2023; Cavalcante et al. 2023).

Carotenoids are terpenoids with 40 carbon atoms in their main chain. The chemical structure of these compounds contains eight isoprene units, forming conjugated double bonds that absorb light. These structural features give carotenoids their characteristic colors, including yellow, orange, and red (Afroz Toma et al. 2023; Lin and Xu 2020). They can be categorized into two main groups: free carotenoids and oxygen-containing xanthophylls. The first group includes α-carotene, β-carotene, lycopene, and torulene, while the second group includes lutein, zeaxanthin, β-cryptoxanthin, and capsanthin (Aziz et al. 2020; Rapoport et al. 2021).

Carotenoids are associated with the cell membrane and are capable of absorbing high levels of light energy, dissipating it harmlessly in the form of heat (Kreusch and Duarte 2021). They are important in thermal adaptation and also involved in photoprotection. Pigments in this class serve an antioxidant role by neutralizing reactive oxygen species (ROS) and protecting cells against deleterious effects of UV radiation (Sandmann 2019; Maoka 2020). Cryophilic habitats, such as polar regions, provide ideal conditions for the production of this pigment by microorganisms due to light exposure (Saijad et al. 2020). The *Rhodotorula* species, in particular, are known to synthesize various types of carotenoids (Cavalcante et al. 2023).

Fungal polyketide pigments consist of tetraketides and octoketides, which are formed by eight C2 units that constitute the polyketide chain. Representative classes of these pigments include anthraquinones and melanins. Anthraquinone is a polycyclic aromatic hydrocarbon derived from anthracene, containing a structure with three fused benzene rings. They impart a yellow color, and the substituents such as $-OH$, $-CH_3$, $-OCH_3$, and $-CH_2OH$ contribute to various hues of the molecules ranging through yellow, orange, red, brown, and violet. Most of the fungi-producing anthraquinones is capable of producing a mixture of pigments from the quinone class, as in *Aspergillus cristatus* capable of synthesizing up to 15 pigments associated with this class, melanins are heterogeneous polymers formed through the oxidation of phenols, followed by the polymerization of intermediate phenolic compounds and their resulting quinones. Fungal melanins are a subset of natural melanins that are polyphenolic and/or polyindolic compounds with high molecular masses and negative charges (Afroz Toma et al. 2023; Lin and Xu 2020; Yadav et al. 2019; Venil et al. 2020).).

Similar to carotenoids, melanins also serve a photoprotective function, especially by absorbing UV radiation. This attribute is manifested through the presence of numerous pigments, resulting in dark colors ranging from brown to black (Kreusch and Duarte 2021). However, melanins can also appear in dark green tones, as observed in *Aspergillus fumigatus*, where the pigment responsible for the gray-green appearance of its conidia is particularly notable (Lin and Xu 2020). The structural and colorational characteristics of fungal melanin can be classified into three types: pheomelanins, which exhibit reddish or yellowish tones; eumelanins, which display black-brown shades; and allomelanins, which present darker hues of black and brown (Saijad et al. 2020). Fungal melanins are further categorized into three types, specifically related to fungal species: 1,8-dihydroxynaphthalene (DHN)-melanin, 3,4-dihydroxyphenylalanine (DOPA)-melanin (eumelanin), and pyomelanin (Perez-Cuesta et al. 2020). The location of this pigment within the organism may vary between species (intra- or extracellularly) and may have distinct functions, including antioxidant activity, which eliminate reactive oxygen species (ROS) and reactive nitrogen species (RNS) in the fungal cell, and may be induced by stressful factors (Perez-Cuesta et al. 2020; Cavalcante et al. 2024), as well as contributing to greater virulence. A notable example is the melanized fungus *Cryomyces antarcticus*, which exhibits radiation resistance in the Antarctic desert environment. This adaptation has been associated with pigments belonging to this class. Melanization is a widely recognized process in microorganisms from extreme environments (Lin and Xu 2020; Pacelli et al. 2020).

Despite the scarcity of works characterizing the pigments produced by Antarctic fungi to determine the class to which they belong, a strategy for the optimization of the production of a new type of red pigment by *Geomyces antarcticus* was developed. In this strategy, it was hypothesized that the synthesis of the pigment most likely occurs via the polyketide pathway (Huang et al. 2020). Furthermore, metabolomic and genomic analyses of *Geomyces* sp. have indicated a metabolic pathway linked to carotenoid biosynthesis (Long et al. 2024). In the dark fungus *C. antarcticus*, which is capable of synthesizing melanin DHN, it was observed that the deletion of the pks1 gene prevents melanization and renders non-melanized mutants with light-pink pigmentation, indicating possible accumulation of carotenoids, while genetic engineering of its material in *Saccharomyces cerevisiae* allows the synthesis of yellowish pigment by coexpression of CaPKS1 or KpPKS1 with a 4′-phosphopantetheinyl transferase (Catanzaro et al. 2024). In contrast, fungi identified as *Pseudogymnoascus* spp. and *Talaromyces cnidii*, isolated from soil samples from the Collins Glacier and capable of producing pigments extracellularly with absorption of yellow, red/orange wavelengths, and pink/violet, are undergoing the chemical characterization process (Cavalcante et al. 2024). This illustrates that the chemical characterization of classes of Antarctic fungal pigments is challenging, as it complicates the classification of the pigments. This is because pigments within the same class often exhibit color ranges that overlap with those of other classes.

13.4 Screening, Production, Recovery, and Structural Characterization of Fungal Pigments

As previously discussed, fungi from polar regions produce a mixture of bioactive secondary metabolites (Barreto et al. 2024), including different types of pigments, which vary in their chemical composition when analyzed at the molecular level. The stages in the study of a pigment-producing fungus, either intra- or extracellularly, generally begin with the environmental screening of pigmented strains, traditionally using a culture recovery method, and after isolation and laboratory preservation, the strain of interest is cultivated to isolate the biopigment (upstream steps) which will be followed by the structural downstream steps (Valenzuela-Gloria et al. 2021).

As pigments are a fermentation product. Methodologies are aimed at producing and extracting them from a polar producer fungus (Fig. 13.4), which when cultivated in a submerged or solid state (Cavalcante et al. 2024). These methods are combined with fermentation optimization through adjustments to the physicochemical parameters of the incubation, respecting time, nutritional composition, temperature, and aeration. Pigments are separated from cell growth and concentrated into the total crude extract (Kalra et al. 2020). Extraction can be carried out using aqueous or chemical solvents, such as acetone, ethyl acetate, dichloromethane, methanol, as well as alternative methods like mannosonication, supercritical CO_2, ultrasound-assisted extraction, microwave-assisted extraction, ionic liquid or pulsed electric field (PEF) or pressurized liquid extraction, for example (Kalra et al. 2020; Valenzuela-Gloria et al. 2021).

Clarifying the pigments of polar fungal taxa from a molecular perspective is a key aspect for the effective characterization and application of these biomolecules (Valenzuela-Gloria et al. 2021). By determining the types of chemical bonds, atoms present, functional groups, and the spatial arrangement of atoms and chemical groups, one can accurately understand the molecular structure of a pigment. Initially, pigments are studied by their functional groups associated with color through comparative colorimetric procedures. However, sophisticated techniques are employed to provide their structural patterns, many based on spectroanalytical methods (Table 13.1), through spectroscopic and spectrometric methods, including ultraviolet and visible spectroscopy (UV-Vis), Fourier transform infrared spectroscopy (FT-IR), nuclear magnetic resonance spectroscopy (NMR), and mass spectrometry (MS) (Valenzuela-Gloria et al. 2021) or by chromatographic couplings such as liquid

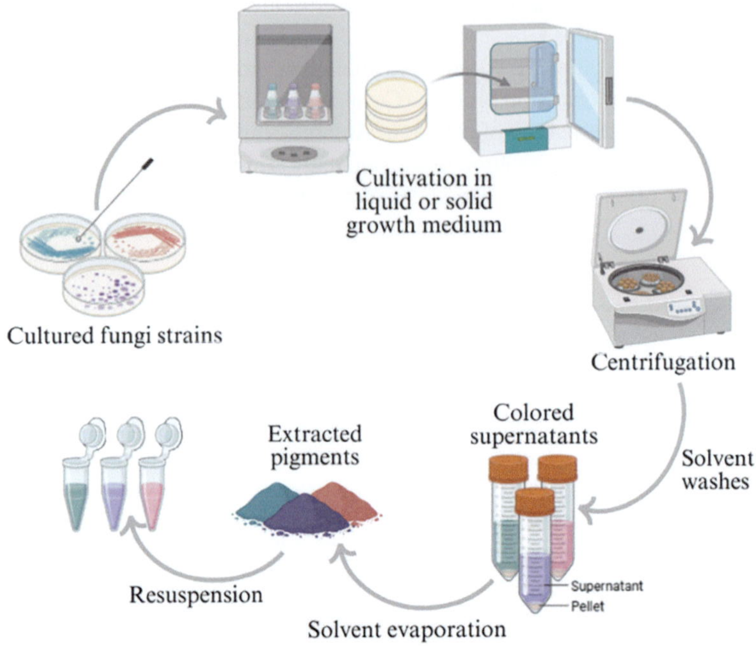

Fig. 13.4 Representation of the stages involved in the pigment production process by fungi from the polar regions

chromatography-mass spectrometry (LC-MS) (Cavalcante et al. 2024), which help to identify the pigment in complex mixtures. This is particularly important when analyzing the pool of metabolites concentrated in the extract of a producing fungus in the upstream stage that precedes structural elucidation.

Each of these tools provides specific chemical information, and, when combined, can be useful for determining the identity of the fungal pigment (Calsteren et al. 1997; Meddeb-Mouelhi et al. 2016) and studying its spectroscopic properties (Anghelone et al. 2018). Before conducting characterization analyses, pigment isolation steps are typically implemented (Rudrappa et al. 2022). This ensures that the spectral information obtained is exclusively related to the compound of interest. The presence of contaminants can generate signals that are misinterpreted as the analyte, potentially leading to its erroneous identification. The acidic or basic behavior of the molecule, determined by its pH, serves as a basis for defining the properties of natural pigments. Additionally, solubility and melting point parameters are identified as critical factors for future applications.

Spectroscopic techniques (FTIR, NMR, and MS) are the most commonly used for identifying the molecular structure of organic compounds and may be sufficient to fully elucidate the structure of a pigment of fungal origin. Although UV-Vis spectroscopy is not a technique specifically designed for the characterization of organic compounds, it can still be useful in the case at hand. UV-Vis spectroscopy is used in

Table 13.1 Techniques for the analysis and chemical characterization of pigments produced by fungi

Technique	Information	Advantage	Limitation
UV-Vis	Absorption spectrum in the ultraviolet and visible	Low-cost instrumentation; useful for explore optical properties	Low diagnostic importance
FT-IR (mid)	Vibrational spectrum of chemical bonds	Speed; high sensitivity; non-destructive analysis	Complex spectra containing broad bands with overlapping bands
NMR	Differentiation of hydrogens and carbons in molecular structure	High diagnostic significance in the structural elucidation of organic compounds; non-destructive analysis	Low sensitivity, requiring considerable amounts of analyte
MS	Molecular weight	High sensitivity; allows determining the molecular formula of a compound; analyzes compounds of different molecular dimensions and polarities	Requires specific ionization methods for certain classes of organic compounds
LC-MS	Molecular weight	High sensitivity; analysis of complex samples	Longer analysis time
GC-MS	Molecular weight	High sensitivity; analysis of complex samples; reproducible spectra	Restricted to the analysis of volatile compounds

various branches of chemical analysis, mainly in laboratory activities for analytical determinations, when the concentration of a particular chemical species in a sample is to be known (Küpper et al. 2007; Domenici et al. 2014). Its use as an analytical resource for identifying organic compounds is quite limited due to the nature of the spectral information obtained, which does not allow much information to be extracted about the molecular structure, such as types of chemical bonds, functional groups and stereochemistry. UV-Vis can also be used to identify the optical characteristics of a pigment and is commonly applied to determine the presence of a pigment in complex extracts, as seen before (Lee et al. 2011; Kumar Shetty et al. 2021).

Some structural characteristics of pigments can generate signals in the molecular absorption spectrum obtained with the UV-Vis technique, allowing certain properties to be inferred, such as the compounds' ability to absorb ultraviolet and visible light (Du et al. 2018). Therefore, its use for qualitative purposes is mainly to find out some spectroscopic properties related to molecular absorption in its operational range of the electromagnetic spectrum (Quijada-Morín et al. 2010; Yoshida et al. 2015), more precisely in the ultraviolet and visible range, from 180 nm to 900 nm.

Among the instrumental techniques mentioned here, UV-Vis spectroscopy is the most cost-effective, with simple optics and an easy sample introduction system. By measuring the absorption of ultraviolet and/or visible light at specific wavelengths, an absorption spectrum is generated in the defined spectral range. The absorption bands present in the spectrum allow us to know the range of the UV-Vis spectrum that a compound is capable of absorbing, generating physicochemical information.

In this sense, pigments are compounds with structural properties that allow them to absorb radiation in the visible spectrum range (350–800 nm). The chemical groups responsible for absorbing UV-Vis radiation are called chromophores. These groups are primarily composed of certain functional groups and conjugated double bond systems (Liu et al. 2020; Nakano and Wiegertjes 2020; Ashenafi et al. 2023), whose molecular orbital electrons are capable of absorbing radiant energy from this range of the electromagnetic spectrum and excited to other orbitals that are energetically more distant (Shabbir and Chauhan 2024).

Similarly, the structural elucidation of organic compounds such as pigments can also be done using FT-IR, which is capable of providing more precise and relevant information for molecular identification than UV-Vis spectroscopy. Recognized for its efficiency in identifying functional groups, its use consists more precisely in identifying the chemical bonds present in a molecular system, conditioning the recognition of certain atoms in its structure (Bhagwat et al. 2024). FT-IR can be used as an additional analytical tool in the pigment characterization process, with the aim of confirming the presence of functional groups noted by other techniques or verifying the presence of new groups (Neha et al. 2017). When irradiated on a sample, infrared rays are capable of producing vibrations in the chemical bonds of molecules. These vibrations are produced because the molecules absorb the radiant energy at specific frequencies or wavelengths, so these absorption bands occur at specific, quantized frequencies (Pavia et al. 2013). Based on this property of organic compounds, it is possible to learn about certain structural characteristics by determining the vibrational states of molecules using FT-IR.

Infrared spectroscopy can be used with two spectral bands, mid-infrared and far-infrared. The mid-infrared range has diagnostic value for identifying organic compounds. By measuring the absorption of infrared radiation in a spectrometer, using frequencies in the range from 400 cm^{-1} to 4000 cm^{-1}, the technique records the frequencies by a molecule. Through the spectrum acquired by scanning the chosen frequency range, it is possible to relate these absorption frequencies to the probable chemical bonds and functional groups present in the natural pigment molecule (Neha et al. 2017), with relevance for its identification (Jeyaram and Geethakrishnan 2020), whose structural characteristics can be recognized through the infrared spectrum, through absorptions that generate visible bands in their pigment spectra.

On the other hand, nuclear magnetic resonance (NMR) is an instrumental technique and one of the most important for characterizing organic compounds, demonstrating high efficiency in determining the structure of pigments (Mateus et al. 2002). It can provide a greater amount of structural information than other techniques for identifying compounds. With NMR it is possible to obtain information such as the number of hydrogens and carbons in a chain, functional groups, chemical bonds, and isomerism, among other information (Stachura et al. 2015).

NMR analysis is based on the determination of spectroscopic properties of molecules that can be obtained from the interaction of radio frequencies with atomic nuclei of certain isotopes, when subjected to a magnetic field of suitable intensity (Pavia et al. 2013). These absorptions are recorded by an NMR spectrometer and are affected by the chemical environment of the nucleus, making it possible to assess the presence of chemical bonds and neighboring functional groups. NMR spectroscopy applied to structural elucidation mainly analyzes hydrogen and carbon nuclei (Dunkel and Wu 2007). There are various NMR methods that can be applied to analyze the different types of carbons present in a chain, in terms of their degree of substitution, for example. It is also possible to determine the position of carbons and hydrogens in a structure, which is conditional on finding the sequence of atoms of this element in the chain (Stintzing et al. 2004). All the information obtained by this technique is extremely useful for identifying a pigment.

While mass spectrometry (MS) is one of the most fundamental characterization techniques in the analysis of natural products, it enables the identification of compounds from numerous classes with a variety of structural properties from different types of samples (Noriega et al. 2021). The technique has the advantage of can be used for the analysis of complex samples, while maintaining high analytical efficiency in identification. It can be coupled with chromatography, an important separation technique, allowing it to be applied to the analysis of different types of samples (Bernardo-Bermejo et al. 2023). When coupled with liquid chromatography (LC) or gas chromatography (GC), it provides two advanced techniques: liquid chromatography coupled to mass spectrometry (LC-MS) and gas chromatography coupled to mass spectrometry (GC-MS) (Vadakedath et al. 2022), suitable for the identification analysis of the components of complex samples, including biological samples, natural extracts, environmental samples, etc. The separation efficiency of compounds determined by chromatography allows mass spectrometry to be used efficiently to identify more than one compound in the same sample.

The identification of a pigment by MS is carried out by analyzing its mass spectrum, which is obtained using a device known as a mass spectrometer, where the compound is ionized and the generated ions are separated and analyzed by the instrument to produce the spectrum of the compound. The mass spectrum is a graphical representation that provides the mass-charge relationships (m/z) of the analyzed ions, useful for determining the molecular mass of each ion (Milman 2015; Christofi and Barran 2023). From this, it is possible to obtain the molecular mass of a pigment of interest.

The use of MS techniques has proven advantageous in the acquisition of structural information from chemical molecules, including fungal pigments. A notable example is GC-MS, which utilizes ionization by high-energy electron impact, thereby generating ions with sufficient instability to undergo fragmentation. Subsequent analysis of these fragments yields a spectrum comprising a substantial number of peaks, corresponding to the m/z values of the fragments. The determined masses of the fragments can be used to propose a fragmentation mechanism and suggest the presence of probable groups in the molecule, which could be confirmed by using an additional technique such as FT-IR (Varga et al. 2016). The use of GC-MS may be limited by some properties of the analyzed compound, which must be volatile under the analysis conditions to be able to make profitable use of the technique (Maji et al. 2023). In the case of polar compounds with low volatility natural products, the employment of MS in conjunction with liquid chromatography (LC-MS) emerges as a suitable analytical instrument.

13.5 Biotechnology of Fungal Pigments: Potential Applications

Fungi exhibit a vast biodiversity and rich metabolism, which result in a wide array of pigments and exhibit a range of biochemical activities, including antimicrobial, immunomodulatory, photoprotective, and stress-tolerant properties (Afroz Toma et al. 2023; Duarte et al. 2019). These characteristics endow them with a broad spectrum of applications, across various sectors in biotechnology (Fig. 13.5) including pharmaceuticals, cosmetics, food production, textile manufacturing, paper production, paint manufacturing, and photography (Elkhateeb and Daba 2023).

Interest in fungal pigments has increased significantly, being a natural alternative to synthetic dyes, as they are sustainable and more likely to be nontoxic to the human body, in addition to being more soluble and stable when compared to plant pigments (Elkhateeb and Daba 2023). In fact, in view of the allergic effects of synthetic dyes, the European Union has started to restrict their use in products food (Molelekoa et al. 2023).

The regulatory standards for the use of dyes in consumer products vary across different countries. Three regulatory authorities have been identified as particularly prominent in this regard: the Food and Drug Administration (FDA) in the United States, the European Commission (EC) in the European Union (EU), and the Ministry of Health, Labor and Welfare (MHLW) in Japan (Guerra et al. 2018). Consequently, there is an absence of a standard that has been formally established for the official analytical methodologies that control concentration limits allowed in each type of product (Guerra et al. 2018; Lores et al. 2016).

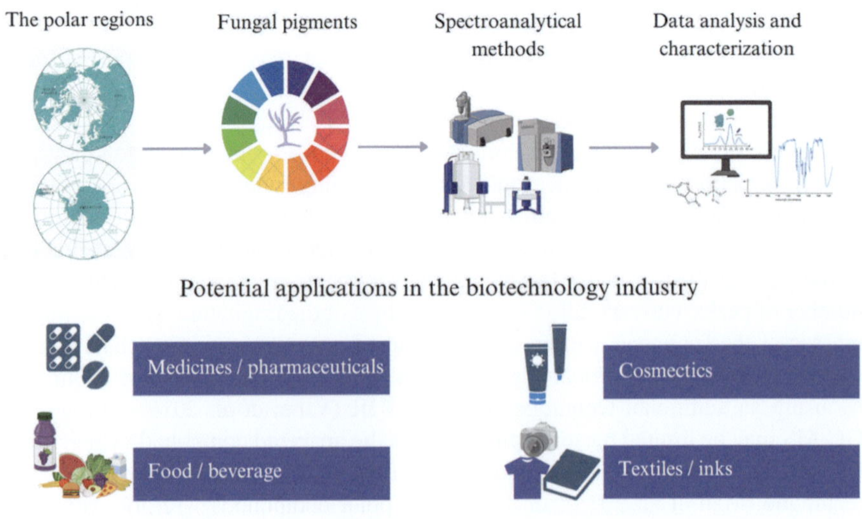

Fig. 13.5 From the polar regions to industry: applications of fungal pigments in different industrial sectors

In 2023, the global pigments market generated US$ 12.36 billion, and it is estimated that by 2032 it will reach US$18.84 billion, growing at a compound annual growth rate (CAGR) of 4.8%, with China, India, and Japan being the main consumers (Mordor Intelligence 2023; Business Research Insights 2023).

The market is undergoing a transformation in response to consumers' escalating demands for sustainability and products that offer added value, characterized by benefits that extend beyond their aesthetic appeal (Kalra et al. 2020). In this context, filamentous fungi emerge as promising solutions, due to their inherent properties as potent antioxidant and antimicrobial agents. These organisms can be readily cultivated from waste, industrial, and low-cost agricultural by-products, a characteristic that enhances their sustainability and economic viability (de Oliveira et al. 2022; Pandey et al. 2023). Moreover, its utilization as a raw material does not pose a threat to the environment; rather, it promotes the conservation and preservation of biodiversity in ecosystems. In contrast to seasonal crops, filamentous fungi are not subject to seasonal variations (Grum-Grzhimaylo et al. 2024).

Emphasizing the therapeutic properties of fungal compounds, the cosmetic industry has increased its use in skin cleansing, makeup, and personal hygiene products, which act to delay skin aging and provide sun protection, reducing unwanted skin pigmentation and damage, due to antioxidant and ultraviolet radiation absorption actions (Alam et al. 2019; Wargala et al. 2021). Furthermore, cosmetics that act to lighten blemishes have antityrosinase activity from natural dyes (Mukherjee et al. 2024).

In the food industry, color plays a pivotal role as a crucial attribute, as it is intimately linked to consumer acceptance of products. This is due to the common association between flavor and color in food and beverages (Mota et al. 2021). A similar importance is observed in the textile industry, where color plays a pivotal role in marketing goods. It is noteworthy that the quality of pigmentation contributes to increased product value. Biopigments, through biotechnology, have the potential to enhance resistance to temperature and light, while exhibiting desirable physical properties (Venil et al. 2020). In addition to their advantageous dyeing properties, biopigments have been observed to possess fungicidal activities, as reported by Mosunova et al. (2021).

The biotechnology industry uses the encapsulation of fungal pigments by biopolymers to increase the stability of these compounds, which occurs mainly by oxidation actions, increased humidity, and other degrading factors (Abel et al. 2023). This allows the color intensity to be maintained for longer and for the color to be released in a controlled manner according to the formulation applied in cases of products pharmaceuticals and cosmetics, for example (Dhanjal et al. 2022).

In recent decades, there has been a marked increase in societal concern regarding the preservation of natural resources. This has led to the adoption of healthier consumption habits, resulting in a lifestyle that is more focused on sustainability (Wargala et al. 2021). Consequently, products of natural origin have gained increased market value, with fungal pigments being a notable example. These pigments contribute significantly to biodiversity preservation due to their biodegradable nature (Kumar et al. 2022).

13.6 Genetically Modified Organism (GMO) Routes for the Production of Fungal Pigments

Genetically modified organisms (GMOs) are integral part of new approaches applied in biotechnology; these are living organisms that have their genetic material remodeled, with the aim of obtaining specifically desired characteristics, such as size, strength, color, and others (Tajima et al. 2022). Genetic engineering with recombinant RNA/DNA technology enables genetic changes that are not possible to happen naturally (Reyes-Avila et al. 2024), making it feasible to perform recombinations of intra- or interspecies genetic materials (Basarkar et al. 2022).

In this context, the term "heterologous" refers to a biotechnological technique that enables the transfer of genes from one organism to another, allowing the expression of specific metabolic pathways in a different host (Meng et al. 2022). Thus, the production of biocompounds can be facilitated or enhanced. For example, overexpression of the AuR1 transcription in *Fusarium graminearum* tripled the production of the mycopigment aurofusarin (Abel et al. 2023). The application of this technique in fungi from polar environments, such as *Antarctomyces* and *Cladosporium*, producers of pigments with antioxidant properties and UV-resistant, may be promising to increase their availability for applications industrial.

Clustered regularly interspaced short palindromic repeats (CRISPR) technology provides evidence that advances in genetic engineering are enabling a remarkable increase in efficiency and effectiveness for the development of innovative products. This genetic editing technology allows the precise alteration of specific DNA sequences (Woodcraft et al. 2022). Furthermore, another eminent strategy is to carry out targeted selection of lines with the aid of metabolomics, since it allows the screening of modified strains with greater yield (Shen et al. 2023).

The utilization of suitable operational criteria for the optimization of fungal cultivation, in conjunction with the regulation of the culture medium and the fermentation process with GMOs, results in the modulation of filamentous gene functions. This, in turn, leads to the enhancement of the production of commercially significant molecules (Kalra et al. 2020).

Furthermore, some pigment-producing fungi are also producers of mycotoxins, which are metabolites that are toxic to humans. Gene editing can overcome this limitation by providing toxicological safety and preserving beneficial effects on the human body (Mosunova et al. 2021). For example, combining heterologous expression techniques with CRISPR/Cas9 resulted in the elimination of the citrinin genomic cluster in *Monascus purpureus*, so that this nephrotoxic mycotoxin was no longer produced while there was an increase in the production of Monascus red pigment by 2–5% (Abel et al. 2023).

This approach enhances the production of molecules of interest, with healthier and more sustainable characteristics, while reducing the environmental impact of production processes, in tune with the concerns of the global market (Choi et al. 2019). It is important to highlight the importance of carrying out toxicological studies and approval regulatory framework for the proper application of GMOs in biotechnology industries, ending with safe marketing for the population (Kalra et al. 2020; Tajima et al. 2022).

It is evident that the main objective of the pigment industry is to identify sustainable and safe sources, particularly for pigments intended for use in food, pharmaceuticals, and cosmetics. In this context, fungi have emerged as promising microorganisms due to their metabolic richness and the ease with which they can be genetically manipulated. These organisms are regarded as "true cell factories," a distinction that signifies their ability to produce high quantities of pigments. Moreover, these pigments are notable for their diverse biological properties and intense coloration (Nielsen 2019).

13.7 Challenges and Future Perspectives

Natural pigments produced by filamentous fungi inhabiting polar environments have emerged as promising compounds due to their significant biotechnological potential and their role as an alternative to the production of synthetic dyes, which are known to be harmful to human and environmental health. Consequently, there is a growing imperative to develop research aimed at encouraging the production, optimization, and characterization of fungal pigments. This research is not only essential for ensuring safe and ecologically sustainable methods, but it is also a key factor in promoting environmental preservation, a matter that has garnered significant attention and support from government authorities.

Acknowledgments The authors would like to acknowledge the National Council for Scientific and Technological Development (CNPq—Process #314775/2023-4) and the Foundation for Research Support of the State Alagoas (FAPEAL—Process #60030.0000000161/2022).

References

Abel G, Amobonye A, Bhagwat P, Pillai S (2023) Diversity, stability and applications of mycopigments. Process Biochem 133:270–284. https://doi.org/10.1016/j.procbio.2023.09.002

Afroz Toma MA, Rahman MH, Rahman MS, Arif M, Nazir KHMNH, Dufossé L (2023) Fungal pigments: carotenoids, riboflavin, and polyketides with diverse applications. J Fungi 9(4):454. https://doi.org/10.3390/jof9040454

Alam MF et al (2019) Assessment of some heavy metals in selected cosmetics commonly used in Bangladesh and human health risk. J Anal Sci Technol 10(2). https://doi.org/10.1186/s40543-018-0162-0

Anghelone M, Stoytschew V, Jembrih-Simbürger D, Schreiner M (2018) Spectroscopic methods for the identification and photostability study of red synthetic organic pigments in alkyd and acrylic paints. Microchem J 139:155–163. https://doi.org/10.1016/j.microc.2018.02.029

Ashenafi EL, Nyman MC, Shelley JT, Mattson NS (2023) Spectral properties and stability of selected carotenoid and chlorophyll compounds in different solvent systems. Food Chem Adv 2:100178. https://doi.org/10.1016/j.focha.2022.100178

Aziz E et al (2020) Xanthophyll: health benefits and therapeutic insights. Life Sci 240:117104. https://doi.org/10.1016/j.lfs.2019.117104

Barreto DLC et al (2024) Phytotoxic and antifungal activity of (−)-penienone produced by *Penicillium palitans* (Ascomycota) isolated from deep sea sediments in the Southern Ocean, Maritime Antarctica. Chem Biodivers 22(3):e202401603. https://doi.org/10.1002/cbdv.202401603

Basarkar V, Govardhane S, Shende P (2022) Multifaceted applications of genetically modified micro-organisms: a biotechnological revolution. Curr Pharm Des 28(22):1833–1842. https://doi.org/10.2174/1381612828666220128102823

Bernardo-Bermejo S et al (2023) Quantitative multiple fragment monitoring with enhanced in-source fragmentation/annotation mass spectrometry. Nat Protoc 18(4):1296–1315. https://doi.org/10.1038/s41596-023-00803-0

Bhagwat SA, Gite AV, Khemnar SY, Phad SH, Rokade VG (2024) A review on IR spectroscopy. Int Res J Mod Eng 4905. https://www.irjmets.com

Business Research Insights (2023) Global pigments market analysis. https://www.businessresearchinsights.com/reports/global-pigments-market-analysis. Accessed 10 Nov 2024

Calsteren M-RV et al (1997) Spectroscopic characterization of Crocetin derivatives from Crocus sativus and Gardenia jasminoides. J Agric Food Chem 45(4):1055–1061. https://pubs.acs.org/sharingguidelines

Catanzaro I et al (2024) Deletion of the polyketide synthase-encoding gene pks1 prevents melanization in the extremophilic fungus *Cryomyces antarcticus*. JUBMB Life 76(12):1072–1090. https://doi.org/10.1002/iub.2895

Cavalcante SB et al (2023) The hidden rainbow: the extensive biotechnological potential of Antarctic fungi pigments. Braz J Microbiol 54(3):1675–1687. https://doi.org/10.1007/s42770-023-01011-4

Cavalcante SB et al (2024) Antarctic fungi produce pigment with antimicrobial and antiparasitic activities. Braz J Microbiol 55(2):1251–1263. https://doi.org/10.1007/s42770-024-01308-y

Chen JL, Kang SC, Wu AD, Hu DDI (2023) Impacts of 1.5 °C global warming on hydrological conditions of navigation along the Northern Sea route and Northwest passage. Adv Clim Chang Res 14(6):904–912. https://doi.org/10.1016/j.accre.2023.11.010

Choi KR, Jang WD, Yang D, Cho JS, Park D, Lee SY (2019) Systems metabolic engineering strategies: integrating systems and synthetic biology with metabolic engineering. Trends Biotechnol 37(8):817–837. https://doi.org/10.1016/j.tibtech.2019.01.003

Christofi E, Barran P (2023) Ion mobility mass spectrometry (IM-MS) for structural biology: insights gained by measuring mass, charge, and collision cross section. Chem Rev 123(6):2902–2949. https://doi.org/10.1021/acs.chemrev.2c00600

Convey P (2011) Antarctic terrestrial biodiversity in a changing world. Polar Biol 34(11):1629–1641. https://doi.org/10.1007/s00300-011-1068-0

de Menezes GCA, Godinho VM, Porto BA, Gonçalves VN, Rosa LH (2017) Antarctomyces pellizariae sp. nov., a new, endemic, blue, snow resident psychrophilic ascomycete fungus from Antarctica. Extremophiles 21:259–269. https://doi.org/10.1007/s00792-016-0895-x

de Oliveira LA, Segundo WOPF, de Souza ÉS, Peres EG, Koolen HHF, de Souza JVB (2022) Ascomycota as a source of natural colorants. Braz J Microbiol 53:1199–1220. https://doi.org/10.1007/s42770-022-00768-4

Dhanjal DS, Mehra P, Bhardwaj S, Singh R, Sharma P, Nepovimova E, Chopra C, Kuca K (2022) Mycology-nanotechnology interface: applications in medicine and cosmetology. Int J Nanomedicine 17:2505–2533. https://doi.org/10.2147/IJN.S363282

Ding MH et al (2024) State of polar climate in 2023. Adv Clim Chang Res 15(5):769–783. https://doi.org/10.1016/j.accre.2024.08.004

Domenici V, Ancora D, Cifelli M, Serani A, Veracini CA, Zandomeneghi M (2014) Extraction of pigment information from near-UV vis absorption spectra of extra virgin olive oils. J Agric Food Chem 62(38):9317–9325. https://doi.org/10.1021/jf503818k

Dotta S, Elias-Piera F, Bassoi M (2022) Antártica, Ártico e Mudanças Climáticas. Interantar 1–33. https://www.researchgate.net/publication/358987964

Du J, Yang R, Fang C, Zhou X, Pan S, Lei W, Su J, Cheng Y, Liu D (2018) Preparation and characterization of organic pigments and their fluorescence properties depending on bulk structure. J Mater Sci Technol 34(11):2218–2224. https://doi.org/10.1016/j.jmst.2018.05.017

Duarte AWF, Dos Santos JA, Vianna MV, Vieira JMF, Mallagutti VH, Inforsato FJ, Wentzel LCP, Lario LD, Rodrigues A, Pagnocca FC, Pessoa Junior A, Durães Sette L (2018) Cold-adapted enzymes produced by fungi from terrestrial and marine Antarctic environments. Crit Rev Biotechnol 38(4):600–619. https://doi.org/10.1080/07388551.2017.1379468

Duarte AWF, de Menezes GCA, e Silva TR, Bicas JL, Oliveira VM, Rosa LH (2019) Antarctic Fungi as Producers of Pigments. IN Fungi of Antarctica Diversity, Ecology and Biotechnological Applications. Springer. https://doi.org/10.1007/978-3-030-18367-7

Dunkel R, Wu X (2007) Identification of organic molecules from a structure database using proton and carbon NMR analysis results. J Magn Reson 188(1):97–110. https://doi.org/10.1016/j.jmr.2007.06.007

Dziurzynski M, Gorecki A, Pawlowska J, Istel L, Decewicz P, Golec P, Styczynski M, Poszytek K, Rokowska A, Gorniak D, Dziewit L (2023) Revealing the diversity of bacteria and fungi in the active layer of permafrost at Spitsbergen island (Arctic) – Combining classical microbiology and metabarcoding for ecological and bioprospecting exploration. Sci Total Env 856:159072. https://doi.org/10.1016/j.scitotenv.2022.159072

Elkhateeb W, Daba G (2023) Fungal pigments: their diversity, chemistry, food and non-food applications. Appl Microbiol 3(3):735–751. https://doi.org/10.3390/applmicrobiol3030051

Grum-Grzhimaylo OA, Shurigina AA, Debets AJM, Aanen DK (2024) Biogeography and uniqueness of filamentous terrestrial fungi in the polar regions. Fungal Biol Rev 49:100382. https://doi.org/10.1016/j.fbr.2024.100382

Guerra E, Llompart M, Garcia-Jares C (2018) Analysis of dyes in cosmetics: challenges and recent developments. Cosmetics 5(3):47. https://doi.org/10.3390/COSMETICS5030047

Huang H et al (2020) Combinatorial strategies for production improvement of red pigments from Antarctic fungus Geomyces sp. J Food Sci 85(10):3061–3071. https://doi.org/10.1111/1750-3841.15443

Jeyaram S, Geethakrishnan T (2020) Vibrational spectroscopic, linear and nonlinear optical characteristics of Anthocyanin extracted from blueberry. Results Opt 1:100010. https://doi.org/10.1016/j.rio.2020.100010

Kalra R, Conlan XA, Goel M (2020) Fungi as a potential source of pigments: harnessing filamentous fungi. Front Chem 8. https://doi.org/10.3389/fchem.2020.00369

Kreusch MG, Duarte RTD (2021) Photoprotective compounds and radioresistance in pigmented and non-pigmented yeasts. Appl Microbiol Biotechnol 105(9):3521–3532. https://doi.org/10.1007/s00253-021-11271-5

Kumar Shetty AV, Dave N, Murugesan G, Pai S, Pugazhendhi A, Varadavenkatesan T, Vinayagam R, Selvaraj R (2021) Production and extraction of red pigment by solid-state fermentation of broken rice using Monascus sanguineus NFCCI 2453. Biocatal Agric Biotechnol 33:101964. https://doi.org/10.1016/j.bcab.2021.101964

Kumar S, Kumar V, Ambika AAA, Nag D, Kumar V, Darnal S, Thakur V, Patial V, Singh D (2022) Microbial pigments: Learning from Himalayan perspective to industrial applications. J Ind Microbiol Biotechnol 49(5). https://doi.org/10.1093/jimb/kuac017

Küpper H, Seibert S, Parameswaran A (2007) Fast, sensitive, and inexpensive alternative to analytical pigment HPLC: quantification of chlorophylls and carotenoids in crude extracts by fitting with Gauss peak spectra. Anal Chem 79(20):7611–7627. https://doi.org/10.1021/ac070236m

Lee SM, Üpping J, Bielawny A, Knez M (2011) Structure-based color of natural petals discriminated by polymer replication. ACS Appl Mater Interfaces 3(1):30–34. https://doi.org/10.1021/am1007968

Lin L, Xu J (2020) Fungal pigments and their roles associated with human health. J Fungi 6(4):280. https://doi.org/10.3390/jof6040280

Liu L, Wu S, Wang W, Zhang X, Wang Z (2020) Sulfonation of Monascus pigments to produce water-soluble yellow pigments. Dyes Pigments 173:107965. https://doi.org/10.1016/j.dyepig.2019.107965
Long H et al (2024) Comparative omics directed gene discovery and rewiring for normal temperature-adaptative red pigment synthesis by polar psychrotrophic fungus Geomyces sp. WNF-15A. Synth Syst Biotechnol 9(4):842–852. https://doi.org/10.1016/j.synbio.2024.07.002
Lopes DDV, Oliveira FSD, Siqueira RG, De Souza JJLL, Schaefer CEGR, Francelino MR (2024) Paisagens e processos pedogenéticos em distintos ecossistemas terrestres Antárticos. Geogr Dep Univ Sao Paulo 44:e210645. https://doi.org/10.11606/eISSN.2236-2878.rdg.2024.210645
Lores M, Llompart M, Alvarez-Rivera G, Guerra E, Vila M, Celeiro M, Lamas JP, Garcia-Jares C (2016) Positive lists of cosmetic ingredients: analytical methodology for regulatory and safety controls – a review. Anal Chim Acta 915:1–26. https://doi.org/10.1016/j.aca.2016.02.033
Maji AR, Roy C, Sinha SK (2023) Gas chromatography-mass spectrometry (GC-MS): a comprehensive review of synergistic combinations and their applications in the past two decades. J Analyti Sci Appl Biotechnol 5(2). https://doi.org/10.48402/IMIST.PRSM/jasab-v5i2.40209
Maoka T (2020) Carotenoids as natural functional pigments. J Nat Med 74(1):1–16. https://doi.org/10.1007/s11418-019-01364-x
Marchetta A et al (2023) A Deep Insight into the Diversity of Microfungal Communities in Arctic and Antarctic Lakes. J Fungi 9(11):1095. https://doi.org/10.3390/jof9111095
Mateus N, Silva AMS, Santos-Buelga C, Rivas-Gonzalo JC, De Freitas V (2002) Identification of anthocyanin-flavanol pigments in red wines by NMR and mass spectrometry. J Agric Food Chem 50(7):2110–2116. https://doi.org/10.1021/jf0111561
Meddeb-Mouelhi F, Moisan JK, Bergeron J, Daoust B, Beauregard M (2016) Structural characterization of a novel antioxidant pigment produced by a photochromogenic Microbacterium oxydans strain. Appl Biochem Biotechnol 180(7):1286–1300. https://doi.org/10.1007/s12010-016-2167-8
Meng X, Fang Y, Ding M, Zhang Y, Jia K, Li Z, Collemare J, Liu W (2022) Developing fungal heterologous expression platforms to explore and improve the production of natural products from fungal biodiversity. Biotechnol Adv 54:107866. https://doi.org/10.1016/j.biotechadv.2021.107866
Milman BL (2015) General principles of identification by mass spectrometry. TrAC Trends Anal Chem 69:24–33. https://doi.org/10.1016/j.trac.2014.12.009
Molelekoa TBJ, Augustyn W, Regnier T, da Silva LS (2023) Chemical characterization and toxicity evaluation of fungal pigments for potential application in food, phamarceutical and agricultural industries. Saudi J Biol Sci 30(5):103630. https://doi.org/10.1016/j.sjbs.2023.103630
Mordor Intelligence (2023) Global pigments market – growth, trends, COVID-19 impact, and forecasts (2023–2028). https://www.mordorintelligence.com/industry-reports/global-pigments-market. Accessed 10 Nov 2024
Mosunova O, Navarro-Muñoz JC, Collemare J (2021) The biosynthesis of fungal secondary metabolites: from fundamentals to biotechnological applications. In: Encyclopedia of mycology, vol 2, pp 458–476. https://doi.org/10.1016/B978-0-12-809633-8.21072-8
Mota IGC, Neves RAMD, Nascimento SSDC, Maciel BLL, Morais AHDA, Passos TS (2021) Artificial dyes: health risks and the need for revision of international regulations. Food Rev Intl 39(3):1578–1593. https://doi.org/10.1080/87559129.2021.1934694
Mukherjee A, Das S, Chakraborty D, Pal N, Das N (2024) Fungi's treasure in cosmeceuticals-a comprehensive chemical approach. S Afr J Bot 166:311–331. https://doi.org/10.1016/j.sajb.2024.01.036
Nakano T, Wiegertjes G (2020) Properties of carotenoids in fish fitness: a review. Mar Drugs 18(11):568. https://doi.org/10.3390/md18110568
Neha T, Shishir T, Ashutosh D (2017) Fourier transform infrared spectroscopy (FTIR) profiling of red pigment produced by Bacillus subtilis PD5. Afr J Biotechnol 16(27):1507–1512. https://doi.org/10.5897/ajb2017.15959

Nielsen J (2019) Yeast systems biology: model organism and cell factory. Biotechnol J 14(9). https://doi.org/10.1002/biot.201800421

Nikitin DA (2023) Ecological Characteristics of Antarctic Fungi. Dokl Biol Sci 508(1):32–54. https://doi.org/10.1134/S0012496622700120

Noriega P, Gortaire G, Osorio E (2021) Mass spectrometry and its importance for the analysis and discovery of active molecules in natural products. Nat Drugs Plants:33–43. https://www.intechopen.com/chapters/76771#:~:text=10.5772/intechopen.97733

Pacelli C et al (2020) Multidisciplinary characterization of melanin pigments from the black fungus *Cryomyces antarcticus*. Appl Microbiol Biotechnol 104:6385–6395. https://link.springer.com/article/10.1007/s00253-020-10666-0

Pandey N, Jain R, Dhakar K, Sharma A, Pandey A (2023) A reduction in temperature induces bioactive red pigment production in a psychrotolerant Penicillium sp. GEU_37 isolated from Himalayan soil. Fungal Biol 127(3):927–937. https://doi.org/10.1016/j.funbio.2023.01.008

Pavia DL, Lampman GM, Kriz G, Vyvyan JR (2013) Introdução à espectroscopia, 1st edn. Cengage Learning

Perez-Cuesta U et al (2020) Melanin and pyomelanin in *Aspergillus fumigatus*: from its genetics to host interaction. Int Microbiol 23:55–63. https://doi.org/10.1007/s10123-019-00078-0

Quijada-Morín N, Dangles O, Rivas-Gonzalo JC, Escribano-Bailón MT (2010) Physico-chemical and chromatic characterization of malvidin 3-glucoside-vinylcatechol and malvidin 3-glucoside-vinylguaiacol wine pigments. J Agric Food Chem 58(17):9744–9752. https://doi.org/10.1021/jf102238v

Rapoport A et al (2021) Carotenoids and some other pigments from fungi and yeasts. Meta 11(2):92. https://doi.org/10.3390/metabo11020092

Reyes-Avila CS, Waldvogel D, Pradervand N, Aubry S, Croll D (2024) Detection of genetically modified organisms using highly multiplexed amplicon sequencing. Food Control 165:110670. https://doi.org/10.1016/j.foodcont.2024.110670

Robinson SA (2022) Climate change and extreme events are changing the biology of Polar Regions. Glob Chang Biol 28(20):5861–5864. https://doi.org/10.1111/gcb.16309

Rosa LH et al (2019) Fungi in Antarctica: diversity, ecology, effects of climate change, and bioprospection for bioactive compounds. In: Rosa LH (ed) Fungi of Antarctica, vol 1. Springer, Cham, pp 1–17. https://doi.org/10.1007/978-3-030-18367-7_1

Rudrappa M, Kumar MS, Kumar RS, Almansour AI, Perumal K, Nayaka S (2022) Bioproduction, purification and physicochemical characterization of melanin from Streptomyces sp. strain MR28. Microbiol Res 263:127130. https://doi.org/10.1016/j.micres.2022.127130

Saijad W et al (2020) Pigment production by cold-adapted bacteria and fungi: colorful tale of cryosphere with wide range applications. Extremophiles 24:447–473. https://doi.org/10.1007/s00792-020-01180-2

Sandmann G (2019) Antioxidant protection from UV-and light-stress related to carotenoid structures. Antioxidants 8(7):219. https://doi.org/10.3390/antiox8070219

Shabbir SW, Chauhan S (2024) A review on use of ultraviolet spectroscopy. Innovare. J Med Sci 12(4):5–11. https://doi.org/10.22159/ijms.2024v12i4.51499

Shen JY, Zhao Q, He QL (2023) Application of CRISPR in filamentous fungi and Macrofungi: from component function to development potentiality. ACS Synth Biol 12(7):1908–1923. https://doi.org/10.1021/acssynbio.3c00099

Stachura M, Gottberg A, Kowalska M, Johnston K, Hemmingsen L (2015) Toward applications of β-NMR spectroscopy in chemistry and biochemistry. Nucl Phys News 25(2):25–29. https://doi.org/10.1080/10619127.2015.1035935

Stintzing FC, Conrad J, Klaiber I, Beifuss U, Carle R (2004) Structural investigations on betacyanin pigments by LC NMR and 2D NMR spectroscopy. Phytochemistry 65(4):415–422. https://doi.org/10.1016/j.phytochem.2003.10.029

Tajima G, Huh S, Schmidt NA, Macdonald JC, Fleischmann T, Wonnacott KM (2022) Impact of genetically modified organism requirements on gene therapy development in the EU, Japan, and the US. Mol Ther Methods Clin Dev 26:74–83. https://doi.org/10.1016/j.omtm.2022.05.012

Vadakedath S, Kandi V, Godishala V, Bharat Kumar Pinnelli V, Sami Alkafaas S, Sami ElKafas S (2022) The principle, types, and applications of mass spectrometry: a comprehensive review. Biomed Biotechnol 7(1):6–22. https://doi.org/10.12691/bb-7-1-2

Valenzuela-Gloria MS et al (2021) Molecular characterization of fungal pigments. J Fungi 7(5):326. https://doi.org/10.3390/jof7050326

Varga M, Berkesi O, Darula Z, May NV, Palágyi A (2016) Structural characterization of allomelanin from black oat. Phytochemistry 130:313–320. https://doi.org/10.1016/j.phytochem.2016.07.002

Venil CK, Velmurugan P, Dufossé L, Devi PR, Ravi AV (2020) Fungal pigments: potential coloring compounds for wide ranging applications in textile dyeing. J Fungi 6(2):68. https://doi.org/10.3390/jof6020068

Wargala E, Sławska M, Zalewska A, Toporowska M (2021) Health effects of dyes, minerals, and vitamins used in cosmetics. Women 1(4):223–237. https://doi.org/10.3390/women1040020

Woodcraft C, Chooi YH, Roux I (2022) The expanding CRISPR toolbox for natural product discovery and engineering in filamentous fungi. Nat Prod Rep 40(1):158–173. https://doi.org/10.1039/d2np00055e

Yadav AN et al (2019) Metabolic engineering to synthetic biology of secondary metabolites production. In: New and future developments in microbial biotechnology and bioengineering. Elsevier, pp 279–320. https://doi.org/10.1016/B978-0-444-63504-4.00020-7

Yoshida K, Tojo K, Mori M, Yamashita K, Kitahara S, Noda M, Uchiyama S (2015) Chemical mechanism of petal color development of Nemophila menziesii by a metalloanthocyanin, nemophilin. Tetrahedron 71(48):9123–9130. https://doi.org/10.1016/j.tet.2015.10.007

Zalar P, Sonjak S, Gunde-Cimerman N (2011) Fungi in Polar Environments. In: Miller RV, Whyte LG (ed) Polar Microbiology Life in a Deep Freeze (pp 79–99). ASM Press, Washington, DC, USA

Chapter 14
Winter Thaws in Polar Terrestrial Environments Under Global Warming: Impacts and Survival Strategies

Anastasiia Kolomiiets, Oleksandr Bren, Pavel Přibyl, Eva Hejduková, Jana Kvíderová, Ekaterina Pushkareva, Burkhard Becker, and Josef Elster

Abstract The Arctic and Antarctic regions are renowned for their harsh environmental conditions, including low temperatures, lack of liquid water and mineral nutrients, and intense solar radiation during the summer solstice period. However, the recent global warming is leading to extreme weather events, including periods of winter thaws, and the widespread melting of ice caps, snowbanks, and glaciers. These environmental changes have a serious impact on local polar communities. The diversity of polar cyanobacterial and microalgal species has recently been the subject of extensive study. Detailed ecophysiological and molecular studies reveal species distributions, either widely distributed (ubiquitous) and/or local (endemic) in a particular polar area and in particular habitat types. To survive, local endemic communities develop a variety of protective mechanisms that differ from those species living in more favourable conditions. In this review, prokaryotic cyanobacteria and eukaryotic microalgae are used as model organisms to discuss the hardening processes, i.e. how they are prepared and/or how they react to the most extreme periods during seasonal cycles, on the ecological, physiological, and molecular lev-

A. Kolomiiets (✉)
Faculty of Science, Department of Experimental Plant Biology, University of South Bohemia, České Budějovice, Czech Republic

Faculty of Science, Centre for Polar Ecology, University of South Bohemia, České Budějovice, Czech Republic

Department of Phycology, Institute of Botany, Czech Academy of Sciences, Třeboň, Czech Republic
e-mail: anastasiia.kolomiiets@ibot.cas.cz

O. Bren · P. Přibyl
Department of Phycology, Institute of Botany, Czech Academy of Sciences, Třeboň, Czech Republic

E. Hejduková
Department of Phycology, Institute of Botany, Czech Academy of Sciences, Třeboň, Czech Republic

Faculty of Science, Department of Ecology, Charles University, Prague 2, Czech Republic

A. W. Fernandes Duarte et al. (eds.), *Polar Microbes and Climate Change*,
https://doi.org/10.1007/978-3-032-07096-8_14

els. While some information on the physiological and morphological adaptations of polar communities is reviewed here, there is a lack of information on the hardening of organisms at the molecular level. It is crucial to determine the mechanisms by which different organisms prepare for the most extreme and stressful conditions during seasonal cycles, involving freezing and desiccation, as well as other important ecological limiting factors.

Keywords Polar areas · Arctic · Cyanobacteria · Microalgae · Winter melts · Starvation · Hardening

14.1 Introduction

Terrestrial parts of the Arctic and the Antarctic are regions which are renowned for their harsh environmental conditions, including low temperatures, which are coupled with predictable annual and diel light availability during the year, and strong short-term variation in their intensities during the day. Temperatures above freezing occur only for a couple of months which is connected to a lack of liquid water and mineral nutrients (Elster and Benson 2004; Thomas et al. 2008). As a result, the polar terrestrial environment contains various stresses for living organisms, including freeze-thaw processes and the persistence of cold temperatures affecting the metabolism of poikilotherm organisms (Vincent 2007). The photosynthetic microorganisms that dominate in these cold polar terrestrial environments are prokaryotic Cyanobacteria and eukaryotic microalgae (to simplify, the term 'microalgae' will include *Cyanobacteria* in the following) (Thomas et al. 2008). They frequently produce visible biomass and affect the mineral and biological development of polar soils.

Seasonal and diurnal variations of polar terrestrial environments present a series of patterns from aquatic (semiaquatic) to dry habitats along a water availability gradient. These patterns differ in periodicity, amplitude, synchrony, and regularity

J. Kvíderová
Department of Phycology, Institute of Botany, Czech Academy of Sciences, Třeboň, Czech Republic

Centre for Biology, Geosciences and the Environment, Faculty of Education, University of West Bohemia, Plzeň, Czech Republic

E. Pushkareva · B. Becker
Department of Biology, Botanical Institute, University of Cologne, Cologne, Germany

J. Elster
Faculty of Science, Centre for Polar Ecology, University of South Bohemia, České Budějovice, Czech Republic

Department of Phycology, Institute of Botany, Czech Academy of Sciences, Třeboň, Czech Republic and initiate a number of different ecological and physiological acclimations and adaptations (Elster 2002). The ecological classification of the polar terrestrial environment is connected to the water gradient and consists of three categories: lacustrine, hydro-terrestrial, and terrestrial. The lacustrine (lake) environment has a limnological character (liquid water is permanently present, similar to marine ecosystems) and is mostly occupied by psychrophilic stenothermic organisms (Newsted and Hüner 1988).

Hydro-terrestrial environments (wetlands) include all types of habitats where liquid water is available for almost the entire period of summer. However, wetlands can dry down during the summer period and/or freeze solid during winter (Hawes et al. 1992; Elster 2002). Polar hydro-terrestrial habitats represent a unique mosaic of unstable extreme environments, where large variations in microalgal species composition and biomass are evident between various types of habitats (Broady 1996).

In the polar terrestrial environments, water is available in liquid form for only a short period (e.g. snow melt, summer rain, or snowfall) or is available as air humidity (vapour absorbed directly from the air). True terrestrial microalgae, including edaphophytes (edaphic, soil microalgae) and aerophytes (lithophytic, aerial microalgae), play an especially important ecological role in polar desert/semi-desert and/or periglacial ecosystems (Elster 2002).

Microorganisms, including microalgae that occupy hydro-terrestrial and terrestrial environments in polar areas, are mostly considered psychrotolerant (Morita 1975). They exhibit maximum growth rates at ≥20 °C but are also able to acclimate, grow, and survive at low temperatures (0–5 °C). Psychrotolerant microorganisms are deemed eurythermal and their survival strategy may provide an adaptive advantage in environments that fluctuate daily and seasonally.

Thermal limits for life are an important parameter of survival in a changing climate. Studies have shown that organisms adapted to stable (stenothermal) and variable (eurythermal) lifestyles are highly affected by climate change (Tomanek 2008). In addition to temperature, photosynthetic microorganisms are able to tolerate desiccation to different levels; they are poikilohydric in that the external water content directly influences their metabolic activity. In addition to poikilohydry, changes in temperature and water status cause a series of adaptive responses such as developing resistance and tolerance to cold, freezing, drought, desiccation, and salinity stress (Elster and Benson 2004).

Winter survival strategies of photosynthetic micro-eukaryotes in polar terrestrial ecosystems have been reviewed by Pichrtová et al. (2020). In this review, the most important adaptive strategies helping to survive polar winters were divided into two categories: (A) formation of specialized stress-resistant (dormant) cells and (B) surviving winter stress conditions in a vegetative stage. Both of these strategies are connected with various acclimation forces and biochemical adaptations, including a shift from phototrophy to mixotrophy (Pichrtová et al. 2020). However, polar hydro-terrestrial and terrestrial cyanobacteria and eukaryotic microalgae have a wide spectrum of winter adaptation and acclimation strategies which have still not been studied.

In this chapter, we review the present knowledge of winter warming extremes, the ecophysiological and molecular-genetic mechanisms for resistance of terrestrial and hydro-terrestrial photosynthetic microorganisms to stresses associated with the harsh seasonally changing Arctic conditions. It is generally expected that hardened and/or starved microalgae are more resistant to Arctic winter, melt-freeze cycles, cryostress and desiccation, compared to those growing under favourable conditions. The physiological differences are accompanied by morphological and ultrastructural modifications and altered metagenomic and (meta)transcriptomic profiles, since the vast majority of adaptation/acclimation mechanisms are connected to metabolic changes.

14.2 Polar Environment and Climatic Extremes

Climate influences a variety of ecological parameters. These effects operate through local weather parameters such as temperature, wind, rain, snow, and ocean currents, as well as interactions among them (Stenseth et al. 2002). Climate change impacts and climatic extremes are not uniform across the Arctic region because interacting factors cause large variations in local ecosystem change (ACIA 2005). Extreme climatic events occur simultaneously against a background of gradual climate warming trends and can redirect ecosystem change along routes that are difficult to predict (Bokhorst et al. 2015).

Hydrographical observatories, which have been operating since 2002 in fjords of the west coast of Svalbard (Kongsfjorden and Isfjorden), show an increase of 2 °C over the 20-year record of observations with the strongest sea temperature increase in winter (Fig. 14.1) (Cottier et al. 2019).

In the later years, there have been episodes of rainfall during winter, mainly around the seacoast (Hanssen-Bauer et al. 2019). The West Spitsbergen Current is particularly important for the shaping of the archipelago's climatic conditions, releasing heat during winter and thus influencing the amount of sea ice (Fig. 14.2) (Walczowski and Piechura 2011; Urbanski and Litwicka 2022; Søreide et al. 2020).

In analysed polar hydro-terrestrial and terrestrial habitats, the freeze-thaw cycles have occurred during winter periods (Solstice dark period) in areas which are most affected by climate warming (Cullather et al. 2016; Hejduková et al. 2020; Jimel et al. 2021; Pushkareva et al. 2024). Freeze and thaw events are accompanied by indirect additional osmotic and mechanical stresses for organisms inhabiting snow, ice, and permafrost-influenced shallow wetlands and soils (Vincent 1989; Tashyreva and Elster 2015). However, organisms in hydro-terrestrial and terrestrial environments can be protected by insulating snow cover and thus they do not experience temperatures far below zero, even during winter (Hejduková et al. 2020; Jimel et al. 2021; Pushkareva et al. 2024; Hejduková et al. under review).

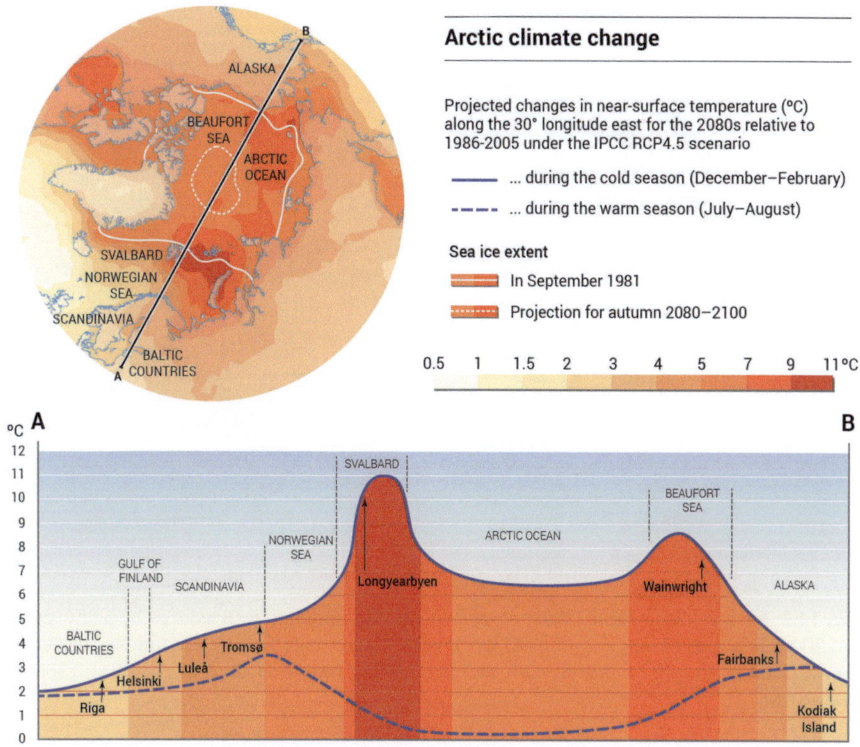

Fig. 14.1 Predicted climate changes (ground surface warming) by 2080 in contrast to 1986–2005 according to the RCP4.5 climate model. (Schoolmeester et al. 2019)

The complicated mosaic of terrestrial habitats evolved in areas located in low elevations close to open water seacoast on the west coast of Svalbard, where winter extreme rain events impacting snow-cover are frequent (Hejduková et al. 2020; Jimel et al. 2021; Pushkareva et al. 2024; Hejduková et al. under review). Here, in some areas, liquid water is soaked up by the snow, developing snow ice layers or ice cones by subsequent freezing. In addition, in areas with thin snow cover, liquid water penetrates down to the soil surface and waterlogs the topsoil layer (watering the surface of soil—biological soil crusts) or can even produce a system of shallow hydro-terrestrial (wetland) habitats (Jimel et al. 2021; Hejduková et al. under review). Snow cover and its properties are principal ecological parameters for the terrestrial polar environment and influence their surface energy balance, water balance, and thermal regimes. The widespread warming and changing winter precipitation amounts in the Arctic and in some areas in Antarctica resulted in different regional snow-cover responses (Callaghan et al. 2011).

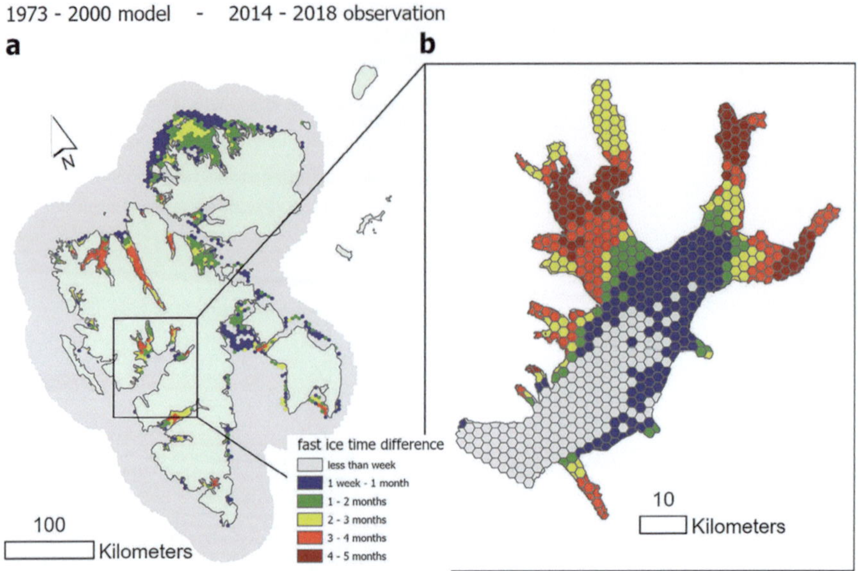

Fig. 14.2 Fast ice extent with a duration of 2 months or more has been reduced by half over the last 30 years (from 12,000 km^2 to 6000 km^2), while the average fast ice duration has decreased from 5 months to only 1 month, with the most severe reduction in West Svalbard. (Søreide et al. 2020)

14.3 Conditions in Low Temperature Fluctuating Environments

During winter, liquid water is usually not available for living organisms; its availability is limited to the solid forms, ice and snow. The amount of liquid water in ice and snow depends on physical, thermal, and mechanical properties, which can vary significantly. Generally, ice has the lowest range of physical and thermal properties (e.g., density or thermal conductivity), while the properties of snow have much larger variations.

The basic properties of ice and snow are modified by environmental conditions (Arenson et al. 2021). However, the amount of available liquid water limits the metabolic processes of living organisms, especially in shallow wetlands and soils due to evaporation during summer (Fig. 14.3) (Everatt et al. 2015).

Recent global warming has increased the melting of snow, ice, and permafrost which can positively as well as negatively impact periods of liquid water availability for local polar organisms (Sawicka et al. 2010). Hydrological models applied to the coast of the Barents Sea predict an increase in the length of growing seasons to 70 days a year by 2100 (Søreide et al. 2020). This change is associated with an earlier start of the growing season of about 3 weeks and a delayed end of 2–3 weeks. The timing of the start of the snow-free period is of critical importance and more important than the timing of autumn snow fall, because sun elevation is high when

Fig. 14.3 The frozen-dry shallow wetlands in late autumn, Endalen, Central Svalbard, High Arctic. (Josef Elster)

photoautotrophs currently start growth. Each extra snow-free day at the beginning of the growing season will enable plants to access high levels of PAR (Crawford 2000). An experiment that manipulated snow conditions by using snow fences at Toolik Lake, Alaska, showed that snowdrift increased winter-time temperatures and CO_2 flux (Walker et al. 1999; Wipf and Rixen 2010; Liu et al. 2023). Temperatures were more constant under the drifts than in control plots. Plants increased their growth despite a shorter growing season, although this was thought to be a transitory response and in contrast to the reduced growth of late snow bed plants (Callaghan et al. 2011).

How the length of the growing season influences the soil surface and shallow wetland organisms is not known yet. However, it can be expected that the growth of biological soil crust and wetland organisms at the beginning and end of the growing season will be supported by earlier melting of the snow cover. It has been shown that temperature fluctuations around the freezing point do not have a fatal effect on photosynthetic organisms in shallow wetlands and biological soil crusts (e.g. Šabacká and Elster 2006; Elster et al. 2008; Jimel et al. 2021; Hejduková et al. under review). However, liquid water availability in the Arctic and Antarctica has a seasonal character, and with the progress of the summer season and the incoming autumn temperatures drops, water levels decrease, and water slowly disappears or becomes less present in the landscape.

14.4 Summer Solstice Period and Hardening

Changes in snow depth and duration are likely to cause hazardous impacts on the frost resistance and avoidance of many vascular plants. Damage to foliage and apical meristems can occur when they are "triggered" to premature bud burst and development by an earlier onset of the growing season. This results from an early snowmelt, when the annual hardening/dehardening is at its most sensitive phase, and when there is a risk of short periods of cold, freezing weather. In addition to frost resistance, frost avoidance may also be disturbed by a thin or even absent snow cover. The risk is likely to be highest at high latitudes, where plants that are genetically adapted to the presence of snow may have lost some potential for frost resistance during their evolution (Niskanen et al. 2019).

Other global change factors might affect frost resistance but only few, and sometimes conflicting reports have been published of studies performed at high latitudes. Nitrogen (or fertilizer) pollutants can impair the frost resistance of plants. Such an effect was demonstrated in Caporn et al. (2000), Shan (2000), Franzaring et al. (2013), Guak and Fuchigami (2002), Taulavuori et al. (2014). However, recent studies with *Calluna vulgaris* (Caporn et al. 1994), *Vaccinium myrtillus* (Zvereva and Kozlov 2005) and *V. vitis-idaea* (Taulavuori et al. 2001) have demonstrated improved frost resistance caused by extra N. Probably, this is because these ericaceous species are adapted to low-nutrient habitats, such as those at high latitudes.

14.5 Seasonal Field Measurements and Reaction of Microalgae Community to Winter Thaws

Seasonal changes in the composition and functional activity of the microbial community in Arctic terrestrial and hydro-terrestrial habitats are strongly influenced by site-specific environmental parameters (Pushkareva et al. 2024). However, our understanding of microbial diversity and microbial activity in hydro-terrestrial and terrestrial habitats across the Arctic is still very limited. The microbial profile and photosynthetic activity within various Arctic latitudes, which are characterized by distinct microclimates, including temperature fluctuations, freeze-thaw cycles, and the duration of the growing season, as well as differences in water content and nutrient availability, were investigated in the central part of West Spitsbergen, Svalbard (Pushkareva et al. 2024; Hejduková et al. under review). Phototrophs and many animals (either invertebrate or vertebrate, mainly herbivores) accumulate energy (amino acids and carbohydrates) during the summer solstice period, and/or during the period of food availability. On the contrary, during the cold, dark winter period, the metabolism depends heavily on catabolic activity of energy which was accumulated during the summer period (Joli et al. 2024; Wang et al. 2024). In addition, the amount of amino acids and carbohydrates decreases significantly during the dark winter period, which is also accompanied by changes in cell morphology in many prokaryotic and eukaryotic photosynthetic microorganisms (e.g. reduction of phycobilisome and chloroplast numbers and size) (e.g. Tashyreva and Elster 2016; Pichrtová et al. 2016; Hejduková et al. 2020), indicating that light starvation during the polar night may trigger rapid autophagic responses, as observed in several starved eukaryotes (e.g. Joli et al. 2024). However, microorganisms in hydro-terrestrial and terrestrial environments may produce cold shock proteins (Csp) or antifreeze proteins (AFP) (Pushkareva et al. 2023, 2024). These proteins enable the cell to adapt and survive under freezing temperatures, maintaining cellular integrity (Keto-Timonen et al. 2016; Pushkareva et al. 2024). In seasonal studies of biocrusts along an altitudinal gradient in Svalbard, the metagenomics and metatranscriptomic functional datasets showed a significant shift in microbial community structure. Cold shock and antifreeze proteins were not highly expressed, whereas genes associated with RNA helicases and heat shock proteins were upregulated, indicating their role in cold acclimation. Differential expression of these genes across seasons suggests a continuous preparation for cold stress, with potential multifunctional properties (Pushkareva et al. 2024).

Simple methods of variable chlorophyll fluorescence measurement techniques are widely used for evaluation of various stresses and stress tolerance determination in plants and microalgae (Büchel and Wilhelm 1993; Maxwell and Johnson 2000; Consalvey et al. 2005). These methods are used in field studies in the polar environment and enable us to measure diurnal and seasonal photosynthetic activities of communities occupying hydro-terrestrial and terrestrial habitats (e.g. Pushkareva et al. 2017; Hejduková et al. under review). More complicated variable chlorophyll protocols for fast (OJIP transient) and slow kinetics (quenching analysis) could be applied for detailed analyses of photosynthetic processes and light energy utilization (Roháček et al. 2008). Combining the variable chlorophyll fluorescence parameters with local microclimatic parameters allows estimation of the photosynthetic activity and limitations. Regular measurements could reveal stress effects by changes in photosynthetic parameters, which are followed by changes in metatranscriptomic performances of photosynthetic-related genes, and further in growth and reproduction.

The study of diurnal and seasonal changes of variable chlorophyll fluorescence measurements of biocrusts along an altitudinal gradient in Svalbard showed that the rate of photosynthesis was driven by irradiance. It was also recorded that there was relatively high photosynthetic activity even in low temperature conditions during autumn (October) (Hejduková et al. under review). Furthermore, winter-thawed biocrusts exhibited the capacity to restore photosynthesis rapidly following thawing, indicating that microalgae are ready to photosynthesize whenever possible (Hejduková et al. under review).

14.6 Response of Shallow Wetlands and Biological Soil Crust Photosynthetic Microorganisms to Winter Thaws

Shallow wetlands and biocrust microorganisms are able to withstand the harsh conditions of polar winters, characterized by extreme cold, prolonged darkness, and limited resources (Tashyreva and Elster 2012, 2015; Pichrtová et al. 2016, 2020; Hejduková et al. 2020; Jimel et al. 2021; Pushkareva et al. 2024). These communities, often dominated by cold-adapted cyanobacteria, green algae, and other microeukaryotes, are capable of surviving in both hydrated and desiccated states. Similarly, the more complex lichens and bryophyte communities which occupy similar habitats have adaptations which help them to survive stresses connected with winter thaw events (Bates and Farmer 1992).

Recent studies have shown that many of these photosynthetic microorganisms can resume metabolic activity, particularly photosynthesis, within just a few hours of thawing (Hejduková et al. under review). This rapid reactivation might be driven by the maintenance of intact photosynthetic machinery and the expression of stress-regulated genes that remain primed during dormancy. Rather than entering a state of full dormancy, these organisms appear to adopt a form of *metabolic pausing*, remaining semi-active and ready to grow when conditions permit (Pushkareva et al. 2024).

However, our current knowledge of how different microbial photosynthetic populations in polar shallow wetlands and biocrusts respond to winter thaws is limited. Understanding these dynamics is critical, as the frequency and intensity of winter thaws are projected to increase under current Arctic climate change scenarios, potentially reshaping microbial community composition and function at the ecosystem scale.

14.7 Hardening and Desiccation and Cryostress

Desiccation stress is a significant challenge for cyanobacteria and microalgae often exposed to fluctuating water availability in their natural environments. To cope with desiccation stress, these organisms have evolved various mechanisms, including hardening. Hardening is a process by which cells become more resistant to desiccation stress through the accumulation of compatible solutes, such as sugars, amino acids, and other osmolytes. For example, the cyanobacterium *Synechocystis* sp. PCC 6803 accumulates trehalose and other osmolytes to protect itself against desiccation stress (Hagemann et al. 2006). Starvation can also play a role in the response to desiccation stress in cyanobacteria and microalgae (Pichrtova et al. 2014; Buitink et al. 2006; Kang et al. 2013). For example, the microalga *Chlamydomonas reinhardtii* enters a state of dormancy when starved of nutrients, which helps it to survive desiccation stress (Buitink et al. 2006; Kang et al. 2013). For surviving desiccation stress, cold acclimation was reported for several *Klebsormidium* (Rippin

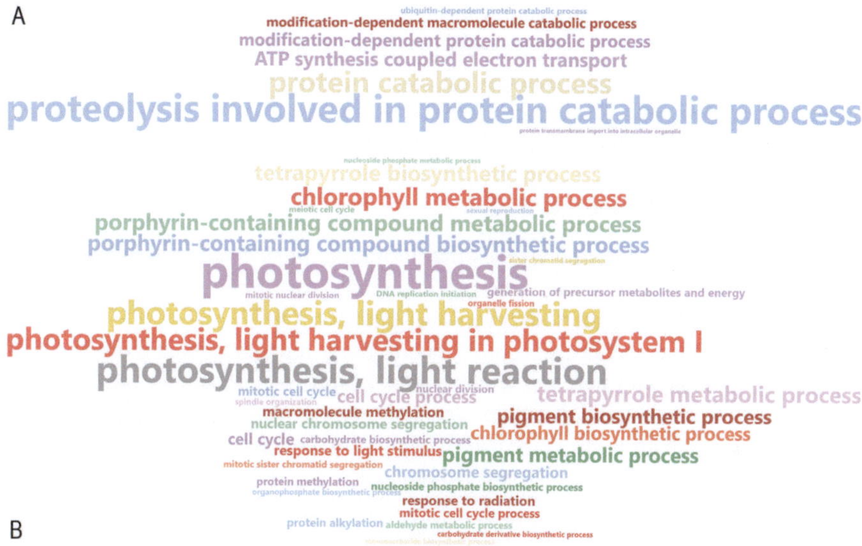

Fig. 14.4 Gene set enrichment analysis shows 50 significantly upregulated (**a**) and downregulated (**b**) enriched gene orthology terms of *Klebsormidium flaccidum* PCC1182 samples cultivated at 22 °C in the presence or absence of nitrogen, with the following desiccation treatment. (Kolomiiets et al. 2025). Upregulated genes indicate their role in cell survival and maintaining physiological processes

et al. 2019; Kolomiiets et al. 2025) and *Chlamydomonas* (Rai et al. 2019) strains, and cyanobacteria (Kang et al. 2011; Lee et al. 2015). In the new study by Kolomiiets et al. (2025), nitrogen starvation had a detrimental effect on desiccation survival (Fig. 14.4).

Cryostress is another significant challenge for cyanobacteria and microalgae, which are often exposed to low temperatures in their natural environments. To cope with cryostress, these organisms have evolved various mechanisms, including cold hardening. Cold hardening is a process by which cells become more resistant to cryostress through the accumulation of cold-induced proteins and other protective molecules. For example, the cyanobacterium *Synechocystis* sp. PCC 6803 accumulates cold-induced proteins to protect itself against cryostress (Hincha et al. 2002). Similarly to desiccation stress, starvation can also play a role in cryostress response of cyanobacteria and algae. For example, the alga *Chlamydomonas reinhardtii* enters a state of dormancy when starved of nutrients, which helps it to survive cryostress (Buitink et al. 2006).

While desiccation and cryostress are different kinds of stresses for cyanobacteria and algae, they share some common mechanisms of stress response. For example, both types of stress can lead to the accumulation of reactive oxygen species (ROS), which can damage cellular components. Plant cells have evolved various mechanisms to cope with ROS, including the production of antioxidants and the activation of stress response pathways (Bolwell et al. 2002). One of the best model systems for studying stress responses is cyanobacteria, and several papers have described their molecular responses to various stressors in detail (Roncero-Ramos et al. 2024; Liu et al. 2017; Padey et al. 2018; Santos et al. 2020). Other well-studied systems are the algae *Chlamydomonas* (Muller et al. 2017; Rai et al. 2019) and *Klebsormidium* (Holzinger et al. 2014; Rippin et al. 2019; Kolomiiets et al. 2025).

In present study of microbial photosynthetic communities, several major research questions are currently addressed:

1. Do well-adapted, endemic species that dominate in selected wetlands and biocrust habitats in the Arctic, where winter thaws are frequent, persist in these environments for long periods, even over glacial cycles?
2. Do cosmopolitan species, which migrate to polar regions, persist more successfully in habitats with frequent winter melts, despite the associated stresses?
3. Do endemic and cosmopolitan species exhibit different ecophysiological responses to winter melts, and if so, what are the underlying mechanisms?
4. Can metagenomics and metatranscriptomics help us to identify significant shifts in microbial community structure in response to winter melts and other environmental stresses?
5. Finally, do species in habitats with frequent winter melts exhibit a lagged response to climatic and environmental changes, or are these extreme conditions sufficient to eliminate these genomes from local environments?

14.8 Conclusions

The polar terrestrial environment is a unique and challenging ecosystem characterized by extreme cold, darkness, and by limited water, energy, and nutrient resources. The photosynthetic microorganisms that dominate in these environments, such as cyanobacteria and microalgae, have evolved various mechanisms to cope with these stresses, including hardening and acclimation. However, our current understanding of how these organisms respond to winter melts and other environmental stresses is limited, and a lot of questions remain to be answered. Further research is needed to investigate the responses of microbial photosynthetic populations in polar shallow wetlands and biocrusts to winter melts and to identify the underlying mechanisms of their ecophysiological and molecular-genetic adaptations. Integrating omics technologies, such as genomics, transcriptomics, proteomics, and metabolomics, into future research can help elucidate the functional and regulatory pathways involved in these adaptations. In addition, long-term monitoring of microbial communities and their environmental contexts is essential to detect gradual or cumulative changes that may result from repeated melt events or other climate-driven shifts. This knowledge will be essential for understanding the impacts of climate change on these ecosystems and for developing strategies to mitigate these impacts.

Acknowledgements This work was supported by the Czech Science Foundation [project 22–08680 L], the Czech Academy of Sciences [long-term research development project No. RVO 67985939] (JE, PP, AK, OB, JK), Grant TAČR TS02030210 (JE, JK, OB, AK), the Deutsche Forschungsgemeinschaft (DFG) BE1779/25-1 (BB, EP). OB was additionally supported by the Researchers at Risk Fellowship (Czech Academy of Sciences) and the IIE Scholar Rescue Fund (IIE-SRF) Fellowship. The authors would like to thank Dr. Keith Edwards for the language correction and the University of West Bohemia, Faculty of Education, Plzeň, Czech Republic for the support as well.

References

ACIA (2005) Impacts of warming climate: arctic climate impact assessment. Cambridge University Press, Cambridge

Arenson L Colgan W, Marshall HP (2021) Physical, thermal, and mechanical properties of snow, ice, and permafrost

Bates JW, Farmer AM (eds) (1992) Bryophytes and lichens in a changing environment. Clarendon Press, Oxford

Bokhorst S, Phoenix GK, Berg MP, Callaghan TV, Kirby-Lambert C, Bjerke JW (2015) Climatic and biotic extreme events moderate long-term responses of above- and belowground sub-Arctic heathland communities to climate change. Glob Change Biol 21(11):4063–4075

Bolwell GP et al (2002) Plant responses to oxidative stress: the role of antioxidant enzymes. J Exp Bot 53(372):1315–1325

Broady PA (1996) Diversity, distribution and dispersal of Antarctic algae. Biodivers Conserv 5(11):1307–1335

Büchel C, Wilhelm C (1993) In vivo analysis of slow chlorophyll fluorescence induction kinetics in algae: progress, problems and perspectives. Photochem Photobiol 1:137–148

Buitink J et al (2006) Desiccation tolerance in plants: from genes to molecules. J Exp Bot 57(4):1055–1066

Callaghan TV, Johansson M, Brown RD et al (2011) Multiple effects of changes in arctic snow cover. Ambio 40:32–45

Caporn SJM, Risager M, Lee JA (1994) Effect of nitrogen supply on frost hardiness in *Calluna Vulgaris* (L.) Hull. New Phytol 128(3):461–468

Caporn SJM, Ashenden TW, Lee JA (2000) The effect of exposure to NO_2 and SO_2 on frost hardiness in *Calluna vulgaris*. Environ Exp Bot 43:111–119

Consalvey M, Perkins RG, Paterson DM, Underwood GJC (2005) PAM Fluorescence: a beginner's guide for benthic diatomists. Diatom Res 20(1):1–22

Cottier F, Skogseth R, David D, Berge J (2019) Temperature time-series in Svalbard fjords. A contribution from the integrated marine observatory partnership (iMOP). In: Orr et al (eds) SESS report 2018, vol 1. Svalbard Integrated Arctic Earth Observing System, Longyearbyen, pp 108–118

Crawford RMM (2000) Tansley review 114. New Phytol 147:257–281

Cullather RI, Lim Y-K, Boisvert LN, Brucker L, Lee JN, Nowicki SMJ (2016) Analysis of the warmest Arctic winter. Geophys Res Lett 43(10):808–816

Elster J (2002) Ecological classification of terrestrial algal communities in Polar environments. Ecol Stud 154:303–326

Elster J, Benson EE (2004) Life in the polar terrestrial environment with a focus on algae and cyanobacteria. In: Fuller BJ, Lane N, Benson EE (eds) Life in the frozen state. CRC Press, Boca Raton, pp 111–150

Elster J, Degma P, Kováčik Ľ, Valentová L, Šrámková K, Pereira AB (2008) Freezing and desiccation injury resistance in the filamentous green alga *Klebsormidium* from the Antarctic, Arctic and Slovakia. Biologia 63(6):839–847

Everatt MJ, Convey P, Bale JS, Worland MR, Hayward SAL (2015) Responses of invertebrates to temperature and water stress: a polar perspective. J Therm Biol 54:118–132

Franzaring J, Holz I, Fangmeier A (2013) Effects of increased autumn temperatures and subacute levels of ammonia on post-winter development of four cultivars of winter oilseed rape (*Brassica napus* L). J Appl Bot Food Qual 85:134–143

Guak S, Fuchigami LH (2002) Foliar applications of urea or ABA affect growth cessation, leaf senescence and abscission, cold acclimation and levels of reserve nitrogen and carbohydrates in nitrogen-treated apple nursery plants. J Hortic Sci Biotechnol 77:137–142

Hagemann M et al (2006) Trehalose and other compatible solutes in the cyanobacterium *Synechocystis* sp. PCC 6803. J Exp Bot 57(4):1071–1081

Hanssen-Bauer I, Førland EJ, Hisdal H, Mayer S, Sandø AB, Sorteberg A (eds) (2019) Climate in Svalbard 2100 – a knowledge base for climate adaptation (NCCS report 1/2019). Norwegian Centre of Climate Services (NCCS) for Norwegian Environment Agency (Miljødirektoratet), Norway, 208

Hawes I, Howard-Williams C, Vincent WF (1992) Desiccation and recovery of Antarctic cyanobacterial mats. Polar Biol 12:587–594

Hejduková E, Elster J, Nedbalová L (2020) Annual cycle of freshwater diatoms in the High Arctic revealed by multiparameter fluorescent staining. Microb Ecol:559–572

Hejduková E, Pushkareva E, Kvíderová J, Becker B, Elster J (under review) Summer and autumn photosynthetic activity in High Arcticbiological soil crusts and their winter recovery. Front in Microbiol

Hincha DK et al (2002) The role of sugars in the protection of plants against freezing stress. J Exp Bot 53(372):1331–1342

Holzinger A, Kaplan F, Blaas K, Zechmann B, Komsic-Buchmann K, Becker B (2014) Transcriptomics of desiccation tolerance in the streptophyte green alga *Klebsormidium* reveal a land plant-like defense reaction. PLoS One 9(10):e110630

Jimel M, Kvíderová J, Elster J (2021) Annual cycle of mat-forming filamentous algae *Tribonema cf. minus* (Stramenopile, Xanthophyceae) in hydro-terrestrial habitats in the High Arctic revealed by multiparameter fluorescent staining. J Phycol 57:780–796

Joli N, Concia L, Mocaer K, Guterman J, Laude J, Guerin S et al (2024) Hypometabolism to survive the long polar night and subsequent successful return to light in the diatom *Fragilariopsis cylindrus*. New Phytol 241(5):2193–2208

Kang JG et al (2011) Cold acclimation and desiccation tolerance in the cyanobacterium *Synechocystis* sp. PCC 6803. J Exp Bot 62(10):3491–3502

Kang JG et al (2013) Starvation-induced dormancy in the microalga *Chlamydomonas reinhardtii*. J Exp Bot 64(10):2751–2762

Keto-Timonen R, Hietala N, Palonen E, Hakakorpi A, Lindström M, Korkeala H (2016) Cold shock proteins: a minireview with special emphasis on Csp-family of Enteropathogenic *Yersinia*. Front Microbiol 7:1151

Kolomiiets A, Bren O, Přibyl P, Kvíderová J, Procházková L, Pushkareva E, Elster J, Becker B (2025) Unveiling the molecular mechanisms of the stress resilience of High Arctic *Klebsormidium flaccidum*: a multifaceted study. Plant Cell Physiol: pcaf091

Lee JH et al (2015) Cold hardening and desiccation tolerance in the cyanobacterium *Synechocystis* sp. PCC 6803. J Exp Bot 66(10):2791–2802

Liu X et al (2017) Desiccation tolerance in the cyanobacterium *Synechocystis* sp. PCC 6803: a proteomic analysis. J Exp Bot 68(10):2811–2822

Liu Y, Hansen B, Elberling B, Westergaard-Nielsen A (2023) Snow depth and the associated offset in ground temperatures in a landscape manipulated with snow-fences. Geoderma 438:116632

Maxwell K, Johnson GN (2000) Chlorophyll fluorescence–a practical guide. J Exp Bot 51(345):659–668

Morita RY (1975) Psychrophilic bacteria. Bacteriol Rev 39:144–167
Muller P et al (2017) Cold stress response in the microalga *Chlamydomonas reinhardtii*: a transcriptomic analysis. J Exp Bot 68(10):2831–2842
Newsted WJ, Hüner NPA (1988) Major sclerotial polypeptides of psychrophilic pathogenic fungi: intracellular localisation and antigenic relatedness. Protoplasma 147(2):162–169
Niskanen AKJ, Niittynen P, Aalto J, Väre H, Luoto M (2019) Lost at high latitudes: Arctic and endemic plants under threat as climate warms. Divers Distrib 25:809–821
Pandey S et al (2018) Desiccation tolerance in the cyanobacterium *Synechocystis* sp. PCC 6803: a comparative study of different osmolytes. J Exp Bot 69(10):2851–2862
Pichrtová M, Kulichová J, Holzinger A (2014) Nitrogen limitation and slow drying induce desiccation tolerance in conjugating green algae. e113137
Pichrtová M, Hájek T, Elster J (2016) Annual development of conjugating green algae *Zygnema* sp. in Arctic hydro-terrestrial environment. Polar Biol
Pichrtová M, Hejduková E, Nedbalová L, Elster J (2020) How to survive winter? Adaptation and acclimation strategies of eukaryotic algae from polar terrestrial ecosystems. Life Extreme Environ:1–25
Pushkareva E, Kvíderová J, Šimek M, Elster J (2017) Nitrogen fixation and diurnal changes of photosynthetic activity in Arctic soil crusts at different development state. Eur J Soil Biol 79:21–31
Pushkareva E, Elster J, Becker B (2023) Metagenomic analysis of Antarctic biocrusts unveils a rich range of cold-shock proteins. Microorganisms 11(8):1932
Pushkareva E, Hejduková E, Elster J, Becker B (2024) Microbial response to seasonal variation in Arctic biocrusts with a focus on fungi and cyanobacteria. Environ Res 263:120110
Rai AK et al (2019) Cold hardening and desiccation tolerance in the microalga *Chlamydomonas reinhardtii*: a review. J Exp Bot 70(10):2871–2882
Rippin M, Borchhardt N, Karsten U, Becker B (2019) Cold acclimation improves the desiccation stress resilience of polar strains of *Klebsormidium* (Streptophyta). Front Microbiol Sec Extreme Microbiol 10:1–19
Roháček K, Soukupová J, Barták M (2008) Chlorophyll fluorescence: a wonderful tool to study plant physiology and plant stress. In: Schoefs B (ed) Plant cell compartments-selected topics. Research Signpost, pp 41–104
Roncero-Ramos B, Savaglia V, Durieu B, Van de Vreken I, Richel A, Wilmotte A (2024) Ecophysiological and genomic approaches to cyanobacterial hardening for restoration. J Phycol 60(2):465–482
Šabacká M, Elster J (2006) Response of Cyanobacteria and Algae from Antarctic Wetland Habitats to Freezing and Desiccation Stress. Polar Biol 30:31–37
Santos R et al (2020) Desiccation tolerance in the cyanobacterium *Synechocystis* sp. PCC 6803: a proteomic analysis. J Exp Bot 71(10):2891–2902
Sawicka J, Robador A, Hubert C et al (2010) Effects of freeze–thaw cycles on anaerobic microbial processes in an Arctic intertidal mud flat. ISME J 4:585–594
Schoolmeester T, Gjerdi HL, Crump J et al (2019) Global linkages/a graphic look at the changing Arctic. UN Environment and GRID-Arendal, Nairobi/Arendal
Shan Y (2000) Phenological disorder induced by atmospheric nitrogen deposition: original causes of pine forest decline over Japan. Part II. Relationship among earlier phenological development, extreme of minimum air temperature, and forest decline of pines over the Japan. Water Air Soil Pollut 117:205–215
Søreide JE, Pitusi V, Vader A et al (2020) Environmental status of Svalbard coastal waters: coastscapes and focal ecosystem components (SvalCoast). In: Moreno-Ibáñez et al (eds) SESS report 2020. Svalbard Integrated Arctic Earth Observing System, Longyearbyen, pp 143–174
Stenseth NC, Mysterud A, Ottersen G, Hurrell JW, Chan KS, Lima M (2002) Ecological effects of climate fluctuations. Science (New York, NY) 297(5585):1292–1296
Tashyreva D, Elster J (2012) Production of dormant stages and stress resistance of polar cyanobacteria. In: Life on earth and other planetary bodies, cellular origin, life in extreme habitats and astrobiology, vol 24. Springer, Dordrecht, pp 367–386

Tashyreva D, Elster J (2015) Effect of nitrogen starvation on desiccation tolerance of Arctic *Microcoleus* strains (cyanobacteria). Front Microbiol 6(278):1–11

Tashyreva D, Elster J (2016) Annual cycles of two cyanobacterial mat communities in hydro-terrestrial habitats of the high Arctic. Microb Ecol 7:1–16

Taulavuori K, Taulavuori E, Niinimaa A et al (2001) Acceleration of frost hardening in *Vaccinium vitis-idaea* by nitrogen fertilization. Oecologia 127:321–323

Taulavuori K, Taulavuori E, Sheppard L (2014) Truths or myths, fact or fiction, setting the record straight concerning nitrogen effects on levels of frost hardiness. Environ Exp Bot 106

Thomas D, Fogg G, Convey P et al (2008) The biology of Polar Regions. 10.1093

Tomanek L (2008) The importance of physiological limits in determining biogeographical range shifts due to global climate change: the heat-shock response. Physiol Biochem Zool 81(6):709–717

Urbanski JA, Litwicka D (2022) The decline of Svalbard land-fast sea ice extent as a result of climate change. Oceanologia 64:535–545

Vincent WF (1989) Microbial ecosystems of Antarctica. Cambridge University Press, Cambridge

Vincent WF (2007) Cold tolerance in Cyanobacteria and life in the Cryosphere. In: Seckbach J (ed) Algae and cyanobacteria in extreme environments, Cellular origin, life in extreme habitats and astrobiology, vol 11. Springer, Dordrecht

Walczowski W, Piechura J (2011) Influence of the West Spitsbergen current on the local climate. Int J Climatol 31:1088–1093

Walker MD, Walker DA, Welker JM (1999) Long-term experimental manipulation of winter snow regime and summer temperature in arctic and alpine tundra. Hydrol Process 13:2315–2330

Wang KJ, Huang Y, Kartzinel T et al (2024) Group 2i Isochrysidales thrive in marine and lacustrine systems with ice cover. Sci Rep 14:11449

Wipf S, Rixen C (2010) A review of snow manipulation experiments in Arctic and alpine tundra ecosystems. Polar Res 29(1):95–109

Zvereva EL, Kozlov MV (2005) Growth and reproduction of dwarf shrubs, *Vaccinium myrtillus* and *V. vitis-idaea*, in a severely polluted area. Basic Appl Ecol 6:261–274

MIX
Papier aus verantwortungsvollen Quellen
Paper from responsible sources
FSC® C105338

FSC
www.fsc.org

If you have any concerns about our products,
you can contact us on
ProductSafety@springernature.com

In case Publisher is established outside the EU,
the EU authorized representative is:
Springer Nature Customer Service Center GmbH
Europaplatz 3, 69115 Heidelberg, Germany

Printed by Libri Plureos GmbH
in Hamburg, Germany